Fundamentals of Composites and Their Methods of Fabrications

Bahram Farahmand

Fundamentals of Composites and Their Methods of Fabrications

PMCs, MMCs, and CMCs

Springer

Bahram Farahmand
Mechanical and Material Department
Former Boeing Technical Fellow
Laguna Hills, CA, USA

University of California
Irvine, CA, USA

ISBN 978-3-031-80200-3 ISBN 978-3-031-80201-0 (eBook)
https://doi.org/10.1007/978-3-031-80201-0

This Springer imprint is published by the registered company Springer Nature Switzerland AG
The registered company address is: Gewerbestrasse 11, 6330 Cham, Switzerland

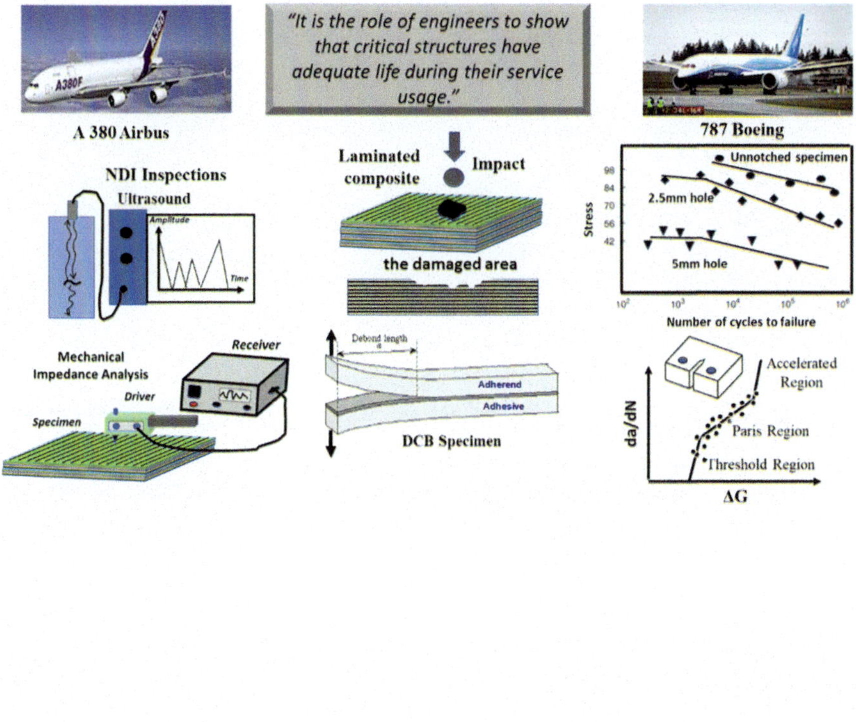

"It is the role of engineers to show that critical structures have adequate life during their service usage."
A 380 Airbus
787 Boeing
NDI Inspections
Ultrasound
Amplitude
Time
Laminated composite
Impact
the damaged area
Unnotched specimen
2.5mm hole
5mm hole
Stress
Number of cycles to failure
Mechanical Impedance Analysis
Receiver
Driver
Specimen
Debond length
Adherend
Adhesive
DCB Specimen
Accelerated Region
Paris Region
Threshold Region
da/dN
ΔG

This book is dedicated to the memory of my loving parents, who instilled in me a profound respect for higher education. I also extend my special thanks to my wife, daughter, and son for their valuable advice throughout my work on this book.

Preface

Our past observation on utilizing metallic materials for industrial use is evident that composite materials are considerably different from our metallic counterparts. To effectively harness the potential of composites in engineering products, new concepts and approaches in areas such as analysis, design, fabrication, quality assurance, testing, inspections, and repairs are essential to instill confidence in their use.

Composite materials are becoming useful to industries because of their superior properties, such as high-specific modulus and strength, good corrosion resistance, high fatigue strength, rapid fabrication process, and many other superior attributes when compared with their metallic counterparts. Recently, composites have been used as the material of choice in aerospace, aircraft, military, electronic packaging, and medical equipment. The goal of this book is to introduce to the readers important aspects of composites, both from their fabrication methods of processing polymer, metallic, and ceramic matrices (PMCs, MMCs, CMCs composites), and methods of obtaining their mechanical properties for failure analysis (Chaps. 2, 4, 5, 6). An additional section is allocated to nanocomposite that provides useful information on parameters that will influence the mechanical properties of nanocomposite parts, such as the interface bonds between the matrix and the nanomaterials (Sect. 4.6).

Composite parts, just like metals, must be designed in such a way that can withstand the service load environment. For that to happen, engineers must perform a comprehensive stress, fatigue, and fracture mechanics analysis of composite primary structures to ensure the part can survive the load environment without failure. These analyses require having material allowable, which are obtainable through testing. Because of the nature of composite materials, which are not homogeneous and not isotropic, testing will be expensive, time-consuming, and labor-intensive. A few sections in this book are allocated to methods of obtaining the mechanical allowable of composite parts by using the rule of mixtures through a micromechanics approach (Chap. 2). In addition, a novel method is presented in this book that will enable engineers to acquire fracture properties of composite parts with the minimum amount of effort through the virtual testing technique. The author has introduced two software programs (FTD and FCG) that can quickly generate the fracture toughness and fatigue crack growth properties of homogeneous, elastic, and isotropic materials with a minimum amount of testing that can be valuable to engineers to perform their structural life analysis (Chap. 10). It is the author's intent to expand this virtual testing approach to composite material as well. In addition, a section is allocated to virtual testing through molecular dynamics and multiscale modeling, which is discussed in Chap. 10. It addresses how one may be able to obtain material properties of composites from the bottom-up approach using molecular dynamics and the coarse grain technique to ease the computer simulations.

Failure of composite parts when subjected to an operating load environment is discussed through the progressive failure analysis (PFA) by using the finite element analysis (FEA) followed by the implementation of a suitable failure theory that can provide information about the damage initiation and structural stiffness reduction (Chap. 8). In addition to the PFA, both the traditional fatigue as well as the linear elastic fracture mechanics (LEFM) approach are discussed from both the Irwin and Griffith energy approach with several example problems that can be useful to the readers (Chap. 9). The failure analysis of a composite bolted joint is presented through the progressive failure analysis approach for engineers to become familiar with the method of structural stiffness reduction and final failure when subjected to applied load.

Because composite parts are susceptible to damage before and during their service usage due to impact, it is important to implement an appropriate nondestructive inspection method to detect the presence and degree of any damage in the part. Different types of nondestructive inspection methods are discussed in Chap. 14, which can be very useful for monitoring damage growth during the composite part's operation. Furthermore, the maximum initial flaw size that can escape detection, in both metals and composite parts, is presented in this chapter, which engineers can utilize for their life assessment of structural parts. If the part has been damaged seriously and engineers believe it should be repaired, then the proper repair method, whether it is a temporary repair or a complete one, must be implemented (Chap. 13). This book will provide several acceptable standard methods of repairing composite parts that can be helpful for engineers to apply when repairing is necessary to implement. Structural integrity assessment of repaired parts for life analysis is also presented with several example problems. The method of applying a composite patch on the thin metallic part that was exposed to fatigue cracking and the life analysis assessment of repaired parts is also presented in this chapter.

Additionally, a short chapter is provided to the reader (Chap. 15) under the 3D printing or additive manufacturing (AM). The direct energy deposition (DED) and powder bed fusion (PBF) techniques, which both fall under laser additive manufacturing (LAM), are discussed in this chapter. Parameters that influence the mechanical properties of the final products are presented in this chapter. Issues related to the time and cost of manufacturing parts, as well as issues related to residual stresses and porosities in the final product and their impact on the mechanical properties, are also discussed. The AM process for both metallic and composite material is considered here. Post-processing operations, such as grinding, surface cleaning, inspections, and annealing, are briefly discussed in Chap. 15.

This book is written for mechanical and materials science engineers who are concerned with the integrity of composite structures under service load conditions, where special attention must be given to the initiation and propagation of damage in the material. The content of this book results from the author's years of academic and industrial experience accumulated throughout his work in various aerospace and engineering firms. However, the content of this book can also be valuable to engineers dealing with the fabrication and analysis of all types of composite materials (i.e., PMCs, MMCs, CMCs). Finally, the author believes that the scope of the material presented here is sufficient for use as a textbook or as a supplement to textbooks for both graduate and undergraduate students.

The author is thankful for the support of his wife, whose sacrifices made it possible to complete this book.

Laguna Hills, CA, USA Bahram Farahmand

Book Description (Summary of Chapters)

Since the 1980s, composite materials have accelerated their applications and become very useful in manufacturing critical parts for many industries, including aircraft, shipbuilding, auto industry, military, aerospace, wind turbine blades, medical equipment, etc. Today, composite material has a special place among other existing materials that engineers can select to fabricate their new products. They have advantages over other materials, especially when compared with their metallic counterparts, in several key areas, such as high specific strength (σ_{Ult}/ρ) and modulus (E/ρ), corrosion resistance, high fatigue strength, and faster assembly. Another advantage of composites over metallic materials is their ability to be custom-tailored to meet specific performance needs and the added benefit of not requiring post-fabrication processes, such as machining or grinding.

The history of composites, from their applications in industries, started after World War II when aircraft companies searched for material with adequate strength, lightweight, durability, and good fatigue and corrosion properties. After 1945, fiberglass became the material of choice and was used mainly for military applications. Soon, polymer matrix fiber-reinforced composites were utilized in several industries. In the early 1950s, manufacturing methods such as pultrusion, vacuum bag molding, large-scale filament winding, and hand layout open mold were developed. The application of filament winding in rocket motor cases consisted of passing fibers through a liquid resin and then winding it on the mandrel with different orientations, which pushed for further space exploration in the 1960s and beyond.

In Chap. 1, the understanding of composite materials is discussed, which are formed from combining two or more basic materials on a macroscale level. After their fabrications, in final form, they exhibit valuable properties, such as good strength, high stiffness, low weight, corrosion resistance, thermal and acoustic insulation, easy to repair, and many other valuable attributes. These desirable qualities make composite material suitable for many industries. The two main elements that form composites are the matrix (a.k.a., binder) and the reinforcement (a.k.a., fibers or whiskers), in which the two are adhesively bonded together (Sect. 1.1). This chapter discusses several types of matrices and reinforcements used in composite materials, which can affect the properties of composites (Sects. 1.2 and 1.3). In addition, the orientation, size, the amount of fibers for each lamina of a laminate, the anisotropic nature of material, the method of manufacturing, the temperature and duration of the cure cycle all contribute to the composite's final mechanical properties are briefly presented in Chap. 1.

In Chap. 2, the concept of micromechanics and understanding lamina through its constituents which are fibers, and the matrix is discussed (Sect. 2.1). The critical point of micromechanics is to look at the material microscopically by focusing on the local stress and strain fields induced by the applied loads upon the constituents. Unlike isotropic materials, experimental evaluation of elastic modulus and strength parameters for a uniaxial lamina is very costly and time-consuming because they are functions of several variables, such as fiber volume fraction (the percent amount of fibers in the total volume of composites), void volume fraction, packing geometry, processing, etc. In this chapter, the material properties of composites are assessed through the rule of mixtures, where one of the key assumptions used here is that even though unidirectional lamina is not homogeneous and anisotropic, it can be assumed the lamina to be homogeneous but not isotropic by focusing on the average response of the lamina to the applied load (Sect. 2.4). If engineers prefer to conduct their analysis through other methods, such as molecular dynamics, nanomechanics, or quantum mechanics, then reliable software with material properties should be available to engineers (Sects. 2.2 and 2.3).

E_1 (longitudinal Young's modulus along fibers), E_2 (transverse Young's modulus), ν_{12} (major Poisson's ratio), ν_{21}, and G_{12} (in-plane shear modulus in plane 1–2) are all calculated through the rule of mixtures with several example problems (Sect. 2.5). To predict properties for any other ply with fibers oriented at some angle θ from the applied load, the transformation matrix can be helpful to obtain both the stress and strain for any other angle with respect to the global axis (Sect. 2.6). Lastly, a brief discussion on several disadvantages and limitations associated with the rule of mixtures for obtaining composite material is discussed in Sect. 2.7.

In Chap. 3, the types of matrices and fibers used in composites are discussed (Sect. 3.1). The matrix material, which holds or binds everything together, is divided into three classifications: (1) polymers, (2) metals, and (3) ceramics. Polymers, metals, and ceramic matrix composites are the three most common types that can be useful for industries, and they will be briefly discussed in this chapter separately. However, a comprehensive description of their manufacturing methods is presented in Chaps. 4, 5, and 6, respectively.

Polymer matrix composites (PMCs) are a type of advanced composite material consisting of a polymer matrix reinforced with high-strength fibers or particles. The polymer matrix, often a resin such as epoxy, surrounds and supports the reinforcing materials, which can be made of materials like fiberglass, carbon fibers, or aramid fibers. PMCs exhibit enhanced mechanical properties, including high strength-to-weight ratio and stiffness, good fatigue, and corrosion resistance, making them suitable for various applications in industries such as aerospace, automotive, and construction (Sect. 3.4). Metal matrix composites (MMCs), with metals used for the matrix, and the reinforcement that can be ceramics or metallic materials are briefly presented in this chapter. From a manufacturing perspective, the MMCs are more expensive than the conventional metallic materials that industries are replacing. As a result, they are budgeted to go forward with the manufacturing of parts where improved properties and performance can justify the added cost (Sect. 3.3). Ceramic matrix composites (CMCs) are a special type of composite material in which both the reinforcement and matrix material are ceramics for the purpose of surviving the high-temperature environment and resistance to oxidation, which is briefly discussed here (Sect. 3.2).

In Chap. 4, polymer matrix composites (PMCs) are discussed in detail. Several methods of fabricating PMC parts are presented, and their advantages and disadvantages are reviewed in this chapter. Despite their significant weight-saving and other advantages that advanced composites could provide, these materials are susceptible to delamination that can shorten the life of the part and must be inspected, analyzed, and monitored during the part's operation.

Several methods of fabricating PMC parts are presented in this chapter. Two types of fabrication methods, the open mold and close mold processes, are presented, which each have problems of their own that may occur during the fabrication process. The spray lay-up technique (an open mold process) is the simplest method of fabricating the PMC parts. This process is a low-cost method and can be finished in a reasonably short time. A detailed description is provided in Sect. 4.3.

The hand lay-up technique for fabricating PMC parts also falls under the open mold process. The composite hand lay-up fabrication process typically involves having some form of molding to shape the resin and reinforcement. In this method, each ply layer is handled only by hand and is stacked layer by layer up with preferred orientations until the desired thickness is reached (Sect. 4.4).

The tape lay-up method also falls under the open mold process, and it works with layers of pre-impregnated fibers or simply prepreg. These prepreg tapes will be applied on the open surface of a mold with the desired orientation. The prepreg tapes can be made as a thermoset polymer matrix material, such as epoxy, or a thermoplastic resin that is already embedded with fibers and is partially cured (Sect. 4.5).

Autoclave molding is an open-mold lay-up process where both fibers and matrix are pre-blended and pre-cured, and each prepreg ply is covered from both sides for protection against a high-humidity environment. The autoclave uses high temperature and positive pressure (above the atmospheric pressure) during curing to consolidate the plies and to provide a path for air and volatiles to escape (detailed description is provided in Sect. 4.6).

Other methods of fabricating the PMCs parts are also discussed in this chapter. Compression molding, also known as press molding, is a closed mold process where the composite material is pressed inside the die (cavity) to form the desired geometry (Sect. 4.7).

Injection molding is another closed mold process for making composite parts by injecting molten polymer or polymer mixed with chopped fibers into a mold cavity. It consists of the high-pressure injection of the raw material into a mold or multiple molds (multi-cavitation), which form the molten material into the desired shape (see Sect. 4.8).

The pultrusion molding process is another closed mold technique in which the fibers impregnated with resin will be pulled through a die with some pulling mechanism to manufacture the part. One of the desirabilities of the pultrusion process is having no limitation on the length of the part. The shape of the cross-section can be mainly rectangular, circular, I-beam shaped, or hexagonal geometry with uniform thickness (Sect. 4.9).

Filament winding is a continuous manufacturing process to produce PMC parts such as tanks, pipes, pressure vessels, pressurized fuel tanks, and other round-form products that require a high degree of structural integrity. The cylindrical composite products that are fabricated by this method have hollow cores in which resin impregnated with high-quality fibers will be wrapped around a rotating cylindrical mandrel mold in the desired pattern or angle to the rotational axis to create different layers (Sect. 4.10).

Even though by using the PMCs parts in the aircraft structure, which can increase both the fuel and maintenance costs by roughly 50% and 20%, nanocomposites have shown promise in the aerospace and aircraft sectors due to their superior E/ρ and σ_{Ult}/ρ over the traditional PMCs parts. Nanoparticles are very small particles with dimensions between 1 and 100 nanometers (nm), where 1 nm $= 10^{-9}$ m. A typical example of nanoparticles is carbon nanotubes (CNTs), which are hollow cylinders made of carbon atoms connected by covalent bonds (see Sect. 4.12 for more discussions related to nanocomposites).

In Chap. 5, the metal matrix composites (MMCs) and the method of their fabrication are discussed. The traditional metal alloys are not good enough to be used when the working environments demand even higher strength, higher stiffness, and better resistance to high temperature and corrosive environments. So, to improve these properties, high-strength MMC composite materials can be used that have excellent reinforcement material together with metallic matrices possessing superb mechanical properties. For example, by choosing metals, such as nickel as the matrix with ceramic fibers as the reinforcement, it can be fabricated to have excellent properties for industries. In this chapter, several MMC's fabrication techniques will be discussed in detail. Both the solid-state (high-pressure diffusion bonding and sintering processes) and liquid-state techniques (the stir casting and infiltration processes) for manufacturing MMC parts are presented.

The high-pressure diffusion bonding process for manufacturing the MMCs parts falls under the solid-state welding technique capable of joining similar and dissimilar metals at the molecular level. The welding can be achieved by employing high pressure, in combination with high temperature, to both materials to be bonded together without applying adhesives. The act of pressure and heat makes the molecules diffuse between the two materials (the matrix and the reinforcement) at the contact areas to take place. The lack of a solid bond might cause premature failure of MMC parts, through the anticipated separation of plates (detailed discussion is provided in Sect. 5.3).

Another MMCs fabrication method is through the sintering and powder metallurgy process. Sintering is a heat treatment process in which a large quantity of loose aggregate material is subjected to a sufficiently high temperature and pressure to cause the loose material to become a compact solid lump for use in industries. The process is involved by mixing the powder of matrix metal with the powder of reinforced material to be sintered and compacting for the subsequent steps, such as heating in the solid state below the melting point. A more detailed description of the MMCs sintering technique is available in Sect. 5.4. A brief discussion on the mechanical properties' comparison of metal matrix composites with the host metal is presented in Sect. 5.5.

The liquid-state stir casting process for fabricating the MMCs parts is presented in Sect. 5.6. In this process, the reinforcement, such as the ceramic particles, is added in the solid form of loose powder to the molten metal matrix. Mechanical force is needed to incorporate the particles uniformly into the molten metal. This is usually performed by stirring the blade while the mixed material is inside the oven (Sect. 5.6).

The liquid infiltration casting process is another process to fabricate the MMCs parts. The concept is based on the infiltration of molten metal alloys through a porous preform of ceramic particles. The liquid matrix metals do not immediately wet the ceramic particles and need external pressure to force the liquid metal into the preform. The external force can be applied using an inert gas or a mechanical device (Sect. 5.7).

In Chap. 6 of this book, the ceramic matrix composite fabrication process is discussed. The process consists of ceramic fibers embedded in a ceramic matrix, and this can be accomplished through several fabrication techniques. Both the mechanical and fracture properties of CMCs are discussed in Sect. 6.2. In Sect. 6.3, the selection of reinforcement and matrix are presented. Methods of fabricating the CMCs are discussed in Sect. 6.4. The SiC/SiC composite, where the silicon carbide fiber is embedded into the silicon carbide matrix, is described in Sect. 6.5. In Sect. 6.6, the sol-gel infiltration approach (wet chemical approach) is presented in detail. The polymer infiltration pyrolysis approach (PIP) and liquid silicon infiltration (LSI) will be discussed in Sects. 6.7 and 6.8, respectively. Finally, the chemical vapor infiltration (CVI), chemical vapor deposition (CVD), and powder processing techniques will be discussed fully in Sects. 6.9 and 6.10, respectively.

The silicon carbide/silicon carbide (SiC/SiC) composite has high resistance capability to heat, and it has application in engine technology where a structure is exposed to high heat. SiC/SiC composite material can withstand temperatures above 2400 °F/1316 °C. Their high heat resistance capability and their reduced weight make them appealing to industries where parts are exposed to high temperature environments (Sect. 6.5).

Sol–gel process is another fabrication technique used to produce CMC products. It is one of the low-cost fabrication techniques that can be used to produce CMC materials. The sol–gel process is a wet-chemical technique where the sol (or solution) evolves gradually from a liquid state toward the formation of the gel.

In this process, a small mass is formed when the solvent evaporates, and the remaining particles begin to join. After drying, gels are converted into the ceramic matrix (see Sect. 6.6).

In Sect. 6.7, the polymer infiltration pyrolysis (PIP) technique is discussed. With this method, the CMCs matrix and fibers can both be fabricated through pyrolysis. As the name implies, the PIP fabrication technique consists of infiltrating a low-viscosity liquid polymer, such as polycarbosilane, into the reinforcing ceramic in the form of fabric preform, where the pyrolysis will follow through heating the polymer precursor. The ceramic matrix is formed from a fluid infiltrating the fiber reinforcement. The PIP method is based on filling the porosity inside the fiber fabrics after the pyrolysis step, where decomposition occurs (detailed discussion is presented in Sect. 6.7).

Another process for fabricating the CMC parts is liquid silicon infiltration (LSI), which is presented in Sect. 6.8. In this process, the ceramic matrix (SiC) is formed due to the chemical interaction between the infiltrated liquid melt (Si) into a porous reinforcing preform where polymer fibers were converted to carbon by the pyrolysis process resulting in the silicon carbide matrix (SiC). It is a fast, low residual porosity, and cost-effective method to manufacture CMC parts.

In Sect. 6.9, the chemical vapor infiltration (CVI) process is discussed. The chemical vapor infiltration method is one of the advanced low temperature techniques that can be implemented by manufacturers to obtain fiber-reinforced ceramic composites for industry use. High-temperature processing techniques can degrade the CMCs constituents. However, with the CVI fabrication technique, damage can be minimized (see Sect. 6.9 for more discussion related to both the CVI and CVD processes).

In Sect. 6.10, the powder processing technique for fabricating the CMC parts is discussed. Both the matrix and fibers must be available in the form of powders. Because of the nature of this process, that is, mixing and consolidation, the size of fibers or whiskers must be small. Even though the reinforcement is small, the process will be efficient, but not to an acceptable degree. However, to make the process effective, the particles in the powder matrix must possess the highest packing density possible (Sect. 6.10).

In Chap. 7, the delamination phenomenon in polymer matrix laminated composites is presented. The failure mechanisms in composites can be due to fiber failure, matrix cracking, fiber/matrix interface failure, or failure because of cracking resulting from the loss of adhesion between two consecutive plies, commonly called delamination. Delamination is the most critical form of damage in laminated composites, where layers that are adhesively bonded together will be separated, grow, and cause structural failure. The initiation and growth of delamination in the polymer matrix composite under service loads have always been a crucial factor, which has the potential to significantly affect the life of the structure and possible loss of life. The cause of delamination due to out of plane load, compressive or bending load, and impacts by an object is discussed in Sect. 7.2. Edge delamination that is considered as a critical failure mode in laminated composites is also discussed, which it refers to as the separation of lamina along the edges of the composite structure (see Sect. 7.2). It typically occurs due to high stresses at the edges of plies, which can be caused by mechanical loading, impact, or environmental conditions. Strategies for minimizing delamination is discussed in Sect. 7.4. The concept of the mode of failures and Griffith energy release rate (including their critical values, G_c) in assessing the delamination growth is presented in Sect. 7.3.

In Chap. 8, failure analysis and life assessment of laminated composites are discussed. In the failure analysis, the typical approach is to look at the stress and strain response of each layer or ply within the stack when subjected to external loads. The assumption is that each layer has its anisotropic properties, and each layer is different from other layers because of their distinct fiber orientations. The individual lamina's failure due to applied load will influence the final failure of the laminate. The failure assessment is possible by conducting a comprehensive finite element analysis and implementing a proper failure theory for each ply within the laminate. However, first, the material allowable must be available to analysts for the material under consideration (Sect. 8.2). To obtain individual lamina properties through the rule of mixtures, mechanical properties of both fiber and the matrix, packing geometry, the relative content of constituents (the volume fraction), the shape of fibers, and the method of processing them must be available. Having all allowable on hand, then the progressive failure analysis (PFA) will be implemented ply by ply. The failure will progress in a stable manner until final structural failure occurs due to reduction of stiffness from E_1, to E_2, ... E_{final}, where $E_1 > E_2 >, \ldots > E_{\text{final}}$ (Sect. 8.3).

Failure analysis of composites requires to have not only a comprehensive material property of both laminate and each individual ply but also a reliable failure theory that can evaluate the local and global failures and be able to compare the analytical results with the available material allowables. It is evident that the theories applied in predicting failure strongly depend on the scale of the analysis, implying that different criteria are used at the macro-, micro-, or nano-scale levels. In the case of nano scale, the failure simulation should begin at the atomistic level and results must be relayed to the micro and macro levels. Subsequently, the continuum mechanics will take over, where finite element technique can be used in the analysis (Sects. 8.4 and 8.5).

Prediction of local failure and damage growth using the well-known polynomial failure criterion, the Tsai-Wu tensor polynomial as well as Hashin's failure criteria are discussed. Hashin's criterion assumes that the composite material consists of fibers and the matrix, and each can fail independently when are subjected to load. The essential disadvantage of these polynomial failure criteria with respect to PFA analysis is that they predict the initiation of failure without providing any information about the lamina failure mode or the damage growth up to the instability (Sect. 8.6).

Both the sudden and gradual degradation scenario are discussed when a lamina fails. The application of gradual property degradation schemes to PFA results in increased computational effort and thus is more complicated compared to the sudden degradation approach. Assigning a constant degradation factor of around 10% applied to both elasticity modulus and Poisson's ratio is acceptable because of their simplicity and the relatively good failure predictions (Sect. 8.7).

Finally, the implementation of the PFA method on a composite bolted joint, made of two different materials, is presented in Sect. 8.8. The PFA analysis of a composite joint is affected by numerous parameters, such as joint geometry, combination of material types of the joined parts, stiffness properties, joint configuration, friction, clamping force, and loading type, which significantly influence the many interacting and complicated failure mechanisms taking place during the bolted joint operational service load. Results of the PFA analysis were compared with the test data, and the successful strength prediction, which occurs due to fiber failure, indicates the accurate detection by the respective fiber failure criterion implemented in the PFA (Sect. 8.8).

In Chap. 9, the application of fatigue and fracture mechanics in composite material is discussed (Sect. 9.1). A brief description of both the Griffith theory of brittle fracture and the Irwin stress intensity factor is presented in Sect. 9.2. Methods of obtaining material fracture toughness and fatigue crack growth through the ASTM standards were presented in Sect. 9.3. The most acceptable equation that can relate the crack growth rate (da/dN) to all the regions of crack growth (regions I, II, and III) provided by the NASGRO computer code (known as the Forman-Newman-deKoning crack growth equation) is discussed in Sect. 9.4.

The energy release rate, the G value, would be a more appropriate parameter to use than the stress intensity factor, K, when delamination growth is required to be assessed. Delamination in composites is generally a mixed-mode fracture process (G_{I} and G_{II}) with interlaminar tensile and shear stresses at the delamination front. The mode I critical value of energy release rate is G_{Ic}, and for mode II is designated by G_{IIc}. For mode III, the critical energy release rate is shown by G_{IIIc}. The double cantilever beam (DCB) test (ASTM D5528) is the most widely used test method for measuring the mode I energy release rate (G_{Ic}) of a unidirectional composite. The fatigue crack growth rate (da/dN) in a lamina as a function of G_{max} is presented in Sect. 9.5. Delamination growth analysis of plies with examples is also provided in this section.

The mode I delamination growth analysis is discussed in Sect. 9.6. The delamination fatigue growth test with several example problems is presented in Sect. 9.7. Analysis of matrix cracking and corresponding crack growth is fully discussed in Sect. 9.8. Here, the primary assumption is the weak bonds between the matrix and fibers that minimize the crack arrest mechanism, and damage will continue to grow within the matrix. Moreover, it is also assumed that fibers will remain intact during the crack growth at the two sides of the crack mouth, and only the separation between the fiber and matrix will occur during the crack growth process. The unbroken fibers at both sides of the crack mouth will be helpful to prevent the total failure of the ply.

Mode II delamination and the related testing standard (ASTM D 7905-14), based on using the End-Notched Flexure (ENF) test specimen, are discussed in Sect. 9.9. The method of obtaining critical energy release rate for both modes I and mode II (G_{Ic} and G_{IIc}) by using the mixed mode beam (MMB) test device is also included in Sect. 9.9.

A final comment on mode I and mode II and providing the variation of mode I versus mode II through conducting several tests on AS4/PEEK composite material is presented in Sect. 9.10. The method of pre-cracking DCB and ENF specimens for obtaining G_{Ic} and G_{IIc} is discussed in Sect. 9.10. The dependency of mode I energy release rate, G_{Ic}, on the amount of pre-cracking is presented in Sect. 9.11. Moreover, a new method of plotting the composite damage growth (the G versus *da*/*dN*) is fully discussed in Sect. 9.12. That is, $\Delta\sqrt{G_I}$ versus the *da*/*dN* where the quantity $\sqrt{G_I} = \sqrt{G_{Imax}} - \sqrt{G_{Imin}}$ supposed to provide a better fit to the test data. An equation similar to the NASGRO is available for composites that can describe the variation of *da*/*dN* as a function of $\sqrt{G_I}$ provided by Hartman-Schijve (Sect. 9.12).

In Chap. 10, obtaining material properties through a virtual testing approach is discussed. Fracture data for metallic materials are currently obtainable through the ASTM testing standards, which are costly and labor-intensive. Difficulties associated with specimen preparation, pre-cracking the notch, recording and monitoring the data, obtaining the variation of fracture toughness versus part thickness, capturing and recording data in the threshold region, repeating the test at least three times, and in some cases, the test must be repeated due to invalid results, they all make the virtual testing technique extremely helpful to overcome the time and cost of testing difficulties. An attempt is made to implement the same methodology on the lamina of a laminated composite by assuming both constituents are linear, brittle, and homogeneous. The virtual testing methodology is briefly discussed in Sect. 10.2. The proposed technique utilizes the basic data from a simple static tensile stress-strain test to estimate the fracture mechanics allowable through the extended Griffith theory. In addition to the fracture toughness value (G_{Ic}), the fatigue crack growth data can also be estimated through the virtual testing concept (Sects. 10.2 and 10.3). In Sect. 10.4, the method of generating the fatigue crack growth rate is discussed, and virtual testing results are compared with several test data available in the literature. In Sect. 10.5, the fracture properties for the epoxy matrix are estimated by the virtual testing method, and the results show good agreement with the test data.

The fracture behavior of adhesively jointed plies under service loading also was assessed by using the virtual testing method (Sect. 10.6). For laminated composites, the fracture toughness, as well as the fatigue crack growth data, must be generated for the analysis assessment of debonding or delamination process. The presence of delamination is considered an essential localized defect that will grow slowly, causing ply separation and final failure. Thus, obtaining both the fracture toughness of adhesive material and the fatigue crack growth rate (how fast the delamination will grow, *da*/*dN*) is vital for a comprehensive life analysis of a laminated composite part. Fracture properties of fibers can also be generated through the virtual testing technique. As an example, fracture properties of Kevlar 49 material were generated and presented in Sect. 10.7. In all cases, the fracture toughness (G_{Ic}) and the fatigue crack growth rate data were compared with the existing test data. Finally, in Sect. 10.8, concluding remarks on virtual testing through molecular dynamics and coarse-grain technique are presented in detail.

In Chap. 11, the high cycle fatigue and behavior of composite material under cyclic loading are discussed. The core of emphasis will be given to the laminated composite material. The fatigue behavior in metals is well understood, and there is no need to elaborate on it in this chapter. However, there are still some concepts in fatigue of metallic material that are necessary to discuss because they are common between both materials. In Sect. 11.2, the reaction of the laminated composite under tension-tension cyclic loading is presented. Under tension-tension fatigue, specimens can achieve as high as 10^6 cycles at load up to 90% of the ultimate strength, indicating that the tension fatigue of composites, even in the presence of holes, is not a major concern. Several S-N data are presented for different PMCs, and results indicate that their behavior is a function of several parameters. The S-N data are strongly stress-ratio, R, dependent. Section 11.3 discusses composite parts under compression-compression cyclic loading with $R > 0$, which results of test data are shown to be different from the tension-tension load environment.

In Sect. 11.4, the influence of both plies orientations and plies arrangements is discussed. The effect of high frequency when composite material is under mean stress (σ_m) > 0 is also presented in this section. Another parameter influencing the S-N curve for the laminated composite material is the amount of fiber volume, V_f, in each ply. The constant life diagram and effect of the notch in laminated composites under cyclic loading are also discussed in Sect. 11.4.

In Sect. 11.5, the formula describing S-N data for composites is presented in detail. In Sect. 11.6, the method of conducting fatigue test data, based on ASTM standards, is shown where all of the fatigue test data scattered in one S-N plot are as a result of several constant amplitude loading tests having the same stress ratio, R. Note that the S-N data for laminated composites will provide lower number of cycles to failure (i.e., shorter life) when the load is highly compressive.

Other parameters affecting material properties, such as the environment and the suitability of whether composite parts should be exposed to that environment when designing the part, are presented in Sect. 11.7. The environmental effects such as moisture, temperature fluctuations, any possible radiation exposure, and probably the presence of chemicals will be discussed briefly in this section. These factors should be fully assessed by engineers before fabrication and during the operation of composite parts if the material is going to be exposed for a long time.

Finally, several analytical models that are able to estimate the number of cycles to failure are presented in Sect. 11.8. Several constants are associated with these models, where they must be obtained by conducting several fatigue tests for the material under consideration. Each fatigue model must establish its own constants through several tests that may not be used for other composite materials with different fiber volumes or orientations.

In Chap. 12, damage accumulation due to cyclic loading in laminated composites is discussed. Under high cycle fatigue, where the load magnitude is low and the number of cycles to failure is high, structural failure seldom occurs rapidly without indicating that the part is failing. In reality, the gradual damage growth that eventually leads to final failure is due to the damage accumulation that starts with damage initiation, damage growth, and final failure. The accumulation of damage can be as a result of several load spectrums of different stress magnitudes. The stronger the matrix and fiber constituents of composite material, as well as their interface, the longer life is expected.

When there is no variation in load magnitude with stress ratio (R) a constant, the damage accumulation can be found directly through the S-N data. However, for variable amplitude load environments, the well-known Miner's rule can be used to determine the total damage accumulation via several S-N data with different stress ratios. Miner's rule for damage accumulation is a linear sum of the contribution of each load environment, as described in Sect. 12.2. The modulus of the elasticity reduction method in the composite material is another method that can be employed for estimating the damage accumulation under cyclic loading. As the fatigue damage accumulates and progresses with load cycling of different magnitudes, the dynamic modulus of the material continuously decreases (Sect. 12.3).

The damage accumulation can also be estimated through the fracture mechanics approach. This approach is similar to the modulus reduction technique described in Sect. 12.3. With this technique, the critical value of the stress intensity factor, K_{Ic}, or the energy release rate, G_{Ic}, can be utilized to measure the incremental growth Δa, from a small preexisting flaw up to the failure for each load spectrum and the total number of cycles associated with it, N_{fi}. Subsequently, Miner's rule can be employed by analysts to find out if the final failure is imminent or not (Sect. 12.4). An example problem related to this approach is also provided to the reader in Sect. 12.4.

In Chap. 13, the repair of damaged composite parts is discussed. Composite laminates are used in a wide range of applications, some of which are utilized for primary structures. When they get damaged during service usage, repairing may be needed to assure structural safety and avoid loss of life. For example, if the fuselage of a passenger aircraft gets damaged, the repair must be done because damage to the fuselage can cause catastrophic failure and harm to passengers during flight. The purpose of any composite repair is to restore the structural loads in the damaged area so that it is adequate for the overall composite structure to perform safely under maximum design load during the service operation without failure. To achieve that, there are a few fundamental steps regarding composite repair that must be followed and are described in Sect. 13.2. In some cases, it is mandatory to conduct a comprehensive, nondestructive inspection after repair, and the residual strength evaluation must show that with the presence of any damage left behind, the remaining damaged structure can withstand the applied load (Sect. 13.2).

There are several ways of repairing damaged composite parts. The tapered scarf joint repair has long been recognized as the best method for repairing laminated composite parts because of its uniform shear properties through the jointed area. In this repairing technique, the already prepared repair material is incorporated into the laminate in place of the material removed due to the damage (Sect. 13.3). Plug or adhesive repair is also a preferred and rapid method of repairing composite structures, especially when externally bonded patches can no longer meet the stiffness, strength, and flushness requirements (Sect. 13.4). In both cases of scarf and plug repair, the extent of damage after inspecting the damaged area determines whether the component needs to be repaired or replaced.

The lap joint and patch repair is another repair technique used in industries. The external patch repair is relatively simple and faster and does not need tapering and removing the parent material; hence it is widely used in aircraft for damaged parts to keep an airplane in operational condition (Sect. 13.5). To ensure that the adhesively patch bonded parts have adequate life, it is required to have mechanical testing, analysis, and post-NDI inspections in the repaired area.

Two examples related to repair analysis that can be helpful to the reader are presented in Sect. 13.6. These examples are provided to assess the integrity of the repaired area and the number of cycles to failure that the repaired area can withstand. Bonded repairs offer many advantages over mechanically fastened doublers. For example, the improved fatigue life, reduced corrosion, in-situ application, and high level of bond durability under operating conditions are benefits of patch-bonded repairs. Analysis of mode I failure, which might occur in a bonded repair of a cracked metallic fuselage skin, is presented in this section. They comprise crack growth, patch debonding, and nucleation of new fatigue cracks in the skin at the patch edges. A comprehensive assessment of damage tolerance of composite patch repair with several example problems is presented in Sect. 13.7.

Parameters that must be considered when conducting patch repair are explained in Sect. 13.8, and finally, a summary of steps for composite repairs on damaged composite parts is provided in Sect. 13.9. These key steps for a successful repair are inspection and damage assessment, removal, and preparation, laminating the repair area, and inspecting the repair and finish.

In Chap. 14, different types of nondestructive testing techniques are discussed. The nondestructive testing (NDT) and inspection of composite structures after manufacturing, to assure the quality of products and for in-service damage detection, have provoked the development and adaptation of several inspection methods and techniques over the years (Sect. 14.1). Different types of NDI methods are currently available to detect flaws in the composite structures prior to, as well as during their service usage. A-scan, B-scan, C-scan (as part of through-transmission ultrasonic), tap hammer device, and the mechanical impedance analysis will all be described here. A brief description of through-transmission ultrasonic (TTU) is presented in Sect. 14.2. In the TTU inspection method, high-frequency sound waves from the probe travel through the tested material to check for anomalies in the material. These sound waves, after bouncing back from the test object, are captured and converted by the probes into an electrical energy pulse that can be assessed for flaw detection. The A-scan (amplitude scan), the B-scan, and the C-scan all use ultrasonic waves and are discussed in Sect. 14.2.

Manual tap test or tap hammer device inspection test utilizes frequencies in the audible range between 10 and 20 Hz. Tap testing is perhaps the most common, simple, and inexpensive inspection technique employed to detect delamination or ply separation. A more detailed description of the tap inspection method is presented in Sect. 14.3.

The mechanical impedance analysis (MIA) is another inspection method for detecting flaws. The MIA method is a well-established NDI technique for detecting a separation between plies because of the lack of adhesive in laminated structures and delaminations inside composite materials up to a depth of a few millimeters. In this NDI technique, the mechanical impedance measurements are aimed at evaluating the damage-induced changes of structural parameters reflected in the frequency response, where the frequency response of a system is the quantitative measure of the magnitude and phase of the output as a function of the input frequency. More explanation on this topic is provided in Sect. 14.4.

The application of NDI in fracture mechanics is to analyze the life expectancy of laminated composite parts from crack initiation to final failure. The purpose of nondestructive inspections in aircraft, aerospace, military, and many other industries is to find out the possibility of finding a sizable crack that, if undetected, would grow in the structure and cause catastrophic failure. Well-known inspection methods, such as standard penetrant inspection, standard Eddy current inspection, standard magnetic particle inspection, standard ultrasonic inspection, and standard radiographic or X-ray inspection, are all briefly described in Sect. 14.5. For each of the abovementioned methods, there is a minimum initial flaw size that will escape detections. These are flaw sizes that analysts can use for their life assessment of their composite parts. Their sizes are determined by well-trained inspection personnel from reliable agencies like NASA. When the standard flaw size is too large because the result of life assessment yields inadequate life, analysts like to have smaller flaw sizes, which can be obtained through the special inspection technique. Both the standard and special inspection flaw size estimations are discussed in Sect. 14.5.

In Chap. 15, the 3D printing process, also known as additive manufacturing (AM), is discussed. It is a relatively new method that creates physical parts from digital models by adding material layer by layer until the desired thickness and shape are achieved. There are several additive manufacturing methods being used for 3D printing of parts that each differ by the materials that may be utilized and the way that layers are added to build composite parts (Sect. 15.1). When a laser beam is used as the source of energy to melt and fuse materials together, it is called laser additive manufacturing (LAM); see Sect. 15.2. There are two LAM techniques that are currently used to create thick composite parts for many industrial applications. They are powder bed fusion (PBF) and direct energy deposition (DED), which are both presented in Sect. 15.3. Several subclasses of DED and PBF are also discussed in this section. Their commonalities and differences between the PBF and DED processes are also included in this section. The two AM techniques utilize a focused energy source, such as laser or an electron beam, as a heat source to melt and fuse wires or powders together layer by layer.

Other 3D printing manufacturing methods, such as binder jetting (BJ) and sheet lamination (SL) methods, are presented in Sect. 15.5, where binder jetting utilizes an inkjet printer to deposit a liquid binder onto powder layers, and the sheet lamination (SL) process is an additive manufacturing technique that creates objects by layering and bonding sheets or foils of material together.

Support structures play an important role in the fabrication process of additive manufacturing parts and are discussed in Sect. 15.6. The support structure is a temporary structure that is created alongside the actual part being manufactured. It mainly provides stability and prevents distortion or collapse of the composite part during the 3D printing or deposition process.

In manufacturing composite parts through 3D printing, it requires a feeding system. The filament feeding, pellet feeding, wire feeding, and powder feeding mechanisms techniques are all discussed in Sect. 15.7. Applications and examples of the DED process in LAM technology are discussed in Sect. 15.8. In this section, the focus is on reducing component failure rates and extending component lifespans through repairing by using the DED technique.

For additive manufacturing, a digitized model of the part you wish to create is required. There are two types of software that can interact with the 3D printers to create the desire object for industry, which are (1) computer-aided design (CAD) and (2) computer-aided manufacturing (CAM). CAD allows users to design new products in 3D, while the CAM software directs the 3D printer to build the composite products in accordance with the CAD model (Sect. 15.9).

Mechanical properties of AM metallic or composite parts, residual stresses, distortions, and porosities are all discussed in Sect. 15.11. Application of 3D printing on fabricating composites is presented in Sect. 15.12. Finally, the additive manufacturing of composite parts with short fibers and long continuous fibers is presented in Sect. 15.13.

Contents

Chapter 1
Understanding of Composite Materials (Constituents of Composites)

1.1 Background

Composites are formed by combining two or more basic materials on a macroscale level, which in final form exhibits valuable properties, such as good strength, high stiffness, low weight, good fatigue properties, corrosion resistance, thermal and acoustic insulation, easy to repair, and many other attributes. These desirable qualities make composite material to be suitable for many industries. Composite materials have emerged as a significant class of structural elements and are now being considered as a replacement for traditional materials currently being used in aerospace, automotive, and others. Unlike metallic alloys (e.g., steel, aluminum, and titanium alloys), composites are heterogeneous mixtures where the combined constituents don't mix and are not soluble, as is in the case of alloys.

In manufacturing industrial parts with composite material, they are usually made of either ceramic, metallics, or polymers as the matrix (aka binder) and high strength reinforcement (aka fibers or whiskers) embedded in the binder acting as the main load-carrying elements. The great thing about composites is that the engineering properties that a part must possess can be achieved by a careful selection of matrix and reinforcement. That is, by properly selecting matrix and fiber materials, with suitable fiber configurations (short, long, straight, woven, braided, laminated, etc.), engineers can tailor their part to the specific requirements set forth by the customers. These two fundamental composite components (i.e., the reinforcement or fibers and the binder or matrix) have distinct chemical and physical properties and are merged to craft a better material with mechanical properties different than the individual constituents. Upon completing the composite material's fabrication, the individual elements maintain their properties, but the mechanical behavior of composites differs from that of each individual component, i.e., fibers and matrix. The matrix provides a medium for binding and holding fibers together through its cohesive and adhesive characteristics. Moreover, the matrix offers protection to the reinforcements from environmental damage, serves to transfer load, and provides durability

B. Farahmand, *Fundamentals of Composites and Their Methods of Fabrications*,
https://doi.org/10.1007/978-3-031-80201-0_1

and functionality to the final products. Note that the fiber constituent part of the composite is of little use by itself and must be adhesively bonded to the matrix in order to be effective as a load-carrying element. It is fibers, not the matrix, that provide the strength, stiffness, and in some cases toughness to the composites. This is especially true for carbon fiber epoxy composites that have excellent mechanical properties and owe it to carbon fiber's outstanding strength and stiffness qualities [1].

Aside from the types of matrices and reinforcements used in composite materials, several other parameters can affect the properties of composites. They are the constituents' relative proportions of matrix and fiber (the volume fraction), the reinforcement geometry, and the nature of the interphase, the type of fiber used (i.e., carbon, glass, aramid fiber, and others), the orientation of the fiber for each ply, the method of fabrication and manufacturing, the temperature and duration of the cure cycle [2]. Regarding the temperature and duration of the cure cycle, they both will contribute to the composite final mechanical properties. Curing is the change in physical and chemical properties of composites by chemical reactions and it is done by applying heat and catalysts with or without pressure. It requires the right amount of time in order to obtain the best properties. Finally, the relative proportions of the fiber in the matrix contribute to the load-carrying capability of the composite material. The greater the volume of fiber in the composite (i.e., the higher the fiber volume fraction), the stronger will be the composite. The volume fraction in structural parts for laminated composites used in the industries is between 60% and 70% [3].

Just like many new alloys being used in industries that have proprietary information about the kind and amount of alloying elements and the level of temperature used during the manufacturing process, composites also have proprietary information on key parameters, such as the curing time, kind of catalyst, number of layers with fibers of different orientations, the stacking arrangement, and the kind and amount of fibers in the matrix. The user may contact the supplier to obtain the best polymer composite material that is most suitable for a particular application and environment under consideration.

Engineers and designers, who are familiar with metallic materials, are very well aware of metals' homogeneous and isotropic nature and should become familiar with composite materials if they are involved with designing structural parts made of composites. An isotropic material is defined as one that has the same properties in all directions. Composites exhibit different properties in different directions since they possess nonhomogeneous and nonisotropic (anisotropic) characteristics. If laminas are made of isotropic homogeneous fibers, and an isotropic homogeneous matrix, the property of the lamina varies from point to point depending on whether the point is in the fiber, the matrix, or the fiber–matrix interface.

One of the advantages of composite over their metallic counterpart is their ability for engineers to select the best combination of each component to provide the best properties tailored to a specific fluctuating load environment that structures will be exposed to during their service life. However, there are a few drawbacks that engineers must be aware of when it comes to selecting composite material for

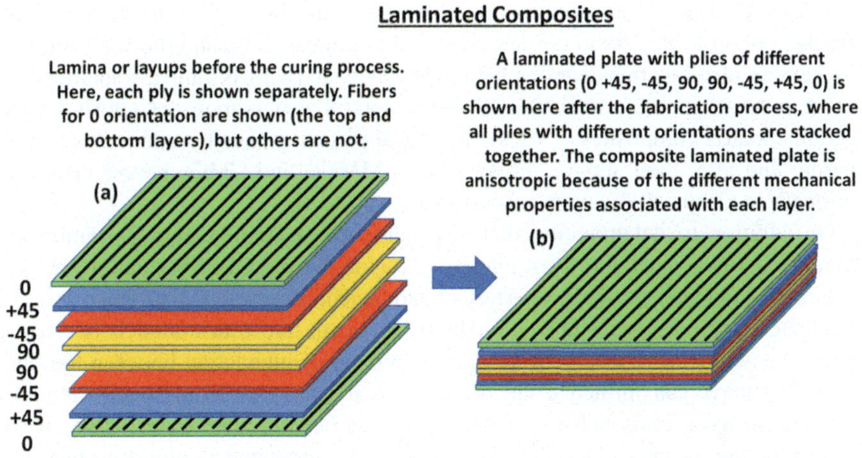

Fig. 1.1 Laminated composite plate before and after final fabrication. (**a**) Layouts before the curing process. Each layer has a different fiber orientation (0 + 45, −45, 90, 90, −45, +45, 0). Only the top and bottom layers are shown here. (**b**) The same plies after the fabrication process

manufacturing critical parts because the failure of critical parts can cause loss of life and, consequently, significant damage to the manufacturer's reputation. For example, in the case of laminated composites (Fig. 1.1), the separation of lamina or plies (aka delamination or debonding) is a common occurrence during the working operation and must be given special attention. In Fig. 1.1a, different plies are shown before the curing process, and in Fig. 1.1b a laminate is constructed by adhesively stacking the same number of plies in the direction of the thickness, where each ply has a different orientation for obtaining the best mechanical properties. As already mentioned, the laminate is composed of several laminas (or laminae) and, in the case of fiber-reinforced composites made of several layers, the buckling stability of the material improves with increasing the number of layers in the laminate. A lamina of a laminated composite material generally having a thickness on the order of 0.005 in (0.125 mm).

The delamination phenomenon is the single most important mode of damage propagation in laminated composites that occurs mainly because of out of plane and compressive loads, that have the tendency to separate plies from each other and cause delamination. Impact load on critical locations of composite structures can also cause delamination within the laminated parts such as hailstones, birds and lightning strikes, runway debris, as well as debris from tires rupture due to landing and takeoff of aircraft, and tool drop during maintenance.

Finally, the anisotropic nature of plies of different orientations, that are adhesively attached to each other, can initiate delamination when plies are subjected to applied loads. Damage initiation must be minimized and prevented, and engineers should give special consideration to those parts that are classified as critical by conducting an appropriate nondestructive inspection prior to and during their service usage. Here, critical parts are defined as those components whose failure during operation can cause catastrophic failure and possible loss of life. The wing and the fuselage of a passenger aircraft are classified as critical parts, and the fuel tank of a rocket can also be considered as being critical. Critical parts can be manufactured using polymer matrix composites (PMCs), metal matrix composites (MMCs), and ceramic matrix composites (CMCs); all critical parts must be given special attention before and after their service usage. PMCs, MMCs, and CMCs are all discussed in-depth in Chaps. 4, 5, and 6, respectively.

In addition to delamination that can cause the separation of plies within the laminated composites, the weak interfacial region between the fiber and matrix has a critical influence on the properties of laminated composites. The interphase of laminated composites is the region in which loads are transmitted between the reinforcement (fibers) and the matrix. The amount of interaction between the fiber and the matrix can influence the design, and it may vary from strong chemical bonding to weak frictional forces. This can often be controlled by using an appropriate coating on the reinforcing fibers. Generally, for a given composite material, with both brittle matrix and fibers possessing high strength properties, it requires to have a strong interfacial bond, but this may lead to a low fracture toughness and fast fracture without having the ability to release energy before fracture. That is, it has low tolerance toward a small preexisting defect. However, by proper control of the physical and mechanical properties of the fiber–matrix interface, high strength characteristics can be combined with high toughness. To fully utilize the potential of such composites without totally reducing the strength, it is necessary to understand the failure mechanisms leading to eventual fracture. Therefore, it is more proper that the design be somehow between the two extremes in order to have both high stiffness as well as some amount of toughness as a result of energy release rate causing stable crack growth.

1.2 Several Types of Composite Material

Composite materials are usually classified by the type of reinforcement (or fibers) and the matrix (or binder) they use. These reinforcements are embedded into the matrix that can bind and hold fibers together to provide strength to the composite. The oldest and simplest example of the application of composite material is the presence of straws in the matrix of mud which has been used for thousands of years since prehistoric times for the purpose of building any kind of earthen construction. Straws are plant material that is left over after grains (like wheat) are harvested and are used as an additive in mud for enhancing strength, i.e., the matrix is the mud, and the reinforcement is the straw. Here, the straw was randomly placed in the mud with various length sizes, both short and long with thicknesses not more than 1–2 mm. The mud without straw has a strength of less than 1 lb./in^2 (psi) and the addition of straw resulted in having strength three times better. Yet disadvantages that a mud earthen construction has when constructed are as follows: (1) they can be easily destroyed by wind, rains, and flooding if it is not protected well enough, (2) they can easily be damaged by dampness or dry weather, and (3) they are not durable, (4) mud flooring gets eroded due to certain activities like constant walking and running. The example above was probably the simplest type of composite that has no application in industries. Actually, the concept of composites is not an invention coming from humans. Wood is a natural composite material consisting of cellulose fibers with good strength and stiffness in a matrix of another polymer. Nature does a much better job of designing and manufacturing composites, where bones and teeth are another natural example of composites, where it is combining hard ceramic reinforcing phases in natural organic polymer matrices.

Nowadays, there are three main types of composites, that have great mechanical and material properties with applications in industries and are designated based on the types of matrices and reinforcements being used. Metal matrix composites (MMCs) [4, 5], ceramic matrix composites (CMCs) [6, 7], and polymer matrix composites (PMCs) [4, 8] are the most popular composite types that have applications in industries like aircraft, aerospace, automobile, and many others. Note that, the term advanced composites are those composites that are traditionally used in aerospace, airplanes, helicopters, automobiles, pipes and tanks, and marine, where these industries alone are consuming a high percentage of advanced composites material in the United States [9, 10]. Advanced composite materials are defined by highly high-strength fibers with unusually high stiffness, or modulus of elasticity properties, where they are adhesively bound together with the weaker matrices. The growth projections for aerospace usage of advanced composites are estimated to range from 8.5% per year to 22% per year. The most common fibers used in advanced composites are glass, graphite, and Kevlar. All three types of composites (PMCs, MMCs, and CMCs) will be discussed later in this book.

1.3 Metallic and Nonmetallic Fibers

Usually, fiber materials are weaker in the bulk format with much lower strength when they are compared with the same material, but with much smaller dimensions. Experimental data show when they are thin (with a high length-to-diameter ratio) their tensile strength is much higher. For example, in the case of metals and their alloys, the bulk materials have much less tensile strength when compared with their metallic filament, that are much smaller in size. These thin filaments known as whiskers or metal wires, usually have around 1–3 mm in diameter with a high length-to-diameter ratio. Generally, whiskers possess high strength and stiffness due to their nearly perfect crystal structure with a minimum amount of micro- or nano-size defects in the material. Their strength is almost close to the theoretical strength of metals [11, 12]. The theoretical strength (or ideal strength) of a solid is defined as the maximum possible stress that a perfect solid (free from flaws) can withstand the applied tensile load. It is often much higher than the strength of materials in the bulk form that possess a reasonable number of microdefects. This is also true with respect to carbon fibers in the bulk form where the strength is much

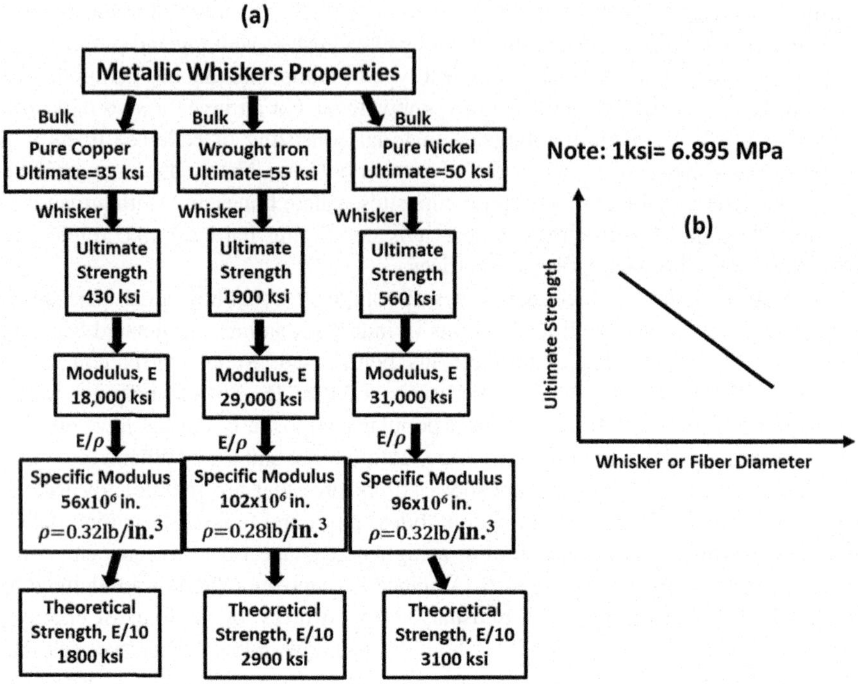

Fig. 1.2 (**a**) Whisker mechanical properties for copper, nickel, and iron metallic materials together with their bulk form values. The theoretical strength values are also presented here. (**b**) Variation of whisker or fiber diameter as a function of ultimate strength [13]. Note that the values mentioned above are average values

lower than the fine thin fiber [13]. In reality, it is impossible for materials to reach their theoretical strength due to internal defects, and also true when they are in the whisker format. The idea that materials in the bulk form are weaker is the probability of finding more defects in the bulk form, where their number is much bigger than when they are thin with a slender shape. These defects are the source of crack initiation and growth that will merge with each other to form a larger crack, leading to structural failure when subjected to the applied load.

The ideal or theoretical strength value of metals is approximated to be equal to ~E/10, where E is the tensile modulus elasticity of the material. Other sources, based on different theoretical strength calculations, provide values ranging from E/4 to E/13. Figure 1.2a shows the strength value of several metallic fibers (whiskers) including their ultimate strength in bulk form and their theoretical value based on the E/10 value [13]. As it is evident from this figure, the tensile strength of metals in the bulk is much lower than when they are in the whisker (filament) form. In the case of copper, the bulk strength has a value of 30.4 ksi, whereas, in the whisker form, the strength is close to 430 ksi, almost 14 times stronger, and the theoretical strength of E/10 ~ 1800 ksi. The same is true with respect to Iron and Nickel as is shown in Fig. 1.2a. Figure 1.2b illustrates the graphical relationship between the whisker

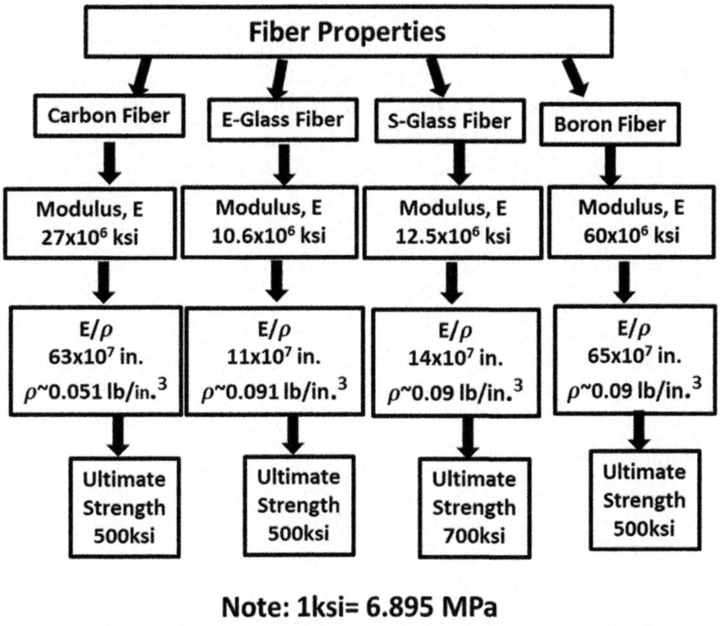

Fig. 1.3 Fiber properties of carbon, E-glass, S-glass, and boron [13]. Note that the values of E and E/ρ mentioned above are average values

(or fiber) diameter and the ultimate strength. The plot illustrates the decline of ultimate strength as a function of whisker diameter, which the difference is due to the inherent flaws in the material, whereby reducing the number of flaws, will help to increase the strength of the material [13].

Some typical metallic and nonmetallic whisker materials are graphite, alumina, iron, copper, silicon carbide, and silicon. Single-crystal whiskers of these materials are known for having very high tensile strength. As it is shown in Fig. 1.2, in addition to the modulus and strength properties of whiskers of copper, iron, and nickel, their specific modulus and strength values are also shown [13], which can be compared with fibers of nonmetallic counterparts shown in Fig. 1.3. These values clearly show that fibers of nonmetallic material have higher E/ρ values when compared with whiskers of copper, iron, and nickel (Fig. 1.2).

In Fig. 1.3, the ultimate strength of several nonmetallic fibers, such as carbon fiber, E-glass fiber, S-glass fiber, and boron are shown [13]. Boron fibers (filaments) have high-strength, high modulus, and lightweight properties that are used in the matrix of ceramics as a reinforcement for advanced aerospace structures. Boron epoxy is also used as a patch on metallic substrates when they encounter cracks during operation to prevent cracks from advancing. E-glass fibers (also known as Electrical glass) are the standard glass composition used for most glass fibers. It is made from silicon oxides (SiO_2) and oxides of aluminum (Al_2O_3), calcium (CaO), magnesium (MgO), and boron (B_2O_3). Note that, in both Figs. 1.2 and 1.3 one of the key parameters used to make a comparison between metallic whiskers (copper, iron, and nickel) and fibers to be used in fabricating composites (carbon, E-glass, S-glass, and boron) was the specific modulus, the ratio of modulus of elasticity and the material density (E/ρ), which its importance with regard to material weight will be described as follow: Assuming the axial elongation of a rod under an axial load, P, is given by $\Delta L = PL/AE$, where L = length of the rod E = Young's modulus of elasticity of the material of the rod, and A is the cross-sectional area. Because the mass, m, of the rod is given by $m = (\rho AL)$, where ρ = density of the material of the rod and AL = volume, therefore, $m = (P\,\Delta L)/(E/\rho)$. From this simple analysis, it can be seen that the lightest beam for specified displacement, ΔL, under a specified load, P, is one with the highest (E/ρ) value. Another important parameter similar to E/ρ is called the specific strength which is defined as the ratio between the strength (σ_{ult}) and the density of the material (ρ). The two ratios (E/ρ and σ_{ult}/ρ) are high in composites and fibers.

For example, the modulus (E) of a graphite–epoxy unidirectional composite (26,300 ksi) could be almost the same as steel (30,000 ksi), but the specific strength is almost ten times that of steel, where the density of steel $\rho = 0.29$ lb./in^3 and carbon fiber of $\rho = 0.051$ lb./in^3 (E/ρ ~10.3×10^7 in vs. 51.5×10^7 in) [4, 13]. To define it differently, for several materials of almost equal modulus of elasticity, E, those with lower density are more attractive to industries. As was shown above, the two materials, steel and graphite–epoxy, have almost the same modulus, E, but different density, ρ, values ($\rho = 0.29$ lb./in^3 and $\rho = 0.051$ lb./in^3, respectively).

In addition to metallic whiskers shown in Fig. 1.2, reinforcement used in the composite matrix are in general classified into several groups as shown in Fig. 1.4. Each group has been categorized into its components. For example, in the case of fiber reinforced, the subclasses are continuous and short fibers which are also categorized into their own classification [14] (Fig. 1.4). Fabricated carbon fibers as well as graphite fibers are originally made of tiny thin fibers which are thinner than the diameter of a human hair (around 10 μm in diameter). Several thousand carbon fibers are bundled together to form a strand or two, which may be used to be woven into a fabric or spools of carbon fibers. Note that carbon and graphite fibers are slightly different from each other. Carbon fibers have 93–95% carbon content, but graphite has more than 99% carbon content. Also, carbon fibers are produced at different temperatures, where carbon fibers are produced at 2400 °F (1316 °C), and graphite fibers are typically produced at more than 3400 °F (1900 °C) [15, 16]. These fibers are not usable by themselves and must be bundled together with a plastic resin to be used for manufacturing industrial composite parts. Otherwise, they are flimsy and cannot hold the applied load other than tension. The carbon fiber strand or tow is the basic form of carbon fiber before it is converted into a finished carbon fiber reinforcement.

Tows are rated according to their fiber, or filament content, commonly referenced as 3 k, 6 k, 12 k, and 15 k. The k stands for "thousand," meaning a 3 k tow is composed of 3000 carbon filaments. Since a single strand of carbon fiber is only about 5–10 μm thick, a 3 k tow comes in at only around 0.125 in thick (3.17 mm). In Fig. 1.5, one can see a solid piece of carbon fabric to be used for a laminated

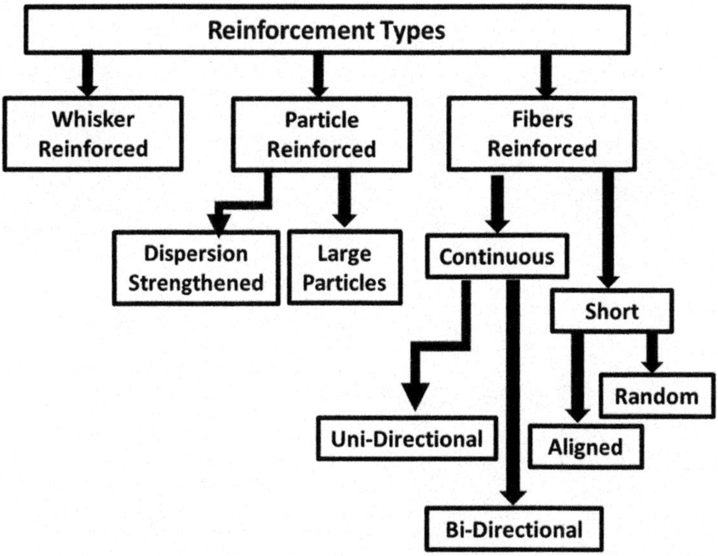

Fig. 1.4 Illustrating different types of reinforcement used in composites (whiskers, particles, and fibers)

composite. Carbon fibers are almost twice as stiff and approximately five times stronger than steel. They are stronger than steel, yet they are much lighter. They are highly chemical resistant with low thermal expansion, however, they are relatively expensive when compared with similar fibers, such as glass fibers. The raw material used to create carbon fiber is almost 90% polyacrylonitrile, and the remaining 10% is made from rayon, which is a fabric made from purified cellulose fibers. Carbon fibers have high strength and high stiffness that are used in a variety of structural applications. Compared to aluminum, carbon fibers are, on average, eight times stronger, yet 1.5 times lighter. Coupled with the proper resins as a matrix, carbon fiber composites are one of the most corrosion-resistant materials available. Moreover, they have superior tension–tension fatigue properties to all known metallic structures. In Fig. 1.5a, tiny carbon fibers (filaments) around 10 μm in diameter are shown. Each strand is almost 100 times thinner than human hair and is made of carbon atoms with covalent bonds, which makes the fiber highly strong. It is possible to bring many of these fibers together in a single ribbon to create a carbon tow. Tows of carbon fiber can be used to manufacture fabric made of woven carbon filaments (Fig. 1.5b).

The polymer by itself is weak and will fail easily when subjected to applied load (Fig. 1.5c), however by implementing fibers in the polymer matrix along the direction of the load (0°), then the strength will increase by several times (Fig. 1.5d). By incorporating fibers in the 0° and 90° orientations, the composite's strength in both directions will significantly increase (Fig. 1.5e). Even though the

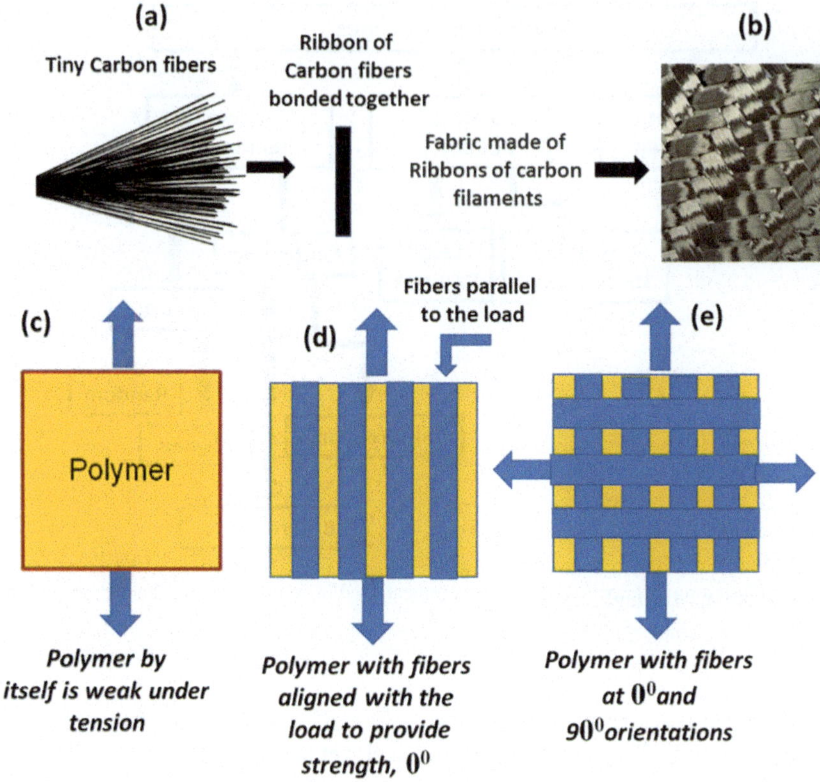

Fig. 1.5 (**a**) Tiny Carbon fibers, (**b**) tiny carbon fibers bonded together, (**c**) polymer without fibers, (**d**) polymer with fibers in **0**° direction, (**e**) fibers in both **0**° and **90**° directions (providing strength in both directions)

presence of fibers will be helpful to bring strength into the matrix and composite, the presence of anomalies that may appear in composite parts during the manufacturing phase, arising mainly from fibers that can weaken the composite material. These are: (1) incorrect total fiber volume fraction, (2) broken fibers, (3) misaligned fibers, (4) inhomogeneous fiber distribution. Other factors that can weaken the composites are: (1) overlaps or other faults in the arrangement of plies, (2) the presence of delamination among plies, (3) mechanical damage around machined holes, (4) local bond failures in adhesively bonded composite components, and (5) high humidity, and a large temperature range environments [7]. For example, as a result of manufacturing defects from fibers, an increase of 1% voids content in composites will lead to a decrease in tensile strength (10–20%), flexural strength (10%), and interlaminar shear strength (5–10%), respectively. These degradations in properties can be eliminated by controlling the processing parameters of manufacturing processes [17].

1.4 Limitations in Using Composites

High cost of fabrication of composite parts and time to manufacture them are two main critical issues for industries when numerous parts are required to have by the customers in a short time. A part made of the composite may cost several times more than the cost of material needed for fabrication. Improvements in processing and manufacturing methods are needed to lower these costs in the future if industries are interested in using their final products. Such a demand to use composite parts is because of their attractive high specific modulus (E/ρ) and high specific ultimate (σ_{Ult}/ρ) properties when compared with metallic material, such as aluminum, steel, and titanium alloys. However, to employ composite materials for industry use, engineers must have material allowable through testing to be able to conduct a meaningful stress and damage tolerance analysis to be sure whether composite parts can survive the load environment or not. These tests must be performed based on well-established guidelines approved by reliable agencies. To obtain composite mechanical allowables, numerous coupon testing is required due to their anisotropic nature that can be costly and time consuming.

From both testing and analysis perspectives, the mechanical characterization of composite parts through testing is much more complex and costly than that of a metallic counterpart. This is because composite materials are not isotropic, and their properties are not identical in all directions. If for an isotropic material only one test is needed to characterize its allowable, for anisotropic material several tests are needed. Therefore, engineers must have more material parameters for their structural integrity assessment. For example, a single layer of a graphite–epoxy composite requires nine stiffness and nine strength constants for conducting a 3D stress analysis. The nine stiffness constants (also they are referred to as *elastic* properties) have three components per direction, which are the modulus of elasticity E, the shear modulus G (also known as the modulus of rigidity), and Poisson's ratio ν, which all are shown in Fig. 1.6a [18]. However, in the case of a typical aluminum alloy that is considered isotropic, engineers require it to have only two elastic properties (E and ν). In Fig. 1.6a, b, the relationship between G and E and their derivation are shown for composite material in 1, 2, and 3 directions.

For composites shown in Fig. 1.6, the stiffness properties of fiber-reinforced composites are highly dependent on the fiber orientation of plies relative to the applied load. In addition to the nine elastic stiffness properties mentioned above for 1, 2, and 3 directions, the nine strength properties of a laminated fiber-reinforced composite are also highly dependent on the material orientation relative to the applied load [19]. In the case of composites, the strength properties can be significantly different under tension and compression loading due to different possible failure modes, including delamination under compressive load (Fig. 1.6a). Both modulus and strength properties, shown in Fig. 1.6, are inadequate to conduct a comprehensive structural integrity assessment when the composite part is under a cyclic loading environment. In addition to static loading, extra data related to fatigue and fracture mechanics must also be available for engineers to determine if the

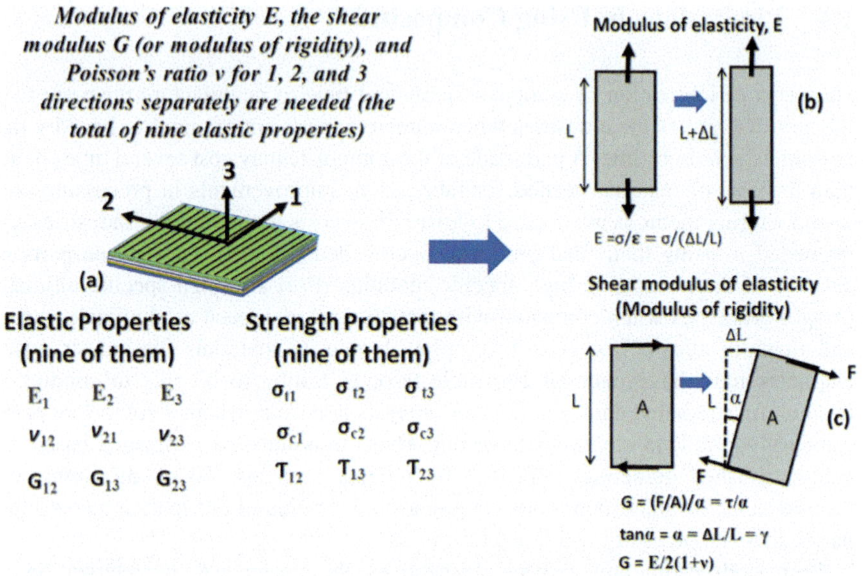

Fig. 1.6 (**a**) Nine elastic modulus and nine strength properties of a laminated composite. (**b**) and (**c**) Simple distinction between E and G of isotropic material. These properties (allowable) are required for a comprehensive stress analysis when the composite part is under load. In part (**a**), t and c are tensile and compressive loads

composite part can survive during its service life or not [20]. These additional properties are also directional dependent due to the anisotropic nature of the material. Figure 1.7 shows a clear description of material properties for isotropic, orthotropic, and anisotropic materials with detailed definitions of their directional dependency. It should be noted that an isotropic material requires only two elastic constants; nine elastic constants for an orthotropic material and 21 elastic constants for an anisotropic materialare needed for engineers to do structural integrity assessment. Therefore, there are cost and time restrictions associated with establishing material properties of the composite material. These properties must be available to engineers, and without them, industries may not show interest in utilizing them for the fabrication of their composite parts.

Another critical point regarding the limitation of using composites material is the uncertainty of their behavior and reduction of their allowable when under various environments, such as exposure to water, water vapor, or other corrosive environments, change in temperature, and long-term physical and chemical stability. To fully be sure about the behavior of a composite part during its service usage, and to prevent structural failure and loss of life, when subjected to a given environment, it will be required to conduct many tests for the environment under consideration.

As was mentioned before, these tests will be costly and time-consuming and will put a financial burden on the manufacturer of parts. Note that, for each allowable, several tests (a minimum of three to five tests) must be conducted and the average

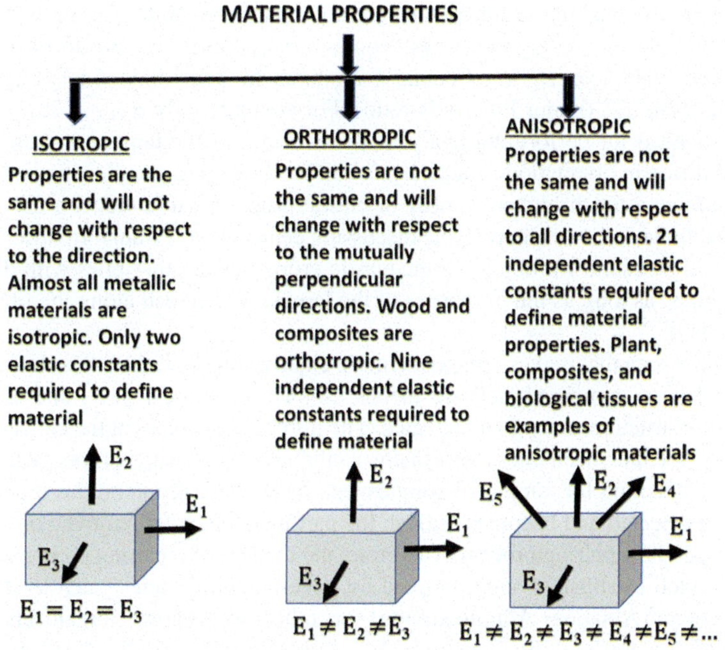

Fig. 1.7 Definition of isotropic, orthotropic, and anisotropic materials. Here, the stiffnesses *E* for isotropic, orthotropic, and anisotropic materials are shown to be different **($E_1 \neq E_2 \neq E_3 \neq E_4 \neq E_5 \neq \ldots$)**

should be used for the analysis as is common in the case of metals. Therefore, there is a need for a reliable alternative, such as an acceptable virtual testing approach that could be able to obtain these properties reliably without relying on many test results [21]. The implementation of a virtual testing technique will be helpful to the industry. It must be available to engineers when they are interested in knowing whether their composite parts can survive the service environment or not. Chapter 9 will discuss a virtual testing technique capable of establishing the fracture properties of metallic materials without the need for costly and time-consuming physical tests. This technique is based on the extended Griffith theory, which can be expanded for composite materials, thereby circumventing the need for expensive and time-consuming testing procedures. Additionally, a section will be dedicated to the molecular dynamics approach to virtual testing. This method involves analyzing the behavior of atoms and molecules when subjected to external forces through computer simulations. Furthermore, the results obtained from molecular dynamics analysis can be interpreted within the context of continuum mechanics to assess failure.

Because of orthotropic or anisotropic properties in composite material, it will make structural analysis more complicated and labor intense. Analysts need to have material allowable available to them to compare their analytical results with the experimental data. In addition, testing, data measurements, and evaluation of data still require establishing standards that are not mature yet. Note that, in the analysis of a single ply only (the plane stress condition), engineers require to have elastic properties limited to E_1, E_2 (Young's modulus in both directions), G_{12} (shear modulus), and ν_{12} (major Poisson's ratio). For strength, only σ_{t1}, σ_{t2}, σ_{c1}, σ_{c2}, and τ_{12} are needed for performing analysis assessment. For the laminates with several plies of different orientations, the analysis will become more complex, and engineers need to have a much higher number of elastic properties to conduct a meaningful analysis. In Sect. 2.6 of Chap. 2, the discussion centers on the transformation matrix, which can be useful for obtaining the elastic properties of other plies with different orientations, as long as the properties of the lamina with a load along the fibers (0°) are available.

Another area that makes composite more challenging to deal with is the repair of damaged parts compared to those metals. Sometimes critical flaws and cracks in composite structures may go undetected. The initiation of cracks at the concentration site and fracture mechanics assessment at the crack tip using proper test data is poorly understood for laminated composites. When damage in composite materials has been detected and becomes critical, the part must be either removed or needs to go through the repairing process. Therefore, the method of repairing composite parts must be well established and approved by government agencies, such as the FAA (the Federal Aviation Administration) or other well-known organizations. In Chap. 13, several repairing methods for composite parts are discussed, including the damage assessment after repairing has been completed. In Chap. 14, the nondestructive inspection methods are described that can be implemented to inspect the parts that have undergone the repairing process and if necessary, the part can have interval inspection during their service usage. The NDI inspections can also be helpful in fracture mechanics analysis to obtain the minimum initial flaw size in the material. Having the initial flaw size in material through the NDI method, engineers will be able to find the total number of load cycles to failure, N_f. Subsequently, the N_f value will be compared with the number of load cycles the structural parts will be exposed to during their service operation [22].

If the traditional linear elastic fracture mechanics analysis on composite parts is required to be implemented, then extensive testing must be initiated to establish fracture mechanics allowable. In addition, conducting the linear elastic fracture mechanics analysis requires a comprehensive nondestructive testing that must be conducted on the part in order to find the minimum initial flaw size that has escaped from detection. That information is necessary as an input for the life assessment of composite parts under cyclic loading [22]. Linear elastic fracture mechanics can be used for a lamina of a laminate if engineers assume the lamina part is homogeneous and linear elastic. However, if this method is not suitable for composite life evaluation, then other analytical tools, such as progressive failure analysis [23], can be implemented, which will be discussed later in Chap. 8. In brief, for progressive failure analysis, a strength analysis proceeds by conducting proper stress or strain analysis of each ply in the laminate and comparing them to the allowable for the ply. Failure often seems to initiate when the induced stress or strain in one of the plies exceeds an allowable stress–strain limit. The first ply failure does not necessarily mean the total failure of the laminate has occurred. The failed ply will transfer some or all its load to another ply in the laminate. When a ply fails, the stiffness of the whole laminate reduces. The analysis will be repeated ply-by-ply until the load-carrying capability of the laminate is depleted, and total failure will occur. This approach is referred to as progressive failure analysis (PFA).

Finally, an additional issue from a fabrication point of few is the ease of processing some kinds of composites, such as polymer-reinforced composites, which are much easier to fabricate when compared with ceramic matrix composites (CMCs). The fabrication of parts made of CMCs is much more costly and time-consuming which may limit industries showing interest in replacing their metallic parts with the CMCs. However, the attractive properties of CMCs, such as their high value of specific modulus and strength, functionality at high temperatures without altering their mechanical properties, and excellent corrosion properties, are forcing industries to improve their manufacturing capability in this area. Chapter 6 will discuss several methods of fabricating CMCs composite parts [24–25].

The immediate future of composite materials may depend more on improving the trustworthiness of present-day composites when subjected to environmental forces rather than on developing and producing new composites. Acceptance of composites in industries will depend very much on the confidence that the end users have in the materials. Therefore, the properties degradation of composite materials by various environmental factors must be understood fully so that corrective measures can be taken to improve their acceptance for various environments and applications in industries. The apparent causes of degradation may involve several factors: (1) loss of strength of the fibers by a stress-corrosion mechanism; (2) degradation of the fiber–matrix interface resulting in loss of adhesion and interfacial bond strength; (3) susceptibility of the matrix material to corrosive agents, such as water vapor, which reduce the properties of composites; (4) sensitivity of matrix modulus of elasticity and strength on time and temperature exposure on the material; (5) faster degradation from the combined action of temperature and moisture; (6) polymer matrix degradation when subjected to a high frequency load environment. As a result of these environmental factors, the effectiveness of composite materials deteriorates when the stiffness is reduced sufficiently to cause structural instability and failure of the composite material.

1.5 Concluding Remarks

The limitations in using composites, such as the high cost of fabrication, the time required to manufacture parts, and susceptibility to unfavorable environments, are the three primary critical issues discussed in this chapter. A part made of composite material may cost several times more than a part made from other materials due to the manufacturing difficulties and associated costs involved in producing consumer goods at competitive pricing. However, composites offer numerous advantages over their metallic counterparts, including superior specific weight and strength, crack arrest features, good fatigue and corrosion properties, and the ability to be tailored to meet specific functional requirements.

Advantages of composites when comparing them to steels in the automotive and transportation industries are: (1) weight reduction between 20% and 40%, (2) around 40–60% savings in tooling cost, (3) reduction in assembly costs and time in joining parts together, (4) resistance to fatigue and corrosion, and scratches, (5) materials and process innovations capable of adding value while providing a cost saving, and (6) safer structure due to the composite material's higher specific energy absorption (how much energy a given mass of material can take under impact before failing).

Undoubtedly, the technology for manufacturing high-quality composite products (like those currently used in aerospace, military, and automotive industries) is available. Still, the equipment, tooling, and running costs are so high that the production of all-purpose items is not feasible. Therefore, there is a need for innovative manufacturing technology of composites so the production time and cost of parts would be more controllable.

Nowadays, three leading composite types have excellent mechanical and material properties with applications in industries and are designated based on the types of matrices and reinforcements being used. Polymer matrix composites (PMCs), metal matrix composites (MMCs), and ceramic matrix composites (CMCs), are the most popular composite types that have tremendous applications in industries. The ceramic matrix composite material (CMCs) has excellent corrosive and high-temperature resistance properties that will make them an excellent choice of material for thermal engine and aircraft blades, automobile, aerospace, rocket, missile, satellite, military, biomedical, electronics, etc. In addition to having excellent static properties, the CMCs possess reasonably good fracture toughness that will overcome the brittle nature of the ceramic matrix, and it is based on the crack arrest mechanism performed by the fibers or whiskers. For that reason, the CMCs can be a perfect candidate for structures where high temperature, lightweight, and corrosive properties features are essential [26].

References

1. A.K. Kaw, *Mechanics of Composite Materials* (Taylor & Francis, 2006)
2. D. Hull, *An Introduction to Composite Material* (Cambridge University Press, 1981)
3. R. Gibson, *Principle of Composite Material Mechanics*, 2nd edn. (CRC Press, Boca Raton, 2007)
4. A.K. Kaw, *Mechanics of Composite Materials*. Chapter 1 (Taylor & Francis, 2006)
5. P. Niskanen, W.R. Mohn, Versatile metal-matrix composites. Adv. Mat. Process (1988)
6. K.K. Chawla, *Ceramic Matrix Composites* (Chapman & Hall, London, 1993)
7. B. Harris, *Engineering Composite Materials*. Chapter 2 (The Institute of Materials, London, 1999)
8. M. Akay, An Introduction to Polymer Matrix Composite. eBooks at bookboon.com (2015)
9. L. Pilato, M.J. Michno, *Advanced Composite Materials*. Chapter 1 and Chapter 2 Matrix Resins (Springer, New York, 1994)

10. OSHA, *Polymer Matrix Materials: Advanced Composites* (U.S. Department of Labor. Public domain content from a U.S. government department, 2009)
11. H. Courtney Thomas, *Mechanical Behavior of Materials* (Waveland Press, 2005)
12. Z. Jin, C. Sun, *Fracture Mechanics* (Academic Press, Waltham), pp. 11–14
13. R.M. Jones, *Mechanics of Composite Materials* (McGraw Hill, 1976)
14. Fiber Composites (Brunel University, London, 2020)
15. T. Johnson, *Carbon Fiber Manufacturing Companies* (ThoughtCo, 2021)
16. A. Mouritz, *Introduction to Aerospace Materials* (Woodhead Publishing Materials, 2012)
17. D.K. Rajak, D.D. Pagar, P.L. Menezes, E. Linul, *Fiber Reinforced Polymer Composites: Manufacturing, Properties, and Applications*, 2019)
18. K.K. Kar, *Composite Material* (Springer., eBook, 2017), pp. 97–104
19. A.K. Kaw, *Mechanics of Composite Materials*. Chapter 2 (Taylor & Francis, 2006), pp. 103–115
20. P.K. Dash, A.K. Chatterjee, Effects of environment on fracture toughness of woven carbon/ epoxy composite. J. Inst. Eng. **85**, 1 (2004)
21. B. Farahmand, *Virtual Testing and Predictive Modeling*. Chapter 1 (Springer, 2009)
22. B. Farahmand, *Fatigue and Fracture Mechanics of High-Risk Parts*. Chapter 5 (Chapman & Hall, 1997)
23. B. Farahmand, *Virtual Testing and Predictive Modeling*. Chapters 4 and 11 (Springer, 2009)
24. Mustafa Akay, An Introduction to Polymer Matrix Composite. Chapter 5, eBooks at bookboon. com (2015)
25. H. Pfeiffer, *Ceramic body flap for X-38 and CRV*, in 2nd international symposium on atmospheric re-entry vehicles and systems, Arcachon, France (2001)
26. F. Raether, Ceramic Matrix Composites – An Alternative for Challenging Construction Task (PDF). Ceramic Applications, pp. 45–49 (2013)

Chapter 2
Understanding Lamina Through Micromechanics (The Rule-of-Mixtures Approach)

2.1 Introduction

The concept of micromechanics can be defined as understanding the mechanical behaviors of composite materials from microstructures of individual constituents. The key point of micromechanics is to look at the material microscopically by focusing on the material locally to evaluate the local stress and strain fields induced by the applied loads in the constituents. It must be realized that laminated composites have heterogeneous properties and consist of distinct components. To be able to apply the micromechanics tool and concepts to the analysis of laminated composites made of plies with different orientations, it will be helpful to use the average lamina properties of both fibers and matrix within a lamina when estimating the laminate properties [1–3]. Furthermore, unlike in isotropic materials, experimental evaluation of elastic modulus and strength parameters for a uniaxial lamina is very costly and time-consuming because they are functions of several variables, such as fiber volume fraction (the percent amount of fibers in the total volume of composites), void volume fraction, packing geometry, processing, etc. Thus, the need for developing an analytical approach to find key mechanical parameters is of great help to engineers. One of the key assumptions used here is that even though unidirectional lamina is not homogeneous, it can be assumed lamina is homogeneous by focusing on the average response of the lamina to the applied load. The lamina can be looked at as a material whose properties are different in various directions at a point (i.e., anisotropic) but homogeneous from one location to another.

B. Farahmand, *Fundamentals of Composites and Their Methods of Fabrications*,
https://doi.org/10.1007/978-3-031-80201-0_2

2.2 Micromechanics and Nanomechanics Concepts

Herein, the focus is on a unidirectional continuous fiber-reinforced lamina because it forms the basic building block of a composite structure. It is a fact that a single laminate plate is nothing but made of several unidirectional laminae adhesively stacked at various angles on top of each other. Therefore, having essential information on a single unidirectional lamina, such as (1) volume fiber and matrix fractions, (2) mass fractions, (3) density, and (4) void content of lamina, and by using the rule of mixtures concept [4–6], it will be helpful to obtain lamina mechanical properties for further assessing the laminated composite. Note that, in a fiber-reinforced composite, the fiber volume fraction must be available to determine the mechanical properties of a lamina. A composite can't have too many fibers, and at the same time, a small amount of it is not desirable and will not provide enough strength to the composite. Therefore, there should be a maximum limit of fiber content in a composite structure to provide the desired strength when they are bonded to the matrix. As the number of fibers in the matrix increases, the strength of the laminate increases. However, it comes to the point that too much fiber in the matrix will bring weakness to the composite [7, 8].

Finally, most computational methods that engineers use in the micromechanics approach are based on continuum micromechanics, where it is assumed that there is a relationship between the macroscopic behavior of materials and the microstructural features on the micro length scale.

If engineers prefer to conduct their computational structural analysis through lower length scale such as molecular dynamics [9, 10], nanomechanics [11], or quantum mechanics computation [12, 13], then reliable software should be available to engineers that could help them to perform their computational analysis accurately. The abovementioned approaches are known as the bottom-up approach (Fig. 2.1). For example, if an analyst likes to use the molecular dynamics approach, then results must be interpreted from the length scale up to the continuum level. Therefore, the bottom-up approach considers simulation from quantum mechanical to molecular mechanical to mesoscale and the bulk scale (aka continuum mechanics). At the bottom of the scale, the size of the computation increases where a greater number of atoms and their random displacement with high velocity necessitates more approximations to be employed to ease the computation; hence, the methods are less numerically precise (Sect. 2.3).

2.3 Nanomechanics Analysis and Limitations

Atomistic simulations are important numerical methods for the investigation of mechanical properties of nanostructures since these modeling approaches can assess atomistic analysis and relate results to the microscale level. All materials have atomistic scale structures that their behavior will affect their macroscopic behavior.

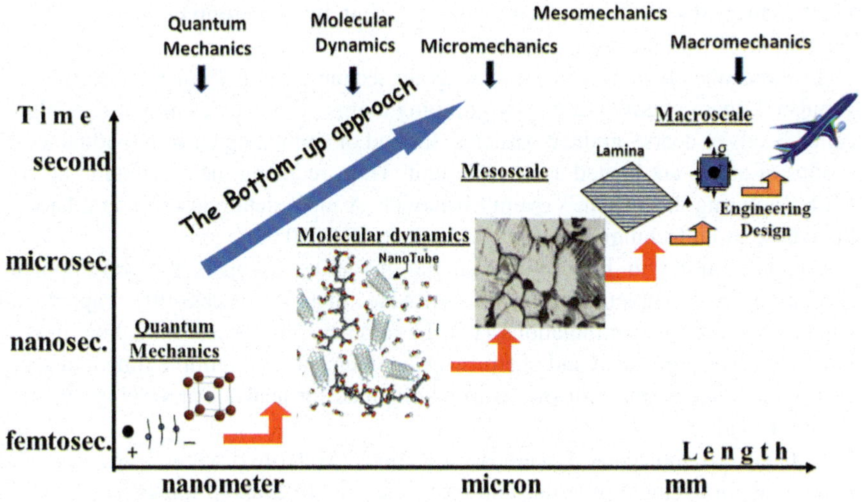

Fig. 2.1 A multiscale modeling that illustrates the type of analysis performed at different length and time scales (from nanometer all the way to millimeter). The smaller scale will make the analysis impossible because of the complicity associated with computational simulation limitation

There are limitations in capturing the exact behavior of atoms under certain load excitation through computer simulation, which will make nanomechanics analysis difficult to implement. For example, the exact modeling of atomic bonds, the interaction of many atoms among themselves (i.e., interatomic potential, Fig. 2.1), and the high velocity associated with vibrating atoms per second, all make the analytical and computational analysis difficult if not impossible to execute [14]. Figure 2.1 illustrates an analysis approach based on different time and length scales starting from atomistic all the way to the continuum level (i.e., quantum mechanics, nanomechanics, molecular dynamics, micromechanics, and continuum mechanics or macro mechanics). If engineers are interested in conducting their analysis by considering analysis to be conducted from molecular or nanomechanics scales upward, then the cost and time of modeling, as well as time of computations, must be taken into consideration. Analysis at the nano level becomes nearly impossible when applied to large structures containing an enormous number of atoms, primarily due to computational limitations. As mentioned earlier, this challenge arises due to numerous variables involved in the analysis. These variables include the substantial quantity of atoms required for a meaningful analysis, the individual atom velocities that can vibrate at rates as high as one cycle per nanosecond under external forces, and the intricate interactions among neighboring atoms. Computation at this lower scale becomes nearly impossible due to these complex parameters. To overcome this problem, analysts attempt to use the coarse grain technique approach. Briefly, in multiscale modeling, the coarse-graining technique involves simplifying a system by reducing the number of degrees of freedom while preserving its important features. The coarse grain process intends to make complex systems more manageable and computationally efficient by representing them at a less detailed level without losing critical information.

In molecular dynamics, the coarse grain technique is helpful to implement to simulate larger system of atoms by grouping atoms or molecules into larger entities called beads or coarse-grained particles. Instead of simulating each individual atom, groups of atoms are treated as a single unit. Thus, reducing the computational cost while capturing the system's overall behavior. A more detail description related to the coarse grain technique is provided in Sect. 10.8–10.9.

Another hurdle in bottom-up material modeling involves the development, validation, and implementation of software that is successfully capable of conducting multiscale simulations [15]. In summary, multiscale modeling aims to integrate diverse physical and chemical phenomena across multiple time and space scales, enabling meaningful results to be obtained for analysts at varying levels of scale [16–18].

In Chap. 10, molecular dynamics simulations (MD) are discussed, and it is used to study the movement of atoms and molecules due to external forces over time. In MD simulations, Newton's equations of motion are solved numerically to simulate the behavior of a system of particles, considering forces between them. The force calculation between atoms in MD involves evaluating interatomic forces based on proper mathematical expressions for potential energies and forces between particles. It represents the attractive and repulsive forces between atoms that govern their behavior and arrangement in different states of matter and is based on quantum mechanical calculations that determine *the* interactions between molecules or atoms in a substance. Later in Chap. 10, the Lennard–Jones potential, that is one of the mathematical models describing the potential energy between a pair of neutral atoms or molecules as a function of their separation distance, will be described in detail.

2.4 Micromechanics Approach to Obtain Lamina Properties (Rule of Mixtures)

2.4.1 Volume, Mass, and Density Fractions

This section focuses on the unidirectional continuous fiber-reinforced lamina because a lamina forms the basic building block of a laminated composite structure. A laminate is generally made of several unidirectional laminae stacked on top of each other, where fibers are oriented at various angles in the matrix. Therefore, the method of obtaining the uniaxial lamina mechanical properties from its constituents is important for further analysis of a laminated composite part when it is subjected to applied service load. The mechanical properties characterization of composite materials generally can be obtained through the rule of mixtures. These are mathematical expressions giving the property of the composite in terms of the quantity and arrangement of its constituting materials. To implement the mathematical rule of mixtures for obtaining material properties, the volume fraction of fibers, matrix, voids, mass fraction, and density fraction must be defined. Engineers must realize that there is a need to conduct a few tests that would be able to verify the lamina mechanical properties estimated through the rule of mixtures if they are going to be used in the stress analysis of composite parts (see Sect. 2.7 on discussions related to weaknesses on using the rule of mixtures).

2.4.2 Volume Fraction of Fibers and Matrix

The amount of fibers in a fiber-laminated composite is directly related to the mechanical properties of the composite. The highest ratio of fibers in the matrix, that can bring optimum mechanical properties, is around 60–70%. Adding too little fiber reinforcement in the composite will weaken the properties of the material. On the other hand, too many fibers in the matrix may also reduce the strength of the composite due to the lack of space for the matrix to fully surround and bond with the fibers [19–22].

In this section, the constituent properties of the lamina, mainly the volume fractions of fibers and matrix, will be examined. The sum of two quantities, i.e., the sum of fiber volume fraction, V_f, and the matrix volume fraction, V_m, is unity, where $V_f + V_m = 1$. The fiber volume fraction, V_f, in terms of the volume of fibers (v_f) and the volume of composite (v_c) will be $V_f = v_f/v_c$. Accordingly, the matrix volume fraction, V_m, in terms of the volume of the matrix (v_m) will be:

$$V_m = v_m/v_c \text{and } V_f = v_f/v_c, \text{ where} \tag{2.1a}$$

$$v_f + v_m = v_c \tag{2.1b}$$

2.4.3 Mass Fraction

Another similar parameter is called the mass fraction (or weight fraction) of fibers (W_f) and mass fraction of matrix (W_m), which they can be written in terms of the mass volume of fibers (w_f) and mass volume of the matrix (w_m) as follows:

$$W_f = w_f/w_c \text{ where } w_c \text{ is the mass volume of composite}$$
$$\text{and accordingly, } W_m = w_m/w_c,$$
$$\text{where the sum of mass fractions is}$$

$$W_f + W_m = 1 \text{ and} \tag{2.2a}$$

$$w_f + w_m = w_c \tag{2.2b}$$

2.4.4 Density Fraction

From the definition of density of the material, where ρ = mass/volume, the mass of composite (w_c) in terms of density (ρ_c) can be written as $w_c = v_c\,\rho_c$. The rule of mixtures can be applied for density and the approach is like the case of the mass fraction. Here, the mass volume of fibers, $w_f = v_f\,\rho_f$, and for the matrix $w_m = v_m\,\rho_m$.

From mass volume fraction equation ($W_f = w_f/w_c$) and replacing w_f with $v_f\,\rho_f$ and w_c with $v_c\rho_c$.

Then $W_f = (\rho_f/\rho_c)V_f$ and $W_m = (\rho_m/\rho_c)V_m$. Finally, the density of composite in terms of its constituents ρ_f and ρ_m can be written by using Eq. 2.2a, where $W_f + W_c = 1$:

$$\rho_c = \rho_f\,V_f + \rho_m V_m \tag{2.3}$$

Example Problem

A lamina consists of a 60% fiber volume fraction. The density of the fiber is ρ_f = 156.1 klb/ft^3 (2500 kg/m^3), and the density of matrix is 74.91 klb/ft^3 (ρ_m = 1200 kg/m^3). Determine (1) the density of lamina, (2) mass fraction of the fiber and matrix, (3) determine the volume of composite lamina if the mass of lamina is 8 lbs., and (4) volume and the mass of fiber and matrix in part.

Solution

From $\rho_c = \rho_f\,V_f + \rho_m V_m$ one can calculate the density of lamina:

$$\begin{aligned}\rho_c &= (156.1)(0.6) + (74.91)(0.4)\\ &= 123.62\ \text{klb}/\text{ft}^3\end{aligned}$$

The mass fraction of the fiber and matrix are:

$$W_f = (\rho_f/\rho_c)V_f$$
$$W_f = (156.1/123.62)(0.6) = 0.7576$$

And for $W_m = (\rho_m/\rho_c)V_m$

$$W_m = (74.91/123.62)(0.4) = 0.2424$$

The volume of composite can be determined as:

$$v_c = w_c/\rho_c$$
$$v_c = 8/123.62 = 0.0647\ \text{ft}^3$$

and finally, the volume and mass of the fiber and the matrix can be obtained as follows:

$$v_f = V_f\, v_c$$
$$(0.6)(0.0647) = 0.0383\ \text{ft}^3$$
$$\text{For } v_m = V_m\, v_c$$
$$(0.4)(0.0647) = 0.0258\ \text{ft}^3$$

The mass of both fibers and matrix:

$$\begin{aligned} w_f &= v_f \rho_f \\ &= (0.0383)(156.1) = 5.978\ \text{lb} \end{aligned}$$

and for matrix

$$\begin{aligned} w_m &= v_m \rho_m \\ &= (0.0295)(74.91) = 2.029\,\text{lb} \end{aligned}$$

2.4.5 *Voids Volume*

During the fabrication of a laminated composite part, voids or porosity is created in the composite. These are generally formed as gas bubbles trapped inside the cured composite materials. Voids are undesirable and have adverse effects on the mechanical properties of composites. As the void content increases, an increase in scatter in the strength data is observed. The role of voids in the initiation and propagation of cracks in composites reinforced with unidirectional carbon fiber has been investigated [22]. Results showed that a void area greater than 0.2 mm in diameter (0.00787 in) with an area greater than 0.031 mm^2, results in the deterioration of mechanical properties. Moreover, an increase in void content increases the composite's susceptibility to water penetration and hence decreases the mechanical properties. In addition to the water penetration, crack initiation and propagation from several voids can cause mechanical properties degradation faster because these growing voids will merge and create a sizable crack. When calculating the density of composite, without considering the presence of voids in material, the calculated value is higher than the actual value. As was noted above, the voids content of a composite can lower the mechanical properties of composites. Voids can act as a source of crack initiation and subsequent growth leading to structural failure. Not much data is available on the effect of void content on the mechanical properties of composite laminates. From a nondestructive inspection perspective, several ways can be used to determine if a composite part contains voids, such as a computerized tomography (CT) scanning device [23] or ultrasound methods [24, 25]. Even a good quality composite part cannot be free from voids and has some small amounts of it is existing in the material when inspected by nondestructive inspections. Usually, their value could be close to an average of 1% of the composite volume.

When composites have voids, with voids volume, v_v, the volume fraction of voids V_v in terms of the total volume of composites (v_c) is defined as, $Vv = v_v/v_c$. The total volume of a composite (v_c), including voids, is given by:

$$v_c = v_v + v_f + v_m \tag{2.4a}$$

or in terms of volume fraction,

$$v_v/v_c + v_f/v_c + v_m/v_c = 1 \tag{2.4b}$$

but, $v_v/v_c = V_v$ or

$$V_v = 1 - V_f - V_m \tag{2.5}$$

If the volume fraction of the fibers (V_f) and matrix (V_m) are known, the volume of voids (V_v) can also be found as was established by Eq. 2.5. Another approach to finding the void volume fraction (V_v) is through the difference in the density ρ_{ct}, which is the theoretical density of the composite without voids, and the measured density of the composite experimentally measured, ρ_{cm}:

$$V_v = (\rho_{ct} - \rho_{cm})/(\rho_{ct}) \tag{2.6}$$

The above equation is derived based on Eq. 2.3 related to the density, mass, and volume. That is,

$$v_c = w_c/\rho_{cm} (\rho_{cm} \text{ is the density from measurements), and}$$
$$w_c/\rho_{ct} = v_f + v_m, (\rho_{ct} \text{ is based on theoretical density}),$$
$$\text{where, } v_c = v_f + v_m.$$

The difference between the two v_c values is the void volume fraction:

$$V_v = v_v/v_c = w_c(\rho_{ct} - \rho_{cm})/(\rho_{cm}\ \rho_{ct}) = v_c(\rho_{ct} - \rho_{cm})/(\rho_{ct}) \tag{2.7}$$

where $w_c/\rho_{cm} = v_c$.

2.5 Evaluation of Elastic Constants of a Unidirectional Lamina (Rule of Mixtures)

For a unidirectional lamina of a laminate (one ply only), there are four elastics constants that must be available for stress analysis assessment, 1) E_1 is longitudinal Young's modulus in direction 1, E_2 is transverse Young's modulus in direction 2, ν_{12} is major Poisson's ratio, where the Poisson's ratio, ν_{12}, is defined as the ratio of the

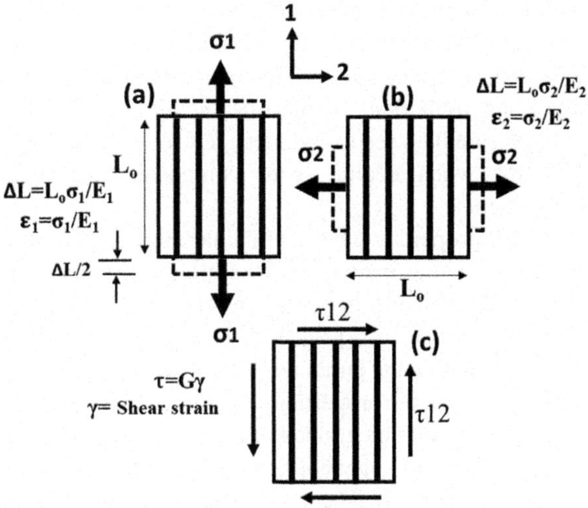

Fig. 2.2 Applied stresses are shown along and normal to the fibers, as well as the shear stresses. Stresses and strains are generally related through engineering elastic constants

negative of the normal strain in direction 2 to the normal strain in direction 1, when the only normal load is applied in direction 1, and G_{12} is in-plane shear modulus (in plane 1–2). The four elastics constants (E_1, E_2, G_{12}, and ν_{12}) that must be available for stress analysis of a unidirectional lamina, are shown in Fig. 2.2. The stress and strains induced in uniaxial lamina are related to each other through the elastic constants mentioned above (i.e., E_1, E_2, ν_{12}, and G_{12}).

To obtain these elastic constants experimentally, tests are conducted by first applying a pure tensile load in direction 1 (Fig. 2.2a). In this case, the stresses in direction 2 and shear stresses must be zero (i.e., $\sigma_1 \neq 0$, $\sigma_2 = 0$, and $\tau_{12} = 0$). In Fig. 2.2a, stress and strains are related by the elastic constant called the modulus of elasticity, E_1, where $\sigma_1 = E_1\varepsilon_1$, but the induced strain in direction 2 is not zero. The strain and stress in the uniaxial direction ($\varepsilon_1 = \sigma_1/E_1$) indicate that the material has been elongated along the direction of the load by the amount of $\Delta L = L_o\sigma_1/E_1$, where L_o is the original length, and ΔL is the total elongation in direction 1. However, based on the law of constant volume, the material has been contracted in the direction 2. The major Poisson's ratio (ν_{12}) is defined as the negative ratio of the normal strain in the transverse direction to the normal strain in the longitudinal direction when a normal load is applied in the longitudinal direction ($\nu_{12} = -\varepsilon_2/\varepsilon_1$) or the transverse strain $\varepsilon_2 = (-\nu_{12}/E_1)\,\sigma_1$, where $\varepsilon_1 = \sigma_1/E_1$. With the same analogy, for direction 2, $\Delta L = L_o\sigma_2/E_2$, $\varepsilon_2 = \sigma_2/E_2$ and the Poisson's ratio $\nu_{21} = -\varepsilon_1/\varepsilon_2$ with the transverse strain $\varepsilon_1 = (-\nu_{21}/E_2)\sigma_2$ (Fig. 2.2b). For isotropic material, the two quantities $\nu_{12} = \nu_{21}$, but this is not the case for composite materials that are not isotropic.

Testing for obtaining mechanical properties of a lamina of a laminated composite, as shown in Fig. 2.2, can be costly, and results can vary because they depend on several parameters, such as fiber volume fraction, void content, packing geometry, processing technique, etc. For those reasons, the rule of mixtures approach can obtain lamina mechanical properties of a laminate for both principal directions without conducting tests. These properties will be the modulus of elasticity (E) for both normal and transverse to the fibers, the maximum strength of composite in both along and transverse to the fibers, and other vital mechanical properties. Disadvantages and limitations of the rule of mixtures are discussed in Sect. 2.7, with an emphasis on the need to conduct a few tests to verify results obtained from the rule of mixtures when dealing with composite parts that are classified as primary components.

When properties for the principal directions were estimated through the rule of mixtures, then properties for other plies in the laminate that are oriented differently with respect to the applied load can be obtained by the transformation matrix described in Sect. 2.6 [26]. Finally, when mechanical properties were obtained for all layups within the laminate, with different orientations, subsequently, they will be used to analyze the properties of a laminated composite through the macromechanics approach. As was mentioned in Chap. 1, a lamina is not homogeneous, but it can be assumed to be homogeneous by concentrating on the average response of the lamina to mechanical loads. The lamina can be looked at as a material whose properties are different in various directions but not different from one location to another (i.e., it is homogeneous). Note that both fibers and matrix of a lamina alone may be homogeneous and isotropic, but when combined, the outcome will not be homogeneous or isotropic, depending on whether the point is in the fiber, the matrix, or the fiber–matrix interface. Considering the above points, it will make any kind of mechanical modeling of the lamina very complicated. For this reason, the macromechanical analysis of a lamina is based on average properties and allows for the lamina to be homogeneous with uniform properties throughout its thickness but not isotropic.

Before modeling the mechanical properties of a unidirectional lamina via the rule of mixtures, it is necessary to remind the concept of the relative fraction of fibers by volume in the composite. This quantity is significant because the formulas for finding the stiffness and strength properties of a unidirectional lamina are a function of fiber volume fraction, which was discussed in Sect. 2.4. Knowing the amount of both constituents based on their mass volume (w_f and w_m), then both fibers and matrix mass fractions must also be defined. Moreover, defining the density of a composite also becomes necessary because its value is used in determining fiber volume and void volume fractions in a composite, which all were discussed in Sect. 2.4. Finally, the value of density is needed because it can be used to determine the two quantities, specific modulus (E/ρ), and specific strength (σ_{Ult}/ρ), which both are valuable parameters to know in the laminated composite materials.

2.5.1 *Longitudinal Modulus of Elasticity,* E_1

In Fig. 2.3a, a lamina is subjected to a uniaxial load parallel and transverse (perpendicular) to the direction of fibers. With the rule of mixtures, the modulus of elasticity, both E_1 (longitudinal modulus) and E_2 (transverse modulus), can be estimated [27]. From Fig. 2.3a, the applied load along the fibers of a lamina will introduce

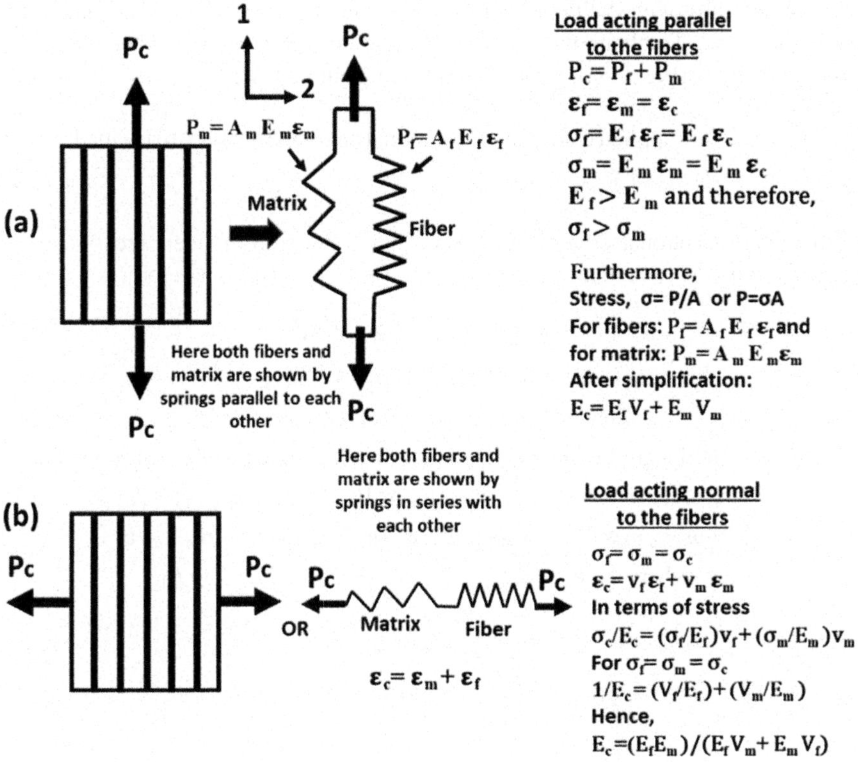

Fig. 2.3 The rule of mixtures are used to obtain a lamina's mechanical properties along and normal to the fibers. Here, the modulus for two cases (**a**) load is parallel to fibers and (**b**) perpendicular to fibers are calculated

the same strain on both fibers and matrix (isostrain—the same amount of strain), if it is assumed the bonding between fibers in the matrix is perfect. The isostrain in this case, where the applied load along the fibers can be written as,

$$\varepsilon_f = \varepsilon_m = \varepsilon_c \text{ (isostrain case)} \tag{2.8}$$

The load, P_c, applied to the lamina is shown in Fig. 2.3a, and both fibers and matrix are represented as springs that can be elongated together, where $P_c = P_f + P_m$. This indicates that the applied load will be taken by both the fibers and matrix. Because both fibers and matrix are assumed to have elastic properties, then the stresses, as a result of applied load, in both fibers and matrix can be written as:

$$\begin{aligned} &\sigma_f = E_f \ \varepsilon_f = E_f \ \varepsilon_c \text{ and } \sigma_m = E_m \varepsilon_m = E_m \varepsilon_c \\ &(\text{Note}: \varepsilon_f = \varepsilon_m = \varepsilon_c) \end{aligned} \tag{2.9}$$

The modulus of elasticity of fiber $E_f > E_m$, and therefore, for the same amount of strain, $\sigma_f > \sigma_m$. To obtain the modulus of elasticity for the lamina when the load is along the fibers (Fig. 2.3a), the following steps will be taken:

$$\text{Stress} = (\text{applied load})/(\text{cross-sectional area}) \text{ or Stress} \times \text{Area} = \text{load}$$
$$\sigma_c = P_c/A_c, \text{ or } P_c = \sigma_c A_c$$

The cross-sectional area of composite, A_c, is the sum of fibers cross-sectional area, A_f, plus the remaining matrix, A_m (i.e., $A_c = A_f + A_m$). Therefore,

$$\sigma_c A_c = \sigma_f A_f + \sigma_m A_m \text{ or}$$
$$\sigma_c = \sigma_f (A_f/A_c) + \sigma_m (A_m/A_c),$$
$$\text{where } (P_c = P_f + P_m) \text{ and } A_c = A_f + A_m$$
$$\text{Because the length of lamina, } L, \text{ is the same, Volume } v_c = LA_c = LA_f + LA_m$$
$$\text{Therefore, the volume } v_c = v_f + v_m \text{ or } V_f = A_f/A_c \text{ and } V_m = A_m/A_c$$
$$\text{Thus, } \sigma_c = \sigma_f V_f + \sigma_m V_m \text{ and } \sigma_c/\varepsilon_c = \sigma_f V_f/\varepsilon_c + \sigma_m V_m/\varepsilon_c$$
$$\text{where } \sigma_c/\varepsilon_c = (E_c)_{\text{Longitudinal}} = E_f V_f + E_m V_m = E_1$$
$$E_1 = E_f V_f + E_m V_m \tag{2.10}$$

For polymer composites, where $E_f >> E_m$. The value of $E_1 \sim E_f V_f$.

2.5.2 *Transverse Modulus of Elasticity,* E_2

In the case of applied load, P_c, perpendicular to fibers of a lamina (Fig. 2.3b), the modulus of elasticity of lamina, E_c, in the transverse direction (E_2) can also be obtained by using the rule of mixtures [27–29]. In Fig. 2.3b, the matrix and fibers are shown as springs in series with each other, where the applied load, P_c, will induce the same stress on both fibers and matrix (i.e., $\sigma_c = \sigma_m = \sigma_f$). The deformation of the entire lamina, because of applied load transverse to fibers, can be written as:

$$\varepsilon_c = \varepsilon_f v_f + \varepsilon_m v_m, \text{ where } \varepsilon_c = \sigma_c/E_c = (\sigma_f/E_f)v_f + (\sigma_m/E_m)v_m$$
$$\text{Because } \sigma_c = \sigma_m = \sigma_f, \text{ then } \sigma_c/E_c = (\sigma_c/E_f)v_f + (\sigma_c/E_m)v_m$$
$$\text{Simplifying it further, } 1/E_c = (1/E_f)v_f + (1/E_m)v_m$$
$$\text{The modulus in the transverse, } E_c = [(1/E_f)v_f + (1/E_m)v_m]^{-1} \text{ or}$$
$$(E_c)_{\text{Transverse}} = (E_f E_m)/(E_f v_m + E_m v_f) = E_2$$
$$E_2 = (E_f E_m)/(E_f V_m + E_m V_f) \tag{2.11}$$

From Eq. 2.11, it can be seen when the fiber fraction volume value is high, $V_f \sim 1$ and V_m is almost close to zero, then the modulus of lamina in the transverse direction (E_2) becomes close to the modulus of fiber (E_f). Accordingly, when V_f value is small, then E_2 becomes equal to the matrix modulus (E_m). A better understanding of E_2 can be obtained by plotting the variation of (E_2/E_m) versus the fiber volume fraction (V_f) for different E_f/E_m values (e.g., $E_f/E_m = 1$, 5, 25, and 125). The results clearly indicated that as V_f value increases, significant changes are not expected as shown in Fig. 2.4, but drastic change will occur when the V_f quantity is above 70% value (E_m becomes small). For the value of $E_2/E_m = 1$, the quantity E_f/E_m remains a constant ~1 (see Eq. 2.11). It is also essential to know that, for polymeric composites (PMC), the fiber-to-matrix moduli ratio is usually high. As an example, for a glass/epoxy polymer matrix composite, $E_f/E_m = 25$. If $E_f >> E_m$, E_2 becomes independent of the property of the fiber ($E_2 = E_m/V_m$), as it is evident from Eq. 2.11.

The E_2 equation shown above (i.e., $E_2 = (E_f\,E_m)/(E_f\,V_m + E_m\,V_f)$) is not a suitable equation to use for practical situations in spite of its occasional agreement with the experimental data. Therefore, better mechanical models have been established by several authors using an empirical model, such as Halpin–Tsai model. The semiempirical Halpin–Tsai equations adjust the transverse modulus equation derived for E_2 with a better agreement to the test data [31, 32]. The model is utilized to determine the stiffness of composites reinforced material, where both the longitudinal and transverse moduli of composites can be calculated. Then, the elastic

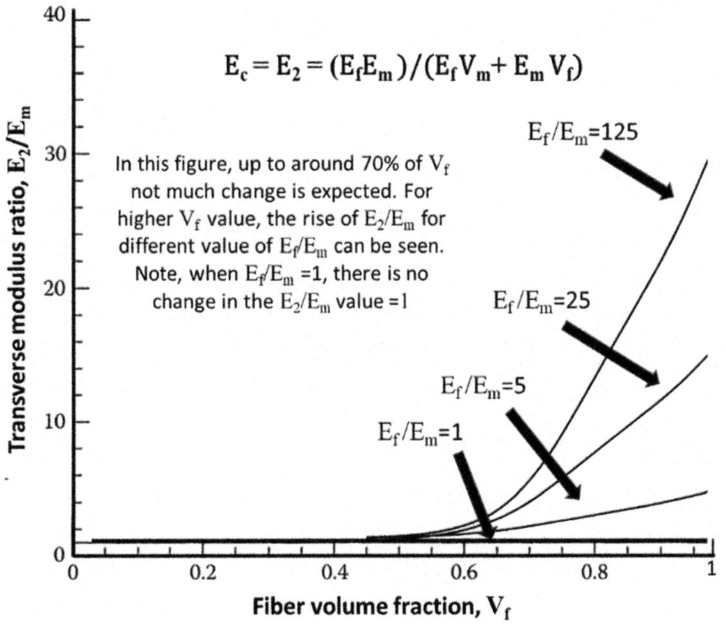

Fig. 2.4 Transverse Young's modulus ratio E_2/E_m, as a function of fiber volume fraction, V_f, for several fiber to matrix moduli ratio, E_f/E_m [30]

transverse modulus of composite is obtained by using the semiempirical modification of Halpin–Tsai equation as.

$$E_2 = E_m(1 + \zeta\eta v_f)/(1 - \eta v_f) \tag{2.12}$$

where, η is a function of fiber and matrix modulus E_f and E_m, and it is given by:

$$\eta = (E_f/E_m - 1)/(E_f/E_m + 2\nu_m)$$

Note that, the quantity η also depends on the shape and aspect ratios of the fibers and fiber packing geometries (hexagonal or squares) and the loading conditions. The quantity ν_m is the Poisson's ratio of the matrix. Moreover, the parameter ζ (shown in Eq. 2.12) in terms of η can be calculated by:

$$\zeta = (2 + \eta)/2(1 + \eta)$$

At this point, both the modulus of elasticity parallel and transverse to fibers, $(E)_{Longitudinal}$ and $(E)_{Transverse}$ or E_1 and E_2 were estimated by using the constituents' properties and the rule of mixtures (see Eqs. 2.10 and 2.11, respectively). Furthermore, the semiempirical Halpin–Tsai equation (described by 2.12), adjusts the transverse modulus equation derived for E_2 with a better agreement to the test data.

Example

Determine the transverse Young's modulus, E_2, of a glass–epoxy lamina with a fiber volume fraction of 65%. The Young's modulus of the fiber glass is $E_f = 12{,}035$ ksi (83 GPa) and the Young's modulus of the matrix is $E_m = 522$ ksi (3.6 GPa).

Solution

Using Eq. (2.12), describing the transverse modulus (E_2) by having both the $E_f = 12{,}035$ ksi and the matrix $E_m = 522$ ksi, the quantity $(E_c)_{Transverse}$ can be calculated:

$$E_2 = (E_c)_{Transverse} = (E_f E_m)/(E_f v_m + E_m v_f)$$

$$E_2 = (12035) \times (522)/(12035 \times 0.35 + 522 \times 0.65) = 6.28 \times 10^6/(4212.3 + 339.3) = 1380.75\,\text{ksi}$$

Comparing the result with Eq. 2.10, where $E_1 = E_f V_f + E_m V_m$, the value of $E_1 = 8005.45$ ksi. Results clearly show that modulus in the direction of fibers, $E_1 = 8005.45$ ksi is much stronger than the transverse direction $E_2 = 1380.75$ ksi.

2.5.3 Major Poisson's Ratio, ν_{12}

The major Poisson's ratio (ν_{12}) is the negative of the ratio of strain in the transverse direction (when a normal load is applied in the longitudinal direction and parallel to the fibers of a lamina) to the normal strain in the longitudinal direction. In Fig. 2.5, both fibers and matrix are represented by rectangular elements, where a representative volume element (RVE) of a material is also shown in Fig. 2.5 [33]. That is the smallest volume over which it includes both the fiber and matrix. Any measurement made on this volume will yield a value representative of the whole material. The term RVE is also used interchangeably in the literature with the representative elementary volume (REV). When a material has a periodic structure, such as aluminum, the face centered cubic unit cell (fcc) represents the volume element of the material and in the case of Iron, it is the body centered cubic unit cell (bcc).

The delta deformations in the transverse direction of the composite (Δ_W), where the width is W, as a result of normal applied load, is the sum of the transverse deformations of the fiber (Δ_f) and the matrix (Δ_m), as shown in Fig. 2.5. In addition, the variation of Poisson's ratio, ν_{12}, versus the fiber volume fraction is plotted in Fig. 2.5b [33]. The derivation of ν_{12} based on the rule of mixtures is as follows:

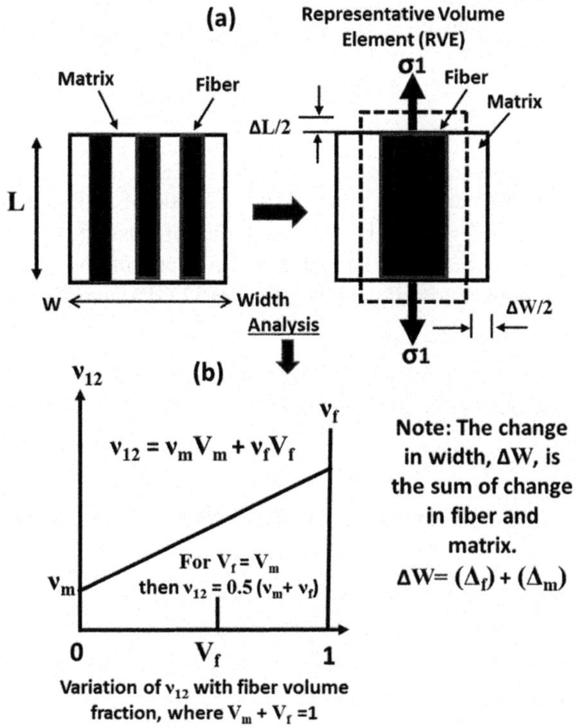

Fig. 2.5 (**a**) The derivation of Poisson's ratio ν_{12} is presented by using volume element in direction 1. (**b**) Variation of Poisson (ν_{12}) with respect to fiber volume, V_f, is plotted

$$\text{By definition, } \nu_{12} = -\varepsilon_2/\varepsilon_1, \text{ where } -\varepsilon_2 = \Delta W/W$$
$$\Delta W = -W\varepsilon_2 = W\nu_{12}\varepsilon_1$$

The change in width, ΔW, is the sum of the transverse deformations of the fiber (Δ_f) and the matrix (Δ_m). Therefore,

$$\Delta W = \Delta_f W + \Delta_m W$$

The contribution of matrix to ΔW is:

$$\Delta_m W = W\nu_m\varepsilon_1 V_m$$

The contribution of fiber to ΔW is:

$$\Delta_f W = W\nu_f\varepsilon_1 V_f$$

Rearranging the above equations:

$$\nu_{12} = \nu_m V_m + \nu_f V_f \tag{2.12}$$

From Eq. (2.12), it can be seen that for $V_m = 1$, the $V_f = 0$ and $\nu_{12} = \nu_m$ and if $V_f = 1$, then $V_m = 0$ and $\nu_{12} = \nu_f$ (Fig. 2.5b). In the case of $V_m = V_f$,

$$\nu_{12} = \mathrm{V_f}\,(\nu_m + \nu_f) \text{ or } \nu_{12} = \mathrm{V_m}(\nu_m + \nu_f),$$

Example

Find the major and minor Poisson's ratio of a glass/epoxy lamina with fiber volume fraction $V_f = 65\%$. Use the properties of glass and epoxy where the Poisson's ratio of glass fiber $\nu_f = 0.2$ and the Poisson's ratio of epoxy $\nu_m = 0.3$. Also, for glass/epoxy lamina material $E_1 = 8776$ ksi (60.52 GPa) and $E_2 = 1504$ ksi (10.37 GPa).

Solution

Using Eq. (2.12) for the major Poisson's ratio ($\nu_{12} = \nu_m V_m + \nu_f V_f$), the calculation will be as follows:

$$\nu_{12} = \nu_m V_m + \nu_f V_f = (0.3)(0.35) + (0.2)(0.65) = 0.235$$
$$\text{Since } \nu_{21} = (\nu_{12})(E_2/E_1) = (0.235)(1504/8776) = 0.040$$

2.5.4 In-Plane Shear Modulus, G_{12}

Apply pure shear stress τ to a lamina, which it represents the volume element, is shown in Fig. 2.6a. The in-plane shear modulus, G_{12}, can be determined by assuming that the shear strain on the fiber (γ_f) and the matrix (γ_m) are equal, where $\gamma_f = \tau/G_f$ and $\gamma_m = \tau/G_m$. Here, the two quantities G_f and G_m are the shear moduli of fibers and matrix, respectively. Furthermore, it is assumed that the relationship between the shear stress (τ) and shear strain (γ) is linear (i.e., $\tau = G\gamma$ analogous to $\sigma = E\varepsilon$). In Fig. 2.6b, the shear deformation on the representative volume element is shown, where the total deformation Δ in terms of both matrix and fibers can be written as:

$$\Delta = \gamma W$$

The total deformation ($\Delta = \Delta_m + \Delta_f$) in terms of fiber ($\Delta_f$) and matrix ($\Delta_m$)

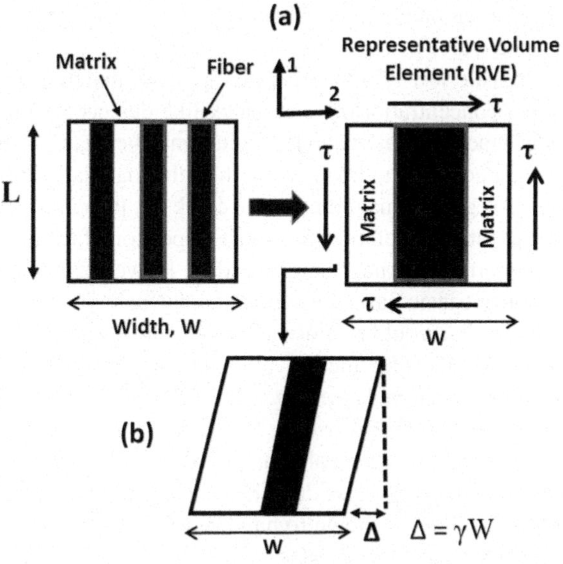

Fig. 2.6 (**a**) Representing volume element in direction 1. (**b**) Using the rule of mixtures concept to derive the shear modulus $G_{12} = (G_m G_f)/(V_f G_m + V_m G_f)$

$$\gamma = V_f\,\gamma_f + V_m\,\gamma_m\text{, but } \gamma = \tau/G_{12}\text{, where } \gamma_f = \tau/G_f$$

$$\text{and } \gamma_m = \tau/G_m$$

$$\tau/G_{12} = (V_f\,(\tau/G_f) + V_m(\tau/G_m)\text{ or}$$

$$1/G_{12} = V_f/G_f + V_m/G_m\text{ or}$$

$$G_{12} = (G_m G_f)/(V_f G_m + V_m G_f)$$

$$\Delta_f = V_f(\gamma_f W) \text{ and } \Delta_m = V_m(\gamma_m W).$$

$$\text{Therefore, } \Delta = V_f(\gamma_f W) + V_m(\gamma_m W), \text{thus, } \gamma = \Delta/W = V_f\gamma_f + V_m\gamma_m$$

$$\gamma = V_f\gamma_f + V_m\gamma_m, \text{but } \gamma = \tau/G_{12}, \text{where } \gamma_f = \tau/G_f \text{ and } \gamma_m = \tau/G_m$$

$$\begin{gathered} \tau/G_{12} = (V_f(\tau/G_f) + V_m(\tau/G_m) \text{ or} \\ 1/G_{12} = V_f/G_f + V_m/G_m \text{ or} \\ G_{12} = (G_mG_f)/(V_fG_m + V_mG_f) \end{gathered} \tag{2.13}$$

2.5.5 *Estimating the Ultimate Strength of Composite*

Both the specific modulus and specific strength ratio are essential parameters to know when both weight and strength play a significant role in manufacturing composite parts. Therefore, it is essential to determine both the modulus and the ultimate strength of the material. The modulus was already derived for both parallel and perpendicular to the fibers using the rule of mixtures (see Eqs. (2.10) and (2.11), respectively).

Tensile test results generally apply to material that does not contain cracks or stress concentrations. When crack-like defects are present either as surface cracks, edge cracks, or internal ones, failure may begin at much lower applied stresses, when compared to the ultimate strength of the material without initial defects. The applied stress is greatly magnified at the crack tip, thus influencing the ultimate strength of a composite. Therefore, a thorough inspection of the test material for ultimate strength is important to know before testing. However, it should be mentioned that microscopic intrinsic flaws always exist in the material. Otherwise, the ultimate strength of the material would be close to its theoretical value. This section briefly discusses two methods of estimating the ultimate strength of a lamina as a part of a laminated composite materials (see Sect. 2.5.5.2).

In Fig. 2.7, the plot of stress versus strain for a typical composite, together with its constituent's mechanical behavior (fiber and matrix) is shown. In most cases, fibers are shown to have less ductility than the matrix, and for that reason, the composite ply as a whole is shown to have less ductility than the matrix, which is the case for polymer matrix composites.

As an example, glass fibers fail at strains of 3–5%, but an epoxy matrix fails at strains above those values (Fig. 2.7).

Two methods will be presented here for obtaining the ultimate strength of composites, the first one will be based on the rule of mixtures method [34], and the second one will be from using the energy release rate, G_{Ic}, approach, where the assumption will be based on required stress needed to fail a small, microscopic crack in the laminated composite [35]. This approach is a fracture mechanics approach, and the critical value of fracture toughness or energy release rate (i.e., G_{Ic}) is required for this calculation. The approach assumes that a critical intrinsic flaw size exists in

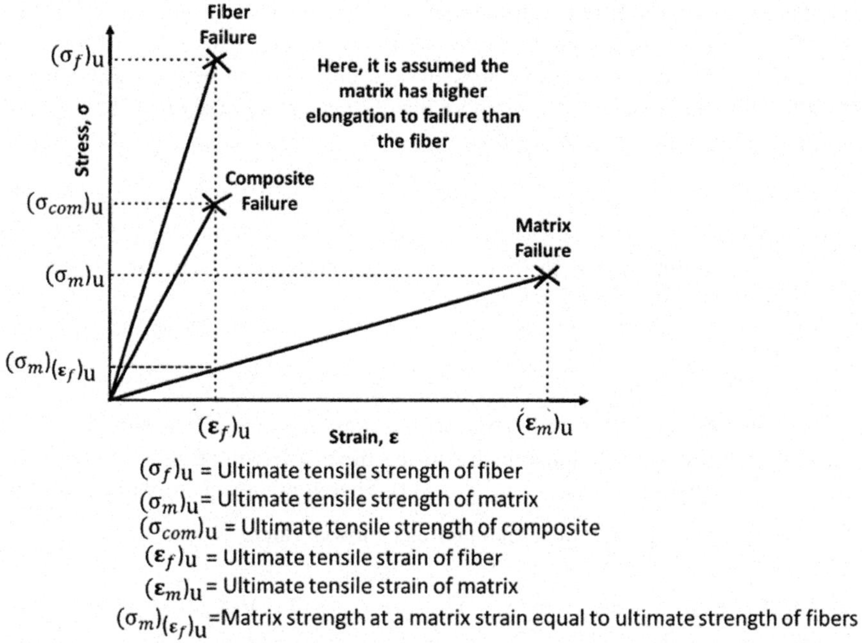

Fig. 2.7 Stress-strain curve for the typical composite, where the matrix has lower strength, but higher elongation to failure

the material that will require an applied stress equal to or close to the ultimate strength of the material to fail it. Note that the calculated ultimate strength by this method can be slightly higher or lower than the actual value because the intrinsic flaws in the material could be slightly smaller or slightly larger than the value assumed by using the fracture mechanics approach.

2.5.5.1 The Ultimate Strength of a Composite Lamina by Using the Rule of Mixtures

Let us assume the ultimate failure strain of the fiber in the elastic range, $(\varepsilon_f)_u$, can be written as:

$$(\varepsilon_f)_u = (\sigma_f)_u / E_f$$

And the ultimate failure strain of the matrix, $(\varepsilon_m)_u$, is

$$(\varepsilon_m)_u = (\sigma_m)_u / E_m$$

The fiber can carry most of the load in polymer matrix composites, therefore, it can be assumed that when fibers in the matrix fail at the strain of $(\varepsilon_f)_u$, the whole composite ply fails, even though the matrix has more strain to failure (i.e., higher ductility). Now if a stress σ applies to the composite lamina, where $\sigma_{com} < (\sigma_{com})_u$ in the direction of the fiber orientation, then both the fiber and matrix act parallel to resist the load. In this case, strain in both fiber (ε_f) and matrix (ε_m) must be the same in the fiber direction and can be written as:

$$\varepsilon_{com} = \varepsilon_f = \varepsilon_m$$

Thus, the ultimate strength of the composite lamina at failure $(\sigma_{com})_u$ can be written from the rule of mixtures as:

$$(\sigma_{com})_u = (\sigma_f)_u V_f + (\sigma_m)_u (1 - V_f) \tag{2.14}$$

As shown in Fig. 2.7, it is obvious that the stress in the matrix will be much less when fibers and composite fail, which it means the matrix will be stressed only up to the maximum strain of the fiber $(\sigma_m)_{(\varepsilon_f)_u}$ [34]. Therefore, the quantity $(\sigma_m)_u$ must be replaced by $(\sigma_m)_{(\varepsilon_f)_u}$, of which the latter has a lower value.

$$(\sigma_{com})_u = (\sigma_f)_u V_f + (\sigma_m)_{(\varepsilon_f)_u} (1 - V_f) \tag{2.15}$$

At the ultimate strain of the fiber, it is assumed the load carrying capability of fiber and as a result, the composite will become zero (Fig. 2.7). However, it is possible that because of fiber failure, the load will be transferred to the matrix, which means that the matrix will withstand the load and be able to take more load before final failure. This case will happen only when the fiber content of the composite is not adequate to prevent ultimate failure. That is, the rule of mixtures for strength calculation is only applicable if the volume fraction of fibers is above a critical value. Because of the strength dependency of composite to the fiber volume fraction, it is possible that at a critical amount of fiber volume $(V_f)_{crit}$, the matrix will be able to handle the load after fiber failure. The critical value of $(V_f)_{crit}$ that allows the fiber to obtain fiber strengthening will be solved [36] by solving Eqs. (2.14) and (2.15) by replacing $(\sigma_{com})_u = (\sigma_m)_u (1 - V_f)_{cr}$ from (2.14)

$$(1 - (\sigma_f)_u (V_f)_{cr} + (\sigma_m)_{(\varepsilon_f)_u} (1 - (V_f)_{cr}) = (\sigma_m)_u ((1 - (V_f)_{cr}) \tag{2.16}$$

Solving for $(V_f)_{cr}$:

$$(V_f)_{crit} = \left[(\sigma_m)_u - (\sigma_m)_{(\varepsilon_f)_u} \right] / \left[(\sigma_f)_u - (\sigma_m)_{(\varepsilon_f)_u} \right] \tag{2.17}$$

Figure 2.8 is the variation of composite ultimate stresses with respect to the amount of fiber in the composite. Equations 2.16 and 2.17 are described through

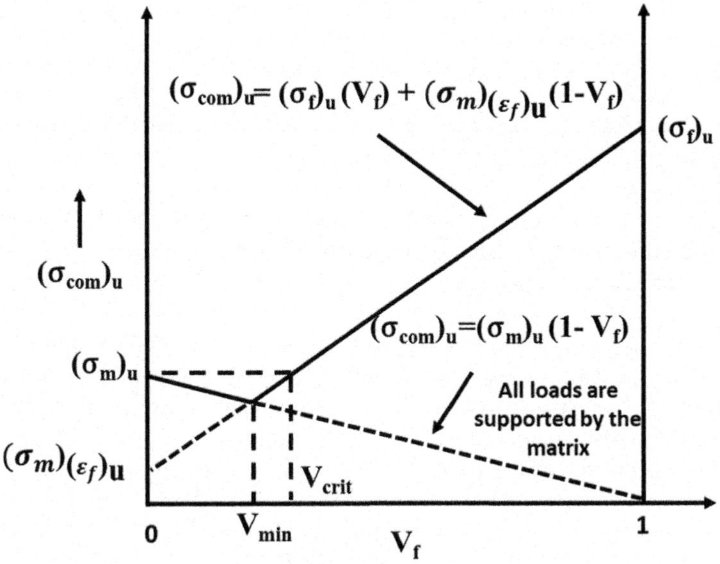

Fig. 2.8 The variation of composite ultimate stress versus the fiber fraction volume. For low fiber volume, the matrix will support the load [34]

Fig. 2.8. Finally, the minimum amount of fiber, $(V_f)_{min}$, and $(V_f)_{crit}$ are shown in Fig. 2.8.

Example

Assume the $(V_f)_{cr} = 35\%$ for the carbon fiber with tensile strength of $(\sigma_f)_u = 71$ ksi (~490 Mpa). The modulus of elasticity for the fiber is $E_f = 10{,}150$ ksi (~70 GPa). Determine the ultimate strength of composite with ultimate strength of epoxy matrix =25 ksi (~172.4 MPa).

Solution

The strain to failure of carbon fiber $(\varepsilon_f)_u$ can be calculated as:

$$(\varepsilon_f)_u = (\sigma_f)_u / E_f = (71 \text{ ksi})/(10,150 \text{ ksi}) = 0.007$$

$$\text{Assuming } (\varepsilon_f)_u = (\sigma_m)_{(\varepsilon_f)_u} / E_m = 0.007$$

$$(V_f)_{cr} = \left[(\sigma_m)_u - (\sigma_m)_{(\varepsilon_f)_u}\right] / \left[(\sigma_f)_u - (\sigma_m)_{(\varepsilon_f)_u}\right] \text{ (from Eq.2.17)}$$

$$(V_f)_{cr} = 0.35 = \left[(25 \text{ ksi}) - (\sigma_m)_{(\varepsilon_f)_u}\right] / \left[(71 \text{ ksi}) - (\sigma_m)_{(\varepsilon_f)_u}\right]$$

$$(\sigma_m)_{(\varepsilon_f)_u} = 0.23 \text{ ksi}$$

This is the stress that the matrix will carry at $(\varepsilon_f)_u = 0.007$ at fibers ultimate. The ultimate strength of the composite from (2.14) will be:

$$(\sigma_{\mathrm{com}})_{\mathrm{u}} = (\sigma_{\mathrm{f}})_{\mathrm{u}}\, V_{\mathrm{f}} + (\sigma_{\mathrm{m}})_{(\varepsilon_f)_{\mathrm{u}}} (1 - V_{\mathrm{f}})$$
$$(71\ \mathrm{ksi})(0.35) + (0.23\ \mathrm{ksi})\ (0.65) = 25.0\ \mathrm{ksi}\ (172.4\ \mathrm{MPa})$$

In Sect. 2.5.5.2, the fracture mechanics approach to finding the ultimate strength of the composite will be discussed.

2.5.5.2 Estimating the Ultimate Strength of the Composite by Fracture Mechanics Approach

The Griffith theory of fracture was established for elastic brittle material, and it is based on the available energy that is consumed at the crack tip to create two new crack surfaces. This theory took the form of an equation relating the energy release rate, G, in terms of breaking stress, σ, young's modulus, E, and half crack length, a. Therefore, the Griffith theory for brittle material and mode I failure can be written as:

$$\frac{\pi\sigma^2 a}{E} = G \tag{2.18}$$

The Griffith theory can be expanded to account for ductile material by adding an additional term that includes the energy dissipated at the crack tip for plastically deforming the material [35]. Therefore, Eq. (2.18) can be modified to account for the plastic deformation. That is:

$$G = \frac{\pi\sigma^2 a}{E} = 2T + \frac{\partial U_P}{\partial a} \tag{2.19}$$

where $2T$ is the energy needed to create two new crack surfaces and the second term is the energy consumed for the formation of plastic deformation at the crack tip prior to final failure. In the analysis of the ultimate strength of the composite with brittle matrix and fibers, the second term will be small and negligible. Therefore, the ultimate strength for brittle material can be written as:

$$\sigma_{\mathrm{ult}} = \sqrt{\frac{GE}{\pi a_{\mathrm{i}}}} \tag{2.20}$$

Note, the quantity $0.01 > 2a_{\mathrm{i}} > 0.005$ in, where a_i = half a crack length. If the material shows some amount of crack tip ductility, then an additional term must be incorporated in Eq. 2.20 to account for the energy consumed for the plastic deformation as shown in Eq. 2.19.

Example 1

Determine the ultimate strength of AS4/PEEK (thermoplastic matrix) with AS4 as fiber of a lamina with $G_{\mathrm{Ic}} = 5.63$ in-lb/in^2 [36] $E = 400$ ksi, and the preexisting flaw in the matrix of $2a_i = 0.005$ inch. Note: the AS4/PEEK is a thermoplastic composite material with 64% AS4 graphite fibers embedded in a thermoplastic matrix of PEEK. The stress needed to propagate the $2a_i = 0.005$ inch crack in the matrix can be calculated as:

$$\sigma_{ult} = \sqrt{G_{Ic}E/\pi a_i} = [(5.63/1000)(400)/(3.14)(0.0025)]^{0.5} = 20.74 \text{ ksi}$$

If the value of initial defect size in the material, $2a_i$, varies between 0.005 to 0.01 inch, then the value of σ_{ult} also varies between 20.74 and 12.0 ksi. This is compatible with the effective stress test value of 36 ksi recorded in [36]. Note that the above assumption is based on a center crack in an infinite plate, and it is a reasonable assumption because the size of the crack is much smaller than the width of the test specimen. Moreover, it is possible to have negligible amount of plasticity at the crack tip that was not considered in the analysis.

The following are two additional analyses conducted for composite materials to estimate and verify the ultimate strength, σ_{ult}, of the material using the fracture mechanics approach. In order to be able to estimate the σ_{ult} value, analysts need to have the fracture toughness, K_{Ic}, or the energy release rate, G_{Ic}, as well as the size of the initial defects in the material, $2a_i$. There are two sources of uncertainty that must be taken into consideration. The first one is the range of initial flaw size in the material, $2a_i$, that already was discussed above. Also, the range associated with the quantity K_{Ic} or G_{Ic} that has been obtained through testing. Herein, the analyses will assume the 5–10% error associated with the fracture toughness measurements. Usually, the average value of five K_{Ic} tests for a given material must be taken into consideration for the G_{Ic} (or K_{Ic}) values. The following examples are for finding the ultimate strength of the epoxy matrix, E-glass fibers with polyamide (nylon) as the matrix, and silicon carbide matrix. In all cases the goal is to obtain the ultimate of the composite matrix value, where crack propagation is more probable through the matrix.

Example 2

Determine the ultimate strength of Epoxy with $K_{Ic} = 1.0$ ksi(in.)^2 [37], and $2a_i = 0.005$ inch. The stress needed to propagate the $2a_i = 0.005$-inch crack can be calculated as:

$$\sigma_{ult} = K_{Ic}/\sqrt{\pi a} = [(1.0)/[(3.14)(0.0025)]^{0.5} = 11.28 \text{ ksi (for Epoxy matrix)}$$

This value is close to the test value of 12 ksi provided by the [38]. If the value of initial defect size in the material, $2a_i$, varies between 0.005 to 0.01 inch, then the value of σ_{ult} also varies between 11.28 ksi to 7.98 ksi [37]. By reducing the value of the initial flaw size to $2a_i = 0.004$ inch, the ultimate value of $\sigma_{ult} = 12.61$ ksi is obtained.

Example 3
Determine the ultimate strength of polyamide (nylon) with $K_{Ic} \sim 2.7$ ksi(in.)^2 [38], and $2a_i = 0.005$ inch.

$$\sigma_{ult} = K_{Ic}/\sqrt{\pi a} = (2.7)/[(3.14)(0.0025)]^{0.5} = 30.47 \text{ ksi (for E-glass fiber)}$$

The actual experimental value is 20 ksi. However, if we use $2a_i = 0.01$ in., the calculated value of ultimate strength will be 21.54 ksi.

Example 4
Determine the ultimate strength of silicon carbide (SiC) with $K_{Ic} \sim 3.0$ ksi(in.)^2 [39], where $2a_i = 0.005$ inch. The stress needed to propagate the $2a_i = 0.005$-inch crack can be calculated as:

$$\sigma_{ult} = K_c/\sqrt{\pi a} = (3.0)/[(3.14)(0.0025)]^{0.5} = 33.85 \text{ ksi}$$

This value is close to the test value of 36.8 ksi provided by the [39]. If the value of initial defect size in the material, $2a_i$, varies between 0.005 to 0.01 inch, then the value of σ_{ult} also varies between 33.25 ksi to 23.94 ksi.

2.6 Transformation Matrix

In Sect. 2.5, it was mentioned that for a unidirectional lamina of a laminate with loads along the fibers, there are four elastics constants that must be available for stress and failure analysis assessment, (1) E_1 and E_2 are the longitudinal and transverse Young's modulus in directions 1 and 2, respectively, where 1 and 2 axes are called the local or the material coordinate system (sometimes they are called longitudinal and transverse directions), (2) ν_{12} is major Poisson's ratio, where it is defined as the ratio of the negative of the normal strain in direction 2 to the normal strain in direction 1, and 3) G_{12} is in-plane shear modulus (in plane 1–2). The four elastics constants (E_1, E_2, G_{12}, and ν_{12}) are essential and must be available for stress and strain analysis of a unidirectional lamina. However, a laminate will not consist of a single lamina only and it requires several of them bonded together. A laminate consists of more than one lamina adhesively bonded together through their thickness. This is because a lamina by itself is too thin with an average thickness of around 0.005 in (0.125 mm), which implies several predetermined selected unidirectional plies will be required to handle the service load. In addition, the mechanical properties of a typical unidirectional lamina are strong along the fibers and weak in the transverse direction. Therefore, this weakness in the transverse direction can be overcome by making a laminate with layers stacked properly on top of each other at different angles to withstand the applied service load.

For a laminate, with plies of different orientations, the stiffness and strength along the direction of fiber and parallel to the load as well as transverse to the fibers were discussed and the rule of mixtures was used to establish mechanical properties in (1) and (2) directions (Sect. 2.5). It is not a good idea to have all plies bonded together in only one direction. Thus, it is necessary to develop the stress–strain relationship for other lamina that are attached together with different orientations. To predict properties for any other ply with fibers oriented at some angle θ from the applied load (Fig. 2.9), the transformation matrix can be helpful to obtain both the stress and strain for any other angle with respect to the global axis or the off-axes (x–y axis). The main assumption used here is the applied load and stresses are in the plane of the ply. In addition, if the laminate has only a few plies, the effect of stacking order becomes important. Unsymmetrical layouts of plies can cause out-of-plane deformations, either twisting or bending.

The procedure for obtaining the mechanical properties for any other plies of a laminate oriented with an angle θ with respect to the load is as follows:

The applied stress in the x–y (σ_x and σ_x) are available as shown in Fig. 2.9. The stresses in the local axis are parallel and transverse to fibers calculated from global to local by tensor equations. Since stresses and strains within the plane of the ply are of concern, only three of each (two normal and one shear) are involved. The first step of

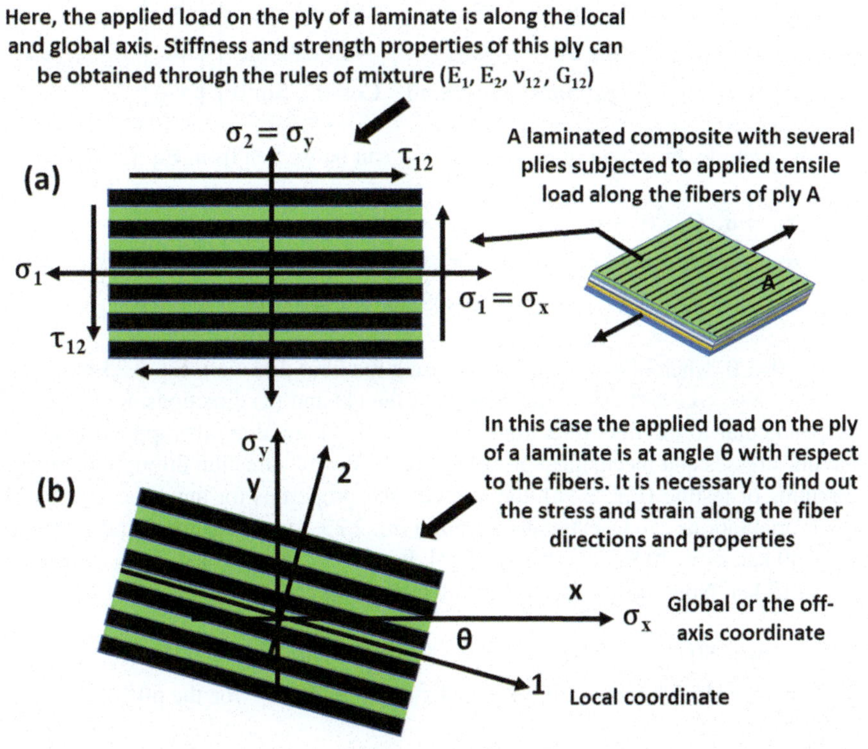

Fig. 2.9 (**a**) The planner mechanical properties of a ply with fibers parallel to the applied load is available. (**b**) By applying the transformation matrix, the mechanical properties of any other ply with fibers at an angle θ from the load can be obtained

resolving the applied stresses, σ_x, σ_y, and τ_{xy}, into components parallel and normal to the fiber axis, σ_1, σ_2, and σ_{12} (see Fig. 2.9b), depends on the angle, θ between the loading direction (x) and the fiber axis (1). The transformation is as follows:

$$\begin{bmatrix}\sigma_1\\ \sigma_2\\ \tau_{12}\end{bmatrix}=[T]\begin{bmatrix}\sigma_x\\ \sigma_y\\ \tau_{xy}\end{bmatrix}\quad \text{OR} \quad \begin{bmatrix}\sigma_x\\ \sigma_y\\ \tau_{xy}\end{bmatrix}=[T]^{-1}\begin{bmatrix}\sigma_1\\ \sigma_2\\ \tau_{12}\end{bmatrix} \tag{2.21}$$

where the transformation matrix is designated by $[T]$ and the inverse by $[T]^{-1}$. Both matrices $[T]$ and $[T]^{-1}$ are a function of θ only. The matrix $[T]$ is given by [40]:

$$[T]=\begin{bmatrix}\mathrm{Cos}^2\theta & \mathrm{Sin}^2\theta & 2\mathrm{Cos}\theta\mathrm{Sin}\theta\\ \mathrm{Sin}^2\theta & \mathrm{Cos}^2\theta & -2\mathrm{Cos}\theta\mathrm{Sin}\theta\\ -\mathrm{Cos}\theta\mathrm{Sin}\theta & \mathrm{Cos}\theta\mathrm{Sin}\theta & \mathrm{Cos}^2\theta-\mathrm{Sin}^2\theta\end{bmatrix}$$

Equation (2.21) in terms of $[T]$ can be rewritten as:

$$\begin{bmatrix}\sigma_1\\ \sigma_2\\ \tau_{12}\end{bmatrix}=\begin{bmatrix}\mathrm{Cos}^2\theta & \mathrm{Sin}^2\theta & 2\mathrm{Cos}\theta\mathrm{Sin}\theta\\ \mathrm{Sin}^2\theta & \mathrm{Cos}^2\theta & -2\mathrm{Cos}\theta\mathrm{Sin}\theta\\ -\mathrm{Cos}\theta\mathrm{Sin}\theta & \mathrm{Cos}\theta\mathrm{Sin}\theta & \mathrm{Cos}^2\theta-\mathrm{Sin}^2\theta\end{bmatrix}\begin{bmatrix}\sigma_x\\ \sigma_y\\ \tau_{xy}\end{bmatrix} \tag{2.22}$$

For example, the value of σ_1, σ_2, and τ_{12}, can be written from Eq. (2.22) as

$$\begin{aligned}\sigma_1 &= \sigma_x\,\mathrm{Cos}^2\theta+\sigma_y\,\mathrm{Sin}^2\theta+2\tau_{xy}\mathrm{Cos}\theta\mathrm{Sin}\theta\\ \sigma_2 &= \sigma_x\,\mathrm{Sin}^2\theta+\sigma_y\,\mathrm{Cos}^2\theta-2\tau_{xy}\mathrm{Cos}\theta\mathrm{Sin}\theta\\ \tau_{12} &= -\sigma_x\,\mathrm{Cos}\theta\mathrm{Sin}\theta+\sigma_y\,\mathrm{Cos}\theta\mathrm{Sin}\theta+2\tau_{xy}\left(\mathrm{Cos}^2\theta-\mathrm{Sin}^2\theta\right)\end{aligned} \tag{2.23}$$

For $\theta=0$, where the applied load is along the fiber direction, Eq. 2.23 will yield $\sigma_1=\sigma_x$, $\sigma_2=\sigma_y$, and $\tau_{12}=\tau_{xy}$. Note that, the (1) and (2) directions are along and perpendicular to the fiber directions and (x) and (y) are along the applied load. By having stresses and mechanical properties in the $\theta=0°$ direction through the rule of mixtures or testing (Fig. 2.9a), the stresses and properties for any other plies with $\theta\neq0$ orientations can be estimated through Eq. 2.23. Let us assume that the stresses $\sigma_x=40$ ksi, $\sigma_y=30$ ksi, and $\tau_{xy}=15$ ksi. For $\theta=0^{\circ}$, $\sigma_1=40$ ksi, $\sigma_2=30$ ksi, and $\tau_{12}=15$ ksi. Now, if $\theta=45°$ or $60°$, then from Eq. 2.23:

$$\begin{aligned}&\sigma_1=50\text{ ksi}, \sigma_2=20\text{ ksi, and } \tau_{12}=-5\text{ ksi} \quad \text{for the }45°\text{ ply}\\ &\sigma_1=45.5\text{ ksi}, \sigma_2=24.5\text{ ksi, and } \tau_{12}=-19.3\text{ ksi} \quad \text{for the }60°\text{ ply}\end{aligned}$$

Furthermore, the relationship between strains and stresses dictated by these elastic constants can be expressed by the compliance tensor, S, which for composite ply, has the form of:

$$\begin{bmatrix} \varepsilon_1 \\ \varepsilon_2 \\ \gamma_{12} \end{bmatrix} = [S] \begin{bmatrix} \sigma_1 \\ \sigma_2 \\ \tau_{12} \end{bmatrix} = \begin{bmatrix} S_{11} & S_{12} & 0 \\ S_{12} & S_{22} & 0 \\ 0 & 0 & S_{66} \end{bmatrix} \begin{bmatrix} \sigma_1 \\ \sigma_2 \\ \tau_{12} \end{bmatrix} \tag{2.24}$$

where the quantities S_{11}, S_{12}, S_{22}, and S_{66} are given as:

$$\begin{aligned} S_{11} &= 1/E_1 \\ S_{12} &= -\nu_{12}/E_1 = -\nu_{21}/E_2 \\ S_{22} &= 1/E_2 \\ S_{66} &= 1/G_{12} \end{aligned} \tag{2.25}$$

Equation 2.24 in terms of 2.25 can be rewritten as:

$$\begin{bmatrix} \varepsilon_1 \\ \varepsilon_2 \\ \gamma_{12} \end{bmatrix} = [S] \begin{bmatrix} \sigma_1 \\ \sigma_2 \\ \tau_{12} \end{bmatrix} = \begin{bmatrix} S_{11} & S_{12} & 0 \\ S_{12} & S_{22} & 0 \\ 0 & 0 & S_{66} \end{bmatrix} \begin{bmatrix} \sigma_1 \\ \sigma_2 \\ \tau_{12} \end{bmatrix} \rightarrow \begin{bmatrix} \varepsilon_1 \\ \varepsilon_2 \\ \gamma_{12} \end{bmatrix}$$

$$= \begin{bmatrix} S_{11} = 1/E_1 & S_{12} = -\nu_{12}/E_1 & 0 \\ S_{12} = -\nu_{12}/E_1 & S_{22} = 1/E_2 & 0 \\ 0 & 0 & S_{66} = 1/G_{12} \end{bmatrix} \begin{bmatrix} \sigma_1 \\ \sigma_2 \\ \tau_{12} \end{bmatrix} \tag{2.24a}$$

Using Eq. (2.24a) allows the strains to be established along the fibers, relative to directions 1 and 2. To establish a relationship similar to Eq. (2.22) relating strains in directions 1 and 2 in terms of strain in the global coordinate (x,y), then the tensor $[T]$ must be replaced by $[T\tilde{}]$. That is:

$$\begin{bmatrix} \varepsilon_1 \\ \varepsilon_2 \\ \gamma_{12} \end{bmatrix} = \left[\tilde{T}\right] \begin{bmatrix} \varepsilon_x \\ \varepsilon_y \\ \gamma_{xy} \end{bmatrix} = \begin{bmatrix} \mathrm{Cos}^2\theta & \mathrm{Sin}^2\theta & \mathrm{Cos}\theta\mathrm{Sin}\theta \\ \mathrm{Sin}^2\theta & \mathrm{Cos}^2\theta & -\mathrm{Cos}\theta\mathrm{Sin}\theta \\ -2\mathrm{Cos}\theta\mathrm{Sin}\theta & 2\mathrm{Cos}\theta\mathrm{Sin}\theta & \mathrm{Cos}^2\theta - \mathrm{Sin}^2\theta \end{bmatrix} \begin{bmatrix} \varepsilon_x \\ \varepsilon_y \\ \gamma_{xy} \end{bmatrix} \tag{2.26a}$$

and the inverse of Eq. (2.26a), where the inverse of $[T\tilde{}]$ is shown by $[T\tilde{}]^{-1}$, can be written as:

$$\begin{bmatrix} \varepsilon_x \\ \varepsilon_y \\ \gamma_{xy} \end{bmatrix} = \left[\tilde{T}\right]^{-1} \begin{bmatrix} \varepsilon_1 \\ \varepsilon_2 \\ \gamma_{12} \end{bmatrix} = \begin{bmatrix} \mathrm{Cos}^2\theta & \mathrm{Sin}^2\theta & -\mathrm{Cos}\theta\mathrm{Sin}\theta \\ \mathrm{Sin}^2\theta & \mathrm{Cos}^2\theta & \mathrm{Cos}\theta\mathrm{Sin}\theta \\ 2\mathrm{Cos}\theta\mathrm{Sin}\theta & -2\mathrm{Cos}\theta\mathrm{Sin}\theta & \mathrm{Cos}^2\theta - \mathrm{Sin}^2\theta \end{bmatrix} \begin{bmatrix} \varepsilon_1 \\ \varepsilon_2 \\ \gamma_{12} \end{bmatrix} \tag{2.26b}$$

Finally, the global strains in terms of global applied stresses for any lamina, with a given orientation, θ, can be written as:

$$\begin{bmatrix} \varepsilon_x \\ \varepsilon_y \\ \gamma_{xy} \end{bmatrix} = \left[\tilde{T}\right]^{-1} [S][T] \begin{bmatrix} \sigma_x \\ \sigma_y \\ \tau_{xy} \end{bmatrix} \rightarrow \begin{bmatrix} \varepsilon_x \\ \varepsilon_y \\ \gamma_{xy} \end{bmatrix} = \left[\overline{S}\right] \begin{bmatrix} \sigma_x \\ \sigma_y \\ \tau_{xy} \end{bmatrix} \tag{2.27}$$

Where $[\overline{S}]$ is the equivalent of multiplication of all matrices shown in Eq. 2.27. That is,

$$\left[\overline{S}\right] = \left[\tilde{T}\right]^{-1} [S][T]$$

Where components of $[\overline{S}]$ are given as:

$$\begin{aligned}
\overline{S}_{11} &= S_{11}\ \text{Cos}^4\theta + (2S_{12} + S_{66})\ \text{Sin}^2\theta\text{Cos}^2\theta + S_{22}\ \text{Sin}^4\theta \\
\overline{S}_{12} &= S_{12}\ \left(\text{Sin}^4\theta + \text{Cos}^4\theta\right) + (S_{12} + S_{22} - S_{66})\ \text{Cos}^2\theta\ \text{Sin}^2\theta \\
\overline{S}_{22} &= S_{11}\ \text{Cos}^4\theta + (2S_{12} + S_{66})\ \text{Sin}^2\theta\text{Cos}^2\theta + S_{22}\ \text{Cos}^4\theta \\
\overline{S}_{16} &= (2S_{11} - 2S_{12} - S_{66})\ \text{Sin}\theta\ \text{Cos}^3\theta - (2S_{22} - 2S_{12} - S_{66})\ \text{Sin}^3\theta\ \text{Cos}\theta \\
\overline{S}_{26} &= (2S_{11} - 2S_{12} - S_{66})\ \text{Sin}^3\theta\ \text{Cos}\theta - (2S_{22} - 2S_{12} - S_{66})\ \text{Cos}^3\theta\ \text{Sin}\theta \\
\overline{S}_{66} &= 2(2S_{11} + 2S_{22} - 4S_{12} - 4S_{66})\ \text{Sin}^2\theta\ \text{Cos}^2\theta + S_{66}\left(\text{Cos}^4\theta + \text{Sin}^4\theta\right)
\end{aligned} \tag{2.28}$$

By using Eq. 2.25 quantities S_{11}, S_{12}, S_{22}, and S_{66} can be calculated and furthermore, from Eq. 2.28, components of $[\overline{S}]$ can be computed for obtaining global strains shown by Eq. 2.27, when σ_x, σ_y, and τ_{xy} are available. The matrix $[S]$ is modulus dependent and the matrix $[\overline{S}]$ is modulus and θ dependent; therefore, the Modulus as a function of θ can be formulated as [41]:

$$\frac{1}{E(\theta)} = \frac{\cos^4\theta}{E_1} + \frac{\sin^4\theta}{E_2} + \left(\frac{1}{G_{12}} - \frac{2\nu_{12}}{E_1}\right) \cos^2\theta\ \sin^2\theta \tag{2.29}$$

And the orientation dependency of the shear modulus, $G(\theta)$, can be formulated also as:

$$\frac{1}{G(\theta)} = 4\left(\frac{1 + 2\nu_{12}}{E_1} + \frac{1}{E_2} - \frac{1}{2G_{12}}\right) \cos^2\theta\ \sin^2\theta + \frac{1}{G_{12}}\left(\cos^4\theta + \sin^4\theta\right) \tag{2.30}$$

Figure 2.10a shows the plot of modulus of elasticity, $E(\theta)$, as a function of orientation angle θ by using Eq. 2.29 and in Fig. 2.10b, the plot of shear modulus, $G(\theta)$, as a function of θ are also shown by applying Eq. 2.30 [42].

Example

A unidirectional ply of graphite–epoxy is subjected to $\sigma_1 = 2$ ksi, $\sigma_2 = -3$ ksi, and $\tau_{12} = 4$ ksi (parallel and perpendicular to the graphite fibers). Determine: (1) compliance matrix $[S]$, (2) minor Poisson's ratio, ν_{21}, (3) reduced stiffness matrix,

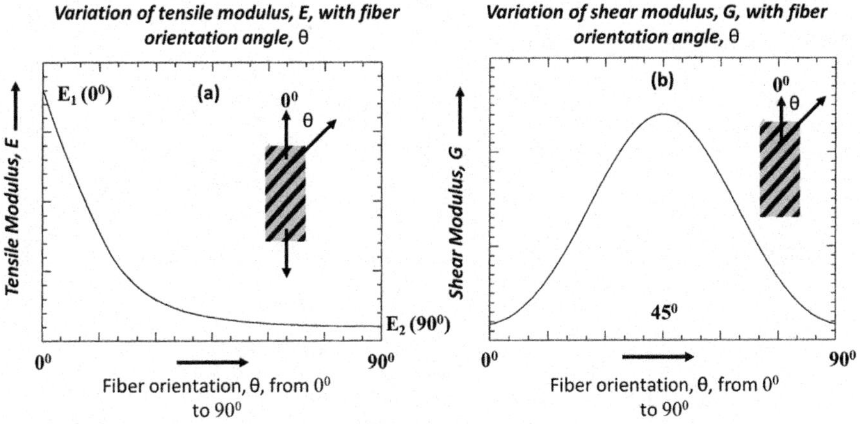

Fig. 2.10 (**a**) Plot of modulus of elasticity, $E(\theta)$, as a function of orientation angle θ. (**b**) Plot of shear modulus, $G(\theta)$, as a function of θ

(4) strains in the 1–2 (the local coordinate system), and (5) the value of E and G for $\theta = 60°$.

Mechanical properties of graphite–epoxy ($\theta = 0°$) are provided by the rule of mixtures as:

$E_1 = 26231 \times 10^3$ ksi, $E_2 = 1492 \times 10^3$ ksi, $\nu_{12} = 0.28$, $G_{12} = 1039 \times 10^3$ ksi with fiber volume fraction $V_f = 70\%$. Using Eq. 2.25, the components of the compliance matrix $[S]$ can be calculated as:

$$S_{11} = 1/E_1 = 1/26231 \times 10^3 = 3.81 \times 10^{-8}\ \text{ksi}^{-1}$$
$$S_{12} = -\ \nu_{12}/E_1 = -0.28/26231 \times 10^3 = -1.06 \times 10^{-8}\ \text{ksi}^{-1}$$

Where $S_{12} = -1.06 \times 10^{-8}\ \text{ksi}^{-1}$ and the minor Poisson's ratio $\nu_{21} = -E_2 \times S_{12}$

$$\left(-1492 \times 10^3\right) \times \left(-1.06 \times 10^{-8}\right) = 0.0158$$
$$S_{22} = 1/E_2 = 1/1492 \times 10^3 = 6.70 \times 10^{-7}\ \text{ksi}^{-1}$$
$$S_{66} = 1/G_{12} = 1/1039 \times 10^3 = 9.6 \times 10^{-7}\ \text{ksi}^{-1}$$

The compliance matrix $[S]$ can be written as:

$$[S] = \begin{bmatrix} S_{11} = 3.81 \times 10^{-8} & S_{12} = -1.06 \times 10^{-8} & 0 \\ S_{12} = -1.06 \times 10^{-8} & S_{22} = 6.70 \times 10^{-7} & 0 \\ 0 & 0 & S_{66} = 9.6 \times 10^{-7} \end{bmatrix}$$
$$= \begin{bmatrix} 3.81 \times 10^{-8} & -1.06 \times 10^{-8} & 0 \\ -1.06 \times 10^{-8} & 6.70 \times 10^{-7} & 0 \\ 0 & 0 & 9.6 \times 10^{-7} \end{bmatrix} (ksi)^{-1}$$

And the inverse of compliance matrix [*S*] is $[S]^{-1}$, where [*S*] $[S]^{-1}$ = [*I*]. The [*I*] is the unity matrix and $[S]^{-1}$ can be written as:

$$[S]^{-1} = \begin{bmatrix} 26.34\times10^6 & 0.42\times10^6 & 0 \\ 0.42\times10^6 & 1.5\times10^6 & 0 \\ 0 & 0 & 1.04\times10^6 \end{bmatrix} ksi$$

Using Eq. 2.24a, local strains (ε_1, ε_2, and γ_{12}) with applied stresses $\sigma_1 = 2$ksi, $\sigma_2 = -3$ksi, and $\tau_{12} = 4$ksi will be calculated:

$$\begin{bmatrix} \varepsilon_1 \\ \varepsilon_2 \\ \gamma_{12} \end{bmatrix} = [S] \begin{bmatrix} \sigma_1 \\ \sigma_2 \\ \tau_{12} \end{bmatrix} = \begin{bmatrix} 3.81\times10^{-8} & -1.06\times10^{-8} & 0 \\ -1.06\times10^{-8} & 6.70\times10^{-7} & 0 \\ 0 & 0 & 9.6\times10^{-7} \end{bmatrix}$$

$$\times \begin{bmatrix} 2\text{ksi} \\ -3\text{ksi} \\ 4\text{ksi} \end{bmatrix} = \begin{bmatrix} 0.044 \\ -1.98 \\ 3.84 \end{bmatrix} (10^{-6})$$

That is, $\varepsilon_1 = 0.044 \times 10^{-6}$, $\varepsilon_2 = -1.98 \times 10^{-6}$, and $\gamma_{12} = 3.84 \times 10^{-6}$. Finally, values of *E* and *G* for $\theta = 60°$ can be obtained by using Eqs. (2.29) and (2.30):

$$\begin{aligned} \frac{1}{E(\theta)} &= \frac{\cos^4\theta}{E_1} + \frac{\sin^4\theta}{E_2} + \left(\frac{1}{G_{12}} - \frac{2\nu_{12}}{E_1}\right)\cos^2\theta\,\sin^2\theta \\ &= (0.0625)/(26.23\times10^6) + (0.562)/(1.49\times10^6) \\ &\quad + \{1/1.03\times10^6 - 2\times0.28/(26.23\times10^6)\}\times(0.75\times0.25) \\ &= 53\times10^{-8} \rightarrow E\ (60°) = 1.86\times10^6\ \text{ksi} \end{aligned}$$

$$\begin{aligned} \frac{1}{G(\theta)} &= 4(\frac{1+2\nu_{12}}{E_1} + \frac{1}{E_2} - \frac{1}{2G_{12}})\cos^2\theta\,\sin^2\theta + \frac{1}{G_{12}}(\cos^4\theta + \sin^4\theta) \\ &= 4\times\{(1+2x0.28)/(26.23x\ 10^6) + 1/(1.49x\ 10^6)\cdot1/(2x1.03\times10^6)\} \\ &\quad \times(0.75\times0.25) + (1/(1.03\times10^6))\times(0.0625+0.562) \\ &= 0.788\times10^{-6} \rightarrow G(60^0) = 1.267\times10^6\ \text{ksi} \end{aligned}$$

Results of above analysis indicated that by using Eq. (2.29), the calculate value of *E*(60°) =1.86 × 10^6 ksi > *E*(90°) =1.49 × 10^6, but smaller than *E*(0°) =2.62 × 10^6 ksi (see Fig. 2.10a). Moreover, the value of shear modulus *G*(60°) = 1.267 × 10^6 ksi > *G*(0°) = 1.03 × 10^6, as it is evident in Fig. 2.10b.

2.7 Disadvantages and Limitations of Rule of Mixtures

In this chapter, it was shown that properties of composite materials can be obtained based on the properties of their individual constituents, which is called the rule of mixtures. Even though the approach is simple, it can provide useful approximations in obtaining composite material allowable. The following is a brief discussion on several disadvantages and limitations associated with this method.

1. *It neglects microstructure*: The rule of mixtures assumes that the mixture of its constituents (both the matrix and fibers) in composite material is uniform, neglecting the effects of several parameters, such as inclusions, voids, or the strength of the interfacial bonding between the matrix and fibers. This simplification can lead to inaccurate estimates in cases where microstructure significantly affects the material's properties. Therefore, there is a need to conduct a few tests that would be able to verify the lamina properties estimated through the rule of mixtures.
2. *It does not' account for reinforcement orientation*: In cases where the reinforcement in the matrix is oriented in different direction, the rule of mixtures may not correctly predict the mechanical properties of composite when transformation matrix is used (Sect. 2.6). Fiber-reinforced composites, for example, can have highly anisotropic properties that the rule of mixtures cannot account for. Moreover, properties of fibers in the matrix are not the same in each ply within the laminate.
3. *It ignores stress transfer mechanisms*: The rule of mixtures assumes that stress is evenly distributed between the constituents, which may not hold true in practice because the strength of interface varies throughout the matrix. In composites, stress transfer mechanisms can vary, and the rule's simplifications may not accurately represent how loads are transmitted between the constituents.
4. *Inaccuracy with nonlinear behaviors*: The Rule of Mixtures is more suitable for linear elastic materials and may not accurately predict the behavior of composites with nonlinear stress-strain relationships or materials that exhibit plastic deformation.
5. *Variation in composite fabrication*: The properties of composites can be significantly influenced by the fabrication process, including factors like curing temperature, pressure, and resin content. The Rule of Mixtures does not consider these factors, making it less accurate when dealing with composites of varying fabrication conditions.
6. *Lack of consideration for environmental effects*: The rule does not account for the effects of environmental factors, such as above room temperature operating conditions, humidity, or chemical exposure, high frequency applied load, which all can lower the properties of composite materials over time.
7. *No consideration of size effects*: For composites with nanoscale or microscale reinforcements, the rule of mixtures may not accurately predict their properties. In many cases the nanoscale reinforcement has poor interface bond with the matrix that upon the application of the load they will not contribute to crack arrest when

crack is advancing. In some cases, the distribution of nanoscale reinforcement is not uniform throughout the matrix, which can contribute to early structural failure during operation.

8. *Limited applicability for complex composites*: Some advanced composite materials, such as laminated composites, have complex arrangements of different constituents. The Rule of Mixtures may not provide accurate predictions for these materials.
9. *Empirical variations*: The rule of mixtures might require empirical correction factors to improve accuracy in specific cases, adding complexity to the approach.

In summary, while the rule of mixtures is a valuable tool for estimating composite material properties, it has inherent limitations, particularly when dealing with complex composites, anisotropic materials, and situations where microstructure, stress transfer mechanisms, and other factors significantly affect the material's behavior. More advanced modeling techniques for obtaining properties may be required to obtain more accurate predictions of composite properties. In practice, even when property values predicted by more advanced micromechanics expressions, it should not be relied upon for design purposes, because of errors introduced by approximations in the models, such as uncertainties in constituents' properties, or by manufacturing variables. Therefore, it is recommended that the ply properties be determined by a limited amount of testing in combination with the modelling technique. Much care is required in the testing process where testing must be conducted based on internationally accepted test standards and should be performed by well qualified technical personnel.

Finally, it should be mentioned that the rule of mixtures for finding the mechanical properties of composites is applicable to all types: PMCs, MMCs, and CMCs. For example, modulus equations derived for both longitudinal and transverse directions ($E1$ and $E2$) are applicable to all three types of composites. However, as mentioned above, even though the approach is simple and provides useful approximations for obtaining allowable composite material properties, a limited number of tests are needed for comparison between analysis and testing.

2.8 Concluding Remarks

By exploring the micromechanics of lamina, it offers invaluable insights into the intricate behavior and properties of composite materials arising from its constituents. By solving the fundamental mechanics of composites at the microscale and understanding the anisotropic nature of the material, engineers can pave the way for not only enhancing the properties of composite materials but also in areas such as design, optimization, and performance, which eventually can drive innovation across several sectors of industries.

The rule of mixtures serves as a fundamental framework for estimating the mechanical properties of composite materials based on the properties of their constituents (fibers and matrix). While it is a simple approach, the broad applicability of it makes it a useful tool for engineers and researchers in designing composite structural parts for a wide range of applications. However, it is essential to acknowledge the limitations associated with predicting elastic constants through the rule of mixtures, and results must be verified through enough tests. Moreover, engineers should explore more sophisticated predictive models to achieve greater accuracy and improve predictions using corresponding data.

References

1. V. Kushch, *Micromechanics of Composites*. Chapter 2 (Elsevier Butterworth Heinemann, 2020)
2. J. Aboudi, S.M. Arnold, B.A. Bednarcyk, *Practical Micromechanics of Composites Materials*, eBook ISBN: 9780128206386 (2021)
3. G. Dvorak, *Micromechanics of Composites* (Springer, 2013)
4. X.W. Yu, *An Introduction to Micromechanics* (Trans Tech, 2016), pp. 3–24. ISBN 9783038357469
5. S.M. Alger, *Polymer Science Dictionary*, 2nd edn. (Springer, 1997)
6. D.R. Askeland, P.P. Fulay, W.J. Wright, *The Science and Engineering of Materials*, 6th edn. (Cengage Learning, 2010)
7. D. Hull, *An Introduction to Composite Materials* (Cambridge University Press, 1981)
8. A. Endruweit, F. Gommer, A.C. Long, Stochastic analysis of fiber volume fraction and permeability in fiber bundles with random filament arrangement, composites. Appl. Sci. Manuf. **49**, 109–118 (2013)
9. D. Pan, K. Zhu, Y. Zhang, L. Sun, X. Hao, First principles and molecular simulation investigation of mechanical properties of the PTFE/graphene. Compos. Part B Eng. **242**, 110050 (2022)
10. J. Zhou, J. Shen, W. Yue, Y. Liu, Z. Chen, Molecular dynamics simulation of reinforcement mechanism of graphene/aluminum composites and microstructure evolution. J. Mater. Res. **23** (2023)
11. B. Farahmand, Chapter 9: Predictive modeling, in *Virtual Testing and Predictive Modeling*, ed. by M. Doyle, (Springer, 2008)
12. V.V. Mitin, D.I. Sementsov, N.Z. Vagidov, *Quantum Mechanics for Nanostructures* (Cambridge University Press, 2010)
13. O. Manasreh, *Introduction to Nanomaterials and Devices* (Wiley, 2012)
14. S. Karabasov, D. Nerukh, A. Hoekstra, B. Chopard, P.V. Coveney, Multiscale modelling: Approaches and challenges. Philos. Trans. R. Soc. **372** (2014)
15. A.G. Hoekstra, P.V. Coveney, S.P. Zwart, Multiscale modelling, simulation and computing. Philos. Trans. R. Soc. **377** (2019)
16. J.T. Oden, K. Vemaganti, N. Moës, Hierarchical modeling of heterogeneous solids. Comput. Methods Appl. Mech. Eng. **172**, 3–25 (1999)
17. Q.H. Zeng, A.B.G.Q. Yu, Multiscale modeling and simulation of polymer nanocomposites. Prog. Polym. Sci. **33**, 191–269 (2008)
18. S.A. Baeurle, Multiscale modeling of polymer materials using field-theoretic methodologies: A survey about recent developments. J. Math. Chem. **46**, 363–426 (2008)
19. R. Gibson, *Principles of Composite Material Mechanics*, 2nd edn. (CRC Press, Boca Raton, 2007)
20. S.-T. Fu, B. Lauke, Y.-W. Mai, *Science and Engineering of Short Fiber Reinforced Polymer Composites* (Woodhead, 2009)

21. N. Pan, Theoretical determination of the optimal fiber volume fraction and fiber-matrix property compatibility of short fiber composites. Polym. Compos. **14**(2), 85–93 (1993)
22. A.R. Chambers, J.S. Earl, C.A. Squires, M.A. Suhot, The effect of voids on the flexural fatigue performance of unidirectional carbon fiber composites developed for wind turbine applications. Int. J. Fatigue **28**, 1389 (2006)
23. Y. Nikishkov, L. Airoldi, A. Makeev, Measurement of voids in composites by X-ray computed tomography. Compos. Sci. Technol. **89**, 89–97 (2013)
24. L. Di Landro, A. Montalto, M. Rigamonti, Detection of voids in carbon/epoxy laminates and their influence on mechanical properties. SAGE J. (2017)
25. A.H. Kite, D.K. Hsu, D.J. Barnard, 34th Annual Review of Progress in Quantitative Nondestructive Evaluation, Golden CO, USA (2007), AIP Conf. Proc. 975 (2008)
26. A.K. Kaw, *Mechanics of Composite Materials.*, Chapter 2 (Taylor & Francis, 2006)
27. M. Akay, An Introduction to Polymer Matrix Composite. Chapter 3, eBooks at bookboon.com (2015)
28. H. Altenbach, J. Altenbach, W. Kissing, *Mechanics of Composite Structural Elements* (Springer, Berlin/Heidelberg, 2004)
29. R.M. Jones, *Mechanics of Composite Materials.* Chapter 3 (McGraw Hill, 1976)
30. A.K. Kaw, *Mechanics of Composite Materials.*, Chapter 3 (Taylor & Francis, 2006), pp. 216–240
31. P.R. Budarapu, P.A. Stephane, Halpin–Tsai model, in *Advances in Applied Mechanics* (2019)
32. J.C. Halpin, S.W. Tsai, *Effects of Environmental Factors on Composite Materials*, Report AFML-TR 67–423, US Air Force Materials Laboratory, Dayton, Ohio (1969)
33. R.M. Jones, *Mechanics of Composite Materials.*, Chapter 3 (McGraw Hill, 1976), p. 94
34. R.M. Jones, *Mechanics of Composite Materials.*, Chapter 3 (McGraw Hill, 1976), pp. 128–131
35. B. Farahmand, *Fatigue and Fracture Mechanics of High-Risk Parts.* Chapter 6 (Chapman & Hall, 1997), pp. 293–294
36. Y. Su, D. Zhang, X. Gong, M. Shen, Mechanical behavior in bending deformation of thermoplastic composite laminates with different stacking sequences. J. Compos. Mater. **50** (2015)
37. A.K. Kaw, *Mechanics of Composite Materials.* Chapter 1, Table 1.10 (Taylor & Francis, 2006)
38. M. Herráez, A. Fernández, C.S. Lopes, C. González, *Strength and Toughness of Structural Fibres for Composite Material Reinforcement* 2016
39. A.K. Kaw, *Mechanics of Composite Materials.* Chapter 1, Table 1.12 (Taylor & Francis, 2006), p. 45
40. A.K. Kaw, *Mechanics of Composite Materials.* Chapter 2, Section 2.5 (Taylor & Francis, 2006)
41. B. Harris, *Engineering Composite Materials.* Chapter 3 (The Institute of Materials, London, 1999), p. 43
42. J. Pabiot, Composite Materials Conference proceedings number 63 (AGARD/NATO, Neuilly, Paris), paper 6 (1971)

Chapter 3
Types of Matrices and Fibers Used in Composites

3.1 Introduction

As already discussed in Chap. 1, the role of the matrix is to have reinforcement embedded inside the matrix, which can provide great properties to the composite like stiffness, strength, and other valuable attributes, such as creep and fatigue resistance, toughness, thermal and electrical conductivity. The matrix material, which holds or binds everything together, is divided into three types as shown in Fig. 3.1. Polymers, metals, and ceramic matrix composites are the three most common types that can be useful for industries, and they will be briefly discussed here separately. However, a comprehensive description of their manufacturing methods is presented in Chaps. 4, 5, and 6, respectively.

3.2 The Ceramic Matrix and Ceramic Matrix Composites (CMCs)

Ceramic matrix composites (CMCs) are a special type of composite material in which both the reinforcement and matrix material are ceramics for the purpose of surviving the high temperature environment and resistance to oxidation (Fig. 3.1) [1, 2]. Current operating temperatures for composites are around 600 °F (~315 °C) for polymers, 1800 °F (~1000 °C) for metals and their alloys, and 2700 °F (~1500 °C) for ceramics. Because both matrix and fibers in ceramic composites possess high melting points, the CMCs have excellent creep resistance properties when subjected to high temperatures and sustain load simultaneously. In some cases, the same kind of ceramic materials are used for both matrix and fibers and if it is needed additional secondary fibers may be added. The word ceramic is a Greek word that means "for pottery" and is made of clay that is brittle, fragile, heat and corrosion resistant, as well as inorganic and nonmetallic materials. Metal oxides (Al_2O_3, FeO, etc.) are

B. Farahmand, *Fundamentals of Composites and Their Methods of Fabrications*,
https://doi.org/10.1007/978-3-031-80201-0_3

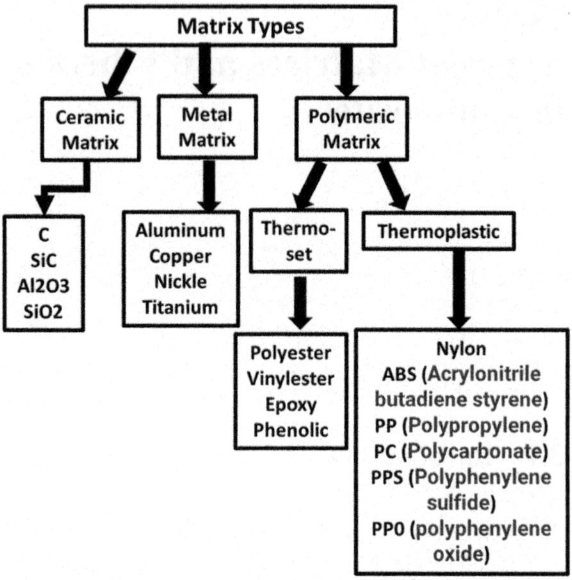

Fig. 3.1 Different matrix types used in composites (ceramic, metals, and polymers): resin matrices for composites, CompositesWorld, CW, 2016

common examples of ceramics, but other compounds such as carbides and nitrides are also included. Porcelain, glass, bricks, and refractory materials are some examples of traditional ceramics. These ceramics have low fracture toughness (low tolerance at the presence of microcracks and lack of ability to resist fracture), and low percent elongation. For those reasons, the existence or formation of a microcrack will lead to fast fracture. That is, they need only the adequate amount of crack tip stress to overcome the surface energy necessary to create two new crack surfaces. The brittleness of ceramics is due to the monolithic nature of the material and because their stress–strain curve does not show yielding or any amount of plastic deformation, therefore, in the presence of microcracks all the available energy at the crack tip would go for generating two new crack surfaces causing the fast fracture. However, in some cases when two ceramic materials are combined (i.e., the ceramic matrix and ceramic fiber), the outcome may be better fracture toughness, percent elongation, greater fatigue, and creep properties. The toughening mechanisms are not due to crack tip plasticity but rather due to crack arrest, deflection of crack, and fiber or whisker pullout, which makes the CMCs to be different from ceramic material without fibers. This enhancement in fracture toughness together with other attractive properties of CMCs can make it to be superior to metals and for those reasons industries are anxious to replace their metallic parts with CMC materials.

To improve the fracture toughness of ceramic materials, a proper type of reinforcement must be embedded in the ceramic matrix. For example, by adding the silicon carbide (SiC) fiber into the SiC ceramic matrix the plane strain fracture toughness, K_{Ic}, will be as high as 23.7 ksi(in)$^{1/2}$ ~ 26 MPa(m)$^{1/2}$. As a second example, applying silicon carbide as a reinforcement to the alumina matrix (SiC/Al_2O_3), will provide fracture toughness of 18.5 ksi(in)$^{1/2}$ ~ 20.3 MPa(m)$^{1/2}$ [3]. This value is equivalent to the lower bound of plane strain fracture toughness, K_{Ic}, of aluminum alloys [4].

3.2.1 Matrix and the Reinforcement Designation in CMCs

As the pressure in commercial aircraft and other related industries for fuel efficiency and a clean environment builds up, ceramic matrix composites are an excellent candidate to be used in the engine hot-zone components. In high temperature environments, both the matrix and reinforcement material must be able to maintain their properties. The reinforcement in the ceramic matrix material can be carbon (C), silicon carbide (SiC), and alumina (Al_2O_3), which all have good resistance to high temperature environments, great creep properties, and lightweight quality. These fibers can take many different forms, including the typical continuous long, short, chopped fibers, or can be particles or whiskers, as well as nanofibers [5, 6]. They all possess a polycrystalline structure like traditional ceramics. Many matrix materials in CMCs are the same as the fiber materials listed above. Therefore, it is more appropriate for the CMCs to be identified in industries and by the users of parts by both the kind of fiber and matrix employed. Because in most cases both the matrix and the fibers are ceramics, the CMCs can be regarded as ceramic material as well. For example, the C/SiC would be carbon fiber reinforced in a silicon carbide matrix, which both are considered ceramic in nature. Other examples are carbon reinforced in carbon matrix (C/C), silicon carbide reinforced in silicon carbide matrix (SiC/SiC), alumina-reinforced alumina (Al_2O_3/Al_2O_3) or oxide-reinforced oxide (OX/OX), alumina and silica oxides (Al_2O_3/SiO_2), and many others. The reason glass, organic, or metallic fibers are not included is that the CMC fibers must remain stable at high temperatures and at the same time must function properly when subjected to the service load environment, which glass, organic, or metallic fibers lack that ability.

The reinforcing fibers are very tiny with the thickness of a human hair or less and in the case of nanofibers, are much thinner. They are often woven into a fabric or tape-like material to be embedded in a CMC part. In a typical CMC process, the fibers are coated with a material such as boron nitride and then passed through a matrix slurry bath, resulting in a prepreg tape or cloth to be used for manufacturing the CMC parts.

3.2.2 Applications of CMCs

Before the application of the CMCs in the manufacturing of turbine blades of aircraft engines, Nickel-based superalloys were the primary high-temperature structural material used for the construction of engine blades. Nowadays, there is an extremely wide range of CMC applications in industries, where high-temperature resistance materials are needed for critical parts. CMCs are also attractive to industries because of their lightweight, almost one-third the weight of the nickel-based superalloys that incorporate chromium, cobalt, and rhenium and are currently being used for aircraft engine blades. The presence of ceramic fibers in the ceramic matrix will provide acceptable fracture toughness for their applications in industries because ceramics alone are brittle and their tolerance regarding the presence of defects is poor. In addition, CMCs can maintain their mechanical strength even at elevated temperatures when subjected to sustain as well as repeated loading.

From the fracture toughness perspective, their resistance to crack propagation is directly related to the presence of reinforcing fibers or whiskers, which could arrest cracks in the matrix and prevent them from growing further that otherwise could cause structural failure. That is, the reinforcements augment the fracture toughness of the CMCs and help the matrix to resist cracks from further crack growth. Their good fracture toughness can be maintained at high temperatures because of their excellent ability to resist heat. Their high-temperature resistance properties mean they have great creep properties when subjected to sustain tensile load at elevated temperatures or in the case of cyclic load environment when the mean stress is positive ($\sigma_m > 0$).

When undergoing tensile tests, the stress–strain curves of fiber-reinforced ceramic–matrix composites (CMCs) exhibit nonlinear behavior when under increasing tensile applied load. The nonlinearity during the application of the load is not due to the yielding material as observed for metallic material that exhibited yielding and plastic deformation, rather it is due to multiple damage mechanisms that take place as damage grows. This behavior is referred to as quasi-plastic, where the matrix cracks in multiple locations, followed by fiber–matrix interface debonding, and fibers break down and final fracture. This is the main reason for the nonlinearity characteristic that will appear in the stress–strain curve of ceramic composite materials as shown in Fig. 3.2a. Both the room and elevated temperature stress–strain curves are shown in this figure that illustrates the properties of CMCs will be maintained at a higher temperature because of the excellent heat resistance that they possess at way above the ambient temperature environment. The standard test specimens (based on the ASTM C1275 standards) used for generating the tensile stress–strain curve is shown in Fig. 3.2b.

Below is the list of parts that are fabricated by the CMC composites where the manufactured part, in most cases, is exposed to high-temperature service environment [7, 8]:

- Heat exchangers.
- Turbine blades.
- High-performance braking systems.
- Immersion burner tubes.
- Bulletproof armor.
- Gas-fired burner parts.

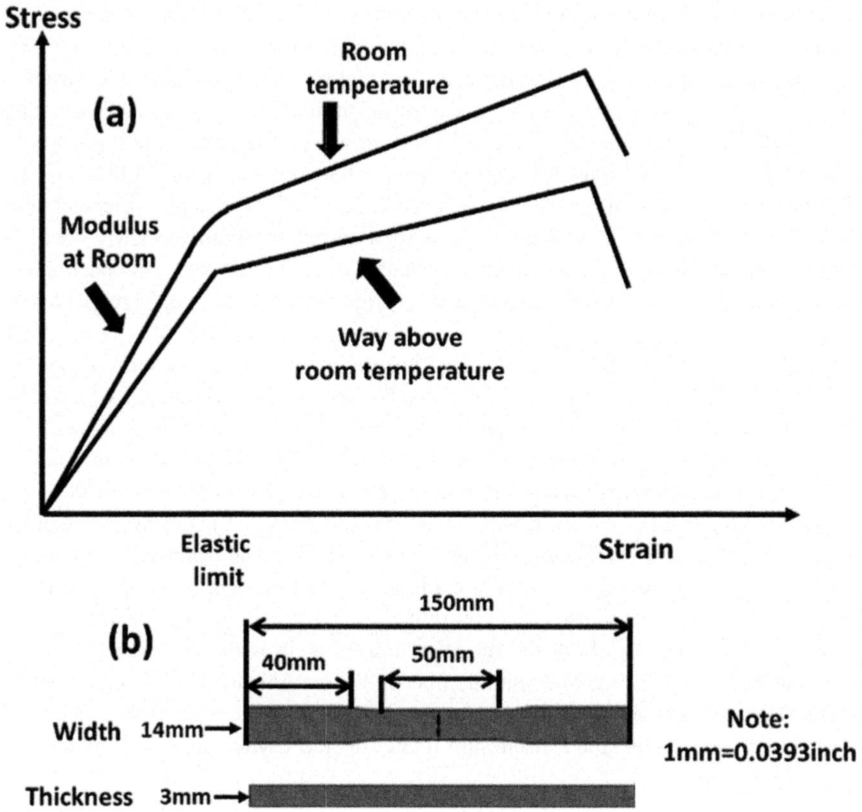

Fig. 3.2 (**a**) A typical stress-strain curve for CMCs at room and above room temperature. Not much degradation in properties is expected. (**b**) A standard specimen is shown for tensile testing based on the ASTM C1275 standards

- Refractory components.
- Hot-pressed dies.
- Thrust control flaps for jet engines.
- Insulation in small arms weapons platforms.
- Multispectral windows.
- Filters for hot liquids.
- Heat shield systems for space vehicles during reentry.
- Burner stabilizers.
- Rocket propulsion components.
- Thermo-photovoltaic burners.
- Turbo jet engine components.
- Combustion liners for gas turbine engines.

In summary, modern structural ceramics bring little resemblance to those traditional materials. Nowadays, they are made from extremely pure, microscopic powders that are fused at high temperatures to yield a dense and durable structure [9]. The U.S. market for advanced structural ceramic composites in 1987 was only $171 million. In about the next 10 to 15 years; however, the market opportunities for structural ceramics expanded rapidly such that by the year 2000, the U.-S. market alone was projected to be an average of $3 billion per year thereafter [10, 11]. The excellent mechanical properties associated with the CMCs will be expanding greatly in the near future, especially in the aircraft and aerospace industries where both the weight, resistance to high temperature, and creep plays a big role.

To have reliable CMC parts for industries are doable but can be more expensive than their metal counterpart that is well established for use in industries for almost more than a century. This is primarily due to the high cost of manufacturing ceramics parts with the hope of being reliable and reproducible. Post-processing, such as grinding and machining processes to give the basic parts their final shape and size, is expensive due to the extreme hardness and brittleness of the material. Nondestructive testing to ensure there are no microcracks in the material is needed, both before and during the operation, which all are a major part of production costs. Therefore, the development of processes that can reliably fabricate a component from start to final shape is crucial to both the manufacturer as well as the customers. In many applications, the presence of a suitable fiber in the ceramic matrix that can meet the customer's requirement both from a cost and performance perspective is important to the end users. A detailed discussion related to the CMC fabrication methods is provided in Chap. 6.

3.3 Metal Matrix and Metallic Matrix Composites (MMCs)

Metal matrix composites (MMCs) are a kind of composite material that have metal as the matrix while the fiber may be a different metal or another kind of material, such as a ceramic or glass fiber [12]. When an MMC (aka metallic reinforced fibers) is composed of three constituents, then it is called a hybrid composite. That is, it is a combination of two or more reinforcement fibers in a given metallic matrix. In some cases, the third material or the second reinforcement may be a different metal, or another material is added, such as a ceramic or organic compound. The aluminum 7075 metal matrix is reinforced with two reinforcements of equal amounts, (1) boron carbide (B_4C) and (2) titanium carbide (TiC), which are fabricated by using the stir casting method (will be discussed in Chap. 5). This is a good example of a hybrid MMC material with two different reinforcements [13]. Hybrid MMCs exhibit outstanding properties and functional responses when compared to reinforced MMCs, and thus have tremendous potential for widespread application in new industrial and engineering applications.

The applications of MMCs in industries are (1) high-performance tungsten carbide cutting tools. It can be fabricated by a fine gray powder, pressed, and formed into shapes through a sintering process for use in industrial machinery such as cutting tools, (2) tank armors, (3) automotive disc brakes, and much more. Usually, the matrix consists of a low-density metal, such as aluminum or magnesium, reinforced with particulate or fibers of a ceramic material, such as silicon carbide ceramic or graphite. Compared with unreinforced metals, MMCs offer higher specific strength ($\sigma ul/\rho$) and stiffness (E/ρ), higher operating temperature, and greater wear resistance, as well as the opportunity to tailor these properties for a particular application by selecting the proper reinforcement. In structural applications, lightweight metallic matrices such as aluminum, magnesium, or titanium, are great for room temperature but in high-temperature applications, cobalt, and cobalt–nickel alloy, are excellent choices to use as a matrix with their properties being maintained at higher temperature environments.

Names of a few MMCs that have a metal matrix with different fibers are aluminum with particulates of SiC, aluminum with fibers of graphite or boron, copper reinforced with graphite, Steel reinforced with boron nitride (to be used in making tank armors), and for high-temperature environments the cobalt–nickel can be an excellent candidate to be used for the metallic matrix composites. There is some overlap between MMCs and CMCs, with the latter typically consisting of less than 20% metal by volume with ceramic as the primary material to be used for the matrix. Another essential point to mention here is the possibility of corrosion where the reinforcement surface of MMCs should be coated to prevent any chemical interactions with the matrix that would cause degradation of the mechanical properties of the metallic matrix. For example, carbon fibers are commonly used in the aluminum matrix to create composites with low density and high strength properties. However, carbon tends to react with the aluminum to produce a brittle compound like Al_4C_3 on the surface of the fiber. To prevent this reaction and the formation of Al_4C_3, the carbon fibers are coated with nickel or titanium boride (TiB_2) [14].

MMCs are fabricated at elevated temperatures, which is an essential condition for diffusion bonding of the fiber–matrix interface. Diffusion bonding is a solid-state welding technique used in metalworking, capable of joining similar and dissimilar metals. Later, when both the metallic matrix and the fiber are cooled down to room temperature, residual stresses are generated in the composite due to the mismatch between the coefficients of expansion of the metal matrix and the fiber throughout the structure. The diffusion bonding process is discussed in detail in Chap. 5.

The manufacturing-induced residual stresses mentioned above can significantly influence the mechanical behavior of the metallic matrix of MMCs. The tensile-induced residual stresses can be detrimental to the structure, which can lead to a premature failure of parts when they are superimposed with the applied tensile load during their service life. On the other hand, the applied compressive residual stresses can be beneficial and increase the fatigue life of the part. In some cases, the thermal residual stresses are high enough to initiate plastic deformation and cracking at the interface between the metallic matrix and the fiber within the composite during the manufacturing process when the matrix is cooling down to ambient temperature.

From a manufacturing cost perspective, the MMCs are more expensive than the more conventional metallic materials they are replacing. As a result, they are budgeted to go forward with the manufacturing of MMCs parts where improved properties and performance can rationalize the added cost. Today these applications are found most often in aircraft engine components and space systems. The scope of applications will undoubtedly increase as manufacturing time, and costs are reduced. However, when compared with the conventional polymer matrix composites, the benefits of MMCs are as follows: (1) their resistant to fire, (2) can operate in a wider range of temperatures, (3) do not absorb moisture, (4) have better electrical and thermal conductivity, (5) are resistant to radiation damage, and (6) do not display outgassing which it is referred to the release of a gas that was dissolved, trapped, frozen, or absorbed in the material. In addition to the attractive properties of metallic matrix in many industries, engineers can use their creativity to mix proper metallic matrix with proper reinforcement to obtain the suitable material with attributes tailored to meet the requirements of a specific application. To do this, there are several variables that engineers need to consider, such as the type and level of reinforcement, the choice of matrix alloy, and the composite processing method. All these factors are interrelated and should not be considered in isolation when developing new material. Chapter 5 provides comprehensive discussions on the MMCs and their method of fabrication.

3.4 Polymer Matrix and Polymer Matrix Composites (PMCs)

Just like both metallic matrix and ceramic matrix composites discussed in Sects. 3.2.1 and 3.2.2, the polymer matrix composite (PMCs) is composed of reinforcement in the form of short or continuous fibers embedded into an organic polymer matrix, which allows the transfer of load between fibers. Polymer means many repeating units of monomers (a small molecule that can bind to other identical molecules to form a polymer) and is often made up of branches of carbon and hydrogen chemically linked together to make a chain. However, it's important to remember that not all polymers are carbon-based and their structures can be more diverse than just simple linear chains. The polymer matrix alone is relatively weak, and it possesses low stiffness and strength, but when it is bonded to the strong fibers, both will be able to transfer loads between them to prevent the structure from failing. In the case of ceramic matrix composites, one of the main functions of reinforcement is to improve the fracture toughness and prevent the matrix from fast fracture. This can happen through the process of arresting cracks by fibers embedded in the matrix. A strong bond between the reinforcement and the polymer matrix will help arrest the growing crack from the matrix. Fibers will deliver excellent strength and stiffness to the composite as a whole that are lacking in the polymer matrix alone as can be seen in Fig. 3.3a, where the strength and stiffness magnitude between the fiber and matrix

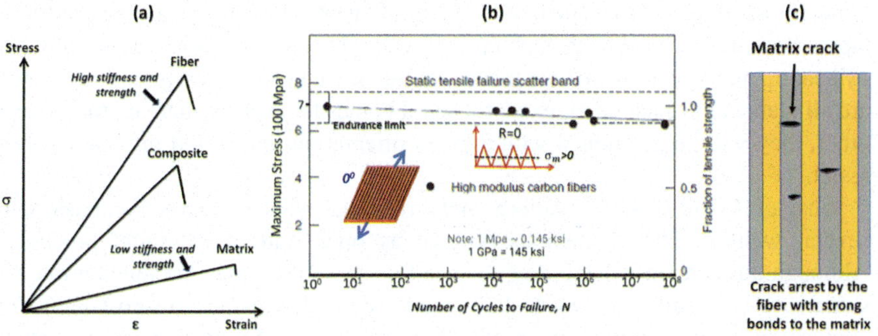

Fig. 3.3 (**a**) Effect of fiber strength and stiffness on composite when load is parallel to fibers. (**b**) Excellent fatigue properties of composites under tensile uniaxial cyclic loading with R = 0. (**c**) Matrix cracks stop growing at the interface between fiber and matrix

are shown. These two valuable combinations, as well as their outstanding lightweight qualities, are the basis of their usefulness in aircraft, automobiles, and other valuable structures. Other desirable properties include superior corrosion and fatigue resistance compared to metals when subjected to tensile cyclic load environments, as shown in Fig. 3.3b, where the S-N curve is almost flat at high applied stress, and the threshold (aka, the endurance limit) value is close to 70% of the material ultimate strength [15]. Even though below the fatigue limit no failure is expected for a large number of cycles, say 10^6 cycles, the presence of humid or other undesirable environments (e.g., slightly higher temperature than it is expected) can lead to degradation of material and lower endurance limit. The fatigue properties of PMCs composites will be discussed later in Chap. 11.

As was mentioned before, the presence of fiber can improve the fracture toughness of the composite as a whole and can prevent the advancing crack that is initiated from the matrix from further growth through the process of the crack arrest mechanism (Fig. 3.3c). It is a fact that the strength of the interfacial bond between fibers and the matrix may make all the difference between a strong composite and a weak one. The stronger the bond between the fiber and the matrix is the key to the crack arrest feature of the composite material.

PMCs are often divided into two types: reinforced plastics and advanced composites. The difference is based on their exceptional strength-to-weight ratio, high stiffness, durability, and resistance to corrosion, chemicals, and environmental exposures, and the cost of manufacturing them. The reinforced plastics, which are relatively low-cost, typically consist of polyester resins reinforced with low-stiffness glass fibers. On the other hand, advanced composites, which have been in use for about 15 years or more, primarily in the aerospace industry, have superior strength and stiffness and are relatively expensive to fabricate [1]. They use advanced fibers like carbon, aramid, or other high-strength, high-modulus fibers. The matrix material in advanced composites might be high-performance polymers or even ceramics or metals. In addition to the limitations that PMCs have concerning their usage, most advanced PMCs today are fabricated by a thorough process called lay-up. This typically involves the placement of sequential layers of polymer-impregnated fiber tapes on a mold surface, with each layer placed with a specific orientation, followed by heating under pressure to cure the lay-up into an integrated structure. These layers are all identical to each other, except when it comes to stacking them on top of each other, they will be positioned with a given orientation that would provide the best result.

Chapter 4 will discuss methods of fabricating polymer matrix composites by several well-established methods, known as open and closed mold processes. Although automation is beginning to speed up these processes, production rates in some cases are still too slow to be suitable for high-volume, low-cost industrial applications such as automotive production lines. New fabrication methods that are much faster and cheaper will be required before PMCs can successfully compete with metals and their alloys in these applications.

Polymers are reinforced with fibers that are 8 to 12 microns in diameter, either continuous or chopped, that are woven into cloth and other types of preformed textiles. The prepared fibers used for reinforcement in the matrix are called fiber preform or simply preform (already shaped or formed beforehand). These fibers are then impregnated into the matrix polymer in liquid form by extrusion, injection pressing, or stamping and then cured to produce the final composite parts. The use of composites in the industry is numerous, and the reader can refer to the literature for more comprehensive applications of PMCs in different industries. However, the typical applications of polymer matrix composites range from tennis racquets, bicycle parts, medical devices, marine, housing items, bridges, as well as aircraft and aerospace parts. A few examples have been taken from a few industries. The military aircraft industry was the first to use polymer matrix composites. In commercial airlines, the use of composites has been conservative because of safety concerns. Because of that, the use of composites was limited to secondary structures, such as rudders and elevators made of graphite/epoxy for the Boeing 767 and landing gear doors made of Kevlar–graphite/epoxy. Composites are also used in panels and flooring of airplanes.

In many cases, the application of PMCs reduced the total weight of the structure but also reduced the number of parts. Recently, the PMCs have been used for the primary structures in 787 Boeing aircraft. Each Boeing 787 aircraft has approximately 70,400 lb. of CFRP composites (CFRP—carbon fiber-reinforced polymer) made with 23 tons of carbon fiber. In addition to weight saving benefit, aircraft engines send out carbon dioxide, Sulphur oxide and nitrogen oxide directly to the upper atmosphere. Such pollutants have a strong effect on atmospheric conditions that can lead to global climate change. Therefore, lighter structural parts imply less fuel consumption, that can be helpful to a cleaner environment.

As mentioned before, several methods can fabricate the PMCs for industrial use. These methods are discussed in Chap. 4 which are Spray lay-up, Hand lay-up, Vacuum molding, Pultrusion, Resin transfer molding, and Filament winding.

3.5 Types of Polymer Matrices

Polymers can be classified as thermosets or thermoplastic (also known as thermosoftening plastic). Both polymers are amorphous, differing from crystalline materials such as metals. An amorphous or noncrystalline solid (from a Greek word meaning without shape) is a solid that lacks the long-range molecular order, which is a characteristic of a crystal or metallic structure having a continuous arrangement of atoms. As an example, aluminum has the face centered cubic unit cells (fcc) and iron the body centered cubic (bcc) crystalline structures, where in both cases of fcc and bcc, unit cells will repeat themselves in three dimensions throughout the body. Thermoset polymers (i.e., epoxy resins) cannot be remolded or reshaped at high temperatures upon curing and cooling (they may burn upon reheating—like charcoal and become useless). On the other hand, thermoplastic polymers, such as PMMA (polymethylmethacrylate), can be remolded upon reheating and are also recyclable and reusable. It has outstanding light transmission characteristics and resistance to weathering. As a transparent sheet, it is used for aircraft windows and many other items, such as lenses and outdoor signs.

On the other hand, thermosetting resins and epoxy resins are the most widely used in aircraft, aerospace, and automotive industries because they have relatively low molecular weight with low shrinkage (1–4%) during cure [16]. They can be cured partially and stored in that state for later use. The cured epoxy resins exhibit good mechanical and thermal properties and high resistance to chemicals and corrosion. The curing agent or hardener is one of the essential ingredients for epoxy. It will control the reaction rate and determine the performance characteristics of the finished epoxy material. During the curing process time, pressure, and temperature also have an influence in completing the process of forming the epoxy resin [16, 17].

Thermoplastic composites consist of a resin matrix and reinforcement material combined to provide specific performance characteristics. They have better properties than conventional engineering materials. On a pound-for-pound basis, they are 10–20 times stronger and several times stiffer than metals. Thermoplastics in the form of composites are not stacked in the form of ply, meaning, the resin is formed into the desired shapes that rely solely on the randomly short, discontinuous fibers from which they are composed to maintain their structure. By having the ability to be reusable upon heating, in many cases, they are preferable over thermosetting polymers. Currently, the primary matrix materials used in aerospace applications are epoxies, however, high-temperature thermoplastics such as PEEK (polyetheretherketone) are considered by many to be the matrices of choice for future aerospace applications because of their lightweight, strength, toughness, and able to survive longer in harsh environments, as well as being recyclable upon heating [16]. However, PEEK is a high price thermoplastic polymer, and due to its complex production process, restricts its use to only the most challenging applications. Other examples of thermoplastic polymers are polyethylene, polystyrene, and polyphenylene sulfide. Finally, thermoplastic matrices, when mixed with proper reinforcement, are being used in a large variety of structures that are subjected to cyclic loading. In aeronautical applications where structures will undergo many cycles, engineers are naturally concerned with design against fatigue, where thermoplastic composite with excellent fatigue properties can be a good candidate to be used in industry.

In the case of thermoset epoxies, they have excellent chemical and mechanical properties. They have low shrinkage and adhere adequately to most types of fibers. Thermosets are usually liquid or malleable (viscous) before curing. They can be molded into the desired shape. As was mentioned before, thermoset composites need to be cured, a process by which the chemical structure of the polymer matrix is permanently changed. Notably, curing involves the application of temperature and a curing agent, pressure, or the addition of a catalyst known as the hardener. It transforms the resin into a plastic (rigid) state, usually by applying heat and, optionally, pressure. Once cured, the thermoset composite has its final properties of rigidity, hardness, and strength and cannot be reverted to its uncured flexible condition (not reusable upon heating) [17].

Thermosetting resins have the propensity to absorb moisture. In typical aircraft environments, they can absorb up to 1% by weight moisture; however, it is not desirable as the moisture softens the matrix resin and degrades its mechanical properties. The effect is more pronounced above room temperature when the matrix is softened. It should be noted that to obtain mechanical properties under these environments, an appropriate knock-down factor must be included if the data is based on room temperature allowable, or allowable should be established through testing based on the environment under consideration. Finally, the application of thermoset polymers as a matrix for manufacturing composite parts must be limited to a specific temperature range when they are used in service. That is, the matrix becomes weak when exposed to a high-temperature environment. Therefore, the application of thermoset composites in industries must be limited to service temperatures below about 450 ° F (~232 ° C).

In summary, epoxy is the most common matrix material used in industry. Although epoxy is more expensive than other polymer matrices, it is the most widely used matrix for PMCs. It is estimated that more than two-thirds of the polymer matrices used in aerospace applications are epoxy based. The main reasons why epoxy is the most utilized polymer matrix material are its high strength, low viscosity, and low flow rates, which allow good wetting of fibers and prevent misalignment of fibers during processing, low volatility during the curing process, low shrinkage rates, which reduce the tendency of gaining significant shear stresses of the bond between epoxy and its reinforcement. Other types of thermoset polymers are polyesters, phenolics, and polyamide [18]. Finally, the differences between the thermoset and thermoplastic polymer to be used for the matrix are, (1) thermoplastics will soften on heating and pressure, and thus be easy to repair, whereas thermosets will decompose upon heating, (2) thermoplastics have high strains to failure and thermoset polymers has low strains to failure, (3) thermoplastics have an indefinite shelf life and can be reprocessed. That is, they can be molded, melted, and remolded without altering their physical properties, whereas thermosets have a definite shelf life and cannot be reprocessed, (4) thermoplastics have short cure cycles and thermosets possess long cure cycles, (5) Fiber volume for thermoset is medium to high (50–70%) whereas for thermoplastic it medium, (6) Fiber length for thermoset is usually continuous and for thermoplastic is in most cases discontinuous, (7) Regarding safety and handling they are good in both cases of thermoset and thermoplastic.

3.6 Types of Fibers and Orientations in Laminated Composites

The fiber in the polymer matrix provides strength and stiffness to composite materials. Different types of fibers already were discussed and their applications in advanced composites include carbon fiber, aramid fiber, boron fibers, high-performance polypropylene fiber, newer fibers such as poly p-phenylene-2,6-benzobisoxazole (PBO), and hybrid combinations, as well [19]. These fibers can be produced by reinforcing continuous and long (Fig. 3.4a) or short and chopped fibers (Fig. 3.4b) in the matrix as well as in the form of particles as shown in Fig. 3.4c [20–24]. Continuous fibers in the matrix of composites have preferred orientation,

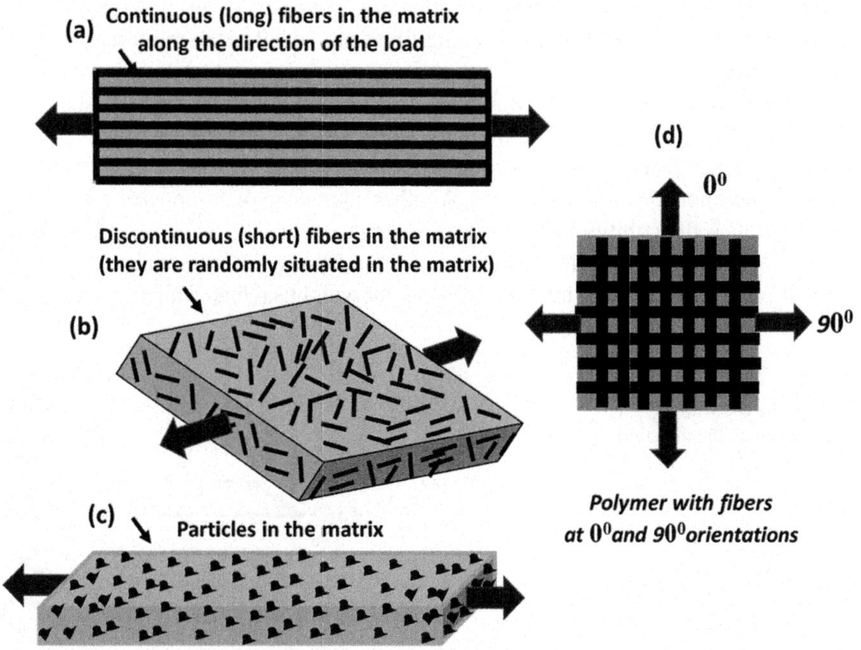

Fig. 3.4 Different types of fiber arrangement in the matrix. (**a**) Long continuous fibers bonded to the matrix. (**b**) Short fibers scattered and bonded in the matrix. (**c**) Particles embedded and bonded in the matrix. (**d**) Polymer with fibers at 0 and 90°

while short fibers are randomly scattered throughout the matrix. Typical examples of continuous fiber composite include unidirectional fibers, cross-plies, and woven fabric. A ply is one single layer of a matrix with fibers aligned in a given direction as shown in Fig. 3.4d, where fibers can have 0° and 90° or any other effective orientations to make the composite valuable to use.

When several of these plies (or lamina) are adhesively arranged on top of each other for manufacturing parts, then it is called a laminated composite. On the other hand, short-fiber reinforced polymer composites consist of slender fibers, relatively short compared to the overall dimensions of the part, and imperfectly (randomly) distributed in the matrix (Fig. 3.4b). The desired thickness for the fabricated part with short fibers can be obtained by stacking several layers on top of each other to achieve the desired thickness. Yet there are other types of composites in that the polymer has small particles in the matrix, which are also randomly bonded in the matrix, Fig. 3.4c. Herein, most of the discussions will be related to the laminated composite and the focus will be on the continuous fiber composites with plies of different orientations, which currently are used in major industries.

Composites are superior to their metallic counterparts in many ways; however, there are certain drawbacks associated with them which will be discussed. Their superiorities are in having high strength-to-weight (σ_{ul}/W or σ_{ul}/ρ) and stiffness-to-weight ratios (E/W or E/ρ), where W is the weight, E the stiffness (the modulus of elasticity), and ρ is the density. Composites are light and strong, the matrix (the epoxy material) is light with low modulus and strength. To bring strength to the epoxy matrix, high modulus, and strength fibers are added as the reinforced material and they are positioned parallel in the direction of the applied load as shown in Fig. 3.5. Here fibers are in the direction of [0°, ±45°, ±90°]2s (the 2 s stands for 2 stacks of 0°, ±45°, ±90°). The Boeing 787 Dreamliner and the Airbus A380, both use composite materials due to their high stiffness to low weight, high tensile strength, and noncorrosive properties. Another advantage of composites over metallic materials is their ability to be custom-tailored to meet specific performance needs, as well as the added benefit of not requiring machining. From the aircraft perspective, the aforesaid qualities translate into greater weight savings which, in turn, result

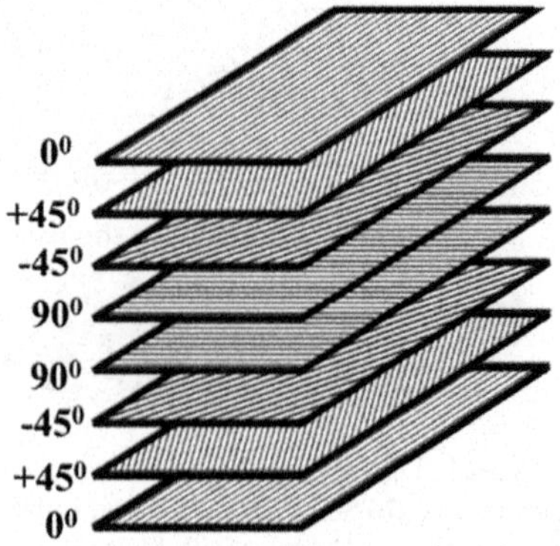

Fig. 3.5 Fiber arrangements in several plies with orientations in the direction of ±0°, ±45°, ±90°

in improved performance, greater payloads, longer range, and fuel savings. However, there are disadvantages regarding the fundamental understanding of failures in the composite material when comparing it with the metallic counterparts. In laminated composites, the damage often initiates in the matrix as matrix cracking, debonding between the matrix and fibers, and eventually the fibers failure. In a ply, the matrix cracking will progress further by failing fibers, debonding them, or going around them.

Matrix cracking and growth will simultaneously occur in other parts of the composite material. They all will grow perpendicular to the application of the load. Eventually, a few of them that are favorably close to each other will merge to produce catastrophic failure of the material. The fracture toughness of composites cannot be assessed through each component in the material since both fibers and matrices are brittle. However, in addition to their outstanding strength and stiffness-to-weight ratios, and good corrosion resistance, they also have some amount of toughness. The toughness in composites does not arise from the inherent characteristics of each individual component of composite materials, but rather from the failure mechanisms that operate during the material fracture process. There are various mechanisms through which fibers provide the composite the proper toughness by stopping the propagating crack straight through the material. This process is more apparent in the ceramic matrix composite and, to some extent in the polymer matrix composites, where the presence of fibers can either stop the crack at the interface, or through the bridging mechanism or cause it to deflect, thus consuming the energy provided to the crack in a safe manner. This topic will be discussed further in Chap. 8.

Another issue with composites when they are subjected to the service load is the delamination of laminates that are stacked on top of each other adhesively, as is shown in Fig. 3.5. The cause of delamination on laminated composites after parts are fabricated has much to do with the nature of the applied load and the weakness of the bond holding plies. The weak adhesive/bond between the plies can cause the separation of plies very quickly under the applied load, causing delamination. Bending, compression, and out-of-the-plane loads can cause the separation of composite layers. Another reason for delamination is the mismatch between the properties of each layer that have stacked together adhesively with different orientations. When the laminated composite is subjected to tensile loading, because each ply has slightly different stiffness, E, and different amount of elongation and Poisson's ratio, they all can contribute to the separation of plies from each other. However, if the extent of orientations between the two plies next to each other is small, then the amount of mismatch and the resultant delamination will be small. This topic will be discussed later in Chap. 7.

Aerospace giants like Boeing and Airbus are among the largest consumers of composite materials in the world. They utilize various materials, from glass fiber to carbon fiber, for manufacturing everything from fuselage panels to seats. The strength-to-weight ratio of these composite materials allows manufacturers to continue building larger aircraft capable of transporting more loads. By weight, the Boeing 787 is 50% composites, 20% aluminum, 15% titanium, 10% steel, and 5%

Fig. 3.6 Different fibers and matrix used in industry. Note that both metallic and nonmetallic materials are included

Matrix Material	Fiber Reinforcement
Acetals	Alumina
Acrylic	Aluminum
Aluminum	Boron
Epoxy	Boron Nitride
Graphite	Beryllium
Nickle	Copper (whisker)
Nylon	Glass
Polybenzimidazoles	Graphite (Carbon)
Polyesters	Iron (whisker)
Polyethylenes	Armid (Kevlar)
Phenolics	Silicon
Polytpropylenes	Silicon Carbide
Polymides	Silicon Nitride
Polyurethane	Titanium
Pheneysilances	Tungsten (whisker)

other elements [25, 26]. Polymer composite materials take up the bulk of the structures. Their fabrication methods for polymer matrix composites (open mold and closed mold processes) will be discussed in Chap. 4. In addition to the abovementioned manufacturing methods, the method of fabricating PMCs and CMCs through the 3D printing is discussed in Chap. 15. Finally, Fig. 3.6 shows different types of matrices and fibers currently being used in industries for manufacturing PMCs, MMCs, and CMCs components [27].

3.7 Concluding Remarks

The selection of matrices and fibers plays a crucial role in influencing the performance, durability, and applicability of polymer matrix composites (PMCs), metal matrix composites (MMCs), and ceramic matrix composites (CMCs). By understanding the unique properties and characteristics offered by different matrices and fibers, engineers and researchers can tailor composite materials to meet specific application requirements across various industries. Engineers must first study the load environment, temperature, and chemical conditions that the part will be exposed to. Results will be helpful for selecting which type of abovementioned composite is more suitable for that environment. As advancements continue to be made in matrix and fiber technologies, the potential for developing next-generation composites with enhanced properties and functionalities remains promising. By leveraging this knowledge and innovation, the future of composite materials looks bright to industries and opens the way for novel solutions in engineering and manufacturing parts.

The selection of matrices and fibers plays a crucial role in influencing the performance, durability, and applicability of polymer matrix composites (PMCs), metal matrix composites (MMCs), and ceramic matrix composites (CMCs). By understanding the unique properties and characteristics offered by different matrices and fibers, engineers and researchers can tailor composite materials to meet specific application requirements across various industries.

Underestanding the environment that the part will be exposed to will help in selecting the most suitable composite material for manufacturing structural parts for the industries.

By continuing this knowledge and innovation, the future of composite materials looks bright for industries, opening the way for novel solutions in engineering and manufacturing parts.

References

1. A.K. Kaw, *Mechanics of Composite Materials*. Chapter 1 (Taylor & Francis, 2006), p. 64
2. K.K. Chawla, *Ceramic Matrix Composites*. Chapter 4 (Chapman & Hall, London, 1993)
3. B. Harris, *Engineering Composite Materials*. Chapter 2 (The Institute of Materials, London, 1999), p. 23
4. B. Farahmand, *Fatigue and Fracture Mechanics of High-Risk Parts*. Appendix A (Chapman & Hall, 1997), pp. 324–356
5. Material & Processes Fibers for Composites, Composite World (CW), published 7/1/2022
6. Aeroengine Composite, The CMC Invasion, Composite World (CW), published 7/3/2015
7. M. Kusay, *Matal and Ceramic Matrix Composites* (Akdeniz University. Chapter 5, 2004), pp. 76–84
8. H. Pfeiffer, *Ceramic Body Flap for X-38 and CRV*, in 2nd International Symposium on Atmospheric Re-entry Vehicles and Systems, Arcachon, France, March 2001
9. F.F. Lange, D.C.C. Lam, O. Sudre, B.D. Flinn, C. Folsom, B.V. Velamakanni, F.W. Zok, A.G. Evans, Powder processing of ceramic matrix composites. Mater. Sci. Eng. A **144** (1991)
10. Ceramic Matrix Composites Market Size, Share & Trends Analysis Report By Product (Oxide, Silicon Carbide, Carbon), By Application (Aerospace, Defense, Energy & Power, Electrical & Electronics), And Segment Forecasts, 2022–2030, Market Analysis Report (2020)
11. Ceramic Matrix Composites Markets in Aircraft Engines, Stratview Research, November (2020)
12. K.K. Kar, *Composite Material – Processing, Applications, and Charecterization*. Chapter 11 (Springer, 2017), pp. 371–388
13. R. Jeyapaul, B. Gugulothu, S. Selvaraj, A. Babu, M. Varatharajulu, R. Jeyapaul, Retrospective investigation on hybrid metal matrix composites: Materials, processing methods and properties of composites. Iranian J. Chem. Chem. Eng. (2023)
14. B. Yang, R. Peng, D. Zhao, N. Yang, Y. Hou, G. Xie, Performance of TiB wettable cathode coating, in *Chemical Engineering and Technology in Mineral Processing and Extractive Metallurgy* (2021)
15. B. Farahmand, *Application of High and Low Cycle Fatigue on Metallic and Non-Metallic Material* (Amazon, 2022), pp. 105–110
16. J. Yin, A. Zhang, K.Y. Liew, L. Wu, *Synthesis of Poly(Ether Ether Ketone) Assisted by Microwave Irradiation and its Characterization* (2008)
17. J.A. Quinn, Chapter 3: Properties of thermoset polymer composites and design of pultrusion, in *Design with Advanced Composite Materials*, ed. by L.N. Philips, (Springer, New York, 1989)
18. A.K. Kaw, *Mechanics of Composite Materials*. Chapter 1 (Taylor & Francis, 2006), p. 29
19. Materials & Processes: Fibers for composites, Compositesworld (CW), Published March 2016

20. S.-y. Fu, B. Lauke, Y.-W. Mai, *Science and Engineering of Short Fiber-Reinforced Polymer Composite* (A volume in Woodhead Publishing Series in Composites and Engineering, 2019)
21. C. Unterweger, O. Brüggemann, C. Fürst, Synthetic fibers and thermoplastic short-fiber-reinforced polymers: Properties and characterization. Polym. Compos. **35**(2), 227–236 (2014)
22. G.M. Jenkins, K. Kawamura, *Polymeric Carbons*, 1st edn. (Cambridge University Press, London, 1976)
23. T.H. Courtney, *Mechanical Behavior of Materials: Second Edition* (Waveland Press, 2005)
24. M. Akay, *An Introduction to Polymer Matrix Composite*. Chapter 3, eBooks at bookboon.com (2015), p. 122
25. 787 Dreamliner Program Fact Sheet. Boeing web page. Boeing. Retrieved July 10, 2007
26. J. Hawk, *The Boeing 787 Dreamliner: More Than an Airplane*, in AIAA/AAAF Aircraft Noise and Emissions Reduction Symposium. American Institute of Aeronautics and Astronautics (25 May 2005)
27. A.K. Kaw, *Mechanics of Composite Materials*. Chapter 1, Table 1.1, 1.2, 1.4. 1.5 (Taylor & Francis, 2006)

Chapter 4
Polymer Matrix Composites (PMCs) and Nanocomposites (Methods of Manufacturing PMC Parts)

4.1 Background

The use of high-performance polymer matrix–fiber composites in aircraft structures has grown gradually since the 1980s, although not as significant as initially predicted. This is despite the significant weight-saving and other advantages that advanced composites could provide. One reason for the slower than anticipated advancement can be the higher cost associated with their fabrications than their metallic counterparts [1, 2]. Other factors include the high cost of certification of new products and their relatively low resistance to impact. Additionally, any low magnitude out of plane loads through the thickness, compressive, or any other situations, such as impact (both before and after part's usage) that can produce separation of plies and cause delamination, must be taken into consideration by engineers. Furthermore, the temperature limitations can introduce creep mechanism when the structure is subjected to sustain in service load or when the cyclic load has a mean stress larger than zero. Thus, in many cases, metals have continued to be favored for designing and manufacturing critical parts in aircraft industries where safety and reliability are essential to consider. Until those abovementioned problems are not fully assessed, the contribution of PMC parts as a primary structure will be limited.

When manufacturing a part, selecting the process to be utilized is strongly related to the (1) material in consideration. That is, whether the selected material by engineers is metallic or nonmetallic (e.g., composites). Note the manufacturing of parts is different depending on the choice of material, (2) The geometry of the part, where engineers must take into consideration whether forming, casting, machining, or molding processes should be used that would be cost-effective and more efficient to the industries, (3) Irrelevant from what material is being used, the number of parts, material costs, and their availability as well as the cost of automation must be taken into consideration. In this chapter, the discussions will be focused on polymer matrix composites, their fabrication techniques, and the advantages and disadvantages associated with each method. Recently, nanoparticles have shown promise in the

B. Farahmand, *Fundamentals of Composites and Their Methods of Fabrications*,
https://doi.org/10.1007/978-3-031-80201-0_4

aerospace and aircraft sectors due to their superior E/ρ and σ_{Ult}/ρ over the traditional carbon fibers used in fabricating PMC parts. In Sect. 4.12, a brief discussion related to nanocomposite is provided to the engineers where weight is an important factor to consider. Nanocomposites are very much like fiber matrix composites except the fibers are now being replaced by nanoparticles that are much stronger than carbon fibers.

Recently, carbon fiber-reinforced polymer (CFRP) composites have emerged as the most preferred and developed composite for aerospace use. Although the raw material costs of this and similar composites are still relatively high, but their advantages over metals in (1) both strength and stiffness to weight ratio (σ_{Ult}/ρ and E/ρ), (2) can be tailored to the desired design, (3) their ease of manufacturing and the post fabrication processes (like machining or grinding) are in many cases preferable over metals. Nonetheless, the rivalry remains intense with developments in structural metals such as aluminum and titanium alloys, which have reasonably good fracture toughness, corrosion resistance, weldable and machinable. Lately, new lightweight alloys such as aluminum lithium alloys, with great ultimate strength and fracture toughness, and the low-cost aluminum casting process are also available for manufacturing aerospace parts that compete with advanced polymer composites. Finally, hybrid metals and composite materials combinations, such as glass laminate aluminum reinforced epoxy (GLARE) material, are good enough, already being utilized on Airbus A380 parts. The GLARE materials that have the best of both metals and composite characteristics are gaining popularity with specific industries. But the GLARE material will not be discussed in this book [3].

4.2 Fabrication of Polymer Matrix Composites

Polymer matrix composites are used extensively across a wide range of industries. Therefore, the design and development of an effective manufacturing process are of great importance. There are numerous methods for fabricating composite parts. Some methods have been borrowed from the plastic and metallic industries (for example, injection molding), but many were developed to meet specific designs or manufacturing when dealing with fiber-reinforced polymers. Selection of a method for a particular part, therefore, will depend on the availability of the material, the characteristic of the design, cost, feasibility of the fabrication of the method, and its usage whether the final part will be utilized for fracture critical parts (also called a primary structure where its failure will cause loss of the whole structure) or as a secondary structure. Typical examples of primary structures of an aircraft structure are wings, control surfaces, flaps, and their attachments, landing gear, engine mounts, and fuselage. Fig. 4.1 shows several methods that are currently being used in manufacturing PMC parts for their applications in industries. These techniques are generally classified as open mold process, closed mold process, and others. For example, in the case of the open mold process, the process includes spray lay-up, hand lay-up, tape lay-up, and the autoclave method. In simple terms, when the plastic resin is exposed to the atmosphere during curing it is known as open

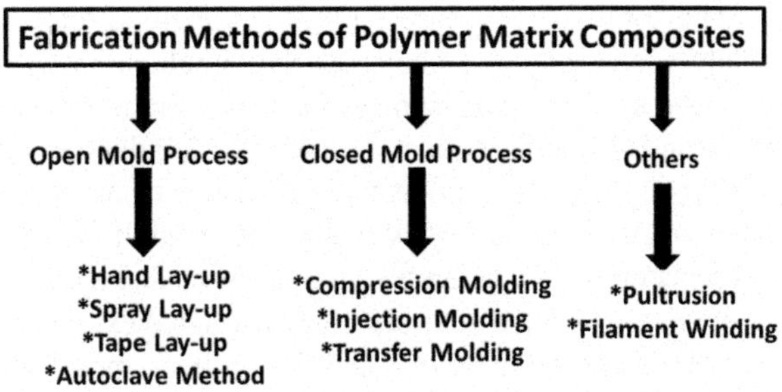

Fig. 4.1 Methods of fabricating the polymer matrix composites are illustrated. All three methods shown here have their own classification

molding, while that which is not exposed to air during curing is known as closed molding [4, 5]. In addition to what was just said, the open molding is associated with the parts that are not closed and don't need to have a mold cavity on both sides. For example, boats, tubs, and many other similar types of designs are falling into the open mold process, which is a low-cost and effective method to manufacture them with composites. Typically, closed mold processes are compression molding, injection molding, and transfer molding, which all involve heating thermoplastic or thermosetting plastic polymers and injecting them into a mold where they fit into the mold cavity, then it will cool and harden. In the case of thermoset molding, time plays a significant role, and an adequate amount of time is necessary for the curing to be completed. Each type of molding has its place within the plastics industry, with advantages and disadvantages for each. It is worth mentioning that the tooling costs for open molds tend to be lower and are generally used for short production runs and are associated with hand lay-up as shown in Fig. 4.1. It's typically used for more oversized products for which automated processes are unsuited or products made in such low volumes that the cost of automating the process is not profitable.

Fabricating parts through both open and closed molds has several different types of problems that may occur during the fabrication and handling of polymer composites. The most common being misalignment and broken fibers, as well as the presence of microcracks in the resin matrix. Others are the presence of voids, porosity, slag inclusions, nonuniform fiber volume fraction, the separation between plies, inadequate curing process, and finally mechanical damage, such as delamination around machined holes. Most polymer molding methods will be discussed here in detail, including the pultrusion and filament winding that both are falling under the others as shown in Fig. 4.1. Even though the filament winding is falling under the others, it can also fall under the open mold category.

4.3 The Spray Lay-Up Technique (An Open Mold Process)

In this section, the spray lay-up technique will be presented, which falls under the open mold process. Later, various closed mold methods will be discussed and in both cases of open and closed mold techniques, the advantages and disadvantages associated with each method will be addressed.

The spray lay-up process [4–7] is a method that uses a handgun to spray resin, catalyst, and chopped fiber material simultaneously. The fibers are short, cut and chopped so they can be sprayed easily with random orientations on the surface of the mold and at the same time, a roller (aka debulked or compacting process) is used to spread and fuse these fibers into the matrix material [4, 5]. The debulking process is used when one layer (a ply or lamina) or multiple layers of composite (e.g., cloth/resin materials) are assembled into a lay-up and a roller or vacuum method is applied to remove unwanted air, voids, and unwanted extra resin within the lay-up. Although debulking can be done by hand with rollers, most fabricators today use a vacuum-bagging technique to remove those mentioned items from the composite part. The process involves placing plastic sheet materials over the lay-up, sealing it at the tool's edges, adding one or more ports for air hoses, and then evacuating air from the space between the sheet and the lay-up using a vacuum pump, which the vacuuming process will insert pressure and removing unwanted air.

As shown in Fig. 4.2a, the purpose of the gun is to spray the three ingredients of composite on the surface of the mold simultaneously and be able to chop the reinforcement into small fibers (chopped fibers). To start the spray lay-up process by using the chopper gun technique (Fig. 4.2), it is important first to choose the most suitable material for the mold itself. There are choices for the mold material to use for this process, but polymer materials have several advantages over others. Polymer-based mold is lightweight, inexpensive, and has good resistance to corrosive atmospheres and humid weather conditions. Furthermore, it is easily workable, reusable, and movable, and it does not induce deformations in the structure to be manufactured. The spray lay-up method is a preferred method to use for big and complex geometry parts to be fabricated, like boats, bathtubs, shower stalls, and tanks (Fig. 4.2b shows a boat fabricated by this process). The spray lay-up process is low-cost and can be finished in a reasonably short time. The stream of the material from the handgun is aimed at the mold and the mixture is left for almost 24 h until it is fully cured under normal temperature and atmospheric pressure. The part

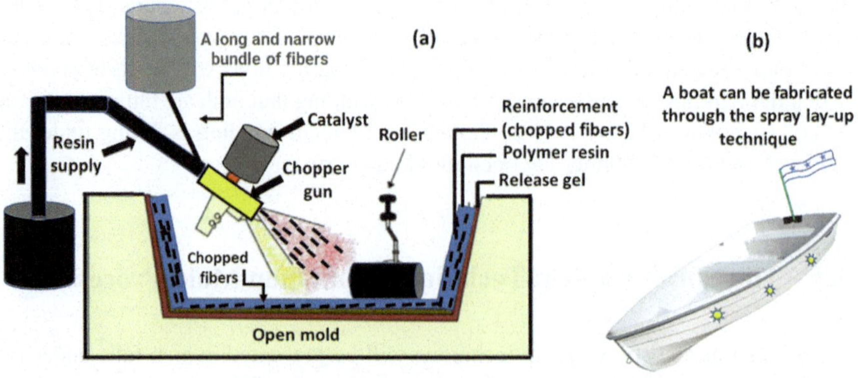

Fig. 4.2 (**a**) Schematic of spray lay-up technique for manufacturing composite parts using the open mold method. (**b**) A boat can be fabricated through the open mold process

thickness can be achieved by simply continuing spraying until the desired thickness has been reached. However, before spraying the three ingredients (reinforcement, catalyst, and resin) on the surface of the mold must be available. In addition, the mold itself must be cleaned by applying the release agent called gel coat to the surface to ease the removal of the final product from the surface. The gel coat is a protective layer that does not allow reinforcement to penetrate inside the mold. Silicon, polyvinyl alcohol, fluorocarbons, water base solvent are the release agents commonly used prior to the spray lay-up process.

4.4 The Hand Lay-Up Technique (An Open Mold Process)

The hand lay-up technique for fabricating PMC parts falls under the open mold process (Fig. 4.3). The composite hand lay-up fabrication process typically involves having some form of molding to shape the resin and reinforcement [8–10]. Just like the spray lay-up technique, using an open mold to fabricate the part in consideration is required for this process. The hand lay-up technique is a wet lay-up method, and it

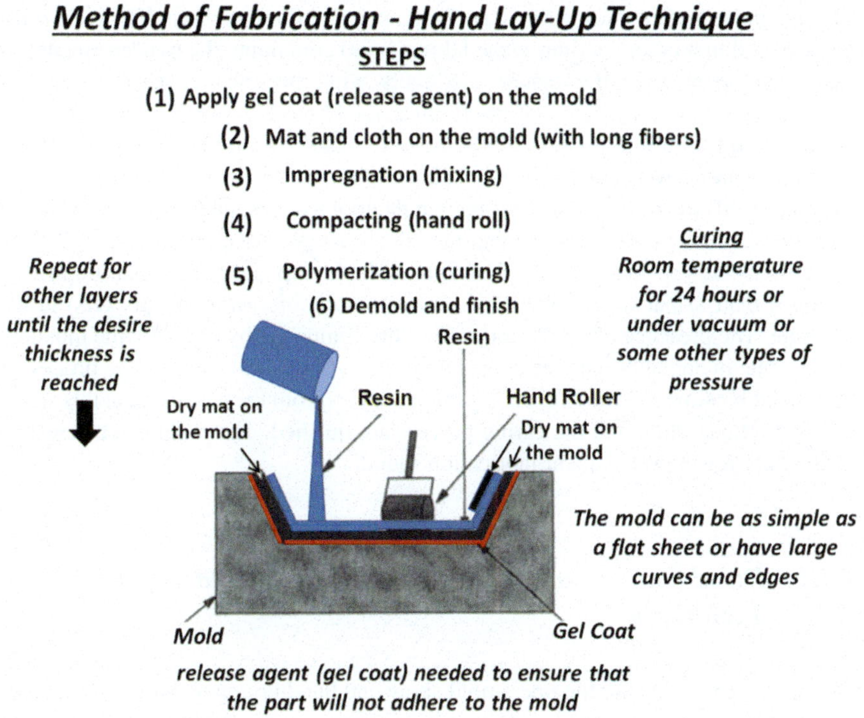

Fig. 4.3 Illustrating the hand lay-up technique for fabrication of polymer matrix composite. This process will repeat and continue until the desire thickness has been reached

is one of the most traditional methods used in the industry. It is a straightforward process, where each ply layer is handled only by hand and is stacked layer-by-layer until the desired thickness is reached. Figure 4.3 shows the fabrication of composite layer-by-layer material and steps. First, the gel coat must apply on the mold. Then, the thermosetting polymer in liquid form is mixed thoroughly in suitable proportion with a prescribed hardener (curing agent or catalyst) and placed onto the surface of the mat (fabric) already placed on the mold. The polymer (resin) is uniformly spread with the help of a brush. The second layer of the mat is then placed on the polymer surface, and a roller is rolled with mild pressure on the mat-polymer layer. The application of pressure helps to remove any air trapped and the excess polymer present. This process will repeat and continue until the desired thickness has been reached. It is up to the designer to organize the type, amount, and direction (orientation) of the fibers being used. The mat and cloth to be used as the reinforcement can have unidirectional and bidirectional orientations with continuous fibers.

With the hand lay-up method, the geometry of the part should be simple and the presence of too many curvatures to be fabricated is not desirable. Finally, the process is completed when the required layers are stacked to obtain the part's thickness demanded by the customer. It should be mentioned that before lay-up, the mold must be cleaned and prepared with a release agent to ensure that the composite part will not stick to the mold when it is ready to be removed. Some of these agents are silicone, polyvinyl alcohol, water-based solvents, and fluorocarbons. The resin must be catalyzed first before adding to the fabrics (mat) containing the reinforcements. A roller can be used to impregnate the fibers with the resin, layer by layer for compacting purposes, which was briefly discussed before.

Although the rolling process can be done by hand with rollers ply by ply, most fabricators prefer to use a vacuum-bagging technique that involves placing plastic sheet materials over the lay-up, sealing it at the tool's edges, and then evacuating air from the space between the sheet and the lay-up using a vacuum pump [11, 12]. In this process the *vacuum-bagging* consolidates the lay-up and removes air trapped in the resin matrix that would otherwise create undesirable *voids* (air pockets) in the laminate. The presence of voids can weaken the composite by allowing the moisture to penetrate in and alter the mechanical properties. Finally, the demolding process is performed by simply removing the developed layers from the mold, and at this stage, the lay-up fabrication and the curing process are finished. The mold can be cleaned to be ready for the subsequent fabrication round.

4.4.1 Disadvantages of Using the Spray and Hand Lay-Up Techniques

Despite the ease of using the open mold spray and hand lay-up techniques, there are two disadvantages associated with both hand spray and the hand lay-up techniques which are, (1) the possibility of low fiber volume fraction and (2) the poor quality of the lay-up process. The poor quality of the lay-up process can lead to uneven ply thickness, and nonuniform distribution of fiber material in the matrix. The time of curing depends on the type of polymer used for composite processing. For example, for epoxy-based material, the normal curing time at room temperature is around 24–48 h. There is a need for a catalyst to be included to accelerate the resin process, which makes the room-temperature curing of the resin possible.

Being an open mold process, it emits a large volume of styrene (phenylethylene) smell which makes the process nonenvironmentally friendly and possible harm to the workers and surroundings. It should be mandatory for the workers involved in the lay-up process to wear masks to be protected from hazardous chemicals.

In some cases, postprocessing fabrication, such as cleaning and polishing the part's surface is required, which makes the process to be time-consuming, causing a lower production rate. All the above factors will force the manufacturers to explore alternatives such as the closed mold process.

4.5 Tape Lay-Up Technique (An Open Mold Process)

The tape lay-up method works with layers of preimpregnated fibers, or simply prepreg, that will be applied on the open surface of a mold in the form of a tape with a specific orientation [13–15]. It is followed by the curing process at the appropriate temperature. The preimpregnated composite is made of fibers, which can be described as a thermoset polymer matrix material, such as epoxy, or a thermoplastic resin, that is already embedded with fibers and is partially cured. The fibers often take the form of a weave, and the matrix is used to bond them together which requires cold storage to prevent the prepreg from becoming fully cured. The tape allows composite parts to be formed by laying down the prepreg tape in a molding tool with the desired orientation and the thickness of the part will be adjusted by stacking several layers of composite tape on top of each other until the desired thickness has been reached. These prepreg layers may have different orientations when stacked on top of each other to obtain the right mechanical properties for the load environment under consideration. Usually, the prepreg ply is covered by special papers which must be removed from the ply before attaching them to the mold or stacking them on top of each other (Fig. 4.4). The automated process of laying down the tapes with different orientations is called automated tape lay-up (ATL). This process can promise a high level of automation and cost savings. The tape lay-up method has been widely used in high-performance industries like aerospace to produce composite laminates from unidirectional prepregs [16, 17].

The stacking process is arranged in such a way as to get the best mechanical properties for the fabrication of the part when subjected to loads. The laid-up components will then unite by a process that involves heating the composite structure, such that the thermoset matrix softens to a sufficient degree to form a single unified matrix. Applying sufficient pressure will then expel any trapped air from the matrix.

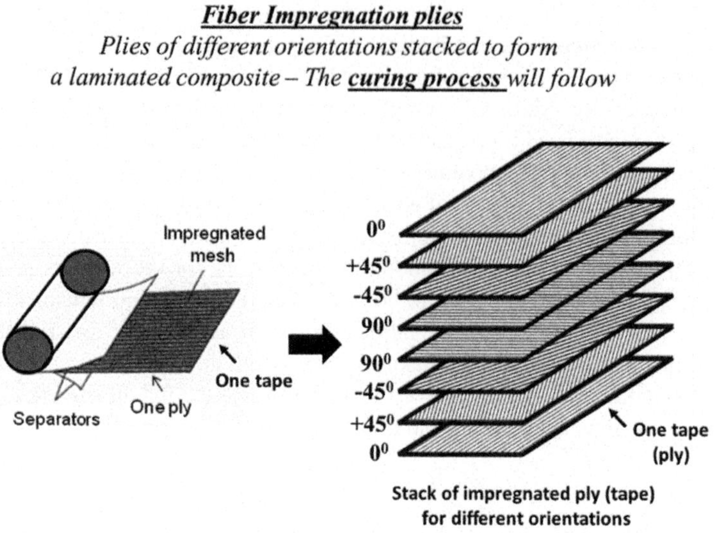

Fig. 4.4 A schematic presentation of several impregnated plies with different orientations to be stacked to form laminated composites

In terms of final structural properties, thermoplastic composites have superior impact and damage resistance properties to those of thermoset composites and are generally tougher and more resistant to chemical attack, all of which are preferable properties within aerospace applications. Furthermore, since thermoplastic composites may be repeatedly reheated and remolded, they are inherently recyclable and more importantly reusable, which is an important point to be considered.

On the other hand, thermoset composite tape has one property, that in relation to the lay-up process, makes it the material of choice for use in aerospace composite parts. This property is part of the thermoset tape that is inherently sticky and adhesive. This stickiness allows the thermoset tape not only to adhere to the complex geometry of mold surfaces but also for separate layers of the thermoset tape to adhere to one another once the initial layer has been applied to the mold surface. For those reasons, the thermoset composite tape makes the lay-up process relatively easy and convenient to physically manage. It should be noted that the separator paper for each ply must be first removed before stacking layers on top of each other, as illustrated in Fig. 4.4.

4.6 Autoclave Lay-Up Technique (An Open Mold Process)

Autoclave molding is an open mold lay-up process that is used mostly in the aerospace industry to produce high-quality composites by using prepregs plies. In this process, both fibers and matrix are preblended and precured, and each prepreg

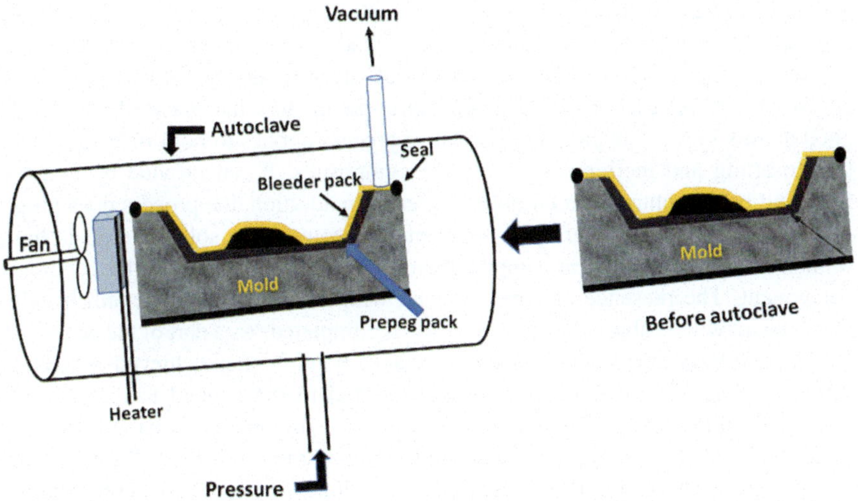

Fig. 4.5 A simplified illustration of the autoclave process. The whole mold assembly will move to the autoclave chamber and the curing process will follow by applying the heater, bleeder, and vacuum will begin

ply is covered from both sides for being protected from the high humidity environment, which can degrade the mechanical properties of manufactured parts. The autoclave cavity uses the right temperature and positive pressure (above the atmospheric pressure) during curing to consolidate the plies and to provide a path for air and volatiles to escape [18, 19]. Most heating devices use a programmable computer control to run the cure cycles. The autoclave molding method is very similar to tape lay-up vacuum bagging except in the case of autoclave, the lay-up is subjected to greater pressures, and therefore denser parts can be better produced, nevertheless, in both processes, they will be able to remove trapped air from the lay-up that can cause voids in the laminate as illustrated in Fig. 4.5 [20, 21].

The autoclave process involves firmly stacking prepreg plies with different orientations in a mold. A release gel is applied to the surface of the mold, prior to the autoclave process, to avoid sticking the polymer to the surface of the mold. After the application of release gel, and stacking prepreg plies, vacuum bagging will follow to remove all possible entrapped air between the layers. Subsequently, the whole assembly is moved to the autoclave machine (Fig. 4.5), where both heat and pressure are applied to aid uniform and effective distribution of fibers throughout the matrix, as well as proper interfacial adhesion or bonding that will require adequate time to reach the final stage. Nonuniform interfacial adhesion or bonding can cause delamination initiation during the service usage of the part resulting in ply separation and possible structural failure.

The autoclave is a cavity or an oven where stacks of prepregs on the surface of mold will be placed inside the oven for subsequent applied heat and pressure (Fig. 4.5). This stage is called curing, which afterward, the composite component is removed from the mold. After cooling the assembly and removal of the vacuum bag, the autoclave process is completed and ready to be separated from the mold. The three cycles of temperature, pressure, and vacuum must be studied experimentally to obtain the best combination of the three parameters that can result in having a good quality composite part. The autoclave process applies to both fiber-reinforced thermosetting and thermoplastic polymer composites. It can provide better inter-layer and better control of both fiber and resin. In addition, the process does proper fiber wetting, and the degree of uniformity in component solidification is high. Moreover, the autoclave can support high fiber volume fraction in the composite component. The presence of voids content, with the help of the vacuum bagging mechanism, will be absent when the process is completed. As a part of the benefits of the vacuum bagging process, a better interfacial bond is often achieved, which can delay the delamination formation when subjected to the applied load. Since this process is falling under the open mold process, it is important to have good quality mold. The mold can be male or female depending on the application. It can be made of different materials, but it should be able to tolerate the shrinkage and the thermal expansion coefficient of the composite material. Additionally, the mold should be stable at the curing temperature, to be able to withstand the operative pressure, withstand wear, and should be resistant to washing solvents and be easy to apply release agents.

Profiting from the abovementioned advantages of the autoclave molding process of manufacturing fiber-reinforced polymer composites, this process has been widely used to manufacture numerous engineering parts such as aircraft components, military, marine, and space crafts, as well as missiles. Despite the benefits of the autoclave molding process mentioned above, it has a few drawbacks such as low production rate, restriction on composite component size, which depends on the size of the autoclave interior space, and finally the involvement of skilled labor, which all the above can make the autoclave process to be an expensive technique for manufacturing the fiber-reinforced polymer composite parts for industries.

The raw materials for the reinforcement material are glass fiber, carbon fiber, aramid fiber which all can be in the form of a mat, bidirectional woven mat, a mat of randomly oriented fibers, and for the matrix material, epoxy, polyester, polyvinyl ester, phenolic resin, unsaturated polyurethane, and thermoplastic resins.

4.7 Compression Molding (A Closed Mold Process)

Several open mold fabrication processes for manufacturing PMC parts were presented in the previous sections. Here, several closed mold processes will be discussed, including the advantages and disadvantages of each method. In this process, the composite lay-up is placed in the open mold cavity, and the mold is closed. Then the force and the heat are applied to fabricate the composite part. It is a closed mold process where the composite material is pressed inside the die (cavity) to form the desired geometry (Fig. 4.6) [22]. The compression molding process is currently being used to produce composite components in high production volume

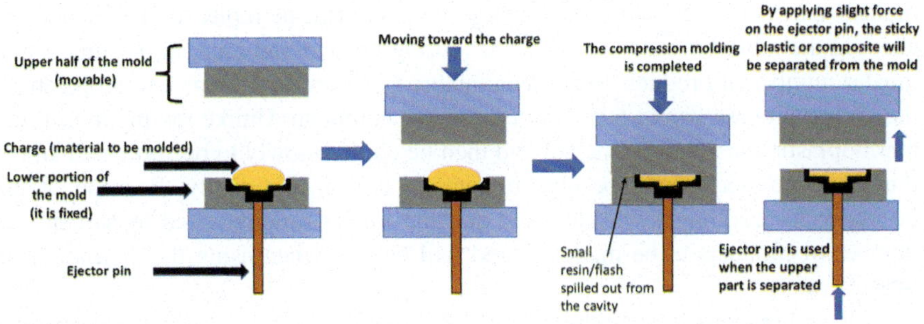

Fig. 4.6 A schematic presentation of a two-dimensional compressive molding. The lower part is stationary, and the upper part will be pressing the charge. After the process is completed and the upper part is separated, the ejector pin will be separating the mold from the cavity [22]

in the automotive industry. It is a simple, fast, flexible, low-cost, and commercially oriented method of fabricating composite components, mostly when the geometry of parts is flat. The compression molding process is one of the low-cost molding techniques when compared with other injection molding methods. Before starting the molding process of the polymer parts, a high-strength metallic mold is needed that must be machined to the exact dimensions required to produce the 3D shape of the part [23]. In addition, some adjustments for shrinkage must be taken into consideration because of the application of heat during the molding process and the subsequent cooling.

In the case of thermosetting matrices, the hot compression molding must be used together with the right amount of pressure to obtain the proper curing.

Therefore, three steps are essential for compression molding to be completed when using the thermosetting matrix. These are, (1) heat the plastic material, (2) deform the plastic material (aka the charge) in the cavity of the metallic mold situated in the lower part where the tool is fixed. Subsequently, exerting pressure on the polymer by bringing the upper half of the mold down toward the lower portion so that the upper and lower parts (matching parts) will meet each other (Fig. 4.6). Under this condition, the material will deform to the shape of the cavity, and (3) allow the cooling to take place. These steps strongly depend on the type of polymer being used. In the case of thermoset polymer, it is necessary to keep the mold hot throughout the process for the curing to happen. In the case of thermoplastic, the mold must cool down gradually for the product to solidify and harden. Immaterial from the type of polymer, the three stages of compression molding must be implemented. It is a high-pressure forming process in which the molten plastic material is squeezed directly into the mold cavity to conform to the designed shape. To summarize, the three steps used in this process that the operator can control are, (1) the heating temperature, (2) the pressure amount, and (3) the holding time in order to obtain the best-quality part.

Engineers should note that the three abovementioned steps can change when the polymer material switches from the thermoset to the thermoplastic. In the case of thermosets, as was mentioned before, the mold remains hot all the way until the molded part is ejected and a new charge material will be replaced. The benefit of keeping the mold hot is to accelerate the curing process and thus shorter time and a greater number of products to be manufactured. Otherwise, at ambient temperature, the curing process would take much longer to complete. Unlike the thermoset, the thermoplastic must first be melted and then be cooled slowly to become hard and at this time the ejector will be able to remove it from the cavity. Because of the necessity for cooling down the mold and the longer time required to harden, the number of products to be manufactured will be less when using the thermoplastic matrix.

Still, in both cases, when the compressive molding process has been completed, and the two parts of the mold are being separated from each other, the bottom part is fixed, and the upper part is movable, then the ejector pin will give slight force to the cavity area to separate the molded material from the cavity, as shown in Fig. 4.6.

Another point to make here is the type of mold design being used in the compressive molding as it is shown in Fig. 4.6. These are (1) the open flash, (2) fully positive, and (3) semipositive mold designs [24]. In the case of open flash mold design (also known as overflow), the amount of charge (i.e., material being used inside the cavity) will be slightly more than what the cavity needs. Upon bringing the upper movable plate down and closing the mold, the excess amount of charge will squeeze out of the cavity which must be removed from the molded part. In the semipositive mold design, no excess amount of charge or flash is allowed in the cavity. That is, if there is any excess charge material present around the cavity, upon bringing the upper movable mold tool down, it will not close, and the upper and lower parts will not mate.

On the other hand, if the amount of charge in the cavity is inadequate (insufficient), then upon closing the mold, the thickness of the part will be less than what it is designed for. Therefore, the amount of charge must be very close to the exact amount, which makes the semipositive mold undesirable to use. Another disadvantage related to semipositive design is that the presence of any excess gas being trapped during the curing process will show as a blister on the surface of the part in the cavity, which is also considered undesirable and must be fixed.

In the compressive molding process described above, composites made of thermosets can be used for fabricating parts through compressive molding. This can be done by placing plies with the desired orientation inside the cavity prior to curing. But the fiber-reinforced resin material is first preformed to the shape of the tool and then proper heat and pressure for the curing will follow. The same is true when fabricating composites made of thermoplastic polymer that is mixed with chopped reinforcement. Of course, the method of heating and cooling for thermoplastic will be different as was mentioned above. Furthermore, there may exist a small amount of resin flash around the edges outside the cavity that needs to be removed, and then the process is finished.

The material to be fabricated in the compression molding process by the thermoset polymer is made of well-mixed, high-quality, unidirectional tapes. Tapes are preimpregnated with resin to be used specifically for compression molding, and then they will be cut and chopped, usually around 2 inches in length. The fiber content by volume in tapes is relatively high and sometimes it is above 50%, which can produce a very high-quality molding compound. The mold is preheated and upon closing the cavity, the proper amount of heat and pressure will apply for the curing process to take place. One can summarize the compression molding for composites as follows: (1) a durable metal tool is made by a toolmaker from metals for mold, usually, steel or aluminum, which must precisely match the measurements needed to produce the desired 3D parts; (2) the desired composite material, usually prepreg, is made before inserting it into the tool; (3) the prepreg composite is inserted into the heated mold cavity. That is, an engineered composite lay-up is placed in the open mold cavity, and the mold is closed. The heat and pressure will apply to the part and the pressure remains on the mold throughout the cure cycle; (4) once the pressure is released, the finished part is removed from the tool using the ejector pin. The combination of heat and pressure can help the curing process to accelerate to produce a composite part with low void content and high fiber volume fraction. If there is any resin/flash left around the edges of the part, it should be removed from the finished part. As was mentioned before, the material utilized for the mold should not have large thermal expansion to create a problem for the manufacturing of composite parts.

4.8 Injection Molding (A Closed Mold Process)

Injection molding is a manufacturing process for making composite parts by injecting molten polymer into a mold cavity [25–28]. It consists of the high-pressure injection of the raw material consisting of polymer mixed with chopped fibers into a mold or multiple molds (aka multicavitation), which form the molten material into the desired shape. It was an American inventor, John Wesley Hyatt, with the help of his brother Isaiah Wesley, who patented the injection molding process in 1892 [29, 30]. It is one of the most used molding processing techniques for plastic parts, and later, the process matured to manufacture thin plastic components of different shapes, sizes, and in some cases with micron-level precision. Manufacturing thick parts through injection molding would be undesirable with this method because of the time required to cool down the thick part, which will slow down the production time considerably. A good metal base injection molding tool should be able to produce excellent-quality plastic parts with tight tolerance and a high production rate.

Injection molding can be performed with various materials, mainly metals, glasses, and elastomers, as well as thermoplastic and thermosetting polymers that can be used for industries (for metals, the process is known as die-casting). Molten material for fabricating the composite part under consideration is prepared into a heated barrel (aka injection chamber) and injected into a mold cavity, where it cools and hardens to the geometry of the cavity (Fig. 4.7).

When the design of the final product is completed by engineers, which is usually done by industrial engineers, the part is ready for the manufacturing phase. But, for the design of the mold itself, it is necessary to make the mold from special metals, either steel or aluminum, and it is done by a mold maker engineer (or toolmaker).

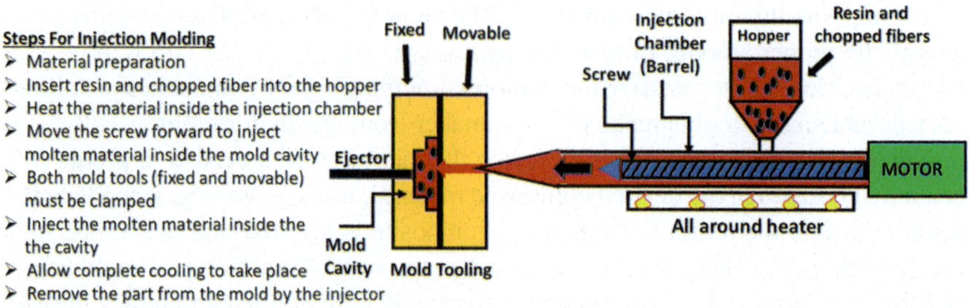

Fig. 4.7 A two-dimensional illustration of the reciprocating screw-type injection molding. The feeding system, motor, heater, and the screw device are all shown in the figure. Steps for injection molding are described

The selected metallic material is carefully machined to the specific design requirements of the part that designers have in mind. There are three types of injection molding design. They are hand injection, plunger type, and reciprocating screw injection molding. Fig. 4.7 illustrates the reciprocating screw type injection molding with the motor, as well as other necessary parts, the hopper, plastic pellet, heater, nozzle, mold tool, mold cavity, and injector.

Other types of injection moldings are very much like the screw type injection, and no need to elaborate on others.

Injection molding is widely used for manufacturing a variety of parts, from the most minor parts to the entire body panels of a car. Parts to be injection molded must be designed such that the outcome of the molding process is entirely acceptable to the customers. For that to happen, the material used for the part, the desired shape and features of the part, the material of the mold, and the properties of the molding machine must all be taken into consideration because they all will contribute to the fabrication of the final product. The flexibility and success of injection molding products are directly related to the parameters mentioned above.

In Injection molding can fabricate composite parts with high precision and at very low cycle times. In a typical injection molding process, resins and chopped fibers in the form of pellets are fed through a hopper, and then they are conveyed by a screw with a heated barrel, as shown in Fig. 4.7. Once the required amount of material is melted in a barrel, the screw injects the material through a nozzle into the mold, which is cooled and acquires the desired shape. Injection molding has been found to be very effective for thermoplastic polymers in many industries.

Adding reinforcement to the polymer matrix in the injection molding process has advantages because of speedy cycle time, low mold clamping pressure, and low scrap rate. With the presence of reinforcement, polymerization shrinkage and thermal expansion are reduced. In addition, fundamental properties such as stiffness, tensile strength, and tensile elongation are increased through this method. Chopped fibers or flakes can be added directly to the resin before reacting in the mixing container. Note that the presence of chopped fibers in the matrix must be uniform, which otherwise can degrade the property of the final products.

The whole process of injection molding for composites can be summarized as follows: (1) resin and chopped fiber are prepared, which will be placed in the hopper; (2) the barrel (or injection chamber) will be heated to melt the material coming down from the hopper; (3) the melted material will be squeezed by the screw into the mold cavity (or cavities) when both the fixed and movable parts are tightly clamped together (Fig. 4.7). The clamping and opening of the mold are done hydraulically; (4) after the injection process has been completed, the cooling process will begin with air and water circulations. With the aid of a cooling system, the final part will be ejected from the mold.

Finally, injection molding is a great process to be used for manufacturing thin parts, however, molding the thick part is possible but can have drastic effects on the cost and production speed of manufacturing because of longer cooling time. The designed wall thickness that can be used for this molding process depends on the size and geometry of the part, structural requirements, flow behavior of the resin, and the effectiveness of the cooling system. The wall thicknesses of an injection molded part generally range from 0.078 in to 0.157 in (2–4 mm). While there are no wall thickness restrictions when using the injection molding process, the goal is usually to choose the thinnest wall possible. Thinner walls use less material which reduces cost and takes less time to cool, thus reducing cycle time.

4.9 Pultrusion Molding

As the name pultrusion implies, it is a molding process that is made of two words, pulls and extrusion, in which the fibers impregnated with resin will be pulled through a die with some pulling mechanism to manufacture the part [31, 32]. One of the desirability of the pultrusion process is no limitation on the length of the part. When it is necessary, based on the design requirement of the length, it can be cut by the machine, as shown in Fig. 4.8. Regarding the shape of the cross section, the shape classification can be mainly rectangular, circular, or hexagonal geometry with uniform thickness [33]. In addition to shape classification, the cross section could be hollow or solid. For example, engineers can have a hollow pipe or a solid shaft with their desired cross-sectional areas and length.

In all the composite manufacturing practices that were discussed in this chapter, the objective for engineers is to choose the best method that can produce a satisfactory and sound product that will be cost-effective for the customer. When the pultrusion becomes the choice of designers for manufacturing parts and because of its automated feature, it can be a low-cost, high-volume manufacturing method where the final product will be having a uniform thickness with no constraint on the length of the product. In simple language, pultrusion is a manufacturing process

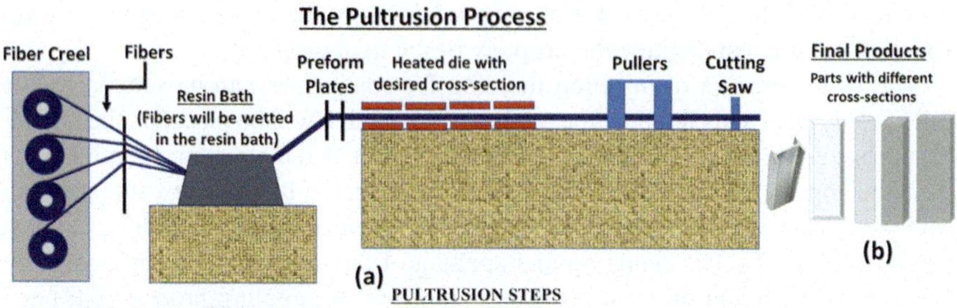

PULTRUSION STEPS

- Material preparation (fibers and resin)
- Fiber Creel or spool, where fibers are coming from
- Resin bath (fibers will be wetted in the resin bath
- Preforming fixture, which are giving shapes before the final shape (also the excess resin and entrapped air are removed),
- Heated die (arrange the metal die with proper heating)
- The impregnated fibers are pulled through a shaping die or mold that gives the composite its desired cross-sectional shape
- Puller where the products are pulled out,
- The cutting saw which each product will be cut with a predetermined length
- Final product

Fig. 4.8 (**a**) Illustrating the pultrusion steps for manufacturing parts. (**b**) Final products with different cross-sections and with no limitation on length can be fabricated through pultrusion process

that can convert reinforced fibers mixed with liquid resin into a fiber-reinforced plastic with low cost, energy-efficient, and no limitation on the size of the product.

The pultrusion process is very much like the case of metal extrusion in which a workpiece, of a certain length and cross-section, is forced to flow through a die, but in the case of the pultrusion process, the resin-wetted fibers are being pulled through the die or the mold. The matrix in the pultrusion process is thermosetting as shown in Fig. 4.8a, where continuous fibers in the form of roving are passed through the resin bath. Fibers can be glass fibers, Kevlar, or can be mat pulled through the resin bath. The resin types can be polyester, polyurethane, vinyl ester resin, and epoxy. Subsequently, the wetted resin bath fibers will be wrapped around the heated dies until the part is cured and then it will be cut by the machine. The operator must be sure that the fibers are fully wetted uniformly, otherwise, the parts may have defects and must be discarded.

In summary, the pultrusion process consists of the following steps: (1) creel or spool (where fibers are coming from toward the resin bath), (2) resin bath and the preforming fixture, (3) removing the excess resin and entrapped air, (4) arrange the metal die with proper heating, (5) puller where the products are pulled out, and finally, (6) the cutting saw with which each product will be cut with a predetermined length (Fig. 4.8a).

As already mentioned, the final products will have several cross-sections, and each can be solid or hollow as shown in Fig. 4.8b. As illustrated in the figure, composites with rectangular, circular, square, L shape, C shape, and I shape, and many others can be produced by this method.

4.9.1 Advantage and Disadvantage of Pultrusion Process

The advantages of the pultrusion process are the high rate of products, the method is a low waste material process, it is able to use a wide range of reinforcement types and many thermosetting resins used as the resin bath, and finally, a thin wall with complex geometry can be fabricated. It also should be mentioned that with this process the final product is the near shape part, which means it is exactly like what the engineers had in mind for use and requires no need for any post-product cleaning. That is, the process will provide a final product with no waste of resin and fibers. Additional two advantages of pultrusion are no limitation on the size of the product that must be manufactured, and extremely large products with no limitation on the width (diameter) and length dimensions are possible to produce. However, there are disadvantages associated with the pultrusion process that must be aware of. One of them is the time of curing. That is, from the time when the resin and fibers are passing through the preform stage and entering the heated die zone, it may not be enough time for the curing and polymerization to take place. Therefore, a quick curing process that is happening only in the heated die zone may not be adequate and the part can have lower mechanical properties than engineers were expecting to have. In general, the curing of thermosetting polymers may take a few hours, whereas, in the pultrusion process, the time that parts are entering the die until their exit from the heated die will be only a few minutes, and no pressure is involved. The quick curing time without pressure may be attractive from a high-volume manufacturing perspective, which can compensate for the low mechanical properties of the final products. In addition to the low curing time and resulting low mechanical properties in the pultrusion process, the lack of fiber volume in some areas can also introduce low mechanical properties. Therefore, special attention must be given to bringing a uniform and adequate volume of fibers into the matrix and a better-quality part. Another drawback associated with the pultrusion process is the uniformity of the parts' thickness. It can be beneficial, but at the same time, it can be an issue when engineers are interested in having nonuniform cross-sectional areas in some areas of the composite part (e.g., design with thickness variation in some areas along the pultrusion part).

4.10 Filament Winding Molding

Filament winding is a continuous manufacturing process to produce composite structures such as tanks, pipes, pressure vessels (with two spherical shape end domes included), rocket motor cases, pressurized fuel tanks, and other round-form products that require to have a high degree of structural integrity [34–36]. The cylindrical composite products that are fabricated by this method have hollow cores in which resin impregnated with high quality fibers will be wrapped around a rotating cylindrical mandrel mold in the desired pattern or angle to the rotational

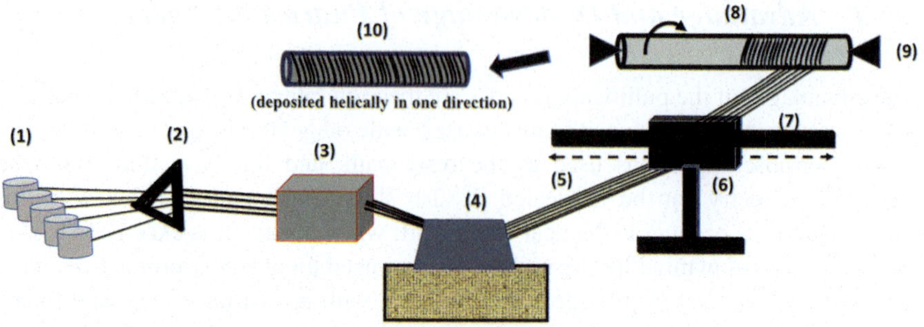

Fig. 4.9 Schematic of the conventional wet filament winding process. (*1*) fiber creels (spools), (*2*) fiber guide, (*3*) tensioning systems and guided pin, (*4*) wet resin bath, (*5*) impregnated fiber bundles for winding around the mandrel, (*6*) traversing carriage (moving horizontally back and forth parallel to mandrel), (*7*) guide rail, (*8*) rotating mandrel, (*9*) spindle, and (*10*) illustrating the final helical winding process around the mandrel in one direction

axis to create different layers. The mandrel will have a circular motion so that upon rotating around its axis, the back-and-forth axial horizontal motion of fibers will be wrapped around the mandrel to create a complete composite structure. Composite products produced through the filament winding automated process display excellent strength because fibers are soaked completely with resins and additionally will wind around the mandrel with proper orientation to resist the induced stresses in the body of the structure during its service life. After the winding process is completed, the entire structure is cured and subsequently, the mandrel is removed. Depending on the resin system and its cure characteristics, often the mandrel is autoclaved or heated in an oven or rotated under radiant heaters until the part is cured [37].

Figure 4.9 illustrates a filament winding process of a cylindrical pipe where wet fibers are used to make the part through the winding operation. But first, dry reinforcements will unwind from the creel, and then fibers will be impregnated in the thermoset polymer bath to be winded around the mandrel in the circumferential and any other desired orientations in order to better withstand the induced circular (hoop) and longitudinal stresses in the body of the part. That is, the wet material will be deposited helically in one direction and on the way back in the opposite direction since the moving rail is traveling horizontally back and forth as shown in Fig. 4.9.

So, in one full cycle (back and forth), the helical winding will be on one side and opposite side. This routine will continue until the desired thickness and proper fiber orientations are achieved. Those fibers that have been soaked in the resin can be wandering in such a way that will provide the best resistance when the load is applied to the structure during service.

Dry fibers that are continuously coming from creel can be glass fibers, carbon/graphite fibers, or Kevlar fibers. After being saturated in the resin bath, fibers will be passed through the nip rollers (aka pinch rolling) to get rid of any excess resin (Fig. 4.9). Nip rolling is a process where rolls are used to press together multiple fibers that were soaked within the risen material. The pressure created from running the material through the rollers helps bond the filaments together, creating a tight bond between the material, and to remove any excess resin. The guide or the feeding port can move back and forth parallel and along the length of the mandrel in such a way that filaments under tensile stress will be rotating over the mandrel with different orientations. The rotating mandrel is made of solid wood or other suitable material and when the filament winding process is over, it can be removed and may be used again.

4.10.1 Advantages and Disadvantages of Filament Winding

The filament winding process starts by first allowing fibers to be wetted by a resin and then uniformly and regularly wound around a rotating mandrel along a predefined path. It is one of the few automated processes for composite manufacturing and can thus produce highly repeatable, high-quality parts at reduced labor content. A big advantage of filament winding is that it uses continuous filaments that have been wetted through the resin bath, which can provide a composite material with high strength and stiffness qualities. There is also the ability to orient filaments to be aligned with the direction of induced stresses. Therefore, the principal advantage of filament winding over other methods for composite fabrication is the possibility of adopting automation and robotic procedures.

High fiber volume fractions are estimated to be between 65% and 80% that can provide a good quality composite part. The filament winding technique can use different filament and matrix types. There are a few major disadvantages of filament winding that should be mentioned here. The major one is the geometric limitation with the inability to wind fibers on negatively curved (dipped) surfaces. That is, the part configuration must facilitate the mandrel path. Also, because the mandrel path is fixed, the ability to change the path of the fiber within the layer is difficult. Finally, the programmer must be sure the machine will be able to create parts per the actual design directed by engineers and free from error. Otherwise, the fabricated parts will be dismantled and that would be a waste of time and money for the company.

4.10.2 A Brief History of Filament Winding

In the early 1960s, the first commercial winding machine was produced. These machines were difficult to work with and in the late 1970s, the first computer-controlled winding machine was used in industries. It was not reliable and difficult to program, and it was a mixture of calculation and trial & error. In the early 1980s, computer numerical control (CNC) winding machines were introduced. These control systems were focused on metal cutting but being general purpose could also be used for filament winding. The need for improved programming methods to take advantage of the new technology was clear to industries and soon the filament winding process became a mature technology and related tools are now becoming standardized and have lower costs. Currently, they are being used in companies of all sizes and are no longer limited to defense, aerospace, and other similar industries [38].

4.11 Other Molding Processes

In previous sections, open mold and closed mold processes were discussed. There are other PMC molding processes used in industries for manufacturing composite parts, which will be briefly discussed in this section. These include sheet molding compound (SMC), bulk molding compound (BMC), and glass mat thermoplastic (GMT), which are commonly used in the automotive industry for manufacturing composite parts. For more in-depth information, readers should refer to the references provided for each molding method.

4.11.1 Sheet Molding Compounds (SMCs)

Sheet molding compound (SMC) is a type of composite material used to manufacture parts consisting of a mixture of resins (typically unsaturated polyester or vinyl ester), glass fibers, fillers, and additives [39–42]. It is commonly used to produce complex composite parts that require precise molding, lightweight properties, high strength, and durability. Since the first molding compounds based on unsaturated polyester and the industrial startup in the 1950s, this class of materials has evolved into a variety of high-performance polymers for various applications, with a broad range of tailor-made performances. A significant portion of the world's thermoset production comes in the form of SMCs [43, 44]. In this section, the composition of SMC in manufacturing composite parts, several advantages of SMC products, and market access for SMC in automotive and other industries will be briefly discussed.

The composition of the SMC matrix includes a resin, typically unsaturated polyester or vinyl ester, acting as the binder that holds the composite constituents together. Reinforcement materials are usually glass fibers, which provide structural strength and stiffness. Longer fibers in the SMC process result in better strength properties. Standard fillers, like calcium carbonate, are often added to improve surface finish. Additionally, additives such as pigments, flame retardants, and other chemicals modify the performance characteristics of SMC parts.

SMC is produced in sheet form, with the resin and fibers laid down in layers. The material is then compressed and cured in a mold under heat and pressure. The preform of the required dimensions is placed on a press, where pressure and temperature are applied for curing, with cycles ranging from a few minutes to a few hours. This makes SMC suitable for high-volume manufacturing processes, such as automotive body panels, electrical components, and other structural parts. The sheet is provided to customers in rolls weighing up to 1000 kg. Alternatively, the resin and related materials may be mixed on-site when greater control over the

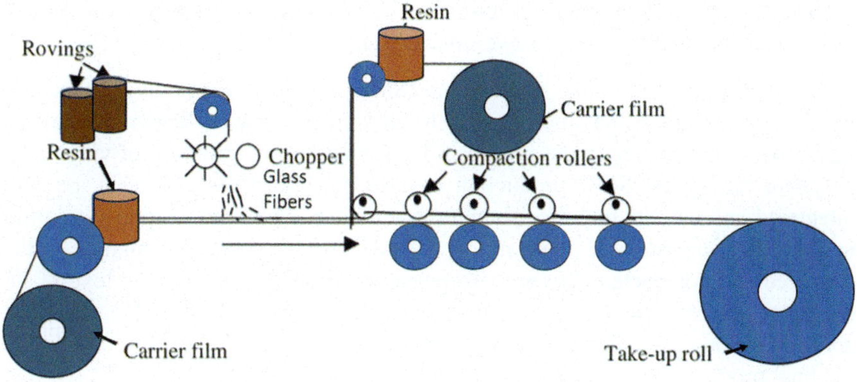

Fig. 4.10 Schematic of sheet molding compound (SMC) manufacturing [7]

chemistry and filler is required. Note that the SMC manufacturing process benefits from very high-volume production capability and excellent part reproducibility. It is a cost-effective molding process with low labor requirements per production level, as illustrated in Fig. 4.10 [45]. The carrier film shown in Fig. 4.10 is a protective layer that serves as a support for the resin paste during the manufacturing process. It typically consists of polyethylene or polyester and is placed on both sides of the resin-fiber mixture to contain the material and prevent it from sticking to the equipment. This material also helps ensure uniform thickness and surface quality of the composite material, which is important for producing high-performance parts.

SMC offers several advantages:

1. *High strength-to-weight ratio*: It is stronger than many light metals and alloys.
2. *Corrosion resistance*: Unlike metals, SMC does not rust.
3. *Dimensional stability*: It holds its shape well under load and temperature changes.
4. *Design flexibility*: Complex shapes can be molded with precision.

Market access for SMC is expanding across various end-use applications and global demand. Suppliers report growing interest in SMC from the industrial, consumer, building and construction, energy generation, electrical/electronic, nonautomotive transportation, and aerospace sectors. However, SMC is most commonly used in automotive parts such as body panels, bumpers, and structural components. It is favored in industries that require high-performance, lightweight, and durable materials. Current automotive and heavy industry applications for SMC materials include [43]:

- Frames for windows/sunroofs
- Air-intake manifolds
- Front-end grill openings
- Battery casings and covers
- Headlamp housings
- Bumpers and bumper reinforcements

- Heat shields (engine, transmission)
- Cylinder head covers (e.g., valve, rocker, cam)

In summary, sheet molding compound (SMC) is a reinforced polyester material featuring glass or carbon fibers, typically an inch or more in length. These long strands of fiber are spread out in a resin bath, made from a mixture of vinyl ester, polyester, or epoxy. SMC production ensures that the fibers and resin are fully integrated and that the interface strength is sufficient. The resin is applied to a film in the form of a paste, and fibers are cut and added to the resin. This mixture is then squeezed and compacted between two pieces of film. Once the correct texture and thickness are reached, it is stored for several days before heat and pressure are applied in a process called compression molding, which cures the SMC and creates the final product.

4.11.2 Bulk Molding Compound (BMC)

Bulk molding compound (BMC) is a compound typically formed into logs or ropes. It is also known as dough molding compound (DMC). BMC is commonly used in the manufacturing of composite materials, with a thermoset as the matrix combined with a mixture of short glass fibers, fillers, and additives. It is typically used for compression molding, injection molding, or transfer molding. The fiber length ranges from 6 to 12 mm. Due to the lower fiber volume fraction and shorter fiber length, BMC composites provide lower mechanical properties than SMC composites. However, BMC offers high strength, durability, and excellent electrical insulation properties [45]. In this section, the composition of BMC in manufacturing composite parts, several advantages of BMC products, and market access for BMC in automotive and other industries will be briefly discussed.

The *composition of BMC* includes:

1. *Resin as the matrix*: Typically made of unsaturated polyester or vinyl ester, providing structural support and flexibility to the final product.
2. *Reinforcement fibers*: Short glass fibers, typically around 6–12 mm in length, providing strength and stiffness.
3. *Fillers*: Commonly calcium carbonate or aluminum trihydrate, added to adjust properties like density and surface finish. Ground limestone ($CaCO_3$) is the most common filler for BMC products.
4. *Additives*: These may include flame retardants, colorants, release agents, and other performance-enhancing chemicals.

The concentration of the mixture can vary depending on the application, but it typically consists of 25–30% glass fibers, 25% filler, and 45% resin, styrene, and initiator. The percentage of fibers can be adjusted based on the customer's desired strength and stiffness for the final product. The glass fibers in BMC improve the strength properties of the product, making it stronger than standard thermoplastic

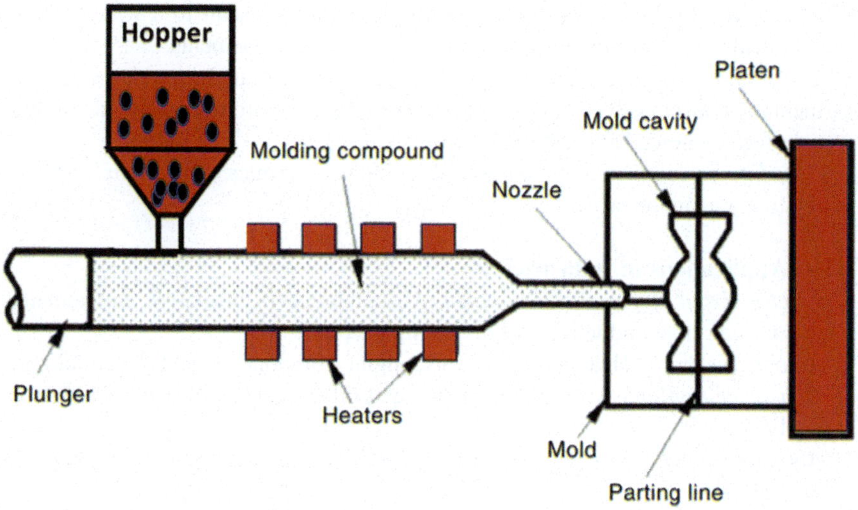

Fig. 4.11 Schematic of the injection molding process for BMC parts [45]

products. The final BMC product is provided in bulk or in logs approximately 4 in in diameter.

BMC is typically made from a thick, dough-like material that is flexible to work with and can be shaped using different molding processes. In the *compression molding process*, a premeasured amount of BMC is placed into a heated mold, which is then compressed and cured under pressure. In *injection molding*, the material is injected into a mold, which is often used for more intricate or detailed parts. In the *transfer molding process*, BMC is transferred into a mold from a chamber, allowing better control over the material during shaping.

Figure 4.11 shows the process of making BMC parts. To make molded parts, BMC is loaded into a hopper and forced by a piston stuffer into the barrel of the molding machine. The material is then conveyed through the barrel by a rotating screw or ram plunger, as shown in Fig. 4.11. The complete injection molding process is automated once BMC is fed into the hopper. The operator enters process parameters into the computer and starts the machine. The operator closely monitors the material flow, part molding, demolding, and cleaning to ensure the process runs continuously without issues [45].

Advantages of BMC Molding Parts

- *High Strength and Rigidity*: BMC offers an excellent balance of mechanical properties, making it suitable for load-bearing applications.
- *Thermal and Electrical Insulation*: Its excellent insulation properties make it ideal for electrical components.
- *Fire and Corrosion Resistance*: BMC can withstand high temperatures and harsh environments without degradation.

- *Chemical and Solvent Resistance*: BMC is highly resistant to a wide range of chemicals and solvents, making it ideal for harsh environments.
- *Eco-friendly and Recyclable*: BMC is a great green solution in today's environmentally conscious world. Its recyclability and low environmental impact make it a preferred choice for sustainable manufacturing.
- *Cost-Effectiveness*: Compared to other high-performance materials, BMC is relatively economical.

BMC Applications in Industries

- *Electrical and Electronic Components*: Circuit breakers, insulators, and housings, due to their nonconductive and insulating properties.
- *Automotive Parts*: Headlamp housings, engine components, and structural parts.
- *Consumer Appliances*: Handles, knobs, and other heat-resistant parts in kitchen appliances.
- *Aerospace and Defense*: Structural parts requiring strength and lightweight properties.
- *Appliance Parts*: Components such as washing machine parts, oven components, and refrigerator linings.
- *Industrial Lighting Housings.*
- *Automotive Components*: Bumpers, engine covers, headlamp housings, and interior components.
- *Aerospace Interior Components*: Brackets and housings.
- *Sports and Recreation Equipment*: Bicycle components, snowboard bindings, and fitness machine parts.
- *Medical Equipment Housings*: Sterilizable components.
- *Building and Construction Components*: Window frames, electrical enclosures, and drainage systems.

Overall, BMC is a versatile, high-performance material used across industries for applications where a combination of strength, heat resistance, and electrical insulation is needed.

4.11.3 Glass Mat Thermoplastic Molding (GMT)

Glass mat molding with polypropylene, also known as glass mat thermoplastic (GMT), is a type of composite material consisting of polypropylene as the matrix material, reinforced with randomly oriented glass fibers. GMT is traditionally available with fiber contents between 20% and 50% and fiber lengths from 20 to 80 mm in the semifinished plates [46]. Due to its lightweight and high-strength properties, GMT is used in various industrial applications for manufacturing composite parts, particularly in the automotive industry. The composition of GMT includes polypropylene and glass fibers. Polypropylene, a thermoplastic polymer, forms the matrix and holds the fibers together, providing flexibility and impact

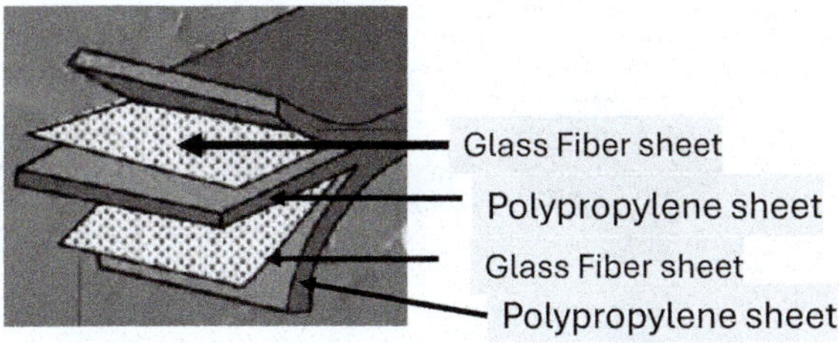

Fig. 4.12 GMT constituents (polypropylene matrix and glass fibers) are shown in the form sheets

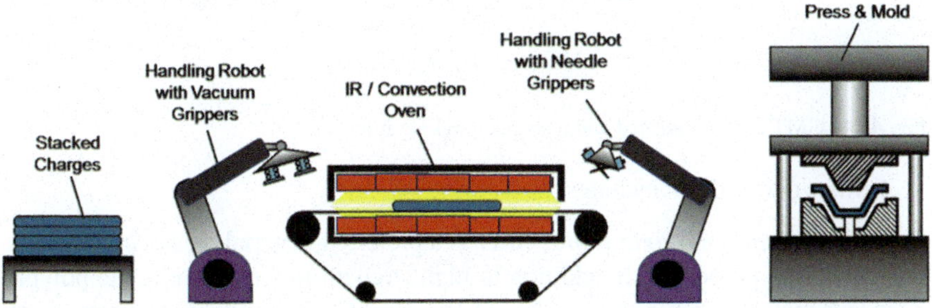

Fig. 4.13 Industrial production setup for GMT compression molding [47]

resistance to the GMT composite parts (Fig. 4.12). The reinforcement fibers are usually randomly oriented in the form of a mat, giving the composite material its strength and stiffness.

The *material preparation for GMT parts* involves:

1. Impregnating glass fibers with molten polypropylene, which is then laid down in the form of a mat.
2. The composite sheet is heated and shaped using various molding processes, such as compression molding. Since polypropylene is a thermoplastic, it softens when heated and hardens upon cooling, making it easy to form complex shapes.
3. Once the material is molded, it is cooled to retain its final shape.

Detailed Description of the Manufacturing Process

The manufacturing process is traditionally split into two separate steps, (1) the manufacturing of the glass fiber mat and (2) following the melt incorporation in a double belt press. GMT materials are traditionally provided in plates and have to be cut and preheated prior to processing. GMT plates are usually cut to the desired shape with plate shears or waterjet cutting depending on the complexity of the shape [47]. The cut plates are then placed into the oven manually or by handling robots (Fig. 4.13). While GMT is still in plate geometry, clamping and vacuum grippers are applied. Since the heating is the most time-consuming step in the compression molding cycle and is a significant cost-driver, there are several heating solutions available for commercial application [47].

Fig. 4.14 GMT can be reheated and reshaped, making it recyclable

Advantages of GMT include:

- *High Strength-to-Weight Ratio*: GMT is a preferred material for having attractive lightweight properties. In addition to high strength-to-weight ratio, the polypropylene matrix, combined with glass fibers, gives GMT excellent impact resistance and toughness.
- *Recyclable*: As a thermoplastic composite, it can be reheated and reshaped, making it recyclable, unlike thermosetting plastics that will be useless after reheating (Fig. 4.14).
- *Resistance to Corrosive Environments*: GMT does not rust or corrode, making it ideal for outdoor or automotive applications.
- *Cost-Effective*: Polypropylene is a relatively inexpensive polymer, making GMT an affordable solution compared to other composite materials.

Applications of GMT in industries include:

1. *Automotive*: Used in structural components such as front-end carriers, door panels, seat backs, and underbody shields due to its high strength and low weight.
2. *Construction*: GMT panels are used in buildings for reinforcement purposes.
3. *Consumer Goods*: GMT is used in items like luggage and protective gear, where durability and lightweight properties are critical.

Comparison to Other Composite Materials:

- *GMT vs. SMC (Sheet Molding Compound)*: GMT is lighter due to the thermoplastic nature of polypropylene, while SMC, made with thermosetting resins, provides higher heat resistance and rigidity.
- *GMT vs. BMC (Bulk Molding Compound)*: GMT offers better impact resistance due to the use of thermoplastics, while BMC typically provides better thermal stability. Note that, GMT material has high stiffness in all three axes and exceptional impact resistance, even at low temperatures (−40 °C/−40 °F).

Overall, GMT is a versatile material that offers a balance of lightweight properties, high strength, great impact resistance, recyclability, and cost-effectiveness, making it ideal for automotive and other demanding industrial applications. Rather than focusing on which manufacturing methods produce superior products (GMT versus BMC and SMC), it is better to recognize that each material has its own strengths and weaknesses, which translate to advantages or disadvantages depending on the application requirements. The more knowledgeable engineers become about these differences, the more successfully they can maximize design-to-cost initiatives by testing and selecting the optimal product and process combination [48]

4.11.4 Concluding Remarks on PMCs

Polymer matrix composites have brought considerable interest to various industries due to their low cost and higher specific strength and stiffness compared to their metallic counterparts. In addition, polymer matrix composites present greater design flexibility and exceptional resistance to corrosion and fatigue, which their use often has associated with financial benefits to the industry immediately or can be foreseen over the service life of the structure. The application of these materials has been recognized since the late 1960s in the aerospace, defense, and automotive industries. In addition to a high-performance polymer matrix, such as epoxy, graphite, phenolics, and polyamides, the high-performance fiber reinforcements are of great interest to military and aerospace because of their capability to handle the load, which they include carbon fibers, silicon carbides, aramids (Kevlar), boron, and silicon nitride.

Currently, polymer matrix composite products are manufactured on a mass scale and their applications are increasing each year because of consumer interest. Hence, there is a need for the government to come up with a way of recycling unwanted composite parts. In the case when the matrix is a thermoplastic polymer, it can be easily melted and remolded again, but information on the quality of recycled material needs to be studied.

Polymer composites are anisotropic materials made up of the matrix (binder) and the reinforcement (fibers or whiskers), which are adhesively bonded to each other. Therefore, their properties will be controlled by factors such as each component's properties, the composition of the individual components, and the interface strength. Although all the abovementioned factors have an influence on the final properties of the polymer matrix composites, the first two can be controlled before processing and the last one can only be determined after the processing and fabrication stage is completed. In order to obtain a good transfer of stress from the matrix to the reinforcements, a strong bond is necessary to be established. A strong interaction such as covalent bonds between the polymer matrix and the fibers can be useful to enhance the mechanical and fracture properties of the composite. From the fracture toughness perspective, a good bond between the matrix and the fibers can act as a crack stopper that would delay the crack growth mechanism in a load-varying environment. Therefore, the matrix part of the composite components serves as a binder to hold fibers in position and to establish the interface but also to assist in transmitting stress within the material. If this is done poorly, that is, the interfacial bond is poor, then it can lead to premature failures in the material.

Finally, engineers should recognize the importance of needed testing and inspection of polymer matrix composite primary structural parts to ensure their integrity during operational use. Testing for mechanical properties should adhere to reliable standards, including ASTM, EN, ISO, and JIS. These standards offer comprehensive methodologies for testing various composite properties, encompassing tensile strength, flexural properties, and impact resistance, among others. They play a crucial role in ensuring consistent material characterization and evaluating performance across diverse industries and regions. Further detailed information on each standard can be found in related literature shown below:

ASTM (American Society for Testing and Materials)

- *ASTM D3039/D3039M*—Determines the tensile properties of composite materials.
- *ASTM D7264/D7264M*—Standard Test Method for Flexural Properties of Polymer Matrix Composites: Evaluates the flexural properties of polymer matrix composites.
- *ASTM D790*—Provides guidelines for flexural testing of plastics, including some composite materials.

EN (European Norms)

- *EN ISO 527*—Specifies the test conditions for the determination of tensile properties of plastics, including composite materials.
- *EN ISO 14125*—Provides guidelines for determining the flexural properties of fiber-reinforced plastic composites.

ISO (International Organization for Standardization)

- *ISO 527*—This standard specifies methods for the determination of tensile properties of plastics, including composite materials.
- *ISO 178*—Specifies a method for determining the flexural properties of rigid and semirigid plastics, also applicable to some composite materials.

JIS (Japanese Industrial Standards)

- *JIS K 7074*—Specifically for testing the tensile properties of carbon fibers often used in composite materials.
- *JIS K 7204*—Addresses the flexural properties of carbon fibers, which are integral components of many composite materials.

In addition to the abovementioned mechanical tests, the significance of nondestructive testing to detect flaws and damage in composite parts before and during their operation cannot be overstated. Chapter 14 of this book thoroughly describes various standard nondestructive testing methods available for this purpose.

4.12 Nanocomposites and Future Work

It is the goal of many industries to minimize the weight of structural parts without compromising their structural integrity and resistance to failure. Recently, composite materials are used extensively in manufacturing many industrial parts, where strength, weight, fatigue, and corrosion are important factors to be considered when designing critical parts. In fiber polymer-reinforced composites, high strength fibers can provide reinforcement to polymers so that they become capable of carrying the high operating load during the service operation. It is estimated that, by using the PMC parts in the aircraft structure, it can increase both the fuel and maintenance costs by roughly 50% and 20%, respectively. However, if it is possible to replace the high strength fibers with another filler type with much higher strength, that can further reduce the fuel and maintenance costs even more. Recently, nanoparticles have shown promise in the aerospace and aircraft sectors due to their superior E/ρ and σ_{Ult}/ρ over the traditional carbon fibers used in fabricating PMC parts. Nanoparticles are very small particles with dimensions between 1 and 100 nanometers (nm), where 1 nm $= 10^{-9}$ meter. A typical example of nanoparticles is carbon nanotubes (CNTs), which are hollow cylinders made of carbon atoms connected by covalent bonds. They have high strength, and great electrical and thermal conductivities since their discovery in 1991 [49–51]. It is the goal of industries to replace the traditional carbon fibers with the CNTs, acting as the reinforcement embedded in the polymer matrix. This combination of polymer matrix and carbon nanotubes is called carbon nanotube reinforced composite or simple nanocomposites [52]. Figure 4.15a shows CNTs in the polymer matrix with their mechanical properties compared with other materials currently being used in

Carbon nanotubes possess exceptionally superior mechanical properties with high elastic modulus and tensile strength. They are an ideal reinforcements for polymer nanocomposites.

(a)

Material	Aluminum	Steel	Glass fibers	Carbon fibers	Silicon carbide fibers	Carbon Nanotubes
Young's modulus, E Ksi (GPa)	10295 (71)	31177 (215)	10440 (72)	43503 (300)	65255 (450)	145011 (1000)
Tensile strength, Ftu Ksi (GPa)	94 (0.65)	130 (0.9)	435 (3)	725 (5)	1450 (10)	29002 (200)

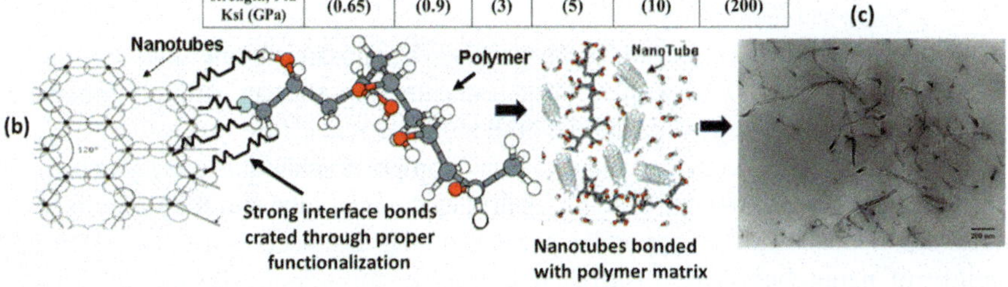

Fig. 4.15 (**a**) Tensile modulus and strength properties of CNTs compared with other aerospace materials (including carbon fibers). (**b**) Illustrating nanocomposites, where nanotubes are embedded in the polymer matrix. (**c**) Distribution of CNTs in the polymer matrix [52, 53]

industries. However, there are two problems of using CNTs in the epoxy matrix and those are (1) poor dispersion of nanotubes in the matrix, which is as a result of CNTs to aggregate and entangle together rather than spread uniformly in the matrix that can cause difficulties in obtaining desired mechanical properties. Their strong tendency to aggregate remains a major obstacle in their applications in industry, and (2) weak interfacial bonding between CNTs and epoxy as the matrix. The lack of strong interfacial adhesion between the matrix and the reinforcement can be overcome by using the functionalized nanotubes (Fig. 4.15b).

By doing that, it can provide multiple bonding sites to the polymer matrix so that the load can be transferred to nanotubes and thus reduce separation between the surfaces of polymer and nanotubes. That is, the functionalization process helps in decreasing the long-range van der Waals forces of attraction between the matrix and the CNTs and gives rise to a homogeneous dispersion and better bonds between the matrix and the CNTs. Finally, Fig. 4.15c shows a photo taken from CNTs embedded in the matrix randomly [53, 54].

To improve the interaction of CNTs particles with the polymer matrix epoxy to develop a strong interface, there is a need to functionalize them to produce a stronger nanocomposite. A weak interface bonding can result in poor load transfer between the two and this substantially decreases the mechanical properties of CNT reinforced composite [55]. To manufacture a robust nanocomposite part, it requires strong covalent chemical bonding between the filler particles and the polymer matrix, rather than the much weaker van der Waals physical bonds which occur if the carbon nanotubes are not properly functionalized.

Many methods for chemical functionalization of carbon nanotubes exist, which will be effective either at the tips or sidewall of CNTs [56]. The chemical covalent bond approach is proved to be very useful to maintain the stable bonds, and significant efforts have been directed toward the establishment of chemical functionalities on the surface of CNTs with a strong stable bond with the polymer matrix. One of them is the amino functionalization of CNTs that has a high reactivity and can react with many chemicals [56].

4.12.1 Summary

The superiority of the mechanical properties of nanotubes alone does not ensure mechanically superior composites. The composite properties are strongly influenced by the mechanics that govern the nanotube–polymer interface (i.e., strong bonds). The structural strength characteristics of nanocomposite materials greatly depend on the mechanical load transfer from the matrix to the nanotube and the strength of the interface bonds. Another important parameter that engineers must consider is the ability of nanotubes to be distributed uniformly throughout the matrix. That is, nanotubes have the tendency to cluster throughout the polymer (rather than uniformly distributed) and will provide little mechanical reinforcement. Ideally the nanotube should be aligned along the direction of maximum loading to resist the

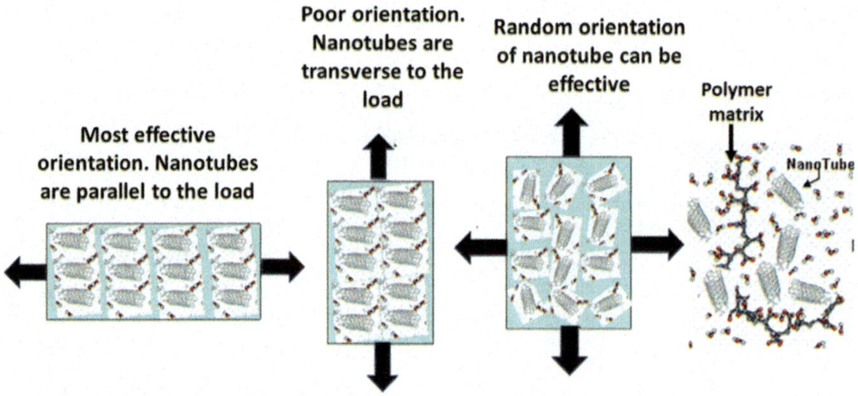

Fig. 4.16 The alignment of nanotubes with respect to the applied load. The most effective orientation is when nanotubes are parallel to the applied load

load, where the probability of failure is maximum. In the case of random alignment of the tubes, most of the nanotubes are not in the direction of loading and when cluster, they act as defects in the composite instead of enhancing the mechanical strength. This may result in degradation of the composite performance. Figure 4.16 shows the proper arrangement of nanotubes in the matrix of nanocomposites. Carbon nanotubes can be aligned during fabrication by mechanical stretching and/or application of electric or magnetic field (exploiting the electrical and magnetic properties of carbon nanotubes). Figure 4.17 shows the ultimate stress value of nanocomposite material with and without the presence of CNTs (0.3 wt% CNTs). In addition, the enhancement of the ultimate after functionalization is shown in this figure [57].

Future work in this area should be concentrated on a few key parameters that can be helpful on having better nanocomposites mechanical properties, (1) a uniform distribution of nanotubes throughout the polymer, (2) a strong interface between the nanotubes and the host material, (3) proper orientation of nanotubes with respect to the applied load, (4) implementation of a cost-effective method of fabricating nanocomposite parts.

4.13 Concluding Remarks

Polymer matrix composites (PMCs) are materials composed of a polymer matrix (either thermoset or thermoplastic) reinforced with continuous or short fibers, or small particles of different geometries. When combined, the reinforcement component enhances the strength and stiffness properties of the final product. The choice of polymer matrix and reinforcement materials allows for tailoring the properties of PMCs to specific applications.

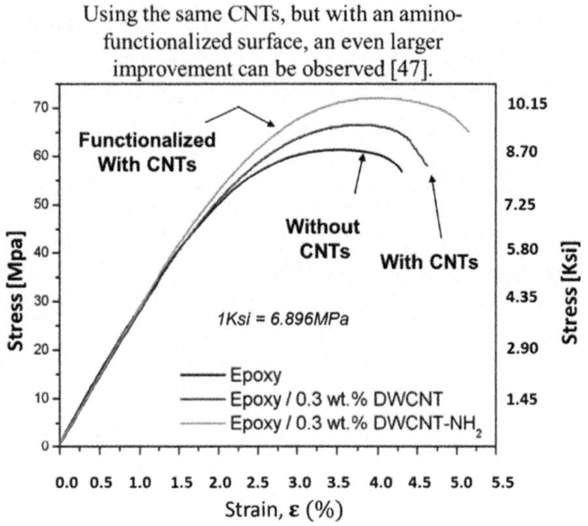

Fig. 4.17 The presence of 0.3 wt% double wall CNTs (DWCNT) leads to an increase in ultimate strength, and strain to failure, of material when compared to the epoxy resin. Functionalizing with CNTs will improve the properties

In this chapter several fabrication techniques that engineers can use for fabricating their parts were discussed. PMCs are attractive in aerospace, automotive, and other industries due to their high both strength and stiffness-to-weight ratio, corrosion resistance, and design flexibility—allowing them to be tailored to meet specific functional requirements. However, their utilization in industry is not as extensive as it could be due to manufacturing difficulties, such as time and cost associated with fabricating consumer goods at reasonable prices. This poses a major issue for industries, highlighting the need for improvements in composite manufacturing technology.

In addition to these challenges, minimizing void content during manufacturing can enhance the mechanical properties of the final product because: (1) voids can be a source of crack initiation and propagation, and (2) voids can allow humidity to weaken the composite.

In conclusion, PMCs have found extensive use in various industries due to their attractive properties and design flexibility. Continued study and improvement in manufacturing processes contribute to further enhancing their properties and lowering cost of these materials, making them important in the development of lightweight and high-performance components for various applications.

The application of nanocomposites in industry is gaining attention due to their unique properties, especially in scenarios where weight is a critical parameter in manufacturing composite parts. Polymer matrix nanocomposites represent a cutting-edge class of materials with promising properties and applications, particularly in aerospace and the automotive industries. These composites consist of a polymer matrix filled with nanoscale reinforcements, such as nanoparticles or nanotubes. The incorporation of nanomaterials into the polymer matrix enhances mechanical properties, opening new possibilities for various applications. While there are challenges related to their fabrication, ongoing advancements, and research in this class of composites are expected to further refine properties and manufacturing processes, ultimately leading to the creation of high-performance and innovative materials for the future.

References

1. Advanced Materials by Design (Part 6 of 18). Princeton.edu. Retrieved 18 Apr 2017. https://www.princeton.edu
2. U.S. Congress, Office of Technology, *Advanced Material by Design*, OTA-E351 (U.-S. Government, Washington, DC, 1988)
3. A. Vlot, *Glare – The History of the Development of a New Aircraft Material* (Kluwer, Dordrecht, 2001), p. 90
4. K. Balasubramanian, M.T.H. Sultan, N. Rajeswari, *Sustainable Composites for Aerospace Applications*, Woodhead Publishing series in composite science and engineering (Woodhead Publishing, Cambridge, 2018), pp. 55–67
5. F. Aird, *Fiberglass & Composite Materials: An Enthusiast's Guide to High Performance Non-metallic Materials for Automotive Racing and Marine Use*, vol 91 (Penguin, London, 1996)
6. S. Mazumdar, *Composites Manufacturing: Materials, Product, and Process Engineering* (CRC Press, Boca Raton, 2001)
7. N. Saba, P.M. Tahir, M. Jawaid, N.A. Ibrahim, Manufacturing and processing of kenaf fibre-reinforced epoxy composites via different methods, in *Manufacturing of Natural Fibre Reinforced Polymer Composites*, ed. by M. Salit, M. Jawaid, N. Yusoff, M. Hoque, (Springer, Cham, 2015)
8. A.F. Avila, D.T.S. Morais, A multiscale investigation based on variance analysis for hand lay-up composite manufacturing. Compos. Sci. Technol. **65**(6), 827–838 (2005)
9. M. Otaka, Hand lay-up molding process. Japanese Patent JP 09,314,686 (1997)
10. K.K. Kar, *Composite Material—Processing, Applications, and Characterization* (Springer, Berlin, 2017), p. 26. Chapter 1
11. Hand Laminate & Vacuum Bag a Simple Carbon Fiber Part, Easy Composites Ltd
12. M. Akay, *An Introduction to Polymer Matrix Composite*, p. 66, Chapter 2 (2015). eBooks at bookboon.com
13. A. Crosky, G. Pearce, C. Grant, D. Kelly, *Advances in Composites Manufacturing and Process Design—Automated Tape Laying* (Elsevier, Amsterdam, 2015), pp. 79–92
14. CW Editorial Team, Fiber placement and tape laying. Composite World (CW), 1 November 2006
15. M. Akay, *An Introduction to Polymer Matrix Composite*, p, 74, Chapter 2 (2015). eBooks at bookboon.com
16. D.H.-J. Lukaszewicz, C. Ward, K.D. Potter, The engineering aspects of automated prepreg layup: History, present and future. Compos. Part B Eng. **43**, 997 (2012)
17. E. Oromiehie, B.G. Prusty, P. Compston, et al., Automated fiber placement based composite structures: Review on the defects, impacts and inspections techniques. Compos. Struct. **224**, 110987 (2019)
18. E. Geoff, *Design and Manufacture of Composite Structures* (Woodhead Publishing Limited, Cambridge, 1994), pp. 273–277
19. H.N. Dhakal, S.O. Ismail, *Sustainable Composites for Lightweight Applications* (CRC Press, Boca Raton, 2021)
20. S.T. Peters, *Handbook of Composites*, 2nd edn. (Chapman & Hall, London, 1997)
21. I. Singh, *Processing of Polymers and Polymer Composites*. An Autoclave Video (Lecture 22)

22. *Introduction to Compression Molding*. eFunda (Engineering Fundamental), March 19, 2013
23. What is compression molding? Coventive Composite, October 1, 2018
24. H.R. Todd, K.D. Allen, L. Alting, *Manufacturing Processes Reference Guide* (Industrial Press, Inc., South Norwalk, 1994)
25. U.S. Patent 133,229, dated 19 November 1872
26. R.K. Meade, H. McCormack, L.T. Clark, G.A. Sclater, L. Lamborn, *Chemical Age* (McCready Publishing Company, New York) Retrieved 27 Apr 2018
27. Y.W. Leong, S. Thitithanasarn, K. Yamada, H. Hamada, Compression and injection molding techniques for natural fiber composites, in *Natural Fibre Composites*, (Woodhead Publishing, Sawston/Cambridge, 2014), pp. 216–232
28. V.M.K. Werner, R. Krumpholz, C. Rehekampff, T. Scherzer, M. Eblenkamp, Thermoplastic encapsulations of a sensor platform by high-temperature injection molding up to 360 °C. Polym. Eng. Sci. **59**, 1315 (2019)
29. J.W. Hyatt, *Fifth of a Series: Pioneer of Polymer Processing*, (Hanser Publishers, Munic, 1837–1920)
30. R.A. Malloy, *Plastic Part Design for Injection Molding* (Hanser Publishers, Munich/Vienna/New York, 1994)
31. D. Verma, G. Joshi, R. Dabral, A. Lakhera, *Mechanical and Physical Testing of Biocomposites, Fibre-Reinforced Composites and Hybrid Composites* (Woodhead Publishing, Sawston/Cambridge, 2019)
32. Y.C. Lam, J. Li, S.C. Joshi, Simultaneous optimization of die-heating and pull-speed in pultrusion of thermosetting composites. Polym. Compos. **24**(1), 199–209 (2003)
33. A. Trivisano, A. Maffezzoli, J.M. Kenny, et al., Mathematical modeling of the pultrusion of epoxy based composites. Adv. Polym. Technol. **10**(4), 251–264 (1990)
34. S.C. Mantell, G.S. Springer, Filament winding process models. Compos. Struct. **27**, 141–147 (1994)
35. C. Hopmann, L. Wruck, D. Schneider, K. Fischer, Automated winding of preforms directly from roving. Lightweight Des. Worldwide **12**, 58–63 (2019)
36. S.W. Beckwith, Filament winding – The string and the glue. Compos. Fab. **14**(1), 8–12 (2004)
37. W. Polini, L. Sorrentino, Winding trajectory and winding time in robotized filament winding of asymmetric shape parts. J. Compos. Mater. **39**(15), 1391–1411 (2005)
38. S. Black, Continuous filament winding: A short history. Composite World (CW), 30 August 2016
39. P. Dumont, L. Orgeas, D. Favier, et al., Compression molding of SMC: In situ experiments, modelling and simulation. Compos. Part A **38**(2), 353–368 (2007)
40. C.L.E. Yen, K.S. Lukas, Preferential curing technique in compression molding of fiber reinforced composites. US Patent 2,006,091,583 (2006)
41. A. Whelan, *Polymer Technology Dictionary* (Springer, Dordrecht, 2012), p. 391
42. AZO Materials, Polyester-SMC; high impact. AZO Materials, 21 August 2001. Retrieved 23 Dec 2018
43. A. Trauth, P. Pinter, K. Wiedemann, Investigation of quasi-static and dynamic material properties of a structural sheet molding compound combined with acoustic emission damage analysis. J. Compos. Sci. **1**(2), 18 (2017)
44. H. William, J. Young, Session 5-C, in *SPI/CI International Conference and Exposition 1998*, (CRC Press, Boca Raton, 1998), pp. 1–9
45. S.K. Mazumdar, *Composites Manufacturing Materials, Product, and Process Engineering* (CRC Press, Boca Raton, 2002)
46. E. Baur, S. Brinkmann, T. Osswald, E. Schmachtenberg, *Seachtling Kunststoff Taschenbuch 30* (Carl Hanser Verlag, Ausgabe/München, 2007)
47. B. Davis, P. Gramann, T. Osswald, A. Rios, *Compression Molding* (Carl Hanser Verlag, München, 2003)
48. CW Editorial Team, Material & processes fibers for composites. Composite World (CW), 1 July 2022

49. S. Iijima, Helical microtubules of graphitic carbon. Nature **354**, 56–58 (1991)
50. D. Qian, G.J. Wagner, W.K. Liu, M.-F. Yu, R.S. Ruoff, Mechanics of carbon nanotubes. Appl. Mech. Rev. **55**, 495 (2002)
51. E.T. Thostenson, T.-W. Chou, On the elastic properties of carbon nanotube-based composites: Modeling and characterization. J. Phys. D. Appl. Phys. **36**, 573–576 (2003)
52. M. Doyle, Chapter 9: Predictive modeling, in *Virtual Testing and Predictive Modeling*, ed. by B. Farahmand, (Springer, New York, 2008), p. 248
53. B. Farahmand, Study of nanoparticles interface with aerospace polymers. 14 Boeing Technical Excellence Conference (BTEC14), February 2008
54. R. Dubey, D. Dutta, A. Sarkar, P. Chattopadhyay, Functionalized carbon nanotubes: synthesis, properties and applications in water purification, drug delivery, and material and biomedical sciences. Nanoscale Adv. **3**, 5722–5744 (2021) Received 21 April 2021
55. J.-P. Salvetat, G.A.D. Briggs, J.-M. Bonard, R.R. Bacsa, A.J. Kulik, T. Stöckli, N.A. Burnham, L. Forró, Elastic and shear moduli of single-walled carbon nanotube ropes. Phys. Rev. Lett. **82**, 944–947 (1999)
56. S. Wang, Z. Liang, T. Liu, B. Wang, C. Zhang, Effective amino-functionalization of carbon nanotubes for reinforcing epoxy polymer composites. Nanotechnology **17**, 1551 (2006)
57. F.H. Gojny, M.H.G. Wichmann, B. Fiedler, K. Schulte, Fundamental aspects of nano-reinforced composites. Compos. Sci. Technol. **66**, 3115–3125 (2006)

Chapter 5
Metal Matrix Composites (MMCs) (Methods of Manufacturing MMC Parts)

5.1 Introduction

Composite materials and related manufacturing techniques are developing rapidly to keep up with demands in various industrial sectors, mainly aerospace, defense, and automotive. Nowadays, new fibers, new matrices, and innovative manufacturing processes are available that can reduce manufacturing time and cost, which are essential to industries for various applications. In this chapter, metallic matrix composites will be discussed. Methods of manufacturing and fabricating parts that can be useful to industries are presented. Later in Chap. 6, the ceramic matrix composite will be described in detail.

Pure metals usually do not have applications in industries because they are soft, not machinable, possess low stiffness, low yield, and low ultimate values. However, when metals are mixed with other elements, such as steel, which is made of iron with a few percent carbon, their mixtures can cause iron's properties to improve significantly. That is, they will have better strength and stiffness, as well as possess better machinability, and weldability, which can be used to manufacture superior products for automobile, space, military, and aircraft structures. However, an extreme service environment presents a challenge that provides an opportunity for engineers to produce products made of unique materials that can survive the harsh environment. For example, in the near-earth orbit, typical spacecraft is exposed to phenomena such as vacuum, thermal radiation, atomic oxygen, ionizing radiation, plasma, and the high reentry temperature, where the friction between the space vehicle and the earth's atmosphere can cause the temperature to rise as high as 3092 °F (1700°C). Other factors, such as high-velocity impact from the incoming space debris, demand more robust material to withstand the damage due to powerful impact in the orbital environment. As an example, the International Space Station (ISS), during its design life of almost 15 years, will undergo many thermal cycles from +257 °F (+125 °C) to

B. Farahmand, *Fundamentals of Composites and Their Methods of Fabrications*,
https://doi.org/10.1007/978-3-031-80201-0_5

−257 °F (−125 °C) and is exposed to the incoming space debris as it moves around the Earth's orbit every 90 min [1, 2]. All the above factors are critical to the structures during their service usage, which demands the engineers design more robust material with better thermal, mechanical, impact, and corrosive properties when compared with metallic alloys, but the cost and time of fabrication of parts are essential factors to be considered by industries.

The traditional metal alloys are not good enough to be used when the working environments demand even higher strength, stiffness, and better resistance to high temperature and corrosive environments. So, to improve these properties, high strength materials can be used both as reinforcements, as well as to be employed for matrices. For example, by choosing metals, such as nickel, as the matrix with ceramic fibers as the reinforcement, it can be fabricated to have excellent mechanical properties for high temperature environments. These newly fabricated materials fall into the category of composites, and just like polymer matrix composites (PMCs), they are called metal matrix composites (MMCs) [3–5]. In the case of MMCs, the metal matrix, such as aluminum, copper, nickel, magnesium, and titanium or their alloys, with strong fibers that can be used as the reinforcement in the metal matrix can have outstanding resistance to high temperature environments. It is worth mentioning that from a matrix perspective, the metal matrix and ceramic matrix differ from each other significantly. However, their properties (i.e., MMCs and CMCs) can be close if the percentage of the metallic matrix and the reinforced ceramics or the ceramic matrix with the metallic reinforcement are close. Figure 5.1 shows the overall properties of the two matrices, indicating they are different. Still, proper reinforcement in each matrix can enhance their properties greatly [6].

In this chapter, the emphasis will be given to metal matrix composites and in Chap. 6, the ceramic matrix composites will be fully addressed.

METALS MATRIX	CERAMIC MATRIX
➢ Crystal structure	➢ Crystal structure
➢ Metallic bond	➢ Covalent bond
➢ Good conductor	➢ Poor conductor
➢ High tensile strength	➢ Poor tensile strength
➢ Good ductility	➢ Poor ductility
➢ Good impact resistance	➢ Poor impact resistance
➢ Moderate corrosion resistance	➢ Great corrosion resistance
➢ Moderate hardness	➢ Extremely hardness
➢ Nonporous	➢ High porosity
➢ High density	➢ Low density
➢ Moderate temperature resistance	➢ High temperature resistance

Fig. 5.1 The differences in properties between metallic and ceramic matrices are illustrated in this figure

5.2 Metal Matrix Composites (MMCs): MMC Fabrication Methods

The main idea of using metal matrix composites (MMCs) is to replace conventional metallic alloys with their inadequate properties with MMCs for increasing their applications in industries. Incorporating various reinforcing materials, such as particulates, whiskers, or short and continuous fibers in the metallic matrix, can be the key to pushing forward the development of MMCs, especially in the aerospace, military, and automobile industries. In addition, metal matrix composites have high specific stiffness and strength which could be used in applications in which saving weight is an essential factor. In order to make MMC materials more commercially attractive, cheaper and faster processing techniques should be given priority.

The MMCs should possess both the ductility and toughness of parent metal but with higher strength, stiffness, and temperature resistance that is provided by the reinforcement materials [7]. In Chap. 4, several different methods were presented related to the manufacturing processes of polymer matrix composites (PMCs). In this chapter, several MMC fabrication techniques will be discussed, and the fabricated products will be able to replace the inefficient conventional metallic materials because of their enhancement in mechanical properties. The manufacturing engineers must conduct a reasonable amount of testing and analysis on the material to verify the superiority of MMCs in strength, stiffness, toughness, temperature resistance, fatigue, fracture, and wear resistance before utilizing them for industrial use.

Utilizing ceramics or high-temperature resistant metallic alloys, as reinforcement, are key to the sustainable development of MMCs. The phase of reinforcement in the metallic matrix composites can be present in the form of high strength metallic whiskers, particulates, or metallic oxides. With respect to the types and dimensions of reinforcements, they can be, (1) continuous or long-fibers, (2) discontinuous or short fibers and (3) particulate (particulates are small particles of metals, ceramics, or polymers). Because of MMCs'usefulness in industries, more attention has been given to their manufacturing processes, where a great process can contribute to better product performance. Currently, Aluminum, Magnesium, Copper, Titanium, Cobalt, and Cobalt-Nickel are used in the form of a matrix, and SiC, TiC, TiO_2, TiN, CBN, ZrO_2, Si_3N_4 are all commercially produced and used as reinforcements [8].

Even though the MMCs appear to be attractive from their superior thermal properties, possessing high modulus, high strength, good wear resistance, and good chemical inertness, other vital factors must also be given a good amount of attention. As was mentioned before, their key essential properties, such as their feasibility of manufacturing parts, the time required to fabricate parts, good processability, and economic efficiency are highly desirable for customers. Indeed, the high production rate of MMC parts provides profit to the company. Figure 5.2 shows several types of manufacturing techniques that are currently being used to produce parts for metallic material. Some of these methods can also be utilized for manufacturing MMC parts. Note that some of these methods already were discussed in Chap. 4 for manufacturing polymer matrix composite parts.

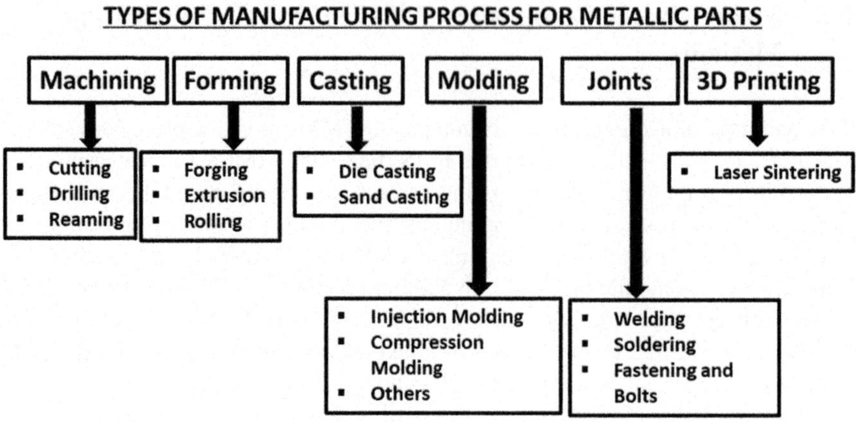

Fig. 5.2 Different types of manufacturing methods for making metallic parts. Some of these methods are also applicable to composites

In Fig. 5.2, machining, forming, casting, molding, joining, and the 3D printing are shown as a method that is currently used in industries for manufacturing metallic parts. Each manufacturing type has its own methods of processing parts. For example, the forming can be performed by forging, extrusion, and rolling. In the case of joining parts together, one can have welding, soldering, fastening, and bolting joints. Each of these methods will have its own subcategories. For instance, welding techniques can be fusion welding, laser welding, friction-stir welding, and several other welding techniques that the reader can become familiarized with the process through the literature. Some of the manufacturing techniques shown in Fig. 5.2, can also be used in the manufacturing of polymer matrix composites. Note that 3D printing, or additive manufacturing is another technique that is currently being used in many industries for fabricating both metallic and composites parts (see Chap. 15 for more detailed discussion on this topic).

In Fig. 5.3, reinforcement types used in the metal matrix composites are shown, which are in the forms of particles, short fibers or whiskers, and continuous fibers (including the sheet as the reinforcements in the metallic matrix) to provide greater properties to the MMCs [9]. Particulates have rough, equiaxial geometry with spherical, angular, or plate-like shapes. They are normally around 1 mm in diameter. The particulates comprising of SiC (Silicon Carbide) or Al_2O_3 (Alumina or aluminum oxide) are the materials of choice for the reinforcement of the metal matrix composites. Materials such as W (Tungsten), WC (Tungsten Carbide), B_4C (Boron Carbide), BN (Boron nitride), TiB_2 (Titanium Boron), TiC (Titanium Carbide), SiO_2 (Silicon Oxide), ZrO_2 (Zirconium Oxide), etc. are also desirable to utilize for reinforcement use in manufacturing MMCs parts [10].

MMCs composite materials become favorable materials that can be used when the working environments are coarse, such as high-load, high-wear/friction, and high-temperature conditions. For those reasons, it is essential to know methods of manufacturing these valuable materials to provide parts for industries.

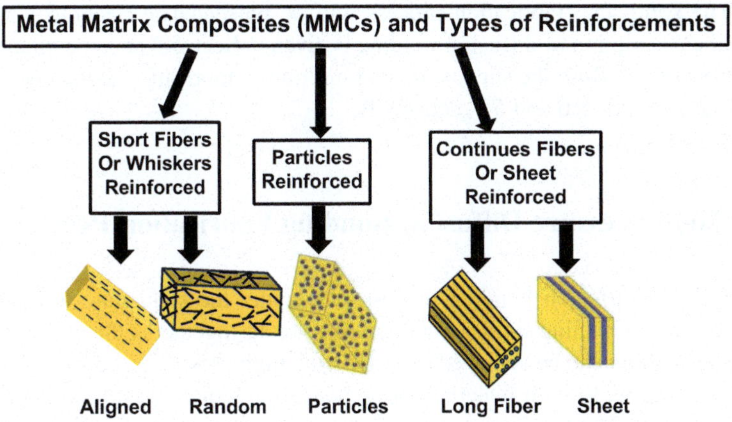

Fig. 5.3 Illustrating metal matrix composite with different types of reinforcements

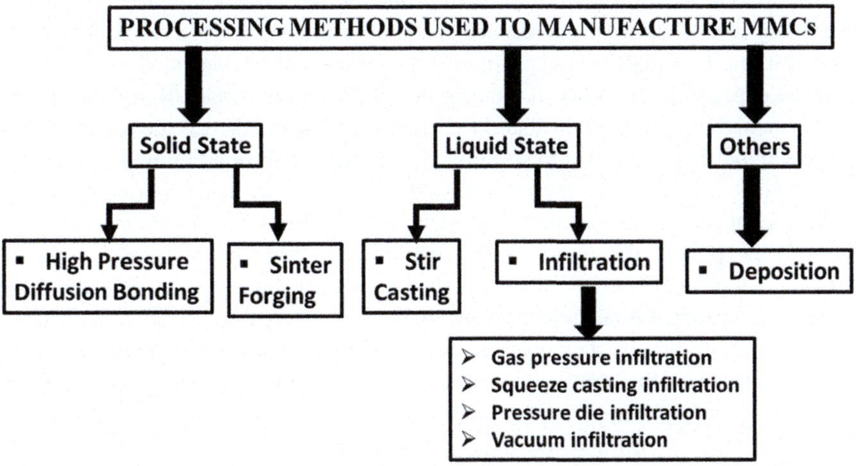

Fig. 5.4 Different methods of manufacturing metal matrix composites

Metal matrix composites can be produced by several methods, but the focus of the selection is to choose the best combination of the host matrix alloy and reinforcement. However, by changing the manufacturing process when using the same metal alloy and reinforcement, it is possible to obtain different mechanical properties that can be below the expectation level for industries' use.

Three processing methods have been primarily used to develop MMCs. They fall into a solid state, a liquid state, and others, as described in Fig. 5.4. The high-pressure diffusion bonding and sinter forging techniques are under the solid method [11–13], and the stir casting and infiltration processes are under the liquid state fabrication process [14]. Finally, the deposition MMCs process falls under others

[15]. The solid-state fabrication process will be presented in Sects. 5.3 and 5.4, where the description of diffusion welding is given in Sect. 5.3 and sintering in Sect. 5.4, respectively. Both the stir casting and infiltration fabrication techniques will be discussed in Sects. 5.6 and 5.7, separately.

5.3 High-Pressure Diffusion Bonding Fabrication Process

High-pressure diffusion bonding process falls under the solid-state welding technique, capable of joining similar and dissimilar metals together at the molecular level [16]. The welding can be achieved by employing high pressure, in combination with high temperature, to both materials to be bonded together without applying adhesives (Fig. 5.5). The act of pressure and heat makes the molecules diffuse between the two materials (the matrix and the reinforcement) at the contact areas to take place. This fabrication process is accomplished mainly by applying pressure at elevated temperatures, around 50–75% of the material melting temperature, and is widely used in the joining of high-strength and high melting point materials, such as refractory metals within the aerospace and nuclear industries. As a result of heat and pressure, the two materials expand, and the amount of expansion depends on the thermal expansion capability of the particles of the two materials. Refractory metals that have high resistance to heat and wear are the best candidate as the reinforcement for metal matrix composites to be fabricated through diffusion bonding. The refractory metals with good temperature resistance are shown in the periodic table as niobium, molybdenum, tantalum, tungsten, and rhenium with a melting point above 2500 °C [17]. They also possess high hardness and good wear resistance characteristics at room temperature. They are chemically inert and have a relatively high density. However, it is necessary to note that when heating the two materials A and B for establishing bonds, their melting point should not be too far apart from each other. Otherwise, the joint will not go through proper procedures when using

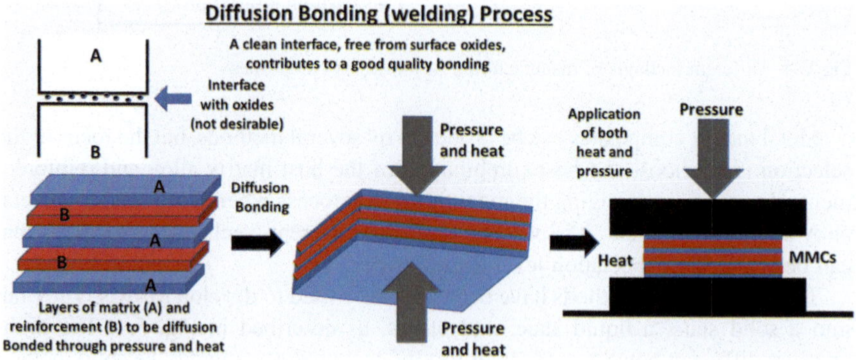

Fig. 5.5 Illustrating the diffusion bonding (welding) for metal matrix composites. Here, the matrixes are shown with (**a**) and reinforcement plies with (**b**)

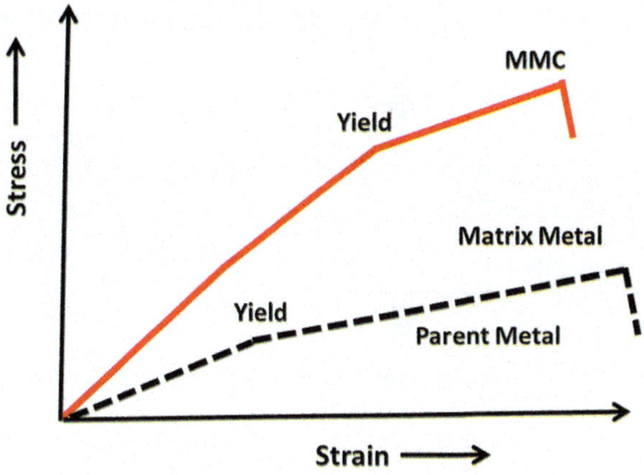

Fig. 5.6 Illustrating the typical stress-strain curves for the matrix metal and the MMC. As it can be seen, the MMCs have better strength and yield, but lower ductility. Fiber has higher strength with much less ductility

diffusion welding. In other words, the 50% melting point of reinforcement should be less than the melting point of the metallic matrix such as titanium or aluminum. That is, the host metal that has a lower melting point should not melt during the diffusion welding process. Finally, when the bonding process is completed, it is a good idea for the joint to be tested for strength using tensile testing along and across the jointed area. The testing is performed because a sound joint in the diffusion bonding process is directly related to the condition of the two jointed surfaces. The testing of the diffusion bonding joint must be performed based on a well documented standard. Note that, the presence of metal oxides at the interface of jointed area can weaken the joint and shorten the life of MMC parts and should be taken into consideration.

The mechanical properties of a good joint should be as good as parent materials or better (Fig. 5.6). This figure illustrates that both the yield and ultimate, as well as the stiffness for the MMCs are higher than the matrix metal.

To create a good bond between the two surfaces at the interface, just like the two plates A and B shown in Fig. 5.5, the interface status plays a big role and must be clean and free from metallic oxides. Metallic oxides will form easily on the surface of the metal matrix when exposed to the air and must be eliminated prior to the diffusion welding process. Therefore, the preparation of the two surfaces before mating with each other for diffusion welding can be effective for having a good joint (Fig. 5.7).

The presence of metallic oxides at the two surfaces, before diffusion welding, can create stress concentration sites and eventual crack initiation and growth as shown in Fig. 5.7, which can weaken MMC materials during their service usage. However, when titanium and aluminum alloys are chosen as a matrix, often oxide particles can be dissolved and diffused away from the bonding surfaces at high temperatures.

For titanium at 1634 °F (890 °C) and aluminum at 977 °F (525 °C) (both below the melting point), when both are under pressure and heat, and with an adequate amount of time, will cause oxide particles to be dissolved and diffuse away [18]. Below this temperature, these two materials may need more time for the oxide particles to dissolve away.

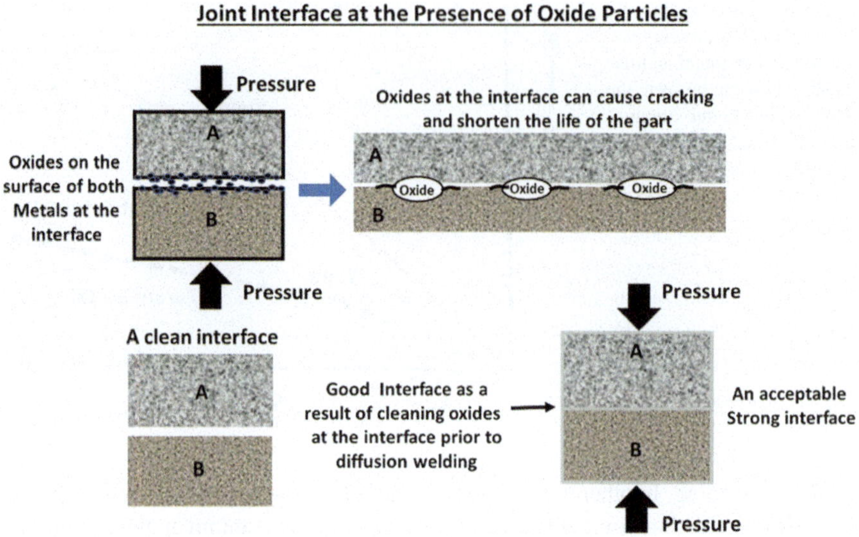

Fig. 5.7 Illustration of the presence of oxides particles between joints. A clean interface can produce a strong interface when subjected to pressure and heat during the diffusion welding process

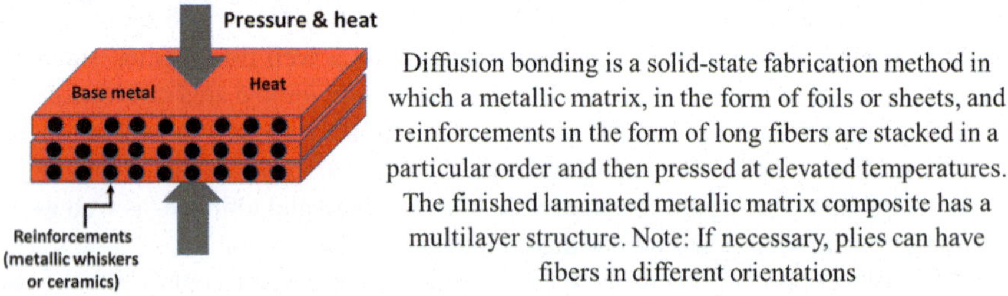

Fig. 5.8 Illustrating diffusion bonding of a base metal with ceramic or metallic whiskers as reinforcements

5.3.1 Interface Bond Between the Metallic Matrix and the Reinforcement

The bonds at the interface between the metal matrix and the reinforcement plates must be solid and undamaged to be able to act as a load-carrying element. In Fig. 5.5, diffusion welding was shown, where several plates were mating with each other by applying both pressure and heat. However, the matrix plate can also have reinforcement in the form of long fibers as illustrated in Fig. 5.8. In this case, an excellent interfacial bonding between the fibers and the metallic matrix should be obtained to have a useful MMC material when parts are under high working stress. It does not matter if the reinforcement is ceramic, metallic, or any other high-temperature resistance material, the interface must be robust. A weak interface will not provide a sound composite, as is expected by the engineers, because the lack of a solid bond might cause premature failure of MMC parts, through the anticipated separation of plates. In Fig. 5.8, the stack of metals with reinforcements is shown to be uniform and the bond between the matrix and fiber is established through pressure and temperature with no adhesive material added. Plies shown in Fig. 5.8 can have fibers in different orientations to have better mechanical properties.

Engineers can design MMC parts in many ways so that the outcome should be best beneficial to the users. For example, the arrangement of plates with fibers can be different throughout the thickness of the part. That is, the first stack could be just a pure metal with no fibers, the second stack could be made of the base metal with reinforcements having a high fiber volume fraction, V_f, with a given orientation, and the third stack could have the base metal with low volume fiber. Or, the engineer may like to have several identical metallic layers diffused together with fibers in different orientations to get the best mechanical properties.

This is similar to the case of laminated polymer matrix composites, where plies had different orientations in order to have greater mechanical properties with respect to the applied load. In addition, engineers may like to fabricate their part through diffusion bonding by choosing metallic plates with reinforcement having different types of metals. For example, the top layer can be an aluminum alloy and the rest of plies with a different metallic material. So, it is up to engineers how they want to manufacture their parts by applying the diffusion bonding process with pressure and heat as the key parameters in order to get the best interface joints and a solid manufacturing result.

In the case of filament in the matrix, a strong interface bond can be achievable by coating the dispersed fiber particles with a proper coating material. The proper coating not only can strengthen the interfacial bond between the two elements but also prevents chemical interaction between the fibers and the matrix, especially at high elevated temperatures. This is called functionalization, where element C (the coating) will provide a strong bond between both matrix A and reinforcement B, such that a good locking between the ceramic particles and metallic matrix will be established. A good bond between the ceramic fibers and the metallic matrix can also be achieved by adding another element that would be friendly to both elements A and B so that it would act as a functionalized element.

5.3.2 Summary of the Diffusion Welding Process

When two or several plates of a similar crystalline structure are bonded together through diffusion bonding (Fig. 5.5), the first step is to clamp plates to each other. Still, before the clamping stage, the interface surfaces must be machined to be as smooth as possible and kept as free from chemical contaminants as possible, such as oxide particles. Any surface oxide material between the two metallic surfaces may prevent proper diffusion of material, hence lower mechanical properties (Fig. 5.7). Once the pieces are clamped, pressure and heat are applied to the components, usually for a predetermined amount of time until the process is completed. The surfaces are heated in a furnace or via the electrical resistance method. Pressure can be applied using a hydraulic device with a proper gage reading. This method allows for exact measurements of applied compressive load on the parts. Alternative methods for applying pressure include dead weights, differential gas pressure between the two surfaces, and high-pressure autoclaves. Diffusion bonding must be done in a vacuum or inert gas environment, especially when using metals that have a strong tendency to develop oxide layers such as copper and aluminum in the presence of oxygen. Improper interface bonding can have influences on composite tensile and compressive strength, stiffness, fatigue and fracture properties, and the inability to transfer load from matrix to fiber, which is an important parameter to be considered. Environmental corrosion can also happen when moisture penetrates through the interface and causes metallic corrosion. Finally, diffusion welding does not require relative motion in any welded parts. The two parts should maintain their position while the pressure is applied to the two plates. This process is a useful technology that enables manufacturers of products to fabricate complex structures that are difficult to produce by machining technique.

5.4 MMCs Fabrication Methods Through Sintering and Powder Metallurgy

Sintering is a heat treatment process in which a large quantity of loose aggregate material is subjected to a sufficiently high temperature and pressure to cause the loose material to become a compact solid lump for use in industries. The process involves mixing the powder of matrix metal with the powder of reinforced material to be sintered and compacting for the subsequent steps such as heating in the solid state, below the melting point. This process involves the atoms in materials diffusing across the particle boundaries of the mating particles and fusing them together into one piece. Figure 5.9a illustrates the sintering process in which particles are first going through pressure (compacting) and subsequently, heat is applied below the melting point of the material to minimize the voids in the processing of the part, Fig. 5.9b. The temperature will be ~50% of the melting point which is considered the sintering temperature point for many materials. Under this condition, the heat and pressure are enough to reduce the porous spaces between the material's particles and squeeze them together into a useful solid chunk. After the completion of the sintering process, the finished products may go through the second operation if it is needed. These secondary operations can be sizing the final product, forging, machining, grinding, and polishing, if necessery (Fig. 5.9).

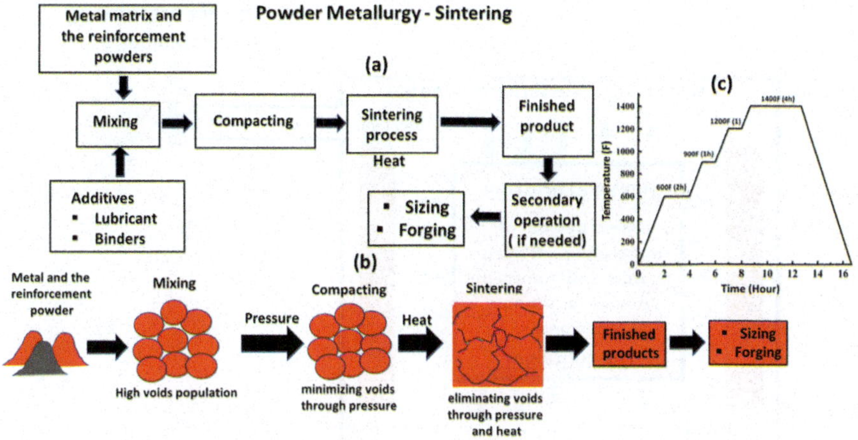

Fig. 5.9 (**a**) Flow chart for the powder metallurgy and sintering process. (**b**) The application of pressure and heat on both the metal matrix and the reinforcement to obtain the finished products. Sintering is used to manufacture MMC parts without being near the melting point of material. (**c**) The stepwise temperature rise and cooling down to room temperature by turning the furnace off as a function of time

5.4.1 Temperature and Pressure Adjustments

If the manufacturer is not interested in going as high as ~50% of the melting point of the material as was mentioned before, then for the product to have acceptable properties, the pressure must be increased. The same is true with respect to pressure. For reducing the pressure, then the temperature must increase above 50% to as high as 70% of the melting point to compensate for the reduction of pressure to minimize the amount of porosity. For powders that are prone to oxidation on the surface, the application of a high temperature sintering can be performed in a nonoxidizing environment.

The temperature rise in the sintering process must be a stepwise process as shown in Fig. 5.9c [19, 20], where the preheating of the powder material is performed, then the temperature rises slowly to its maximum value near the sintering temperature of the material, and finally, the cool-down step will take over by turning the furnace off and letting it cool slowly to room temperature. Otherwise, fast cooling may introduce residual stresses in the part and cause cracking when the load is applied to the structure.

To obtain uniform hydrostatic pressure all around the powder material that is supposed to be used for the sintering process, that is, to transmit pressure equally to the part in all directions, a special device is needed to apply pressure while the heat is implemented on the particles (Fig. 5.10). In this figure, the inert gas, such as helium, is used to pressurize the tank, where the container with powder material will be

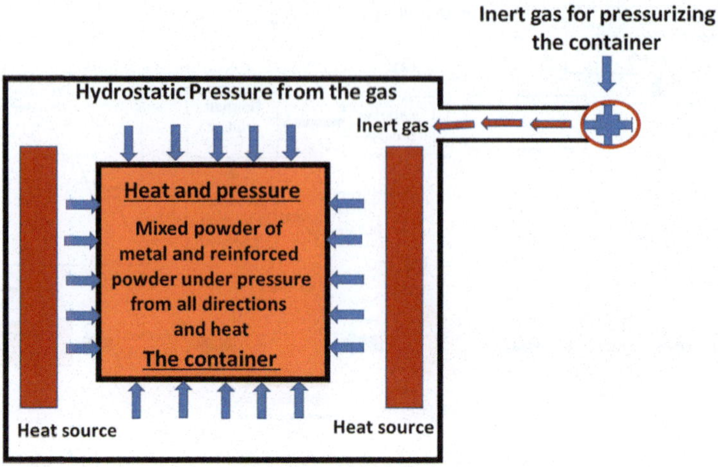

Fig. 5.10 A sintering container. Mixed powder material (matrix metal and reinforced powder) inside the container under pressure and heat. Inert gas is used for hydrostatic pressure (pressure from all directions)

subjected to hydrostatic pressure from all directions. Inside the tank, heat is generated that would allow the temperature to rise slowly to its maximum value near the sintering temperature of the material set by the engineers (Fig. 5.10).

The material involved in the sintering process can be metals, ceramics, plastics, or other desirable materials suitable for the products. Because some materials have a high melting point when going through the casting process, such as the refractory metals with a high melting point, the sintering process can be an ideal alternative manufacturing method, mainly because the process does not require reaching its melting point, only up to 50% of the melting point of the material. Note that melting metals are energy and labor-intensive, especially for materials with a higher melting point. Even though in the sintering process both heat and pressure are required, the total energy, labor, and time consumption can be less for sintering when compared with melting material where the casting process is used.

5.4.2 *Post Sintering Process*

There are several types of post-sintering processes that are available after parts have been manufactured and are ready for application. The post-sintering process for fabricating parts depends on the material being joined, whether they are metallics, ceramics, or plastics.

In the case of MMCs post sintering process, the material can be deformed to the desired shape either by forging, pressing, drawing, or extrusion process (Fig. 5.9). These post processes could be implemented at low temperatures, when the material is below the recrystallization temperature, or hot at above the recrystallization temperature. In Sect. 5.5, an example of a processing method of fabricating aluminum metal matrix with the silicon carbide particles (SiC) from powder material to the final part ready for application in industries will be presented and results will be compared with the host metal.

5.5 Comparison of Metal Matrix Composites with the Host Metal

In manufacturing parts with metallic matrix composites using the mixed powder techniques, the matrix is mostly chosen as aluminum, copper, magnesium, or titanium, where these metals show the best to be used in making structures when high resistance material is the material of choice. In this discussion, the emphasis will be given to aluminum as the host metal. Once aluminum is reinforced with fine silicon carbide particles (SiC), with particles of less than a millimeter in size, the mixed powder is ready for the sintering process as illustrated in Fig. 5.11 [21]. The MMCs are produced via a powder route by simply blending the silicon carbide with metal-based aluminum.

The mixed powders are consolidated in the solid state by applying heat around 50% of the aluminum melting point (around 330 °C) under isostatic pressing to fuse small particles into a dense bulk (Note: Both hydrostatic and isostatic pressures

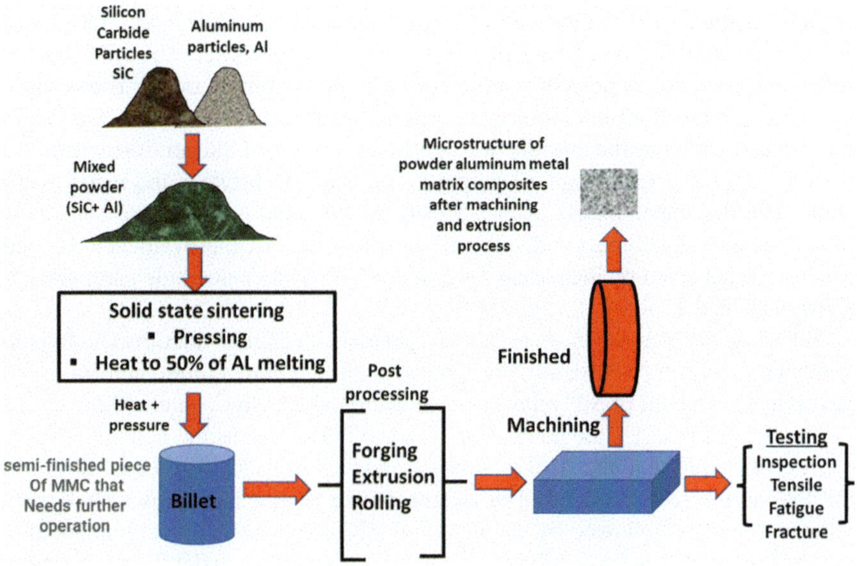

Fig. 5.11 Aluminum metal matrix composites process route. From raw powdered material to final manufacturing operation, post processing, and testing for a reliable service usage

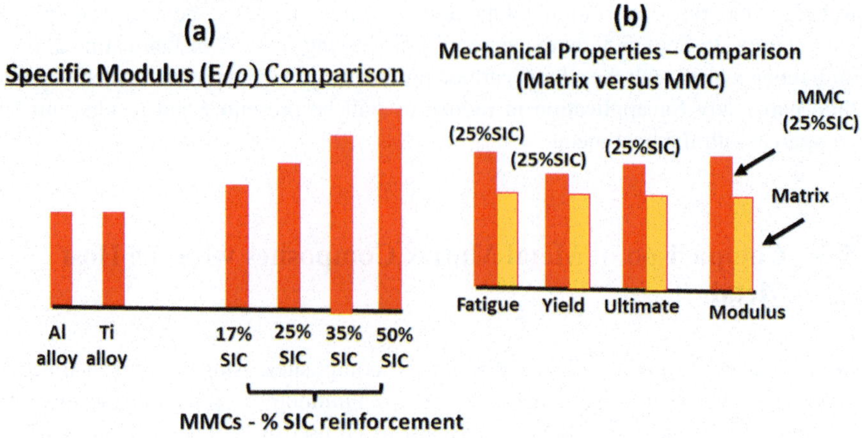

Fig. 5.12 Comparison of (**a**) specific modulus and (**b**) mechanical properties of the metallic matrix and the MMC. The presence of SiC will enhance the MMCs mechanical properties [22]

contribute to the overall densification and quality of the final sintered product). After sintering operation, the composite material in the form of the billet is produced, which can be processed further by conventional means such as forging, rolling, or extrusion and subsequent machining as shown in Fig. 5.11. Prior to machining process, nondestructive tests, and mechanical testing, such as tensile testing, fatigue, and fracture related tests, all through the ASTM testing standards, should be conducted in order to check the integrity of the final products. In Fig. 5.12a, the specific modulus (E/ρ) comparisons between aluminum and titanium alloys with different percent reinforcement (SiC) in the aluminum matrix are shown. From this figure, the presence of percent reinforcement in the aluminum matrix is causing to elevate E/ρ ratio. For aluminum metal matrix composites, the ratio of E/ρ can be more than doubled by the introduction of 50% by volume of SiC reinforcement [22].

In Fig. 5.12b, the mechanical properties comparison between the metal matrix alone and the metal matrix with 25%SiC as the reinforcement are shown for properties such as fatigue, yield, ultimate, and modulus. An improvement in strength can also be achieved by increasing the amount of fine silicon carbide particulate, as shown in Fig. 5.12b.

However, the increase in strength may lead to a reduction in ductility and fracture toughness, so a compromise must be reached on the amount of reinforcement, where the optimum amount of SiC reinforcement is around 25% by volume [22].

Finally, in Fig. 5.13, the number of cycles to failure versus the maximum stress, σ_{max}, is plotted for two matrices (aluminum and titanium alloys) and the MMC with the SiC as the reinforcement. The fatigue failure of titanium alloy shows better fatigue properties, followed by the aluminum SiC composites. The aluminum alloy used as the matrix has less resistance to a cyclic environment under the stress ratio of $R = -1$. Therefore, the introduction of fine silicon carbide particulate to the aluminum matrix can significantly increase the fatigue life and other mechanical

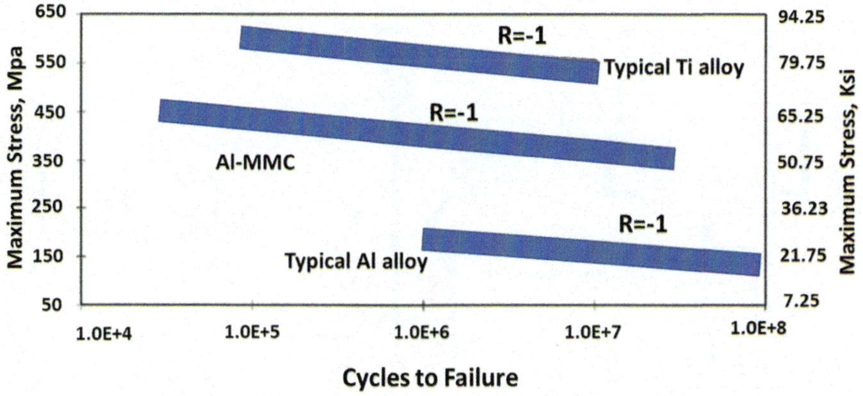

Fig. 5.13 Comparison of fatigue properties between the Al-MMC with 25% SiC and the typical aluminum and titanium alloys under the stress ratio of $R = -1$ [22]

properties as shown in Fig. 5.12. The main reason for the increase in the life of MMCs over the host metallic matrix is the presence of strong fibers in the matrix that would act as the crack stopper, hence an increase in the number of cycles to failure. However, it should be also mentioned that the increase in fatigue life has a direct relationship with a strong bond between the matrix and the reinforcements. The data presented in Fig. 5.13, corresponds to the average values, and the scatter is not shown here.

5.6 Liquid State Stir Casting: Method of Fabricating MMCs

In this metal matrix composite fabrication process, a flow chart is shown in Fig. 5.14a, where the reinforcement (i.e., ceramic particles, also known as the dispersed phase) is added in the solid form of loose powder to the molten metal matrix. When ceramic is used as the reinforcement, it exhibits poor wetting (i.e., poor interfacial bonding) in the liquid metal, so the mechanical force is needed to incorporate the particles uniformly into the molten metal, and this is usually performed through stirring blade while the mixed material is inside the oven, Fig. 5.14b. The melt stirring method is currently the most inexpensive and cost-effective method and leads to the conventional production of material by utilizing secondary processes such as casting followed by extrusion, rolling, forging, and machining for the final fabrication of desired parts [23, 24]. The final liquid composite material is cast and then solidified in a specific mold or ingot by conventional casting methods for further applications.

In the preparation of MMCs by the stir casting method, three key factors contribute to the quality of final products: (1) uniform distribution of reinforced

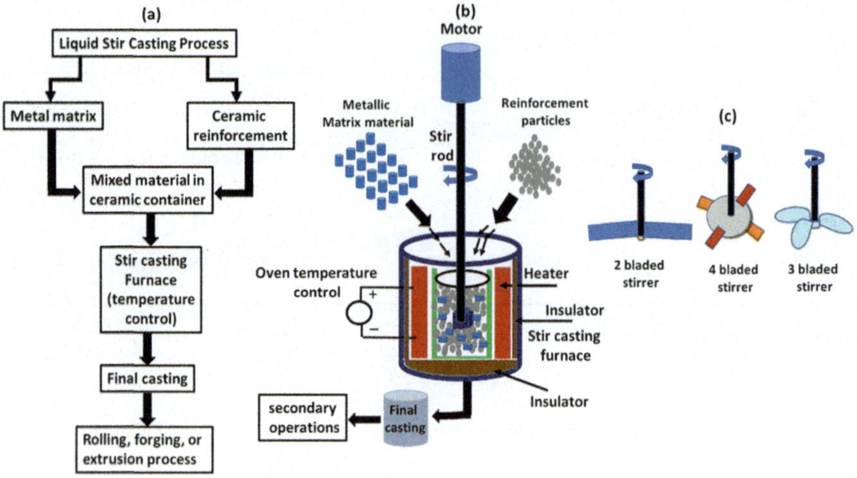

Fig. 5.14 (**a**) Flow chart for the liquid stir casting. (**b**) Device for melting the metal matrix and stirring the reinforcement material. (**c**) The homogeneity of MMCs depends on the stirring speed, geometry, and time

material in the matrix, (2) wettability between ingredients, and (3) minimization of voids between the matrix and the reinforcement material should be given attention so that good mechanical and metallurgical properties can be obtained.

Processing parameters for obtaining good quality MMCs are (1) the stirrer design (the device that does stirring), (2) the rotational speed of the stirrer, (3) stirring time, (4) preheat temperature of reinforcement, (5) the temperature of mold (preheat), and (6) pouring of mixed melt should be properly controlled. Figure 5.14c shows examples of three different types of stirrers having different numbers of blades, which can be used for stirring the mixed melt for better wetting and having a more significant quality product.

To provide a high level of mechanical properties of the composite, good interfacial bonding (wetting) between the dispersed phase and the liquid matrix should be obtained. Wetting is the ability of a liquid to maintain contact with a solid surface, resulting from intermolecular interactions when the two are brought together. Wetting improvement may be achieved by coating the dispersed phase particles (i.e., ceramic fibers). The proper coating prevents chemical interaction between the dispersed phase and the matrix.

Stir casting is characterized by the amount of dispersed phase that can be incorporated into the molten matrix, which is limited to around 30% of the total volume. Three problems must be given special attention in using the stir casting process, which they are (1) the distribution of dispersed phase throughout the matrix is not uniform and homogeneous, (2) the cluster of fibers in the metallic matrix can occur and should be prevented, and (3) the segregation of the dispersed phase on the bottom of the casting container, which can happen due to a difference in the density of the ceramic versus the metal matrix. These three problems can be eliminated through a good design in the stirring process.

5.7 Liquid State Infiltration Casting

The liquid infiltration casting processes are used to fabricate MMC parts. Its concept is based on the infiltration of molten metal alloys through a porous preform of ceramic particles [25–27]. A preform is simply the collection of short fibers, which could act as infiltrating sites for the liquid metal. Infiltration has been used for many years in powder metallurgy to strengthen porous steel parts with a suitable ally such as copper. However, in MMC fabrication through the liquid state infiltration process, there is a need to force the liquid metal into the reinforcement or vice versa. That is, liquid matrix metals do not immediately wet the ceramic particles (to be used as the reinforcement) and need external pressure to be applied to force the liquid metal into the preform. The external force can be applied using an inert gas or a mechanical device. This is called the infiltration technique, which will be discussed later. Another method of bonding the dispersed phase and metallic matrix liquid to achieve the required mechanical properties is through wetting agents. It means that ceramic particles should get wet with molten metal adequately not to expel the particles from the molten metal. Since most ceramics are not getting wet by molten alloys, adding wetting agents or coating the ceramic particles before mixing seems essential.

The parameters that can contribute to good-quality infiltration products are the initial composition of the matrix and the reinforcement, volume percentage of reinforcement, temperature of the reinforcement, and the magnitude of the external force applied to the metal, noting that in some cases, there is no need to force the liquid metal into the preform. All it may be necessary is a good stirring of the mixture to produce a good-quality MMC material, where the molten metal wets the reinforcement with no difficulty.

Infiltration is classified according to the nature and magnitude of the motive force applied during the infiltration process. The external forces can be: (1) no external force, (2) vacuum-driven infiltration, (3) pressure-driven infiltration, and (4) other forces. In the following sections, only the no force, gas pressure, and force-driven infiltration techniques will be discussed.

5.7.1 Liquid State Infiltration Casting: No External Force

In some specific cases, the molten metal wets the reinforcement with no difficulty at all. The motive force of an infiltration process in this case is the capillary force of the dispersed phase (i.e., spontaneous infiltration). Examples are infiltration of copper as the metal matrix into tungsten as the reinforcement [28, 29], which is driven by the direct wetting of metals or the infiltration of reinforcements into molten copper by direct coating of reinforcements (tungsten) by nickel is done without applying force [30]. The infiltration by melting without the application of external force is called no force infiltration method. In melt infiltration, reinforcements (as a solid) are first placed in the die, and the molten alloy is poured, then penetrated on to it and permitted to solidify without any external force.

5.7.2 *Liquid State Infiltration Casting: Gas Pressure-Driven Infiltration*

Pressure driven infiltration is used to overcome poor wetting conditions, which use external work to force the metal into a preform, i.e., the reinforcement. Pressure is applied either by an inert gas or mechanically. In the case of gas pressure, inert gas from the outside, such as argon, is used for applying pressure and to push the metal into the preform dispersed phase. The inert gas does not interact with the material to be used for the manufacturing of MMC parts, and it will prevent oxidation formation. A typical sketch of the process is given in Fig. 5.15. The gas Pressure Infiltration method is used for manufacturing large composite parts. In contrast to the methods using mechanical force, the gas pressure infiltration process results in low damage to the fibers. Usually, pressure ranges from 1 to 10 MPa (1.45 ksi) are used, but the use of higher pressures up to 17 MPa (2.47 ksi) has also been investigated for aluminum matrix composites. To compare this value with 1 atmospheric pressure, the pressure exerted by the weight of the atmosphere, which at sea level has an average value of roughly 14.7 pounds per square inch,

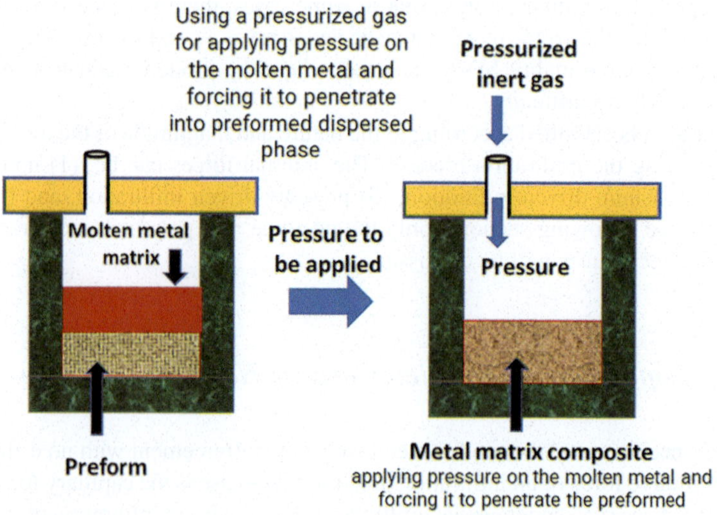

Fig. 5.15 The process of manufacturing MMCs by using the gas pressure infiltration process. Pressure on the molten metal and forcing it to penetrate the preform

1 psi = ~0.069 MPa. However, safety issues and possible damage become significant factors in this case when the pressure is too high.

The pressure used for the infiltration process allows the bonding of ceramic reinforced components directly to the molten metal during the infiltration stage. However, the application of high pressure might cause preform damage during infiltration, which is also observed experimentally [30].

5.7.3 Liquid State Force-Driven Infiltration (Squeeze Casting Infiltration)

The squeeze casting infiltration method is a process of applying a force to the molten metal on the preform [30]. Aluminum as a host metal is in a molten state and infiltrates the preform from the top end to the bottom end under the squeeze pressure. The method utilizes mechanically applied pressure using the piston of a hydraulic press that applies a force on the molten metal subsequently maintained during solidification (Fig. 5.16). In this case, the pressures are higher than they are in gas driven processes as already discussed above, which in this case ranges from 10 to 100 MPa (1.45–14.5 ksi). This type of infiltration process is the most widely investigated for commercial applications since it has been directly adapted from established processes used to cast unreinforced metals such as squeeze casting in which the piston represents part of the mold.

In Fig. 5.16, the molten metal is poured on the preform and then the assisted force is applied for the metal to penetrate the reinforcement, which will result in improving wettability and homogeneous distribution of reinforcement in the metal matrix. As

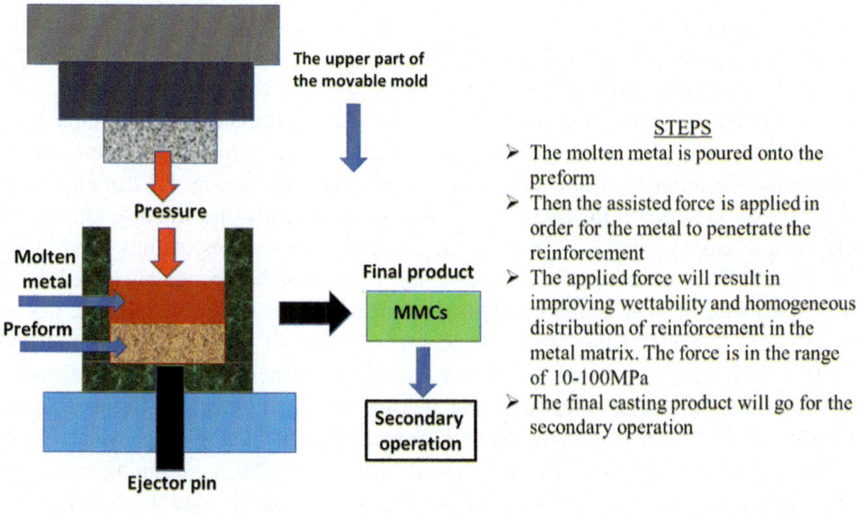

Fig. 5.16 Illustration for the squeeze casting infiltration process by applying force onto the preform

was mentioned before the applied force ranges from 10 to 100 MPa (1.45–14.5 ksi), almost 10 times higher than the gas driven pressure as was in the case of the inert gas pressure infiltration technique (Fig. 5.15). The metal matrix composite products produced by the external applied force method (Fig. 5.16), generally has a porosity-free matrix as a result of high applied force (a complete infiltration). That is, the molten metal will be penetrating inside the voids due to the applied force and the final casting product will be almost without porosity.

In Chap. 6, the ceramic matrix composite will be discussed. As was in the case of the metal matrix, in this case, the matrix is ceramic and the reinforcement in most cases is ceramics in order to maintain their properties at high service temperature environments. Several methods of fabricating ceramic matrix composites will be presented in Chap. 6.

5.8 Parameters Influencing the MMC Properties

Parameters that can influence properties of MMC materials are: (1) the properties of reinforcement and metal interface, (2) volume fraction of reinforcing filler, (3) geometry arrangement of the reinforcing filler, (4) properties of reinforcing filler, (5) properties of host metal, (6) residual stresses developed during manufacturing phase, (7) degradation properties of both the matrix and reinforcement during manufacturing process at elevated temperature.

5.9 Advantages and Disadvantages of MMCs over PMCs

A comparison between the MMCs fabricated parts and parent material (the host metal) was made in Sect. 5.5. The mechanical properties comparison between the metal matrix alone, and the metal matrix with 25% SiC as the reinforcement, were demonstrated and properties such as fatigue, yield, ultimate, and modulus were assessed. It was shown that the improvement in strength can be achieved by increasing the amount of fine silicon carbide particulate (see Fig. 5.12b). However, the increase in strength may lead to a reduction in ductility and fracture toughness. Here, an attempt is made to compare the two types composites, mainly the MMCs and PMCs. Both advantages and disadvantages between the two are presented. Note that in the case of MMCs, the matrix is metallic and in the case of PMCs, the matrix is polymer. The advantages of MMCs composites are: (1) better temperature capability, (2) not easily burnable, (3) higher transverse stiffness and strength, (4) moisture resisting, (5) better electrical and thermal conductivities, (6) Better fracture toughness, (7) higher resistance to crack growth, (8) no creating outgassing.

Regarding disadvantages of MMCs over PMCs, (1) MMCs require high temperature processing, (2) they have higher density, and (3) metals tend to corrode.

5.10 Concluding Remarks

Different methods of fabricating MMC parts were discussed in this chapter. Manufacturers of metal matrix composites will select the proper technique suitable for fabricating parts for specific applications. Both solid-state and liquid-state methods were presented here in detail. High-pressure diffusion bonding, sinter forging, stir casting, and infiltration techniques were discussed, and the advantages and disadvantages of each method were presented.

For engineers interested in having composite parts that can withstand harsh environmental conditions, the choice of using MMCs to fabricate their structural components is preferred even though the cost and time of manufacturing exceed traditional methods of fabricating metallic alloys. The attractive properties of MMCs include a high strength-to-weight ratio, high fatigue and fracture properties, high stiffness, and high corrosion resistance. The application of these composites in defense, aerospace, automotive, and other advanced industries can be useful to resolve many major problems that cannot be solved by conventional material.

The high cost of fabricating parts with MMC techniques is attributed to the cost of reinforced materials, matrix metals, and, above all, fabrication processing expenses. These costs can deter manufacturers of MMCs from day-to-day applications.

Finally, a proper standard testing technique must be available to manufacturers so that they can obtain material properties. Engineers can then implement these properties (allowables) for structural integrity and failure assessment. These properties encompass static, fatigue, and fracture allowables, and they should be based on ASTM testing standards.

It's important to note that metal matrix composites can exhibit varying degrees of isotropy and homogeneity depending on their manufacturing processes and specific composition. Therefore, when conducting tests, the degree of anisotropy must be taken into consideration when dealing with MMC materials properties and related analysis.

References

1. M. Garcia (ed.), *International Space Station Facts and Figures* (NASA, Washington, DC, 2023)
2. S.P. Rawal, Metal-matrix composites for space applications. J. Mater. **53**, 14–17 (2001)
3. C. Legzdins, I. Samarasekera, J. Meech, MMCX—An expert system for metal matrix composite selection and design. Can Metallurg Q **36**, 177–202 (1997)
4. K.K. Kar, *Composite Material—Processing, Applications, and Characterization* (Springer, Berlin, 2017), pp. 382–420. Chapter 1
5. M. Gopi Krishna, K. Praveen Kumar, M. Naga Swapna, J. Babu Rao, Bhargava, Metal-metal composites – An innovative way for multiple strengthening. Mater. Today Proc. **4**, 8085–8095 (2018)
6. M. Rosso, Ceramic and metal matrix composites: Routes and properties. J. Mater. Process. Technol. **175**(1–3), 364–375 (2006)

7. J.K. Katiyar, A.S. Mohammed, *Encyclopedia of Materials: Composites* (Elsevier, Amsterdam, 2021)
8. K.U. Kainer, Chapter 1: Basics of metal matrix composites, in *Metal Matrix Composites: Custom-Made Materials for Automotive and Aerospace Engineering*, (Wiley, Weinheim, 2006)
9. K.K. Kar, *Composite Material—Processing, Applications, and Characterization* (Springer, Berlin, 2017), p. 384. Table 11.1
10. A. Kumar, O. Vichare, K. Debnath, M. Paswan, Fabrication methods of metal matrix composites (MMCs). Mater. Today Proc. **46**, 6840–6846 (2021)
11. K.B. Jiju, G. Selvakumar, S. Ram Prakash, Study on preparation of Al – SiC metal matrix composites using powder metallurgy technique and its mechanical properties. Mater. Today Proc. **27**, 1843 (2020)
12. A.B. Susula, I. Suguhartono, M.A. Marpaung, Study on mechanical properties of metal matrix composites (MMCs) Al-Cu-Mg/SiC_p with powder metallurgy. J. Phys. **1402**, 044109 (2019)
13. C. Zweben, *Metal Matrix Composites*. Contractor report for OTA (January 1987)
14. D.K. Sharma, D. Mahant, G. Upadhyay, Manufacturing of metal matrix composites. Mater. Today Proc. **26**(Part 2), 506–519 (2020)
15. F.L. Matthews, R. Rawlings, *Composite Materials: Engineering and Science* (Chapman & Hall, London, 1994)
16. T. Barrett, The future of metal is in matrix composites. Mach. Des. (2017)
17. A.A. Shirzadi, E.R. Wallach, A new method to diffusion bond superalloys. Sci. Technol. Weld. Join. **9**(1), 37–40 (2004)
18. A.A. Shirzadi, H. Assadi, E.R. Wallach, Interface evolution and bond strength when diffusion bonding materials with stable oxide films. J. Surf. Interf. Anal. **31**(7), 609–618 (2001)
19. H. Lee, K. Dongkeun, I. Lee, Y.-J. Moon, J.-Y. Hwang, K. Park, S.-J. Moon, Stepwise current electrical sintering method for inkjet-printed conductive ink. Jpn Soc. Appl. Phys. **53**, 05HC07 (2014)
20. J. Li, X. Ren, Y. Zhang, H. Hou, Silicon carbide hot pressing sintered by magnesium additive: Microstructure and sintering mechanism. J. Mater. Res. Technol. **9**, 520–529 (2019)
21. A.D. Chandio, M.B. Ansari, S. Hussain, M.A. Siddiqui, Silicon carbide effect as reinforcement on aluminum metal matrix composite. J. Chem. **41**(04), 650 (2019)
22. M. Kusay, *Metal Ceramic Matrix Composite* (Institute of Physics Publishing, Bristol and Philadelphia, 2004)
23. J. Hashim, L. Looney, M.S.J. Hashmi, Metal matrix composites: production by the stir casting method. J. Mater. Process. Technol. **92–93**, 1 (1999)
24. K.K. Kar, *Composite Material—Processing, Applications, and Characterization* (Springer, Berlin, 2017), p. 395
25. L.H. Hihara, A. Bakkar, Corrosion of metal matrix composites, in *Materials Science and Materials Engineering*, (Elsevier, Oxford, 2016)
26. R. Etemadi, B. Wang, K.M. Pillai, B. Niroumand, E. Omrani, P. Rohatgi, Pressure infiltration processes to synthesize metal matrix composites—A review of metal matrix composites, the technology and process simulation. Mater. Manuf. Process. **33**, 1261 (2018)
27. K.M. Sree Manu, L. Ajay Raag, T.P.D. Rajan, M. Gupta, B.C. Pai, Liquid metal infiltration processing of metallic composites. Metall. Mater. Trans. B **47**, 2799–2819 (2016)
28. Jiří Matějíček, Preparation of W-Cu composites by infiltration of W skeletons. 2021 materials science, METAL 2021 conference proceedings (2021)
29. L. Dong, M. Ahangarkani, W.G. Chen, Y.S. Zhang, Recent progress in development of tungsten-copper composites: Fabrication, modification and applications. Int. J. Refract. Met. Hard Mater. **75**, 30–42 (2018)
30. Dmitri Kopeliovich, "Liquid State Fabrication of Metal Matrix Composites," SubsTech (Substances & Technologies), Royal Societty of Chemistry, (2012)

Chapter 6
Ceramic Matrix Composites (CMCs) (Methods of Manufacturing CMC Parts)

6.1 Background

As briefly mentioned in Chap. 3, ceramic matrix composites (CMCs) are categorized as composite materials, just like the metal matrix and the polymer matrix composites discussed in the previous Chaps. 4 and 5, respectively. They consist of ceramic fibers embedded in a ceramic matrix through several fabrication techniques that will be discussed here. The fibers can take many different forms, including the more traditional continuous fiber as well as short fibers, particles (spherical or irregular shapes), whiskers, layers, and nanofibers [1–3]. The reinforcing fibers are less than a millimeter thick and, in the case of nanofibers, are even much smaller. The selection of ceramic whiskers in the fabrication of CMCs is typically high-strength single crystals with a length at least ten times greater than their diameters. However, they still fall into the short reinforcement category. Silicon carbide, SiC, is probably the most common nonoxide whisker material used as the reinforcement in the CMC matrix.

In most cases, the same kind of ceramic is used for both the matrix and fibers (e.g., SiC/SiC). Additional secondary fibers (aka fillers) may also be added to enhance the CMC properties. When both the matrix and reinforcement are ceramics, CMCs could be categorized as a subgroup of both composite materials and fully ceramics.

From a chemistry perspective, ceramics alone have both metallic and nonmetallic atoms, such as silicon carbide (SiC), titanium carbide (TiC), silicon nitride (Si_3N_4), and aluminum oxide (Al_2O_3). From a failure perspective, when ceramic fibers are entrenched in the ceramic matrix, the failure strain of the matrix is lower than the failure strain of the fibers, which makes the CMCs to be different from both polymer and metal matrix composites, where the host matrix has good ductility and fracture toughness. In both the PMCs and MMCs, their matrices are less brittle than the reinforcement material. To overcome the brittleness of ceramic matrix, it is a good idea that the reinforcement to be chosen as a metallic whisker that possesses high

B. Farahmand, *Fundamentals of Composites and Their Methods of Fabrications*,
https://doi.org/10.1007/978-3-031-80201-0_6

ductility and good resistance to temperature, like the refractory metals, such as tungsten, molybdenum, niobium, tantalum, and rhenium. In a situation like this, where the reinforcement is of refractory metals, it can provide high fracture toughness and high resistance to temperature to the CMC. These advanced CMCs with metal reinforcements are called cermets (ceramics and metals). The negative aspect of cermets CMCs is the weight issue that can always be an essential factor in aircraft, aerospace, and automotive industries. In addition, the mismatch between the ceramic matrix and the metallic fiber can introduce residual stresses, causing premature failure of the part when it is subjected to a service load and temperature environments. Because of that, the refractory metallic fibers may not be a suitable candidate to implement as a reinforcement when compared with the ceramic fibers, and they will not be discussed here. Another important point from the industry's perspective is the fact that ceramics are usually lighter and harder than metals with a much lower density as well as having lower thermal and electrical conductivity. For those reasons, industries are interested in utilizing them since they can lead to much better fuel efficiency due to their lightweight and, in turn, less pollution when used in the auto and aircraft industries. It is estimated that reducing 1 lb (0.453 kg) of mass in a commercial aircraft by switching to a lighter material can reduce fuel consumption by 360 gal of fuel per year [4]. So, if it is assumed that the weight has been reduced by 100 kg due to employing lighter material, the fuel consumption can be reduced by 79,470 gal of fuel per year. General Electric estimated that a 1% reduction in fuel consumption could save more than $1 million a year for commercial air carriers. This next-generation CMC material technology produced by GE Aviation will improve fuel efficiency by 1–2% [5].

Because of all that, it is highly under consideration by manufacturers to replace metallic parts with ceramic matrix composites in both room and elevated temperature environments. Another benefit of replacing the metallic material with the CMCs is their ability to operate at high temperatures without the need to use any cooling systems, and that will result in lowering the cost of electricity. From a mechanical properties' perspective, ceramic properties are superior to their metallic counterpart. However, their resistance to cracking is poor, with low fracture toughness properties. Below is a comparison of flexural strength, modulus of elasticity, and fracture toughness between three ceramic matrices (Al_2O_3, SiC, and Si_3N_4) at room temperature.

Flexural strength (unit of MPa): $Al_2O_3 = 358$, $Si_3N_4 = 700$, SiC = 380 (Note that 1 MPa = ~0.145 ksi.)

Elastic modulus (unit of GPa): $Al_2O_3 = 300$, $Si_3N_4 = 310$, SiC = 410 (Note that 1 GPa = ~145 ksi.)

Fracture toughness (unit of MPa $(m)^{1/2}$): $Al_2O_3 = 4$–5, $Si_3N_4 = 6$, SiC = 4.6 (Note that 1 MPa $(m)^{1/2}$ = ~0.91 ksi$(in)^{1/2}$.)

Aside from the ceramic matrices being hard, brittle, and fracturing easily under mechanical or thermo-mechanical loads, they possess excellent corrosion and wear-resistant properties. Their good corrosion resistance characteristic is due to their high chemical stability in a harsh environment, which enables them not to react to other chemical substances and materials during their service usage. This is directly related to the type of chemical bond in the ceramic material, i.e., the covalent bond. The covalent bond, which is the strongest bond in the material, is because of electrons being shared between the two neighboring atoms. Whereas, in the case of metals, the bonds between atoms are not of the sharing type. Instead, many electrons are free to move throughout the metal so that each electron can interact with many of the fixed nuclei of atoms. Those free electrons in the metallic material are responsible for excellent electric conductivity in metals. In the case of ceramics with covalent bonds, it requires lots of energy to separate atoms from each other. For those reasons, ceramic materials have much higher modulus and strength when compared with metals but no ability to conduct electricity. Moreover, they have high temperature oxidation stability compared to metals. Note that a ceramic container that was made hundreds of years ago, is still intact even though it was exposed to harsh environments, which the same environment could cause severe damage to metallic and polymer materials.

CMCs are useful for structures under load and simultaneously subjected to high temperature service conditions. They have shown increasing potential for their applications in aerospace, automobile, biomedical, and electronic. Finally, one of the most common applications of the CMCs is in the cutting tools used in industries where high wear and high resistance to temperature are essential.

CMC materials can make good candidates for the hot sections of turbine engines, primarily since the demand for greater jet engine efficiency and power has driven the need to withstand heat near the melting point of nickel and titanium alloys. The presence of CMCs in manufacturing the turbine engine can help to have higher blade frequencies, longer service life, and higher thrust. While these ceramic base materials offer the greatest payoff in jet engine hot sections, they also benefit engine nozzles, tail cones, and other components not subjected to as much intense heat and stress during usage. For all the above qualities of ceramic matrix composite over its metallic counterpart, they can be of great use for aerospace companies.

The fact that these materials are highly in demand in a certain sector of our industries, the problem that is facing the manufacturer of ceramic composites is the development of a suitable, cost-effective method for combining the ceramic matrix and their reinforcement to obtain the required shape with properties that tailored to fit the design requirements lay down by the customers [6]. To achieve the desired properties in any target application, the selection of manufacturing technique, ceramic matrix, and reinforcing part, along with its orientation and amount of volume fiber content, are crucial parameters to have an acceptable part.

Finally, the manufactured CMC parts may need to undergo a limited number of tests by the American Testing for Testing Composite Material to establish the required properties and to be sure these properties are acceptable or not [7]. These tests are performed to obtain mechanical, fatigue, and fracture properties, where data can be used for further analysis, such as structural integrity assessment and whether the part can withstand the design life. Properties such as the modulus of elasticity, the ultimate strength and strain, high- and low-cycle fatigue data, and plane strain and stress fracture toughness, as well as the fatigue crack growth data, must be all available to see whether the designed part can withstand the load environment throughout its service life. Since ceramic matrix composites are anisotropic (and sometimes orthotropic), with properties different in other directions, the most robust mechanical properties will be along the length of the fibers. This allows them to be tailored to a specific application's needs if fibers are continuous and not randomly oriented in the matrix. There is a wide range of international and national test standards (ASTM, EN, ISO, JIS) were developed to test composite materials, along with several historical standards that are still in use, such as those published by the SACMA (Suppliers of Advanced Composite Materials Association) and CRAG (Composites Research Advisory Group). The following standards are important for engineers to follow. It provides consistency and reliability in the testing and characterization of ceramic composites, covering mechanical properties, fracture toughness, and other essential aspects of these materials.

ASTM C1275—Evaluates the tensile properties of continuous fiber-reinforced ceramic composites at room temperature.

ASTM C1341—Determines the flexural properties of advanced ceramic composites at ambient temperature.

EN 12011—Determination of flexural strength: Provides guidelines for determining the flexural strength of monolithic ceramics at room temperature. Note: EN Standards is the European Norms Standards.

EN ISO 14704—Specifies a method for determining the fracture toughness of monolithic ceramics.

ISO 14704—Like the EN standard, this ISO standard specifies the method for determining fracture toughness in monolithic ceramics. Note: ISO Standards is the International Organization for Standardization.

ISO 20750—Provides guidelines for determining the flexural strength of monolithic ceramics.

JIS R 1601-3—Covers the determination of tensile properties of ceramic matrix composites at room temperature. Note: JIS Standards is the Japanese Industrial Standards.

JIS R 1601-4—Specifies the method for determining flexural properties of ceramic matrix composites at room temperature.

Finally, applying nondestructive inspection on CMC parts before their usage and during their operation based on a well-established method is essential not only to engineers but also to industries. Chapter 14 is allocated to applying nondestructive inspections based on several acceptable methods to be used in industries.

6.2 Mechanical Properties of CMCs

In the case of CMCs' structural failure, it is the matrix that fails first due to its brittle behavior. That is, the existence of small microcracks in the ceramic matrix can cause instability and fast fracture due to the applied monotonic or fluctuating load. But

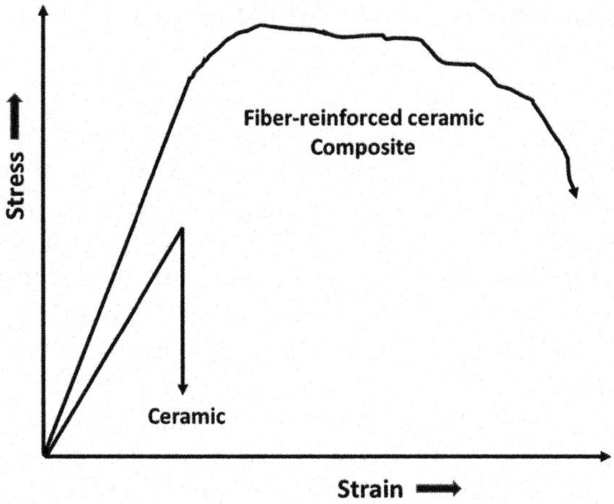

The stress-strain behavior of CMCs can be explained as the bilinear behavior caused by preexisting micro cracks that can advance and stop repeatedly as the external load increases beyond a proportional limit. That is, the presence of a strong fiber-reinforced ceramic matrix interface will help to stop the crack from growing, causing a bilinear behavior, which provides better fracture toughness than the ceramic matrix alone with brittle behavior.

Fig. 6.1 Sketches of mechanical behavior of ceramic composite (with fiber reinforced) versus the ceramic alone that has brittle fracture behavior

with the presence of ceramic fibers, crack growth will be halted and prevent it from catastrophic failure. The failure of the ceramic matrix alone under the monotonically increasing load is shown in Fig. 6.1. In this figure, both the stress–strain curves for the monolithic ceramic matrix and the reinforced ceramic matrix are plotted. The ceramic by itself shows brittle behavior with no elongation, however, by adding reinforcement, the new material can absorb much more energy prior to its failure due to the presence of fibers (or whiskers), which very much behaves like metallic material. The reinforced fibers in the matrix will prevent fast fracture and allow elongation to take place before final failure, which is because of the bridging mechanism by the reinforcing fibers. In addition to having a superior stress–strain curve, another mechanical property of CMCs is the enhancement of its fracture toughness over the ceramic matrix alone when ceramic is without being reinforced with fibers. Whether the fracture toughness superiority of CMCs is a result of fibers in the matrix, it is still necessary to obtain a reasonable amount of useful test data to be sure about the applicability of CMCs in different working environments. For example, it is highly beneficial to generate test data that can depict the degree of degradation of CMCs mechanical properties, such as fatigue and fracture toughness, as a function of temperature or fiber volume in the matrix ceramic. This variation of fracture toughness as a function of temperature can be useful to engineers when designing a CMC part for a given load and temperature environment to which the part will be exposed to.

6.2.1 Fracture Mechanisms Behavior of CMC Materials

The presence of reinforcements in the ceramic matrix will be helpful to stop the crack from growing which provides better fracture toughness than the ceramic matrix alone. While cracks are originating and growing from the matrix, its crack tip stress intensity factor, K, is smaller than the fracture toughness of fiber material ($K < K_{\mathrm{Ic}}$), and for that reason, the crack will be stopped when facing the fibers bonding with the matrix. If the crack continues to grow in the matrix through the debonding mechanism, without failing fibers, then the crack mouth is to be bridged by the unbroken fiber. So, the fiber will provide toughness to the otherwise brittle matrix. In addition to the reinforcement in the matrix, sometimes a third element called filler can be also added to enhance other properties such as electrical conductivity, or any other properties that can be improved to have a more useful composite.

Both fibers and fillers can be useful to bring better toughness and better electrical conductivity properties to the CMCs. Regarding the enhancement of fracture toughness due to the presence of reinforcement, an adequate bonding between the matrix and fibers is useful to have a sound mechanical property. Usually, the stress–strain curves for the CMC material at room and elevated temperatures under tensile tests show nonlinear behavior, which looks as if the material has been deforming plastically like their metallic counterparts, Fig. 6.2a. This behavior is called quasi-plastic because the effect is the result of the microcracks in the ceramic matrix, which are formed, grow, arrested, debonding fibers from the matrix, and keep growing in a stable manner with increasing load up to the failure during the full stress–strain test [8–11]. The occurrence of multiple damage and toughening mechanisms, such as, matrix multiple cracking, crack bridging, fiber–matrix interface debonding, and fibers fracture, are the main reasons for the nonlinear characteristic seen in the

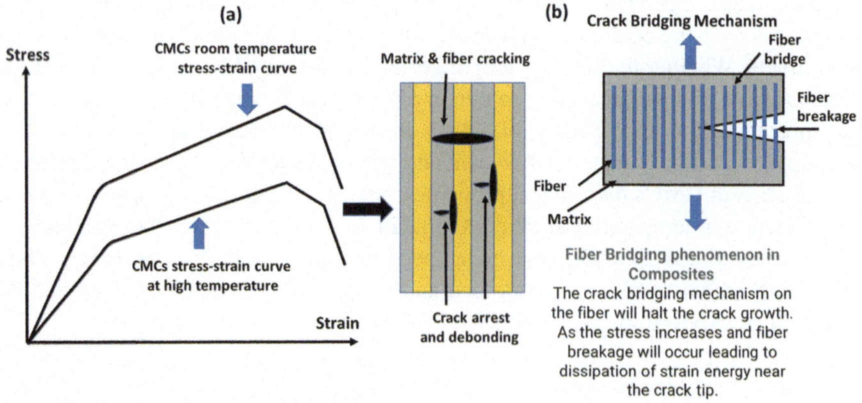

Fig. 6.2 (**a**) Sketches of nonlinear stress-strain curves, which look as if the CMCs material has been deforming plastically. Both room temperature and higher temperature stress-strain curves are shown. (**b**) Matrix multiple cracking fiber/matrix interface debonding, fiber bridging, and fiber fracture is responsible for the nonlinear behavior

CMCs as shown in Fig. 6.2a. In Fig. 6.2b, the crack bridging, one of the toughening mechanisms for the ceramic matrix composite, is shown. The crack flank bridging by the ceramic fiber can slow down the crack from advancing and it causes gradual failure of the composite rather than a catastrophic one. As the stress increases further, fiber will eventually break which leads to partial dissipation of strain energy release near the crack tip. In some cases, the debonding occurs and the crack can continue growing in the matrix or change its path from being perpendicular to the applied load to parallel to the load. In this situation, the growing crack would change from mode I (when the crack is under the tensile load, it is called the opening mode or mode I) to become parallel to the applied load (mode II where the crack is under the shear stress or sliding mode). It is well established that crack growth and final failure occur under mode I and seldom happen under mode II. When the reinforced material is left behind and undamaged, and forms a bridge across the crack flanks, it comes to a point that the reinforced fibers will eventually fail since the applied stress intensity factor, K, rises above the critical value, K_c due to an increase in crack length or applied load [12]. Here, the overall (total) energy release rate (G_{total}) is a function of both the critical energy release rate for crack arrest and trapped (G_{ct}) and the energy release rate for crack bridging (G_{cb}), which can be written as $G_{total} = G_{ct} + G_{cb}$ [13, 14].

Another point to note is the strength of the interfacial bond between the fibers and the matrix which plays a big role in obtaining a satisfactory composite part with long life when subjected to monotonic increasing load or fluctuating load environments. If the interfacial bond is inadequate and weak, it will result in a shorter life. Therefore, a healthy bond between the matrix and the fiber helps the load sharing between the two elements to happen, which will lead to a larger number of cycles to failure for the part with the crack. Some also argue that the weaker interface may have a better fracture toughness mechanism because the energy release rate, G, will happen gradually from one fiber to the next. That is, a strong interface will dissipate all the available energy at once because G will become large at the interface and cause a catastrophic failure ($G_I = G_{Ic}$), not an incremental energy release rate, $G + \Delta G$. In the case of $G_I = G_{Ic}$ the applied stress increases at the interface, while in the case of $G + \Delta G$ crack length is increasing in a stable manner until $G = G_{Ic}$.

Finally, the uniform distribution of fibers embedded in the ceramic matrix will be helpful to enhance the fracture toughness of the combined material system. Engineers must recognize that the inherent properties of the ceramic matrix by itself that have high strength, high Young's modulus, and good creep is good for the CMCs to have a longer life, both at room and elevated temperatures when subjected to monotonic or fluctuating load environments during their service usage. However, when fibers are not uniformly distributed in the matrix, the fracture toughness and other mechanical properties of CMCs can be degraded greatly because of agglomerations of fibers in one location that will act as stress concentration sites in the matrix. Therefore, the uniform distribution of fibers in the matrix is important to get a sound and uniform mechanical property throughout the composite especially when dealing with crack arrest at the interface region.

6.3 Selection of Matrices and Reinforcements for CMCs

The most common examples of ceramic materials are clay, porcelain, brick, and roof tiles. Clay is a form of soil made up of very small particles of aluminum silicate (Al_2SiO_5) and it is the earliest ceramics item made by humans. The manufacturing of ceramic matrix composites is through the lay-up and fixation of the ceramic fibers. That is, different types of fibers are embedded in the ceramic matrix using some of the techniques that are already discussed in polymer matrix reinforced composite materials (PMCs), such as lay-up of fabrics, filament winding, and braiding. The ceramic fibers have different geometry and size such as particulates, whiskers, and short and long fibers, that are arranged and scattered in the matrix in such a way that could withstand the load environment during the part's design life. The use of long fibers may result in better toughness because parallel fibers can hold the CMC together even after the ceramic matrix has begun to crack. They can significantly slow down crack propagation.

Ceramic fibers must stay stable at high temperatures, normally above 1000 °C (1800 °F). In working environments, materials such as fibers of alumina, mullite, and SiC, can be used because of their resistance to high operating temperatures. Silicon carbide fibers (SiC), have an elongation capability above 2%, much larger than conventional ceramic materials with elongation of around 0.05–0.10% [1]. Figure 6.3 illustrates different types of ceramic matrices and reinforcements that can be good candidates for manufacturing purposes. Any of these ceramic matrices, shown in

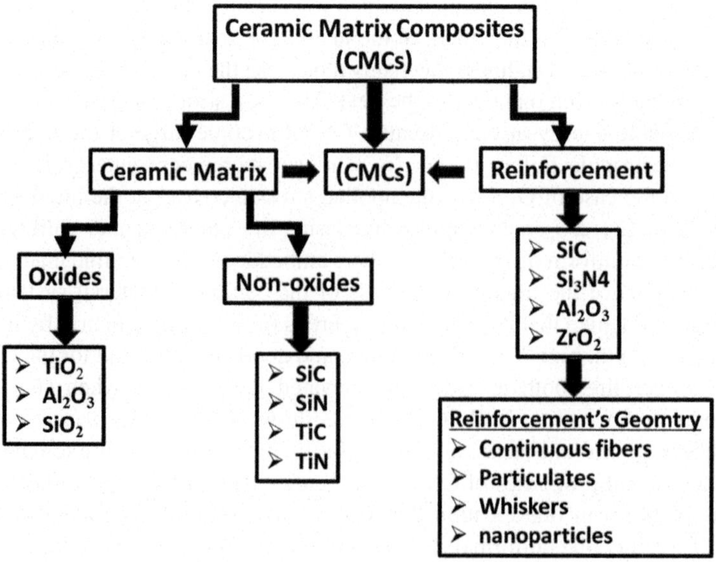

Fig. 6.3 Types of matrices and reinforcements used in ceramic matrix composites

Fig. 6.3, can be blended, or mixed with these reinforcements to fabricate CMC parts. For reinforcements, there are particulates, whiskers, and nanoparticles that must establish a good bond with the matrices in order to have a reasonable amount of toughness due to the crack arrest behavior described above.

Currently, there are ceramic oxides (TiO_2, Al_2O_3, SiO_2) and nonoxide matrix materials (SiC, SiN, TiC, TiN) that can be used in industries for the high temperature environment and with the right reinforcement to manufacture the desired CMC parts (see Fig. 6.3). Recently, ultra-high temperature resistance ceramics were investigated for their applications as a matrix in the CMCs for their usage in high temperature environments and are discussed in these references [15–17]. One of the important points to remember when fabricating CMC parts is the way fibers are distributed in the matrices, which already was briefed before. In the case of small whiskers or small fibers, the agglomeration of fibers in the matrix is a possibility and must be avoided because the uniform distribution of reinforcement in the matrix will cause uniform mechanical properties. The agglomeration of fibers in the matrix means in some areas of the matrix there will be much fewer fibers and the possibility of fast fracture can occur. This is because of the lack of crack arrest feature of fibers that are missing in the matrix which can stop the advancement of a crack growing in the matrix. In addition, the accumulation of fibers in one location can cause stress concentration in that locality, which is not a desirable thing to have. Engineers must try their best to avoid it and instead have a uniform distribution of fibers in the matrix through trial and error and figure out what is the best way of obtaining uniform distribution of reinforcement in the ceramic matrix. Finally, for CMC parts to be acceptable for industry use, certain requirements must be followed, such as fabrication processes that could yield CMC components not only with the proper performance at high temperature environments but also with the proper shapes, sizes, and functional features. Thus, the following parameters, such as ceramic fiber types, fiber geometry, functional fiber coatings for the interphases, matrix constituents, and their fabrication processes, must be carefully selected and tailored for specific tasks. Any post fabrication processing must also be considered by the manufacturer, such as the feasibility of the grinding process or final machining if needed [18, 19]. Note that the machining process is the final operation after the manufacturing process to give the final shape to a part and to reach the required tolerances. Machinists must be extra careful not to damage the part; otherwise, it will be a big economic loss to the manufacturer, as well as the end user of the part. If possible, the nondestructive inspection should be applied before parts usage to check for the possible presence of cracks during the post fabrication process. The nondestructive inspection of CMC parts should be a requirement because the failure of parts during their operations, due to undetected flaws, can be catastrophic to the user of the part. Ultrasonic, radiography, thermography and acoustic emission are used for inspecting flaws in the CMC parts and they are discussed in Chap. 14.

6.4 CMCs Fabrication Methods

The mechanical properties of structural parts made of CMC material strongly depend on the kind of ceramic matrix, the reinforcement, as well as the method of manufacturing them. Several techniques are currently available that engineers can select for fabricating the CMC parts. These CMC processing techniques are different from each other and have a strong influence on the final products. Only those important ones widely used for manufacturing CMCs will be discussed in detail. The advantages and disadvantages of each method will also be addressed here.

In manufacturing ceramic matrix composites, there are three steps that manufacturers are required to take for fabricating CMC parts. In step (1), the preparation of both the ceramic matrix and the reinforcement materials separately must be available before the fabrication process. In step (2), mix the two quantities by a selected technique to be determined by the manufacturer to obtain the best CMC products. Because ceramic materials have a high melting point, it is difficult to melt them and cast them into different shapes, as is a common practice among metals or polymers. In step (3), we must realize that in any methods engineers use to manufacture the CMC parts, the post machining process is cumbersome because of the high hardness associated with the ceramics and the possibility of creating scratches on the surface. For that reason, it is much better to produce parts that are very close to the actual shape to minimize the machining process [19]. Some other post processing items may also need to be implemented in ceramic composites when parts are completed. They are grinding, lubrication, enhancing the surface hardness, and glazing processes (glazing process involves applying a thin, glossy, transparent coating to the ceramic surface). Additionally, it may be necessary to conduct a comprehensive nondestructive inspection which could be necessary to check for the presence of embedded or surface defects. As was mentioned before, if the manufacturing process allows producing the near shape casting products, then the amount of post processing work on the part can be minimized. But, for some critical parts, the NDI inspections, and a limited number of tests, could be necessary based on the customer's request.

One of the challenges that CMCs may encounter during the fabrication process is the high temperature environment that it will go through. The reinforcement must be able to withstand the high-temperature environment because of its smaller size, which contributes to having a slightly lower melting point. Differences in the thermal expansion coefficient between the reinforcement and the matrix material must also be given special attention because they can lead to thermal residual stresses when the material is cooled down to room temperature. The tensile residual stresses will shorten the part's life when additional tensile stress is imposed on it during its usage, causing premature failure. Therefore, to check the integrity of CMC parts by a given technique, mechanical tests are necessary to conduct and find out the severity of the residual stresses built up in part and how one can minimize it. In addition to mechanical testing, proper nondestructive inspections will be helpful to

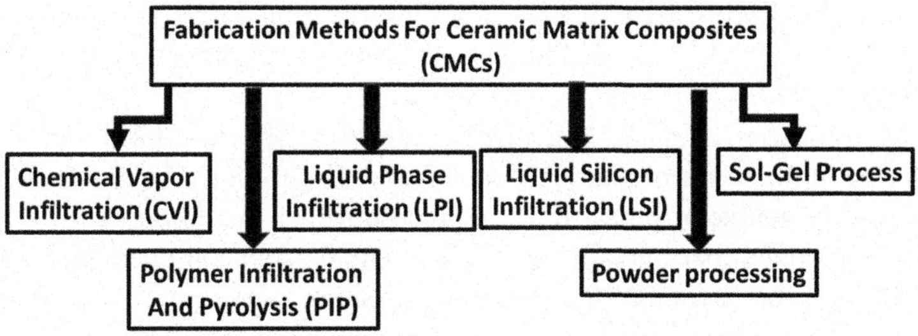

Fig. 6.4 Different methods of fabricating ceramic matrix composites

check for crack-like defects after fabrication and interval inspection throughout the structural usage.

Figure 6.4 shows different manufacturing techniques currently available for making CMC parts. In this figure, the CMC manufacturing processes are categorized into five groups. These are the sol–gel, the polymer infiltration and pyrolysis (PIP), the chemical vapor infiltration (CVI), the liquid phase infiltration (LPI), liquid silicon infiltration (LSI), and sintering methods [20].

One of the points that industries are interested in is the implementation of an automation process that will be able to produce CMC parts in a short time. The automation process can be helpful to save time and increase production, which can bring profit to the industries.

For reinforcement preparation (fibers or whiskers), it will take many different forms, including continuous fibers, short fibers, particles, whiskers, and nanofibers. These fibers all have a polycrystalline structure (crystals that are variously oriented) like traditional ceramics possess. Fibers are often woven into a fabric or tape-like material to be embedded in the matrix. In a typical CMC process, the fibers are coated with a material, such as boron nitride, and then passed through a matrix slurry bath, resulting in a prepreg tape or cloth. Other CMC fibers like glass, organic, or metallic fibers are not stable at temperatures above 1800 °F (982 °C) and, for that reason, are not good enough to be part of reinforcement for manufacturing CMC parts that must withstand higher temperatures unless they are being precoated with high temperature material before being embedded in the matrix [21, 22].

For producing the prepreg CMCS tape, the fibers would first be passed through a slurry bath of binder's solvents, which is the matrix material being used as the binder (e.g., silicon carbide). The various layers of prepreg tape are laid up and shaped to form the final part. In most cases, the part then goes through pyrolysis to convert the matrix material to a ceramic. The pyrolysis process [23] is the thermal decomposition of a complex organic substance, such as carbon, into one more straightforward structure by means of heat. The decomposition will take place at high temperatures in the presence of an inert gas atmosphere and in the absence of oxygen to avoid oxidation. Note that different products are released during the pyrolysis process, some of which are dangerous. These products may include CO, H_2, CH_2, H_2O, and CO_2. Upon conducting pyrolysis, the part will introduce porosity, which later must be infiltrated by appropriate material for the densification purpose and elimination of voids (porosity) as much as possible. The pyrolysis and infiltration technique will be discussed later in Sect. 6.7.

The identification of ceramic matrix composites is based on both the fiber and matrix (type of fiber/type of matrix). For example, C/SiC would be a carbon fiber in a silicon carbide matrix or others like C/C (carbon fiber in carbon matrix), SiC/SiC (the silicon carbide fiber into silicon carbide matrix). As another example, the fiber of aluminum oxide is embedded into the aluminum oxide matrix (Al_2O_3/Al_2O_3). Sometimes the manufacturing process is also included in labeling the CMC before the fiber/matrix type. For example, the C/SiC composite manufactured through the liquid polymer infiltration (LPI) process and is abbreviated as C/SiC-LPI or SiC/SiC fabricated by the chemical vapor infiltration method and it is abbreviated by SiC/SiC-CVI.

A few methods of fabricating CMCs are discussed below. The SiC/SiC composite, where the silicon carbide fiber is embedded into the silicon carbide matrix, is described in Sect. 6.5. In Sect. 6.6, the sol–gel infiltration approach (wet chemical approach) is described in detail. The polymer infiltration pyrolysis approach (PIP) and liquid silicon infiltration (LSI) will be discussed in Sects. 6.7 and 6.8, respectively. Finally, the chemical vapor infiltration (CVI), chemical vapor deposition (CVD), and powder processing techniques will be discussed in Sects. 6.9 and 6.10.

6.5 Method of Fabricating the SiC/SiC CMCs

The silicon carbide/silicon carbide (SiC/SiC) ceramic matrix composite has high resistance capability to heat is essential in industries, primarily in engine technology where the structure is exposed to high heat. SiC/SiC CMCs can withstand temperatures above 2400 °F/1316 °C. Their high heat resistance capability and their reduced weight (almost 30% the weight of nickel superalloys) make them appealing to engine manufacturers that are looking to reduce weight in the engine hot zone areas in quest for greater fuel efficiency. Because SiC/SiC possess more outstanding durability to heat and can maintain its mechanical properties at elevated temperature, for that reason, it requires less cooling air than current materials (e.g., nickel-based superalloys) when being used in manufacturing gas turbines and jet engines. These gains will improve combustor efficiency and reduce fuel consumption. In addition to excellent heat resistance properties, the SiC fibers have good ductility, strength, and fracture toughness. In the form of whisker reinforcement, the young modulus and ultimate tensile values are $E = 68{,}571$ ksi (480 GPa), almost twice as much as steel, and $F_{tu} = {\sim}300$ ksi (2070 MPa). In the form of the matrix, it has a young modulus and ultimate tensile values of $E = 57{,}971$ ksi (400 GPa) and $F_{tu} = 43.4$ ksi (300 MPa), respectively. The fracture toughness of the SiC matrix is around $K_{Ic} \sim 4.2$ ksi$(\text{in})^{0.5}$, which is still much lower than metallic material. For example, the plane strain fracture toughness of typical aluminum alloys falls in the range of

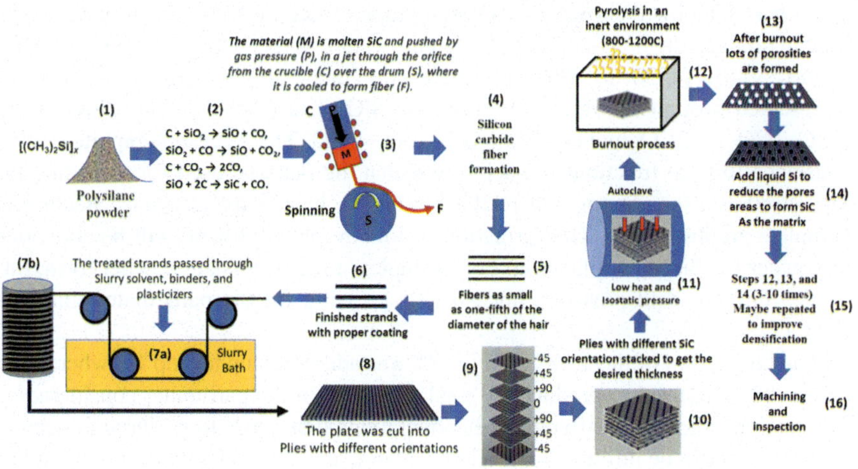

Fig. 6.5 Steps 1–16 toward making SiC/SiC plies and CMCs by melting the polysilane powder

$K_{Ic} \sim 25$–30 ksi(in)$^{0.5}$, and the value of K_{Ic} (currently being used for steel in aerospace industries) is above 80 ksi(in)$^{0.5}$. Note that 1 MPa(m)$^{1/2} \sim 0.91$ ksi(in)$^{1/2}$.

The preparation for fabricating SiC/SiC CMCs starts from (1) making the SiC fibers, (2) preparing the SiC matrix, and finally, (3) incorporating reinforcement into the matrix. One of the methods of fabricating the reinforcement is by using polysilane powder to produce the SiC fibers [24–27]. The following example provides a detailed description of one of the methods of fabricating SiC/SiC by starting from using the polysilane powder, as shown in Fig. 6.5. Polysilane is mainly used in industries as a precursor to form the SiC by first decomposing it through heat; the process is called thermolysis (a chemical decomposition caused by heat). The steps are as follows (Fig. 6.5):

In step 1, the polysilane powder is prepared and then converted to SiC through decomposition as shown in step 2. Note that the SiC can also be obtained by combining silica sand (silicon dioxide, SiO_2) and carbon in an Acheson graphite electric resistance furnace [28] at a high temperature range, between 1600 °C (2910 °F) and 2500 °C (4530 °F). Then by spinning the melted SiC over the drum, (step 3), it is possible to turn the melting SiC into high strength, low density, corrosion resistance silicon carbide fiber strands of SiC (the drum is shown in Fig. 6.5 by S). Each fiber is as thin as one-fifth of the human hair diameter, and a bunch of them together forms a strand that can be useful for further processes (steps 4–6). Strands go through several special coatings to enhance their strength and other properties. Subsequently, they will pass through the slurry bath of solvent, binders, plasticizers, and then wet winding around the drum (steps 7a and 7b). Subsequently, transforming it into tapes with fibers aligned and oriented in one direction, acting as a wide unidirectional ply (steps 8–9). The large ply that has been processed will then be cut by a special cutting machine into smaller plies and will stack on top of each other with different orientations, as illustrated in Fig. 6.5 (step 10).

As was mentioned above, the Acheson graphite electric resistance furnace can be used to produce the SiC matrix at a high temperature range, between 1600 °C (2910 °F) and 2500 °C (4530 °F). The chemical reaction in the SiC formation process can be represented through four steps: (1) $C + SiO_2 \rightarrow SiO + CO$; (2) $SiO_2 + CO \rightarrow SiO + CO_2$; (3) $C + CO_2 \rightarrow 2CO$; and (4) $SiO + 2C \rightarrow SiC + CO$. The Acheson graphite electric resistance furnace functions in such a way that the material used for providing the SiC, itself is a conductor, where the product is heated by resistance heating to complete graphitization. The formation of SiC in the solid form will be converted to powder for further application. For example, in the solid-state powder form, SiC can go through cold pressing, hot isostatic pressing for the particulate-reinforced CMCs.

The fabrication of the CMCs of Fig. 6.5 was addressed up to step 10, where plies of step 10 were used to produce the final geometry of the part under consideration. The stack of plies with different fiber orientations first needs to go through decomposition so it can be infiltrated with liquid silicon and subsequently turn into the SiC matrix. Therefore, the plies of step 10 will be placed into the autoclave with high pressure and low temperature to be adhered together (step 11). Afterward, as was mentioned above, they will go through the burnout pyrolysis process (step 12). Pyrolysis is a chemical decomposition caused by heat without oxygen, which results in the formation of porosities in the plies (step 13). Therefore, densification through melt infiltration is needed to increase the density of the final product by filling porosities introduced due to pyrolysis by silicon [26]. The burnout process of step 13 will be done by converting all the organic material into carbon. The densification will be performed by infiltrating the molten silicon into the pores plies after the pyrolysis process is performed (step 14). The process should be repeated several times to reduce the porosity amount to an acceptable level (step 15). Here, the molten silicon and the carbon will be combined to produce the SiC as the matrix with SiC as the reinforcement.

Finally, when the part is fabricated, the final machining, grinding, mechanical testing, and nondestructive inspection (NDI) will follow (step 16 of Fig. 6.5).

6.6 Sol–Gel Infiltration Approach (A Wet Chemical Approach)

Sol–gel process is one of the low-cost fabrication techniques that can be used to produce CMC materials from the formation of small particles in the presence of a precursor (it is defined as something that comes before and leads to the development of something else. In this case the precursor is SiO_2) in a solvent solution [29–32]. This technique is used to prepare the SiC matrix of CMCs from small size solid particles that are so small that even the gravitational force is not adequate to bring them down and hence are floating in the water (i.e., floating ultrafine particles in a liquid). Water acts as a catalyst, which solid particles first soak and then convert into

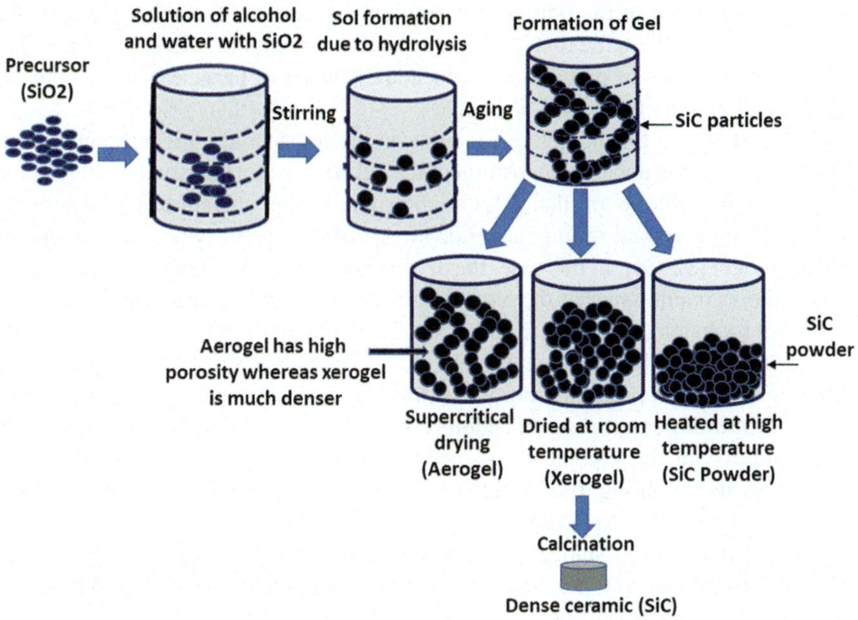

Fig. 6.6 The sol-gel process for fabrication of SiC begins by using the SiO_2 as a precursor going through hydrolysis. The process must be repeated several times in order to obtain an adequate amount of SiC ceramic

a solid (gel) immersed in the solvent. Not only SiC matrices but also other CMC matrices, such as carbon (C), alumina (Al_2O_3), and many others, can be fabricated by the sol–gel process.

The sol–gel process is a wet-chemical technique where the sol (or solution) evolves gradually from a liquid state towards the formation of gel having two phases, which basically is a solid-state network in a liquid. In this process, a small mass is formed when the solvent evaporates, and the remaining particles begin to join. After drying and firing, gels are converted into the ceramic matrix. In brief, the sol–gel process involves the following steps: (1) hydrolysis and polycondensation, (2) gelation, (3) aging, (4) drying, (5) densification, and (6) crystallization, where the hydrolysis of the precursor is done primarily on the acidic or basic mediums (Fig. 6.6).

In the case of silicon oxide, SiO_2, as a precursor, it should be dissolved in water (acting as a catalyst) and used as a starting material. In the sol–gel process, the precursor must be chosen such that it will tend to form gels, but the silicon oxide is not soluble in the water and the question of how to dissolve SiO_2 in the water can be done by adding alcohol into the solution. This addition will cause the precursor to be solved gradually in alcohol. To accelerate the process, the act of serious stirring will be helpful for the water molecules to react with the substances. This process is called hydrolysis, where the reaction between the catalyst and the precursor (the silicon oxide, e.g., SiO_2) will take place. Therefore, the sol solution contains tiny solid ionic particles suspended in the liquid. This solution should stay for a few days, so the aging will occur where the time and the heat will be helpful to change the sol to gel. That is, the network of particles starts to shape (cluster of particles) throughout the liquid medium, and thus gel is formed.

The next step is for the gel to dry either in the room or at a higher temperature, where densification and decomposition of the gel will occur. The drying will result in the complete collapse of the gel into powder. If the undisturbed gel network undergoes supercritical drying, also known as critical point drying, the outcome is called aerogel [32–34]. In this case, the drying step is performed inside an autoclave, which allows overpassing of the critical point of the solvent. Furthermore, if the gel is dried at room temperature, it is called xerogel (from the Greek word "xero," which means dry). Xerogels and aerogels (see Fig. 6.6) can be described as dried gels that maintain (at least in part) their porous texture after drying. Later the porosities must be minimized through the densification process. Finally, if the gel is heated to high temperature with the complete collapse of the gel, it will be converted to SiC powder, as shown in Fig. 6.6. As it is shown in this figure, the degree of shrinkage experienced by the xerogel after drying is higher than aerogel.

The gel that is dried and formed aerogel has greater porosity and is less dense when compared with xerogel. Since the ceramic amount in the gel is relatively low, it is due to significant shrinkage occurring after drying. Therefore, the sol–gel process must be repeated several times to obtain an adequate amount of ceramic for the SiC matrix. Figure 6.7 shows the flowchart associated with the sol–gel process starting from precursor (here it is SiO_2) to the formation of aerogel, xerogel, and powder for fabrication of SiC or other ceramic matrices mentioned above (carbon (C), and alumina (Al_2O_3)). This figure is analogous to Fig. 6.6, showing the steps necessary to obtain the SiC matrix from the sol–gel technique. The SiC powder can be used for sintering process, where particles under pressure and heat will produce a solid mass of material. In brief, in the sintering process, the atoms in the materials diffuse across the boundaries of the particles due to the applied heat and pressure, fusing the particles together and creating one solid mass of material.

6.6.1 Advantages and Disadvantages of Sol–Gel Process

Some of the advantages of the sol–gel process are (1) less damage on the reinforcing fibers due to relatively low temperature processing, which usually is performed at about room temperature or slightly above room temperature, (2) controllable matrix composition, (3) it can create fine powders for further sintering process by applying heat and pressure, (4) low equipment costs, (5) low machining cost due to near net shape fabrication, (6) complex parts can be fabricated, (7) it produces compositions not possible by solid-state fusion, and (8) high-quality materials with homogeneity and purity at lower temperatures than conventional methods. Products formed from the sol–gel process have many applications, such as optics, electronics, energy, space, sensors, and medicine.

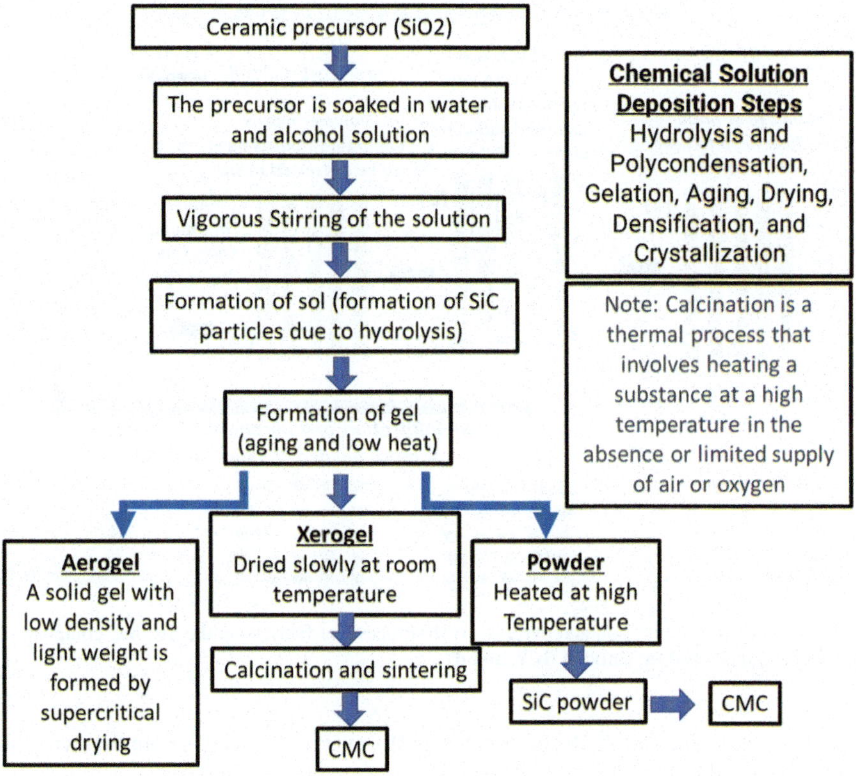

Fig. 6.7 Chart illustrating the steps for the fabrication of SiC ceramic matrix through sol-gel process

Regarding disadvantages of the sol–gel process, there are a few limitations: (1) possible matrix cracking because of considerable shrinkage and porosity, (2) multiple infiltration–gelation cycles are needed to increase the ceramic density, (3) low mechanical properties due to the presence of voids, and (4) a very time-consuming process especially when multiple infiltration–gelation cycles are needed.

6.7 Polymer Infiltration Pyrolysis Approach (PIP)

The sol–gel process for obtaining the CMCs was discussed in Sect. 6.6. In this section, the polymer infiltration pyrolysis (PIP) technique will be discussed. With this method, the CMC matrix and fibers can be fabricated through pyrolysis. The term pyrolysis means thermal decomposition of materials at elevated temperatures in an inert or low-oxygen atmosphere. The PIP fabrication technique consists of the infiltration of a low viscosity liquid polymer, such as polycarbosilane, into the

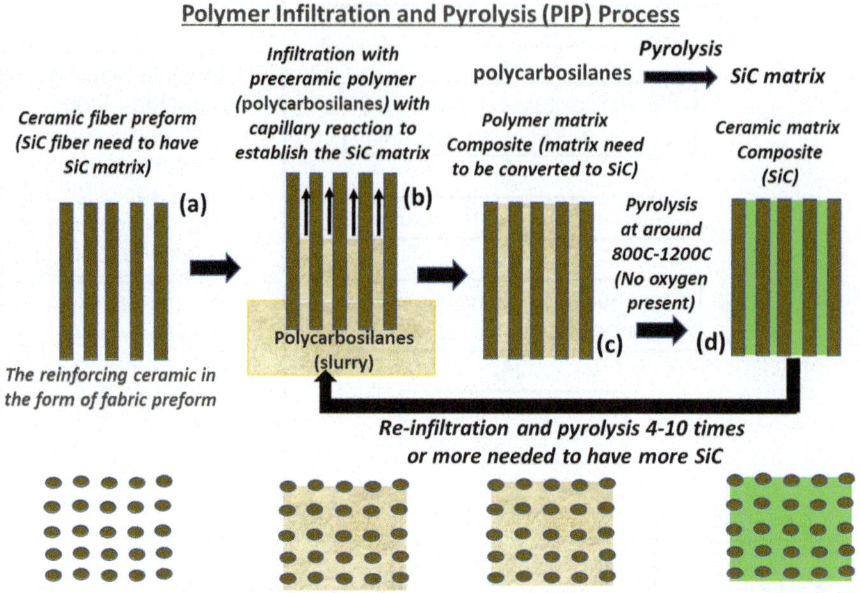

Fig. 6.8 Polymer infiltration and pyrolysis (PIP) process for fabricating the SiC/SiC composite is shown and described (**a**) through (**b**, **c**, and **d**)

reinforcing ceramic in the form of fabric preform (Fig. 6.8a, b), where the pyrolysis will follow through heating the polymer precursor. The precursor is the initial material that is used as the starting point for the processing of the final material. The ceramic matrix is formed from a fluid infiltrating the fiber reinforcement [25–27]. In another word, the thermal decomposition of an organic substance occurs at high temperatures in the presence of an inert atmosphere to avoid oxidation (Fig. 6.8c) and the infiltrated polycarbosilane will be changed to silicon carbide, SiC, upon pyrolysis process at a temperature between 1472 °F and 2192 °F (800–1200 °C) (Fig. 6.8d). To have more SiC in the matrix, the process should be repeated several times to fill the porosities until adequate densification has been achieved.

The PIP method is based on the concept of filling the porosity inside the fiber fabrics after the pyrolysis step, where decomposition occurs. In Fig. 6.8d, the SiC ceramic matrix, resulting from the decomposition of preceramic polymer precursors (polycarbosilanes) will be repeated several times to fill up most of the porosity created as the result of pyrolysis. Molecules of preceramic polymers generally contain carbon (C) and silicon (Si) but may also contain boron (B), aluminum (Al), and titanium (Ti). Because the liquid polymer is used for infiltration, the process is also called liquid polymer infiltration or simply LPI. Both the PIP and LPI involve the infiltration of a polymer precursor into a porous ceramic matrix, which is subsequently converted into a ceramic phase.

The PIP and LPI techniques were briefly mentioned in Fig. 6.5, where polymer infiltrated in plies of different orientations that have already undergone pyrolysis at a temperature between 1472 °F and 2192 °F/800 °C and 1200°C (step 12 of Fig. 6.5). A subsequent densification process took place through liquid silicon infiltration. Unstable products (some of which can be volatile), such as CO, H_2, CO_2, CH_2, and H_2O are released and as a result of pyrolysis, it created a porous ceramic matrix. In Fig. 6.8, the steps of infiltration and pyrolysis are demonstrated, and to reduce pores areas, the pyrolysis process should be repeated a few times to achieve an acceptable level of matrix densification by allowing the polymer to infiltrate into the open pores and leading into more SiC in the matrix. A typical number of PIP repeats is between 6 and 10 or more, depending on the final densification and desired properties.

Moreover, the level of porosity should fall below an acceptable level to avoid premature cracking of the part because ceramics are susceptible to cracking in the presence of defects. For that reason, porosity should be minimized. Therefore, nondestructive inspections prior to the part usage are essential to check for porosity and other types of defects in the CMC parts.

The pyrolysis approach is preferred over other high temperature CMC processing methods such as sintering and hot pressing. Both sintering and hot pressing can result in high porosity, mechanical damage to the fibers, and fiber ruining as chemical reactions occur at the interface between the fibers and the matrix.

The preparation of fiber reinforcement in the PIP method includes not only shaping fibers into desired shapes but also requires having a strong interface to the matrix by coating the interface. The interfacial coating helps to transmit the load from the matrix to the fiber to improve the mechanical properties. In addition, the presence of coating can protect the interface and fibers from damage due to the high-temperature molten materials during the pyrolysis process. Therefore, proper coating material that can withstand high temperatures will be helpful.

6.7.1 Advantages and Disadvantages of PIP Method

The advantages of polymer infiltration and pyrolysis (PIP) can be summarized as follows: (1) fiber damage is prevented due to the fabrication process that takes place at a relatively low temperature. By using other processes it may require high temperature near the melt, where damage to reinforcement can be possible; (2) different types of fibers (particulate, fibrous in the form of short and long) may be used; (3) ceramic matrices of different compositions (silicon carbide, silicon nitride, silicon carbonitride) may be obtained through the PIP fabrication method, (4) The densification process can be repeated until desire matrix densification can be obtained, and (5) The remaining silicon in the matrix will be minimum after multiple infiltrations, pyrolysis, and densification cycles. However, the disadvantages of the PIP are as follows: (1) the fabrication time is relatively long due to the multiple infiltration–pyrolysis cycles in order to achieve adequate infiltration and densification (usually four to ten or more may be necessary); (2) there can be residual porosity left behind that can degrade the mechanical properties of the composite, as well as, the possibility of crack growth and final cracking when subjected to load environment; (3) telatively high production cost, higher than liquid silicon infiltration (LSI) method, which will be described below.

6.8 Liquid Silicon Infiltration (LSI) Method

Currently, there are four processing methods for fabricating CMC parts that are used for space and other applications. Both the sol–gel method and polymer infiltration pyrolysis (PIP) were already discussed in Sects. 6.6 and 6.7, respectively. The CVI approach, including the chemical vapor deposit (CVD) techniques, and powder processing, will be discussed in Sects. 6.9 and 6.10, separately. For nonoxide CMCs, the liquid silicon infiltration (LSI) process of fabricating the CMCs will be discussed in this section, where it is a fast, low residual porosity, and cost-effective method performed through liquid silicon (Si) infiltration [35, 36]. In this process, the ceramic matrix (SiC) is formed due to the chemical interaction between the infiltrated liquid melt (Si) into a porous reinforcing preform where polymer fibers were converted to carbon by the pyrolysis process resulting in the silicon carbide matrix.

Figure 6.9 illustrates the LSI fabrication process. Here, the polymer impregnated with SiC fibers (Fig. 6.9a) are first going through pyrolysis at high temperature

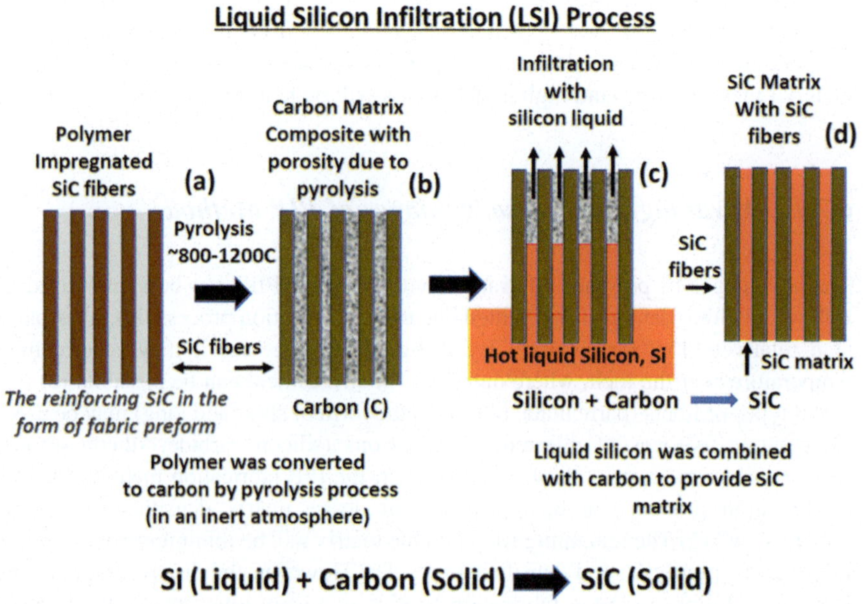

Fig. 6.9 Liquid silicon infiltration (LSI) process is shown. Here liquid silicon is infiltrated into the carbon matrix to form SiC matrix. The process is illustrated from step (**a**) through (**d**)

(Fig. 6.9b), where the thermal decomposition of an organic, containing carbon, will take place. Then the polymer will turn into a carbon matrix, as shown in the figure. The silicon carbide matrix will be formed due to the chemical interaction between liquid silicon and carbon (Fig. 6.9c). This process has been extensively used in the fabrication of SiC matrix composites. After the pyrolysis process, the carbon porous preform will be infiltrated with molten silicon (Si) at a temperature above 1200°C through the capillary forces. The surface of carbon preform is wetted by Si liquid. The melt reacts with the carbon matrix to form a silicon carbide matrix (SiC) according to the reaction shown in Fig. 6.9d. The LSI process was also shown briefly in Fig. 6.5, wherein step 13 pyrolysis was conducted and in step 14, the molten silicon was infiltrated to fill the porous areas. The process was repeated several times for the desired densification to be reached.

In the LSI process, large pores because of pyrolysis help to get great silicon melt infiltration. However, it may also cause pores areas filled with free silicon (Si) where incomplete chemical interaction with carbon can take place. That is, pores areas will be infiltrated with silicon only without the chemical reaction going on with carbon. On the other hand, if the amount of pores areas after the pyrolysis process is not enough for infiltration (due to the lack of pores), it can lead to poor properties.

That is, free available silicon without interaction with carbon can occur because of a lack of pores availability and not enough pyrolysis. It should be noted that, in this process, the ceramic matrices (SiC) developed by the LSI process are much denser (have low residual porosity) as compared to the composites fabricated by other methods such as polymer infiltration and pyrolysis (PIP) or the CVI (will be described in Sect. 6.9).

6.8.1 Advantages and Disadvantages of the LSI Process

Advantages associated with this method are as follows: (1) the LSI process is considered as a low-cost CMC fabrication method; (2) it has a short production time when compared with other methods; (3) it has very low residual porosity; (4) the LSI process is a near net shape process without repeating the infiltration steps as was the case for the PIP and gel-sol techniques. Disadvantages of the LSI process are as follows: (1) the high heat from liquid silicon may damage fibers. To overcome this drawback, the fibers are coated with weak interphase layers using the CVI method (Sect. 6.9); (2) the presence of residual silicon (extra silicon) in the matrix can influence the mechanical properties; (3) poor mechanical properties will result if the process is not fully completed [37].

In summary, the LSI process provides a dense matrix, which means less open porosity leading to less cracking when parts are subjected to monotonic or cyclic loading. In addition, it has greater thermal conductivity, higher proportional limit, lower oxidation, less environmental degradation, and it has significantly lower processing or fabrication time when compared with other methods.

6.9 Chemical Vapor Infiltration (CVI) and Chemical Vapor Deposition (CVD)

Manufacturing structural parts using ceramic matrix composite material is not an easy task. It requires the preparation of both matrix and fibers, and in some cases, it requires applying pressure and high temperature, as it was shown in the case of LSI and PIP fabrication techniques. Engineers must realize that in selecting any given approach, the CMCs mechanical properties are strongly dependent on the properties of ceramic matrix materials, fibers, the volume fraction of fibers in the matrix, and the interface strength between the matrix and the reinforcement. In the case of the ceramic matrix alone that is highly brittle, the mechanical behavior is very sensitive to small defects (i.e., of the order of a few micrometers) and thus should be processed very carefully to minimize the formation of small flaws and surface scratches as much as possible. That is, the tolerance of ceramics in the presence of flaws is very low, and very little energy is needed to propagate the crack (i.e., low fracture toughness, K_{Ic}, or low energy release rate, G_{Ic}). In addition, engineers must also consider the fact that ceramic matrix alone has high melting points 4532 °F (~2500 ° C), which can cause damage to fibers in the matrix if the fabrication technique requires going through high temperature process. In some cases, it may be necessary to coat the fibers to avoid any damage to fibers and to have a strong interface between the matrix and fibers [38]. Any fabrication technique that can be avoided by the manufacturer by not going through the high temperature process will be useful to the industries that are interested in using the CMC material for their applications. Chemical vapor infiltration (CVI) is a chemical vapor deposition method, one of the advanced low temperature techniques, that can be implemented by manufacturers to obtain fiber-reinforced ceramic composites for industry use [39–42]. In this case, the CVI method will minimize the loss of mechanical properties of the fibers owing to the use of much lower temperature and pressure than both the LSI and PIP processes. However, it should be noted that the CVI process is quite slow and can take from a day to more than a week to fabricate the CMC parts and has poor duplicability. That means the reproducibility of fabricating the same part with the same properties is low.

Chemical vapor infiltration is a CMC process that allows the fabrication to take place at a much lower temperature, whereby the matrix material will infiltrate into the porous fibrous preforms in the form of reactive gases to form the ceramic matrix composites. As shown in Fig. 6.10a, fibers are initially free from infiltration, and after being infiltrated by the chemical vapor, the ceramic matrix will be deposited on the fiber (Fig. 6.10b). The deposition of gases will continue with time until the ceramic matrix grows around the fiber, with some small amount of porosity in the matrix left behind (Fig. 6.10c).

In the CVI process, the infiltration of matrix on the fibers through deposition of gases will gradually fill the space between fibers. In effect, the matrix material is considered as the deposited material and dispersed phase is the fibers preformed (Fig. 6.10). In Fig. 6.11, the chemical vapor infiltration process is shown, where the

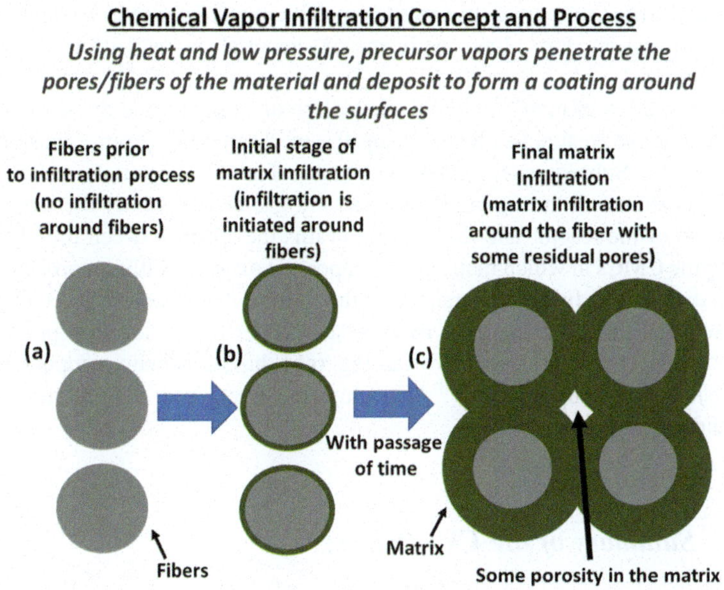

Fig. 6.10 Chemical vapor infiltration concept and process are shown through steps (**a**) through (**c**). Note: the matrix grows around the fibers. The process leaves some porosity in the matrix

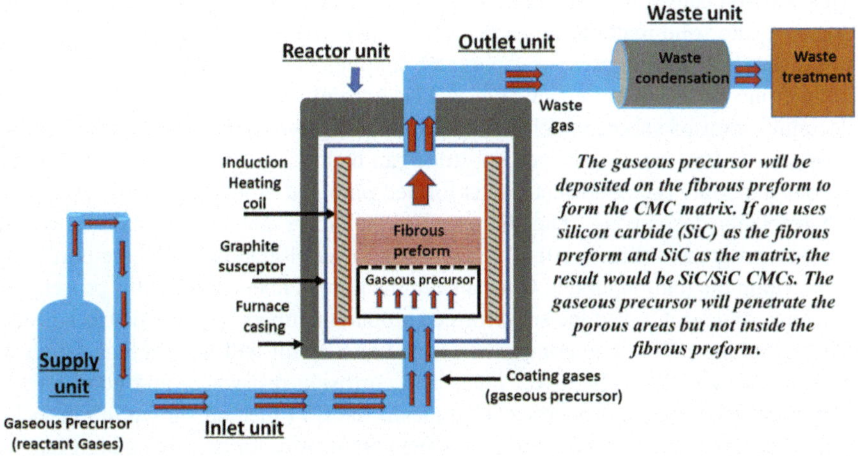

Fig. 6.11 A simplified illustration of the CVI concept where precursor gases will be deposited on the surface of the porous fibers to form the CMC matrix. The supply, inlet, outlet, and waste units are shown

fibrous preform is supported on a porous metallic plate (substrate) through which a mixture of carrier gas along with matrix material is passed at an elevated temperature.

The infiltration of matrix material takes place in a reactor, and the waste material will undergo the treatment plant to keep the work area free from exposure to the hazard waste. Subsequently, gases and residual matrix material are chemically treated. As shown in the figure, the matrix and gas material enter the reactor from the bottom of the reactor and the deposition fills the space between the fibers, thus forming the CMCs in which matrix is the deposited material. Filling porosities by the CVI method results from mass transfer in the form of the deposit of reactant gas into the pores preform, which occurs by diffusion under a constant low pressure and temperature. The CVI process shown in Fig. 6.11, has three main components: (1) a feed system or the supply unit, (2) a heating chamber, where the preform has been kept, and (3) an effluent system.

6.9.1 Summary of the CVI Process

Chemical vapor infiltration is a process in which a solid material is deposited in a porous structure by the disintegration of vapors. Inside the heated chamber, the gaseous precursors decompose at high temperatures. This decomposition releases the elements they contain, which then react with the surface of the porous structure. This reaction causes the deposition of solid material onto the structure's surface. In this case, the solid material is the ceramic matrix, the porous material is reinforcement, and the reactant gases (or vapors) are supplied to the furnace, flow around, and diffuse into the porous preform (Fig. 6.11). A solid material is formed because of decomposition of vapor, which is deposited from gaseous reactants onto a heated substrate. In this case, the heated substrate can be the reinforcement material or the porous material. The solid material is formed onto and around the porous fibers as the reactants decompose. The diameter of the fibers gradually increases as the reaction progresses (Fig. 6.10). The deposited material gradually fills the pores and gaps within the reinforcement. One can assume porous fibers used in this process are silicon carbide (SiC), and the solid formed around the fibers (i.e., the matrix), is also silicon carbide (SiC). The composite formed as a result will have a high strength, high modulus of elasticity, great creep and corrosion resistance, and good fatigue and fracture toughness properties. That means the CVI method shown in Fig. 6.11, can fabricate the SiC/SiC composite suitable to industries for producing high-quality CMC parts. The equation describing the formation of the SiC matrix, due to the decomposition of methyltrichlorosilane (CH_3SiCl_3) gas is shown as:

$$CH_3SiCl_3(\text{gas}) \rightarrow SiC(\text{solid}) + 3HCl(\text{gas})$$

Here, the 3HCl is the waste material and must be removed.

6.9.2 The Chemical Vapor Deposition (CVD) Process

A similar technique called chemical vapor deposition (CVD) is also used for deposition, where it involves depositing the gaseous precursor on the substrate. The main difference between CVI and CVD is being that the deposition of CVD is on hot bulk surfaces, while the CVI deposition is on porous substrates. The CVD process is a deposition method performed in the vacuum and used to produce high-quality solid materials or it can be used to enhance the surface properties of the substrate material through the deposition process for prevention of surfaces from oxidation and corrosion that can degrade their mechanical properties (Fig. 6.12). In a typical CVD process, the wafer (or the metallic substrate) is exposed to one or more precursors, which react and/or decompose on the substrate's outer surfaces to produce the desired deposit. In general, the process involves the deposition of a layer or layers of a substance (coating) onto a surface [43, 44]. That is, the CVD is used to create a coating or a thin layer of a different material on the surface of a substrate to alter its properties like conductivity and corrosion resistance. The protective coating is frequently used in aerospace for turbine blades and other industries, which run at very high service temperatures and in a very oxidizing atmosphere. Other applications are the thin film coating shown in Fig. 6.12, which can be utilized for surface coating of material that needs to be waterproofed (aka

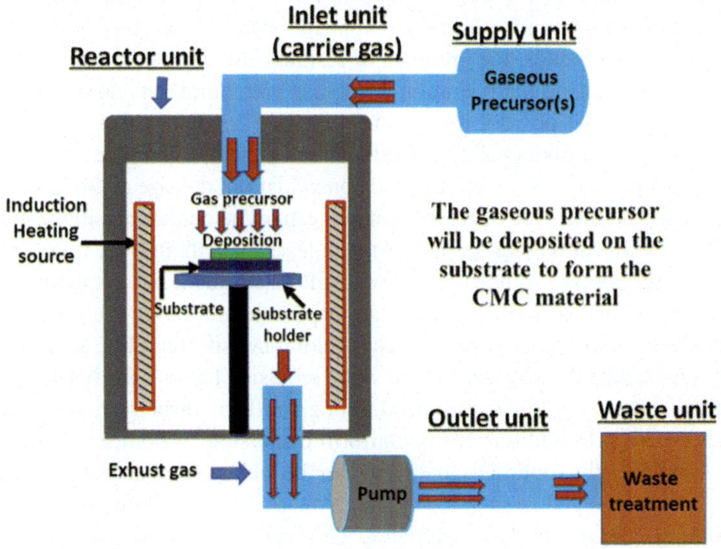

Fig. 6.12 A simplified illustration of the CVD concept where the precursor gases will be deposited on the surface of the substrate. The CVD process can be employed to modify the surface properties of the composite, enhance its performance, or add specific functionalities to the CMC part. The supply, inlet, outlet, and waste units are shown

water hydrophobic). That is, a water drop will not be absorbed on the surface by the surface treatment provided by the CVD process. Also, they can be applied to the surface of parts that need surface protection from damage, such as scratches or other physical wear. Finally, another important application of CVD is the infiltration of precursor gases on the surface of reinforcements to fill the pores areas.

Frequently, in the CVD process, volatile by-products gases are produced, which are removed by gas flowing through the reaction chamber. The volatile by-products are connected to a waste treatment plant and subsequently, the waste gases and residual matrix material are chemically treated as shown in Fig. 6.12.

6.10 Powder Processing CMCs: Hot and Cold Pressing

In this chapter, the sol–gel, polymer infiltration pyrolysis (PIP), liquid silicon infiltration (LSI), and chemical vapor infiltration (CVI) fabrication methods were discussed in detail. Some of these processes are mostly required to have high-temperature fabrication settings, where both the matrix and the reinforcement would be exposed to a high temperature environment. As was mentioned before, the presence of high temperature can damage the fibers or whiskers that are bonded to the matrix, which will result in the breakage of fibers and degradation of the mechanical properties of composites as a whole. In addition, the difference in coefficients of thermal expansion between the matrix and reinforcement can lead to residual stresses upon cooling the composite material to room temperature. Just like in the case of metals, the presence of residual stresses can be superimposed on the applied load during the operation and cause premature structural failure.

In the case of the powder processing technique for fabricating the CMC parts, both the matrix and fibers must be available in the form of powders. Because of the nature of this process, that is mixing and consolidation, the size of fibers or whiskers will be small and therefore the reinforcement cannot be continuous in the matrix and the effect of fibers as a crack stopper is much less. Even though the reinforcement is very small, the process of crack arrest will still be efficient, but not to an acceptable degree. Moreover, to make the fabrication process effective, the particles in the powder matrix must possess the highest packing density possible. It is known that the highest packing density is achieved when repulsive forces exist between particles and voids content should be minimized through uniform dispersion of the reinforcement and matrix. In addition to the uniform dispersion of both reinforcement and matrix material, the optimum packing can be achieved when the particle size distribution is arranged in such a way that around 30% by volume is small particles and 70% by volume of large particles. This percent distribution apparently can provide maximum packing arrangement when using the powder processing technique [45].

6.11 A Brief Summary of CMC Materials and Difficulties

Even though there are fabrication issues related to ceramic matrix composites for industry use, such as the time of manufacturing parts, high temperature involvement, and the possibility of damage done to the reinforcements, but CMCs offer a breakthrough in performance in extreme environments like their outstanding resistance to high service temperature, corrosion, and wear. From a mechanical properties' perspective, they are attractive due to their high specific modulus (E/ρ) and specific ultimate (σ_{Ult}/ρ) properties, when they are compared with metallic counterparts, such as aluminum, steel, and titanium alloys. In addition, due to the presence of fibers or whiskers, they can delay crack growth and prevent the possibility of catastrophic failure and loss of life. In this chapter, SiC/SiC composite was examined and several methods of manufacturing it, its superior mechanical properties, its application at high service temperatures, and its applications in industries were discussed. Similarly, other types of CMCs, such as Al_2O_3/Al_2O_3 or C/C and C/SiC, can also be fabricated through the abovementioned methods.

These ceramic matrix composites must be well established before their applications in aerospace and auto industries because catastrophic failures must be avoided at all costs. Therefore, a comprehensive amount of testing, analysis, and inspections are needed to be conducted by engineers to be able to assess the integrity of parts during their operational life. From testing and analysis perspectives, the mechanical properties characterization of composite material through testing is much more complex and costly than that of metallic material. This is because composite materials are not isotropic, and their properties are not the same in all directions. Static, fatigue, and fracture mechanics data for ceramic matrix composites must be available based on a reliable standard for engineers to be able to assess the life of the parts when subjected to an in-service load environment. Due to the anisotropic nature of these materials, gathering comprehensive static, fatigue, and fracture mechanics allowables for analysis purposes is not an easy task and would be costly, labor intensive, and time-consuming. In addition, more tests are needed if the exposed environment is different than room temperature. A slightly higher or colder working environment will necessitate establishing more test data to find out if mechanical properties will be affected or not by the environmental temperature under consideration. This data, together with a reliable computational tool, must be provided to engineers to conduct their structural integrity assessment. Following testing standards (i.e., ASTM, EN, ISO, JIS) provides consistency and reliability in the testing and characterization of ceramic composites. They cover mechanical properties, fracture toughness, and other essential aspects of CMC materials. Another possible approach to overcome the difficulty associated with testing is to establish an acceptable virtual testing technique, that would be greatly helpful to engineers to obtain CMC material allowables with a minimum amount of testing and much less time. The virtual testing approach will be helpful to both engineers and industries that need to have this data in a short time. Chapter 9 briefly describes the virtual testing methodology based on the extended Griffith theory. The proposed technique can estimate the fracture toughness, G_{Ic}, and fatigue crack growth of metallic material by having only the full stress–strain curve of the material under study. The virtual testing technique described above can be extended to composite material. Another approach to virtual testing is through molecular dynamics simulation and it is described briefly in detail Chap. 9.

Another area that makes ceramic matrix composite more challenging to deal with is the repair of CMC-damaged parts compared to that for metals. This area is still in its early stage and requires a great amount of effort by engineers to bring it to a mature stage that could be reliable enough for the industry to apply when repairing composite parts is needed. Chapter 13 describes the method of repairing composite parts that engineers believe the part needs to be repaired and be able to use it again for service operation.

Finally, structural failures can cause loss of life, which can damage the reputation of the industry under consideration. Expensive lawsuits due to loss of life and high cost associated with the loss of structure can economically damage companies involved in manufacturing ceramic matrix composite parts. For all the above reasons, manufacturing CMC parts must be expensive, but their unique performance in high temperature environments and their lightweight properties make them extra useful for industrial applications. It is up to the user to decide if the ceramic matrix composite parts are economically beneficial to be used or not. However, as customers and engineers begin to be comfortable with the type and nature of CMC materials, other industries will begin to come forward and costs will eventually become more feasible.

6.12 Concluding Remarks

In summing up discussions on ceramic matrix composites (CMCs), it is important to emphasize their significance in various industries, particularly when considering ceramic parts exposed to high-temperature service environments. These advanced materials, comprised of ceramic fibers embedded in a ceramic matrix, exhibit exceptional strength, high temperature resistance, corrosion resistance, and lightweight properties. Several techniques were discussed in this chapter for fabrication of CMC parts and advantages and disadvantages of each process were presented. These materials hold great promise for applications in aerospace, aircraft, automotive, and other industries where CMC parts are exposed to elevated temperature.

Ongoing research and development aim to improve fabrication techniques and enhance both mechanical and thermal properties, paving the way for innovative solutions in engineering and technology. As we move forward, the exploration of new manufacturing techniques and materials will undoubtedly contribute to the evolution and widespread adoption of ceramic matrix composites in diverse industrial sectors. Furthermore, there must be more emphasis given to the stress analysis of CMC parts. However, there is a need to have material allowables available in order to go forward with a comprehensive structural integrity and failure analysis.

These analyses are necessary for customers to verify whether critical CMC parts can withstand the operating environment during their service usage. It should be noted that material allowables must be obtained through standard testing that would be acceptable to the customers. Moreover, due to the anisotropic nature of CMCs, numerous tests should be conducted for material allowables, such as ultimate tensile strength, ultimate compressive strength, shear strength, and modulus of elasticity, all in the longitudinal and transverse directions to the fibers. Otherwise, the results of the analysis for CMC parts could be unconservative, which can lead to the premature failure of structural composite components.

References

1. T.R. Cooke, Inorganic fibers – A literature review. J. Am. Ceram. Soc. **74**(12), 2959–2978 (1991)
2. K. Kumagawa, H. Yamaoka, M. Shibuysa, T. Ymamura, Fabrication and mechanical properties of new improved Si-M-C-(O) Tyranno fiber. Ceram. Eng. Sci. Proc. **19A**, 65–72 (1998)
3. F. Zivic, N. Busarac, S. Milenković, N. Grujović, General overview and applications of ceramic matrix composites (CMCs), in *Encyclopedia of Materials: Composites*, (Elsevier, Amsterdam, 2021)
4. G. Daugherty, General Electric Aviation building U.S. blueprint to industrialize CMCs, by GE Aerospace (22 February 2017)
5. Aerospace Manufacturing and Design, "GE Aviation is building to industrialize CMCs". Posted by Eric Brothers, 28 February 2017
6. R.R. Naslain, M.R. Pomeroy, Ceramic matrix composites: matrices and processing. Materials Science and Materials Engineering (Elsevier 2016), pp. 106–122
7. X. Fan, X. Yin, Progress in research and development on matrix modification of continuous fiber-reinforced silicon carbide matrix composites. Adv. Compos. Hybrid Mater. **1**(4), 1–11 (2018)
8. B. Farahmand, *Application of High and Low Cycle Fatigue on Metallic and Non-Metallic Material* (Amazon Self Publisher, Seattle, 2022). Chapter 12
9. G. Xiguang, L. Long, S. Yingdong, A temperature-dependent constitutive model for fiber-reinforced ceramic matrix composites and structural stress analysis. Int. J. Damage Mech. **23**, 507 (2013)
10. S. Zhang, Y. Feng, X. Gao, Y. Song, F. Wang, S. Zhang, Modeling of fatigue failure for SiC/SiC ceramic matrix composites at elevated temperatures and multi-scale experimental validation. J. Eur. Ceram. Soc. **42**, 3395–3403 (2022)
11. A.G. Evans, F.W. Zok, R.M. McMeeking, Fatigue of ceramic matrix composites. Acta Metall. Mater. **43**(3), 859–875 (1995)
12. X. Kong, Y. Qiao, Fracture in ceramic-matrix composites reinforced with strongly bonded metal particles. Mech. Compos. Mater. **41**, 205 (2005)
13. K.K. Kar, *Composite Material—Processing, Applications, and Characterization* (Springer, Berlin, 2017), p. 471
14. Z. Poniżnik, Z. Nowak, M. Basista, Numerical modeling of deformation and fracture of reinforcing fibers in ceramic-metal composites. Int. J. Damage Mech. **26**(5), 711 (2015)
15. L. Zoli, D. Sciti, Efficacy of a ZrB 2–SiC matrix in protecting C fibers from oxidation in novel UHTCMC materials. Mater. Des. **113**, 207–213 (2017)
16. L. Zoli, A. Vinci, L. Silvestroni, D. Sciti, M. Reece, S. Grasso, Rapid spark plasma sintering to produce dense UHTCs reinforced with undamaged carbon fibres. Mater. Des. **130**, 1–7 (2017)

17. P. Galizia, S. Failla, L. Zoli, D. Sciti, Tough salami-inspired C_f/ZrB_2 UHTCMCs produced by electrophoretic deposition. J. Eur. Ceram. Soc. **38**(2), 403–409 (2018)
18. H.-T. Lin, J. Hemrick, *Advanced and Refractory Ceramics for Energy Conservation and Efficiency*, Ceramic transactions, vol 256 (The American Ceramic Society, Westerville, 2016)
19. O.G. Diaz, G.G. Luna, Z. Liao, D. Axinte, The new challenges of machining ceramic matrix composites (CMCs). Int. J. Mach. Tools Manuf. **139**, 24–36 (2019)
20. F. Zivic, N. Palic, Z. Jovanovic, N. Grujović, Processing routes for ceramic matrix composites (CMCs), in *Encyclopedia of Materials: Composites*, vol. 2, (Elsevier, Amsterdam, 2021), pp. 20–36
21. Committee on Advanced Fibers for High-Temperature Ceramic Composites, *Ceramic Fibers and Coatings – Advanced Materials for the Twenty-First Century* (National Materials Advisory Board, National Academy Press, Washington, DC, 1998)
22. B. Harris, *Engineering Composite Materials* (The Institute of Materials, London, 1999), p. 171. Chapter 7
23. D. Kopeliovich, *Fabrication of Ceramic Matrix Composites by Polymer Infiltration and Pyrolysis (PIP)* (SubsTech (Substances & Technologies, 2012)
24. GE Aerospace, *GE's Super Material Sic/SiC. How the CMC Process Works [Video]*. YouTube. https://www.youtube.com/watch?v=mM-XM1ObH04YouTube (2018, September 12)
25. Z. Luo, X. Zhou, Y. Jinshan, K. Sun, F. Wang, Mechanical properties of SiC/SiC composites fabricated by PIP process with a new precursor polymer. Ceram. Int. **40**(1, Part B), 1939–1944 (2014)
26. R.R. Naslain, SiC-Matrix Composites: Nonbrittle Ceramics for Thermo-Structural Application. Int. J. Appl. Ceram. Technol. **2**, 75–84 (2005)
27. L. Jin, K. Zhang, T. Xu, T. Zeng, S. Cheng, The fabrication and mechanical properties of SiC/SiC composites prepared by SLS combined with PIP. Ceram. Int. **44**, 17 (2018)
28. What is Acheson graphitization furnace? Advantages of Acheson graphitization furnace, By Yvonne Mar 29, 2021
29. D.A. Hanaor, H. Chironi, I. Karatchevtseva, I. Triani, G. Sorrell, Single and mixed phase TiO_2 powders prepared by excess hydrolysis of titanium alkoxide. Adv. Appl. Ceram. **111**(3), 149–158 (2012)
30. C.J. Brinker, G.W. Scherer, *The Physics and Chemistry of Sol-Gel Processing* (Academic Press, Boston, 1990)
31. L.L. Hench, J.K. West, The sol-gel process. Chem. Rev. **90**, 33–72 (1990)
32. L. Klein, *Sol-Gel Optics: Processing and Applications* (Springer, New York, 1994)
33. C. Jeffrey Brinker, G.W. Scherer, Civil & environmental engineering, in *Sol-Gel Science: The Physics and Chemistry of Sol-Gel Processing*, (Academic Press, San Diego, 1990)
34. J. Zha, H. Roggendorf, *Sol-Gel Science: The Physics and Chemistry of Sol-Gel Processing* (Academic Press, Boston, 1991)
35. W. Krenkel, F. Berndt, C/C–SiC composites for space applications and advanced friction systems. Mater. Sci. Eng. **412**, 177–181 (2005)
36. K.K. Kar, *Composite Material—Processing, Applications, and Characterization* (Springer, Berlin, 2017), p. 482. Chapter 14
37. K. Raju, Y.-H. Seong, S. Kim, S.-H. Kim, I.-S. Han, H.-J. Bang, Fabrication of SiC_f/SiC composites through hybrid processing via chemical vapor infiltration, electrophoretic deposition, and liquid silicon infiltration. J. Asian Ceram. Soc. **9**, 1031–1037 (2021)
38. L. Li, *High Temperature Mechanical Behavior of Ceramic-Matrix Composites* (Wiley, Weinheim, 2021)
39. D.R. Petrak, *"Ceramic Matrices", Composites, Vol 21, ASM Handbook* (ASM International, Detroit, 2001), pp. 160–163
40. K.-H. Bang, G.-Y. Chung, H.-H. Koo, Preparation of C/C composites by the chemical vapor infiltration (CVI) of propane pyrolysis. Korean J. Chem. Eng. **28**, 272–278 (2011)
41. I. Singh, *Mod-06 Lec-04 Chemical Vapour Infiltration*. NPTEL YouTube Channel. National Programme on Technology Enhanced Learning. Archived from the original on 21 December 2021. Retrieved 21 January 2014

42. M. Balasubramanian, *Composite Materials and Processing*. eBook, pp. 417–412 (2013)
43. R. Sadri, Controlled physical properties and growth mechanism of manganese silicide nanorods. J. Alloys Compd. **851**, 156693 (2021)
44. L. Sun, G. Yuan, L. Gao, J. Yang, M. Chhowalla, M.H. Gharahcheshmeh, K.K. Gleason, Y.S. Choi, B.H. Hong, Z. Liu, Chemical vapour deposition. Nat. Rev. Methods Prim. **1**, 6 (2021)
45. F.F. Lange, D.C.C. Lam, O. Sudre, B.D. Flinn, C. Folsom, B.V. Velamakanni, F.W. Zok, A.G. Evans, Powder processing of ceramic matrix composites. Mater. Sci. Eng. **144**, 143–152 (1991)

Chapter 7
Delamination in Laminated Composites

7.1 Introduction

The failure mechanisms in composites can be due to (1) fiber failure because of tensile fracture, or local compressive fiber kinking, (2) matrix cracking, (3) fiber–matrix interface debonding, (4) failure because of damage caused by the loss of adhesion between two consecutive plies, which generally is called delamination, where the cause may be due to manufacturing anomalies during lay-up or curing or out-of-plane load. However, there is a strong acceptance among engineers that delamination is the most critical form of damage in laminated composites, where layers that are adhesively bonded together will be separated and cause structural failure [1]. The inadequate adhesion between layers will initiate ply separation or interlamina disconnection, leading to gradual growth and eventual failure during the part's usage. However, the delamination will not grow much if the applied load is below a certain value, but under cyclic loading, where the load magnitude is low, it can grow slowly until structural failure occurs. Delamination or separation of plies represents one of the most occurring failure modes of laminated composites [2]. The damage induced because of ply separation takes place mainly in the polymer matrix composites (PMCs). For this reason, the delamination related discussions are allocated to this class of composites, where plies are adhesively attached and arranged next to each other with different orientations.

The fabrication process of polymer matrix composites was already discussed in Chap. 4. The process involves the placement of sequential layers of polymer-impregnated fiber (prepreg tapes) on a mold surface, followed by heating under pressure to cure the lay-up into an integrated structure. Thus, the lack of understanding of the importance of interlayer regions, which can lead to delamination and subsequent separation of plies under applied service load, has a significant impact on the part's structural integrity. Although several decades have passed since the recognition of the importance of delamination, it remains an essential factor that limits the application of structural elements made of laminated composites in

B. Farahmand, *Fundamentals of Composites and Their Methods of Fabrications*,
https://doi.org/10.1007/978-3-031-80201-0_7

industries. More standard testing as well as computational modeling and analysis, are needed under different loading conditions for engineers to have a better understanding of the delamination initiation, growth, and possibly repair in laminated composite parts [3, 4]. For designers to avoid delamination growth for composite parts is to bring the applied load magnitude low enough that the number of cycles to failure would exceed the number of cycles that the structure is exposed to.

The growth of delamination in the polymer matrix composite under service loads has always been a critical factor, which has the potential to significantly affect the life of the structure and possible loss of life. The demand and application of PMC composites in aerospace, commercial aircraft, helicopters, large military transport aircraft, and auto industries are growing rapidly. For that reason, the delamination issue must be fully assessed by engineers through testing, analysis, inspection, and a better fabrication process to prevent any catastrophic failures during their operational life. The understanding and assessment of delamination will be more crucial for primary structures. Noting that the failure of a primary component of a structure can cause the catastrophic failure of the whole structure. An example of a primary structure is shown in Fig. 7.1. The failure of the vertical stabilizer of an American Airlines Flight 587 passenger airplane, which was made of carbon fiber-reinforced polymer (CFRP), led to the loss of 260 people. The failure occurred at the connection of the fuselage to the vertical stabilizer with six attaching points, in which the joint is a critical region of an aircraft. The accident was attributed to the plane's tail breaking off from the fuselage under the force of swirling wakes generated by another jetliner flying at a path around 1.2 km away. A comprehensive review of the failure of Flight 587 is given in [5, 6].

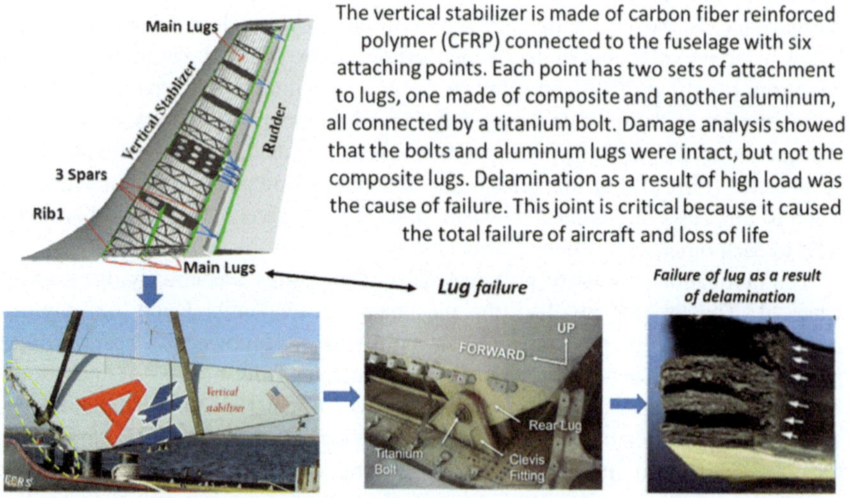

Fig. 7.1 Failure of the vertical stabilizer of American Airlines Flight 587 at its attachment point to the fuselage. It is a critical joint, and delamination was the cause of failure that led to 260 fatalities on board and 5 on the ground [5, 6]

Knowledge of the internal stresses at all points in the body is needed through finite element modeling. Based on the finite element results, engineers must always check the induced maximum flight stresses that must be less than the strength of the material. However, the presence of preexisting defects or defects initiated during the service flight load environment can significantly degrade the material properties and structural life of the components.

Here, the damage investigation analysis of Fig. 7.1 showed that the bolts and aluminum lugs were intact, but not the composite lugs. The presence of initial delamination, because of a lack of enough bonding between plies and high environmental loads due to swirling wakes generated by another jetliner right after takeoff, was the cause of structural failure and loss of life [5]. Composite structures can be damaged by the initiation of delamination through lightning strikes; tool drops; service vehicle collisions; impacts by hail, runway debris, and flying birds. Other sources of damage initiation include erosion, abrasion, manufacturing defects and design, excess heat exposure, fluid infiltration due to humidity, and the possibility of preexisting delamination because of inadequate adhesive between plies. The extent of the damage by those abovesaid factors determines whether the component needs to be repaired or replaced. Visual and nondestructive inspections may be required to assess the extent of the damage further. In Chap. 13, the method of composite repair will be addressed and in Chap. 14, the nondestructive inspection is presented.

As was mentioned in Chap. 6, regarding the fabrication of ceramic matrix composite (CMC), the reinforcement in the ceramic matrix is used mainly to improve the fracture toughness of composite parts. Reinforcements would act as a crack stopper, whereas fibers in the PMC provide features such as high strength, and great stiffness, and arrest the damage initiated in the brittle matrix from growing further. In the PMC fabricated parts, the applied mechanical load carried by fibers and a strong interface bond between fibers to the matrix will enable the transfer of load between them. The debonding of laminas of different orientations occurs because of weak adhesive bonds between them, which will be fully discussed in this chapter. Both the experimental data and the analytical tools are needed to assess the growth of delamination under fluctuating load in order to estimate the number of cycles to failure from initiation to final failure. Therefore, the topic of initiation and growth of delamination in polymer-matrix fiber composites under fatigue loading, using fracture-mechanics approach or progressive failure analysis, is necessary to be presented for engineers to have a better understanding of the way delamination initiates and accelerates. The number of cycles to failure is important for both engineers, and the users of the composite part to know, because it can be compared with the total load cycles that part is exposed to during its operations.

7.2 Cause of Delamination and Damage Assessment

Delamination initiation and its stable propagation (or, in some cases, unstable growth) are one damage mechanisms among others that can occur in cross-ply laminated composites under mechanical loading. The delamination in composites

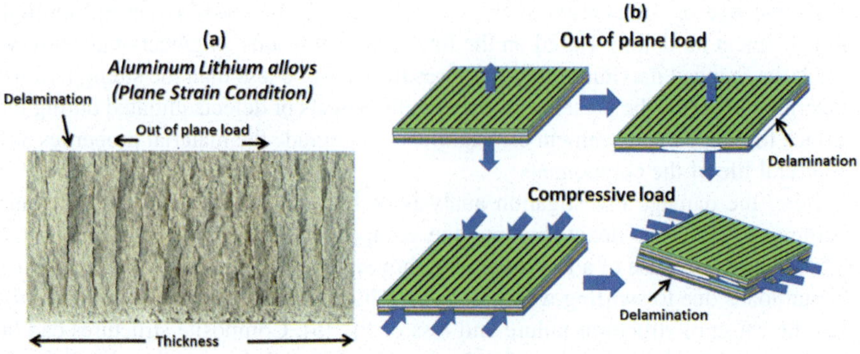

Fig. 7.2 (**a**) Delamination in metallic aluminum lithium alloy subjected to out-of-plane load. (**b**) Polymer-reinforced composites under both out-of-plane and compressive loads

is a mode of failure where the composite will fracture into plies and can be initiated as the result of lightning strikes; tool drops; service vehicle collisions; impacts by hail, runway debris, bird strike; as well as the existence of anisotropy between bonded plies. Still, delamination is most critical under pure compressive, bending, or the out-of-plane loads. Surprisingly, in some metallic materials, such as aluminum–lithium, in the plane strain condition, delamination also occurs, where the induced out-of-plane stress or residual stresses during the manufacturing can separate plates from each other [7]. In metal forming processes, such as steel, formed by the rolling process, can also cause layer separation like delamination. Figure 7.2a shows the delamination in aluminum–lithium alloy under the plane strain condition, where out-of-plane load is responsible for layer separations. Aluminum–lithium alloy is lighter and stronger than traditional aluminum alloys and has recently been given lots of attention in the aerospace industry. However, the delamination issue associated with them must be resolved. Another issue related to aluminum–lithium alloys arises when welding parts together. The cracking occurs during welding due to residual stresses in the heat affected zone. In the case of PMCs, the delamination occurs under compressive and out-of-plane loads as shown in Fig. 7.2b.

In addition to out-of-plane and compressive loads that can cause delamination initiation and growth in composites, the mismatch in Poisson ratios and modulus of elasticity between plies can result in the initiation of delamination when plies are subjected to tensile load. The anisotropic nature of individual plies that have different mechanical properties, because of their fiber orientations, can lead to delamination initiation.

In Fig. 7.3, two specimens are shown, and both are subjected to uniaxial tension. In Fig. 7.3a, the test specimen was fabricated from an isotropic material, and its mechanical properties are the same in all directions. That is, the modulus of elasticity in both directions 1 and 2 ($E_1 \sim E_2$), as well as their strength, are the same, including an identical stress–strain curve up to the failure. However, in Fig. 7.3b, the ply specimen has anisotropic properties with fibers oriented at some angle α from the main axis, that is, axis 1. In this case, the modulus of elasticity and other mechanical

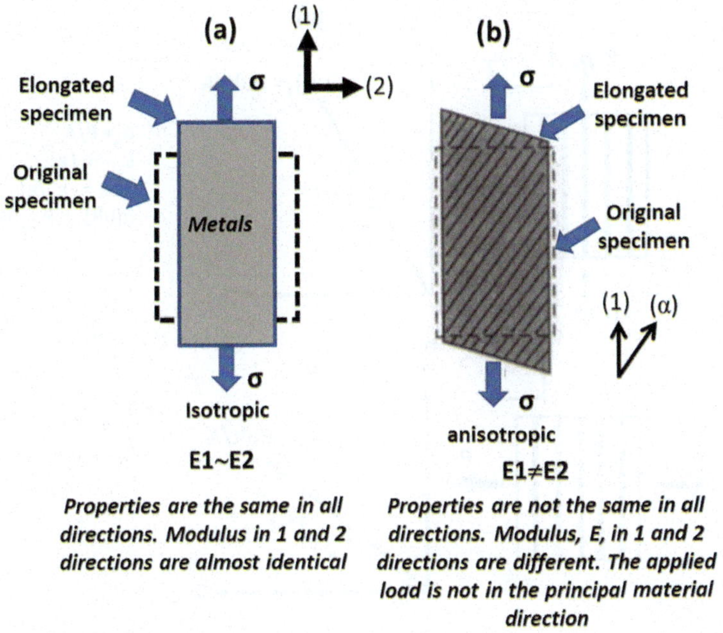

Fig. 7.3 The behavior of (**a**) isotropic ($E1 \sim E2$) and (**b**) anisotropic materials ($E1 \neq E2$) when subjected to uniaxial tension

properties are different in directions 1 and 2 ($E_1 \neq E_2$) [8–10, 12]. Because of a mismatch in elastic moduli between the layers of a laminated composite that are adhesively attached to each other, interlaminar stresses between layers will develop. These normal stresses and shear can be high enough to cause edge delamination between the layers.

Figure 7.4a illustrates a simple example of a uniaxial tension test along the fiber direction (direction 1) for a flat lamina. This is a continuation of Fig. 7.3a. In Fig. 7.4a, the stress–strain plot is shown with, $\sigma_1 = \mathrm{P/A}$, $E_1 = \sigma_1/\varepsilon_1$, $\sigma_{\mathrm{ult}} = P_{(\mathrm{ult})1}/A$, and the Poisson's ratio $\nu_{12} = -\varepsilon_2/\varepsilon_1$. The quantity A is the cross-sectional area and P is the applied uniaxial load. Now consider the uniaxial load, P, is in direction 2 and applied on the same specimen shown in Fig. 7.4a. Similarly, $\sigma_2 = P/A$, $E_2 = \sigma_2/\varepsilon_2$, $\sigma_{(\mathrm{ult})2} = P_{(\mathrm{ult})2}/A$, and the Poisson's ratio $\nu_{21} = -\varepsilon_1/\varepsilon_2$ as shown in Fig. 7.4b.

Both the stress–strain curves generate the E_1 and E_2 quantities that are not equal, as depicted in the figure. The two quantities, ν_{12} and ν_{21} would be equal for isotropic material, and therefore the material can be described by the linear elastic relations ($\nu_{12}/E_1 = \nu_{12}/E_2$). However, if the two moduli are not equal ($E_1 \neq E_2$), as is expected due to the application of the load, then the material does not obey Hooke's law.

In Fig. 7.5, two identical plates with different fiber orientations are adhesively attached to each other. The mismatch in properties can result in the formation of shear stress for both plies in the x–y plane. To maintain equilibrium, the induced

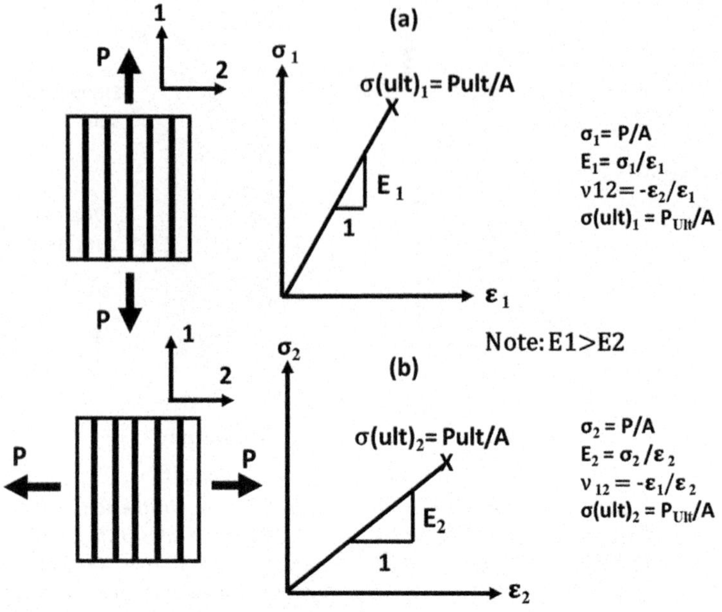

Fig. 7.4 Uniaxial testing in 1 and 2 directions, where E_1 and E_2 are not equal. Here, both $\boldsymbol{\varepsilon}_1$ and $\boldsymbol{\varepsilon}_2$ are measured at the failure (instability). (**a**) For tensile along the fibers and (**b**) for tensile transverse to the fibers

shear stress will have opposite directions on each ply. These shear stresses produce a bending moment that is balanced by stress in the z-direction (like the out-of-plane load) [8]. For some lay-up sequences, substantial out-of-plane tensile stresses occur at the edge of the panel (maximum), which leads to the formation of delamination and separation of plies. When attached plies have very small orientation differences with respect to each other, the chances of having delamination because of mismatch are much smaller.

In Fig. 7.5a, b, the two plies are shown as having two different orientations and not loaded. In this case, no deformation is expected. The same two plates are now subjected to uniaxial tensile load separately as illustrated in Fig. 7.5c, d. Due to the nature of their anisotropy, it is expected the two plates will go through some elongation in the normal direction along axis 1 and be tilted by some angle α due to the shear stresses as illustrated in the figure. To avoid distortion, shear stresses are applied as illustrated in Fig. 7.5e, f. Imagine the two plies shown in Fig. 7.5a, b are now glued together and will be subjected to axial tensile load. Due to the anisotropic nature of the two attached plies with different orientations, delamination between the two plies is expected if the applied tensile load is adequate (Fig. 7.5g).

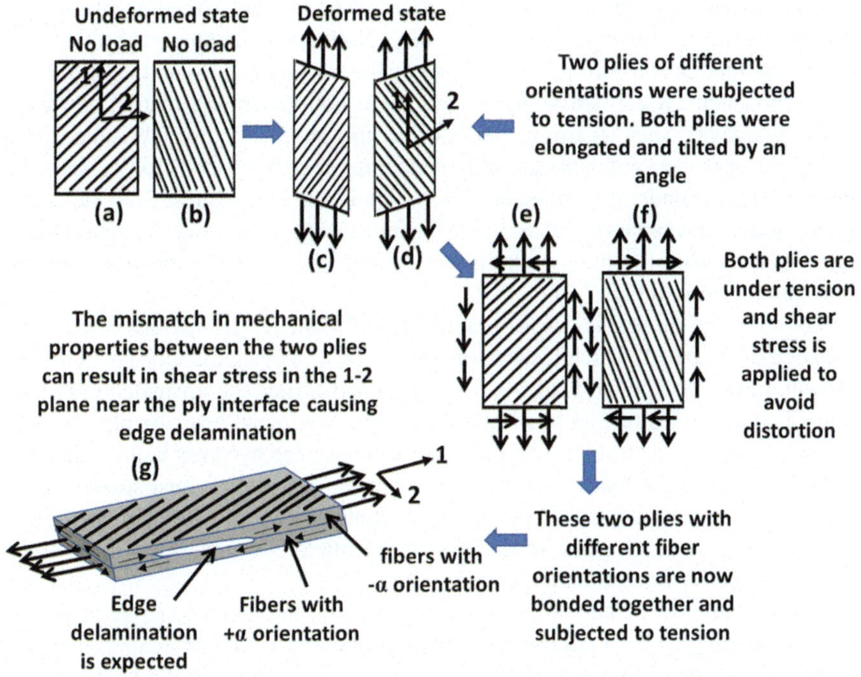

Fig. 7.5 Edge delamination initiated as a result of a mismatch in mechanical properties of two plies that are adhesively bonded together and subjected to uniaxial tension (**a**, **b**) in undeformed state, (**c**, **d**) in deformed state, (**e**, **f**) shear forces applied to avoid distortion, and (**g**) edge delamination

7.3 Application of Griffith Fracture to Delamination

Laminated composite materials are fundamentally not homogeneous regarding their mechanical properties. Their properties are not the same at each point in the body when subjected to load. When applying a load on composite parts with several plies of different orientations stacked on top of each other, a nonuniform distribution of stress is set up within the composite. Induced stresses and strains, as a result of mismatches between mechanical properties of adjacent plies, can overcome the strength and stiffness of the composite material allowables, which can lead to structural failure. Much effort has been devoted to understanding and predicting this stress distribution. By knowing the stress distribution in composite parts through inspections or finite element approach, it determines how the material will distribute the load, where the high stresses are located, and how engineers can change the design to eliminate the stress risers. In Chap. 2, the micromechanics approach of obtaining laminate material properties from a unidirectional lamina was estimated through its constituent properties (i.e., fibers and matrix), where the rule of mixtures was implemented. Because a laminated composite part is made of several plies with different orientations stacked on top of each other, the understanding of each ply through their constituents can be of great help. Noting that the physical properties of composite material are strongly influenced by the microstructures and the amount of each constituent, which are volume fractions of both fibers and matrix, as well as other parameters, such as arrangements of lay-ups and method of fabrications. The fracture assessment of delamination in the laminated composites is fully assessed in the next chapter under the failure analysis of laminated composites. Both the out-of-plane mode I delamination, by using the Griffith energy release rate concept (G_I) and the in-plane shear of failure, mode II delamination (G_{II}) [12], will be discussed later in Chap. 8 (see also Fig. 7.7).

For some lay-up sequences, substantial out-of-plane tensile stresses can occur at the edge of the panel, which leads to the formation of a delamination/separation and growth during the applied load. As was mentioned earlier, the mismatch in Poisson ratios between plies results in shear stress in the x–y plane near the ply interface. These shear stresses produce a bending moment balanced by stress in the z-direction (like the out-of-plane load). That is, near the edges, the normal shear stress drops to zero, and the out-of-plane shear stress becomes large causing ply separation. Figure 7.6 shows two plies with different orientations can develop interlaminar stresses, delamination initiation, and growth.

The initiation of delamination observed based on the nondestructive inspection (NDI) and the growth of delamination through the service life of the part using the crack growth rate is discussed in Chap. 8. The mode I ASTM-D5522 DCB test [13] corresponds to interlaminar delamination due to out-of-plane loading, and its critical value, G_{Ic}, is the most performed testing technique for obtaining fracture toughness

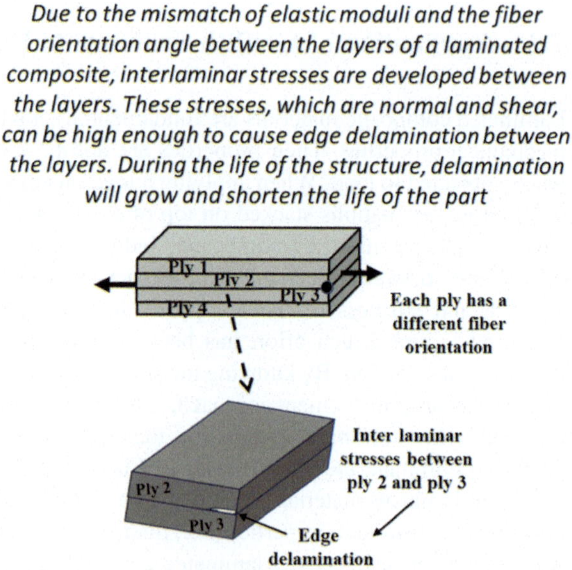

Fig. 7.6 High interlaminar stresses, as a result of mismatch between plies, are the cause of ply separation, which cause edge delamination, growth, and ply separation

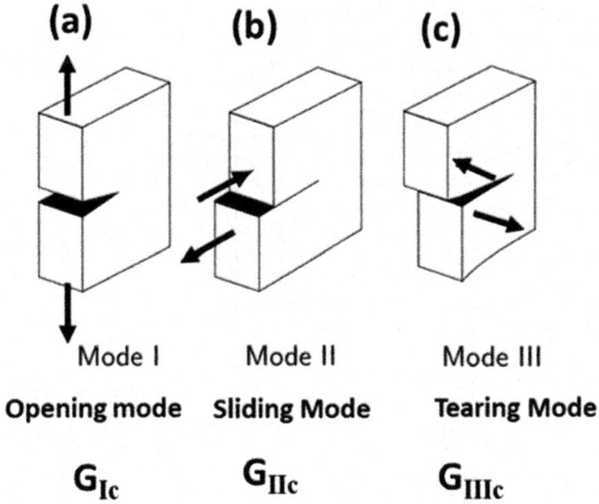

Fig. 7.7 Three modes of failure. (**a**) Mode I or the opening mode, (**b**) mode II or the sliding mode, (**c**) mode III or the tearing mode

results for laminated composite materials (Fig. 7.7a). Mode II ASTM test corresponds to interlaminar shear loading, and its critical energy release rate value, G_{IIc}, is also commonly used when predicting delamination growth and final failure (Fig. 7.7b) [14]. Mode III corresponds to less commonly tested interlaminar (aka scissoring) shear loading (Fig. 7.7c) [15]. In actual composite structures, delamination occurs when it is subjected to a combination of loading modes, mainly modes I and II, known as mixed-mode delamination. In Chap. 8, therefore, the focus will be on mode I, mode II, and the mixed-mode (mode I and II) testing, for which standardized tests exist.

The application of linear elastic fracture mechanics and the stress intensity factor, K, is well established for isotropic materials such as metals and their alloys. However, the application of fracture mechanics to composite materials, that are homogeneous (i.e., considering the average properties of both fiber and matrix) and anisotropic, makes the application of the stress intensity factor, K, or the Griffith energy release rate, G, difficult to use. However, Wu [16] found out that for homogeneous and anisotropic materials, the stress intensity factor, *K*, as a function of half a crack length, a, can be expressed by $K_I = \sigma(\pi a)^x$, where the exponent x ~ 0.5, the same as metals. Wu's result was obtained by testing several composite crack plates subjected to tension (Fig. 7.8). At the onset of failure, a plot of applied stress versus critical crack length was plotted in a Log-Log plot. The straight line shown in Fig. 7.8 indicated that the slope is ~0.5. Therefore, for a crack subjected to mode I loading, the fracture mechanics assessment of a composite lamina with homogeneous anisotropic properties also shows the same trend at the onset of failure for the stress intensity factor relating the far field applied stress to the critical crack length. This relationship has been verified by test data [16, 17]. The same can be true for mode II, where $K_{II} = \tau(\pi a)^{0.5}$ [16]. In both the K_I and K_{II} cases, the concept of linear elastic fracture mechanics (LEFM) remains valid until the onset of failure, as

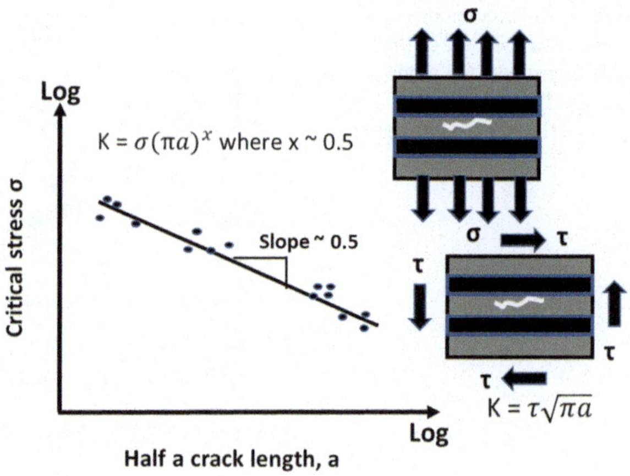

Fig. 7.8 Failure tests on crack plates under mode I and mode II. The slope of stress versus crack length indicated that $K = \sigma\sqrt{\pi a}$. (By Wu, Ref. [16, 17]). Here, both σ and τ are far field stresses

demonstrated by Wu. However, engineers are also interested in whether the LEFM concept can be extended to assess crack (or delamination) growth in laminated composites.

7.4 Strategies for Minimizing Delamination

To minimize delamination in laminated composite parts, engineers must consider the influence of the following: (1) material selection plays a big role in avoiding the delamination. That is, choosing the right combination of matrix and fiber materials with good interfacial bonding and compatible coefficients of thermal expansion can reduce the occurrence of delamination, (2) minimizing large fiber orientation difference between two adjacent plies to avoid delamination, which otherwise can lead to significant interlaminar stresses and strains. Induced stresses and strains can overcome the strength and stiffness of the composite material, causing it to fail or crack, (3) incorporating toughening materials between plies to avoid cracking and delamination. This includes using toughened resin systems, inserting toughened films or layers, to enhance bonding between plies, (4) better manufacturing process, such as proper curing, adequate pressure during lay-up, and maintaining consistent fiber alignment, helps prevent defects that could lead to delamination, (5) ensuring proper surface preparation before adhesively bonding layers together is crucial. This involves cleaning the surfaces to be bonded to enhance adhesion between layers, (6) quality control and inspection, using nondestructive inspection can detect potential delamination early, allowing for corrective action before it becomes a significant issue, (7) finite element analysis can be helpful to identify potential delamination areas, (8) proper handling and service conditions can protect composite parts from premature delamination. By incorporating these tactics, engineers will be able to reduce the likelihood of delamination and ensure the structural integrity of composite components [9–11].

7.5 Summary of Delamination in Laminated Composites

In a laminated composite material composed of multiple layers (plies) that are adhesively held together, delamination refers to the separation of these layers due to anisotropy associated with each layer when subjected to tensile load. That is, when a tensile load is applied to the laminate, the layers experience forces pulling them apart along their interface. If the bonding between the plies or the adhesion with the matrix material is weak, this stress can cause the layers to separate at their interfaces, leading to delamination or partial separation of layers. Tensile loading creates stresses, which are normal and shear *st*resses along the edges of the composite material and are concentrated at the boundaries, where the layers meet. Delamination can be further taking place due to (1) inadequate curing, (2) due to out-of-plane as well as bending and pure compressive loads, (3) Sudden forces or impacts can initiate delamination, (4) the presence of moisture (moisture ingress) can weaken the bonding between layers, leading to delamination, and (5) delamination can cause reduction in structural stiffness when plies separation occur. Therefore, engineers must find a way to reduce the interlaminar stresses and ways to counter their effects to avoid structural failure due to delamination during parts' operational life. One way of approaching it is to arrange the stacking sequence influences in such a way to minimize the effect of delamination. Other approaches are to improve resistance to delamination by using toughened resin and incorporating systems to prevent separation of plies [18–20]. Engineers should note that delamination is a critical concern in industries where composites are widely used, such as aerospace, automotive, and sporting goods, hence by detecting and mitigating delamination is crucial for ensuring the reliability and safety of composite structures. If delamination was detected through nondestructive inspection prior or during structural usage, repairs may be required and that involves injecting adhesives, applying patches, or using other methods to restore the integrity of the composite.

7.6 Concluding Remarks

Concluding remarks on delamination in composites are essential for pinpointing key findings to engineers, discussing consequences of delamination taking place for critical composite parts, and proposing future methods of preventing it. Some potential points on delamination in composites are discussed here: (1) to emphasize the importance of preventing delamination and the role it plays in total failure of composite structure. The failure is as a result of gradual reduction in strength and stiffness during the operational life when delamination grows slowly during service operation; (2) engineers must determine factors that influence delamination and attempt must be made to minimize its occurrence. Various factors that contribute to delamination occurrence and its growth, including manufacturing processes, loading conditions, and environmental factors should be taken into consideration; (3) implement advanced detection and monitoring techniques, such as nondestructive methods like ultrasonic testing, infrared thermography, and acoustic emission can be helpful to detect and stop delamination growth and related failure. Early detection of delamination and preventing its growth is helpful to prevent catastrophic failure and possible loss of life; (4) methods of resistance to delamination initiation and growth can be provided to manufacturer of composite parts, such as interlaminar toughening mechanisms, more robust bonding techniques, and implementation of a better design are all contribute to resistance to the initiation and growth of delamination in laminated composite.

In addition to all above parameters related to delamination growth, it is essential to highlight the application of computational modeling such as a 3D finite element to simulate areas that will be susceptible to delamination initiation under different loading conditions. Finally, it is important to emphasize the necessity of experimental verification to the theoretical models and simulations to be sure results of analysis are acceptable to engineers. Moreover, in conducting computer simulations on delamination, the impact of environmental factors, such as moisture, temperature, and aggressive chemicals, should be included in the final analysis.

References

1. W.J. Cantwell, J. Morton, The impact resistance of composite materials – A review. Composites **22**(5), 347–362 (1991)
2. M.R. Wisnom, The role of delamination in failure of fiber-reinforced composites. Philos. Trans. R. Soc. A Math. Phys. Eng. Sci. **370**(1965), 1850–1870 (2012)
3. C.G. Ddvila, D.R. Ambur, *Numerical Growth in Simulation of Delamination Composite Materials* (Langley Research Center/NASA, Hampton, 2001)
4. J.G. Bjeletich, F.W. Crossman, W.J. Warren, *The Influence of: Stacking Sequence on Failure Modes in Quasi-Isotropic Graphite-Epoxy Laminates*, in Failure Modes in Composites IV, AIME (1977), p. 118
5. B. Farahmand, History of Structural Failures (Failures of Airlines, Fighter Jets, Rockets, Space Shuttles, and Pressurized Tanks (Amazon eBook, 2021)
6. B. Farahmand, Application of High and Low Cycle Fatigue on Metallic and Non-Metallic Material (Application of Fatigue for Engineers) (Amazon eBook, 2022)
7. B. Farahmand, *Fracture Properties Estimation of Aluminum Lithium Alloys Subjected to Exposure Time (Analytical Approach Versus Physical Testing)*, in Conference: 45th AIAA/ASME/ASCE/AHS/ASC Structures, Structural Dynamics & Materials Conference (2004)
8. R.M. Jones, *Mechanics of Composite Materials* (McGraw Hill, 1976), pp. 210–220
9. R.B. Pipes, N.J. Pagano, Interlaminar stresses in composite laminates under uniform axial extension. J. Compos. Mater. **4**, 538 (1970)
10. N.J. Pagano, R.B. Pipes, The influence of stacking sequence on laminate strengths. Int. J. Mech. Sci. **15**, 679 (1973)
11. S.S. Wang, Edge delamination in angle-ply composite laminates. AIAA J. **21**, 256 (1984)

12. A.A. Griffith, The phenomena of rupture and flow in solids. Philos. Trans. R. Soc. Lond. A **221**(582–593), 163–198 (1921)
13. D 5528–01, Standard Test Method for Mode I Interlaminar Fracture Toughness of Unidirectional Fiber-Reinforced Polymer Matrix Composites, Approved 2007
14. ASTM D7905/D7905M, Standard Test Method for Determination of the Mode II Interlaminar Fracture Toughness of Unidirectional Fiber-Reinforced Polymer Matrix Composites, Last Updated: Nov 06, 2019
15. P. Zdenk, Bazant, P.C. Prat, Measurement of mode iii fracture energy of concrete. Nucl. Eng. Des. **106**, 1–8 (1988)
16. E. Wu, *Fracture Mechanics of Anisotropic Plates* (Technomic, 1968), pp. 20–43
17. R.M. Jones, *Mechanics of Composite Materials*. Chapter 6 (McGraw Hill, 1976), pp. 291–298
18. N. Sela, O. Ishai, Interlaminar fracture toughness and toughening of laminated composite materials: A review. Composites **20**, 423 (1989)
19. J.G. Williams, T.K. O'Brien, A.J. Chapman III, *Comparison of Toughened Composite Laminates Using NASA Standard Damage Tolerance Tests*, in NASA CP 2321, ACEE Composite Structure Technology Conference, Seattle, WA, August 1984
20. S.F. Chen, B.Z. Jeng, Fracture behavior of interleaved fiber–resin composites. Compos. Sci. Technol. **41**, 77 (1991)

Chapter 8
Progressive Failure Analysis (PFA) of Laminated Composites

8.1 Introduction

The high specific strength and stiffness of composite materials have been of great importance in their structural applications where weight is a critical factor to be considered in many industries, such as aerospace, automobile, defense, and many more. However, parts made of laminates exhibit their own problems, for instance, their susceptibility toward the initiation and growth of defects between the layers during their service usage, and engineers must be given special attention to it in order to have a safe structure. In the failure analysis of laminated fiber-reinforced composite parts, the common approach is to look at the stress strain response of each layer or ply within the stack when subjected to external loads. The assumption is that each layer has its own anisotropic properties, and each layer is different from other layers because of their distinct fiber orientations. The individual laminas failure due to applied load will influence the final failure of the laminate. That is, the failure assessment of a laminated composite part demands having a good knowledge of the stresses and strains in the lamina by conducting a comprehensive finite element analysis and the implementation of proper failure theories for each ply within the laminate. Any available failure criteria that are being used for both cases of isotropic, orthotropic, or anisotropic material must be first validated by testing before being used by the analyst. For isotropic material, the failure theories used for both brittle and ductile material are well established [1–5]. The maximum stress and strain theories are used for brittle materials that show little or no ductility before their final failure. The maximum shear stress or the von Mises stress distortion energy criteria are used for a ductile material. None of these failure criteria are applicable to anisotropic materials and are of no use for predicting failures in the composite lamina. However, several alternative failure theories have been developed [6, 7] which can be used to predict the failure of the composite lamina, which will be briefly discussed here. The following methodology is used when engineers attempt to assess and analyze the failure of laminated composites by focusing on a lamina first.

B. Farahmand, *Fundamentals of Composites and Their Methods of Fabrications*,
https://doi.org/10.1007/978-3-031-80201-0_8

8.2 Method of Approach to Failure Assessment of Laminated Composites

A simple and common approach to assess the mechanical properties of fiber-reinforced laminated composite structures is implemented in this section through the micromechanics approach, where the material allowables of constituents must be available for any failure analysis assessment. However, it is vital for engineers to choose their composite material first with the proper amount of fiber content in the matrix, so as to arrive at proper stiffness and strength suitable for the part when subjected to a load environment that the structural material will be exposed to. After all, it is the fiber that provides strength and stiffness to the composite and has the load carrying capability when they are adequately bonded to the matrix. When the volume of fibers in the matrix is determined by the engineers, then allowables must be established for further analysis pertaining to structural failure. Having allowables available, engineers could be able to compare their calculated stresses and strains, which were obtained through the finite element method, to material allowables, and be able to assess if the failure is imminent or not.

Because of many parameters that are involved, when trying to obtain material properties of composites, such as modulus, ultimate strength, and other key properties that are dependent on the fabrication process, fiber content, and their orientations, both experimental and analytical approaches are difficult to conduct because of a large number of tests and detailed 3D analyses, which both can be costly and time-consuming. Therefore, other feasible approaches should be used by implementing simple and satisfactory assumptions. For example, by taking the average ply properties from the individual properties of the constituents (i.e., fibers and the matrix materials) and by considering the ply to be homogeneous, both assumptions will be of great help to estimating allowables. That is, the mechanical properties of a lamina are the same at any point in the lamina, but they are different in various directions [8, 9]. This is the general assumption that is used in micromechanics of a lamina (Fig. 8.1a). In effect, the stiffness and strength of the lamina vary from point to point depending on whether the point is in the fiber, the matrix, or the fiber–matrix interface. Because of these variations in properties, any kind of in-depth mechanical modeling of the lamina from a micromechanics perspective will be complicated, and computationally impossible. Therefore, the above assumptions of homogeneous and using average properties of constituents are reasonable to implement.

Methods to find these average properties of a lamina are based on the individual mechanical properties of the fiber and the matrix, the relative content of constituents (the volume fraction), the shape of fibers, and the method of processing them. For example, the modulus of elasticity for a lamina parallel to fibers, E_1, can be expressed by a simple rule of mixture for continuous fibers as [10, 11], which was derived in Chap. 2 (Sect. 2.5.1) as,

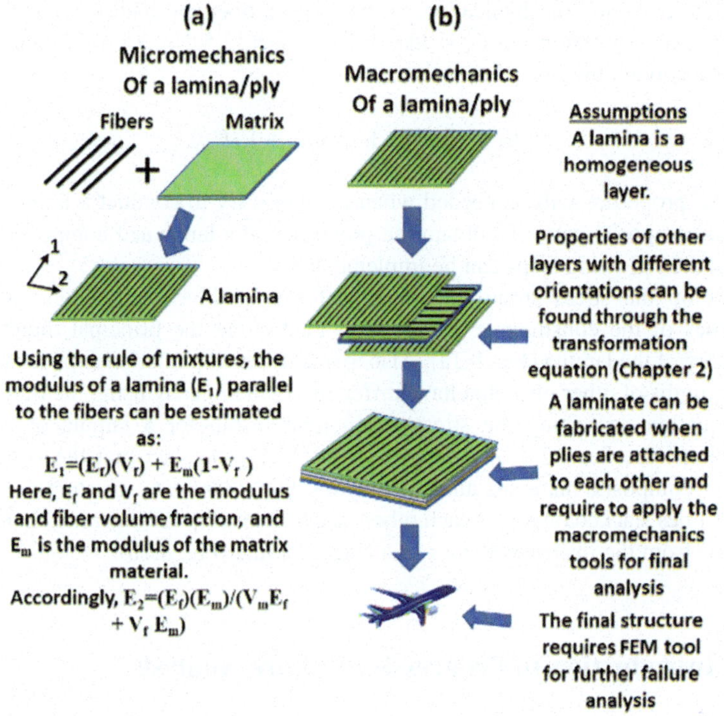

Fig. 8.1 Illustrating (**a**) using the micromechanics approach to find the properties of a lamina from its constituents. (**b**) Knowing the properties of a lamina, properties of a laminate can be obtained through macromechanics approach

$$E_1 = (E_f)\ (V_f) + E_m\ (1 - V_f) \tag{8.1}$$

where E_f and V_f are the modulus and volume fraction of fibers, and E_m is the modulus of matrix material in the uniaxial direction parallel to the fibers. Here, the quantity V_f = (Volume fraction of fibers)/(total volume of composites) and the volume fraction of matrix V_m = (Volume of the matrix)/(total volume of composites). Note that the sum of volume fractions of fibers and matrix is unity. That is,

$$V_f + V_m = 1 \tag{8.2}$$

If the void volume fraction (V_v) in the lamina is also of interest to include in the micromechanics analysis of the rule of mixtures, then it should be added as an additional parameter in such a way that,

$$V_v + V_f + V_m = 1 \tag{8.3}$$

In addition to elastic modulus, E_1, other properties of a lamina, such as strength in the axial direction, modulus and strength in the transverse direction can all be estimated by having mechanical properties of both fiber and matrix available to the analyst. The expression for E_2 is like E_1 (see Sect. 2.5.2 of Chap. 2) and can be obtained through the rule of the mixtures as,

$$E_2 = (E_f)(E_m)/(V_m E_f + V_f E_m) \tag{8.4}$$

These properties will be needed when engineers try to conduct a failure assessment of composite parts. To obtain the properties of a laminated composite from a lamina, the following steps can be implemented:

Develop the stress–strain relationships for a unidirectional lamina from the properties of the constituents by applying load along the principal directions of symmetry of the lamina (Fig. 8.1a). Also, one can develop relationships for stiffness and strengths of other plies that have different orientations by using the transformation equation describing the stress–strain relationship for a lamina of arbitrary orientation (see Sect. 2.6 of Chap. 2). By knowing the fact that the structure is made of composite materials and by knowing that a laminate structure is made of various plies stacked on top of each other, one can develop the macromechanics of a laminate from the micromechanics of a single lamina (Fig. 8.1b).

8.3 Introduction to Progressive Failure Analysis

The method of obtaining composite properties from their constituent was briefly mentioned in Sect. 8.2. Having material allowables available, a strength analysis proceeds by conducting proper stress or strain analysis of each individual ply in the laminate and comparing them to the material allowables for the ply. Failure is often seemed to initiate when the induced stress or strain in one of the plies exceeds an allowable stress and strain limits or by applying a suitable failure theory to check for the damage initiation, subsequent damage growth, and the final ply failure. The failure of composite structures is a progressive series of events that usually starts in the matrix that is much weaker than the fiber, at a weak matrix–fiber interface, and finally the failure of fibers.

The first ply failure does not necessarily mean the total failure of the laminate [12]. The failed ply will transfer some or all the load it carried to another ply in the laminate. When a ply fails the stiffness of the whole laminate drops as shown in Fig. 8.2, where the recalculate stiffness, with different slope, accounts for the local changes in material stiffness. The analysis will be repeated ply-by-ply until the load-carrying capability of laminate will be depleted and the total failure will take place (Fig. 8.2). In this figure, the stiffness of the structure is reduced from E_1 to E_f, where E_1 is the original (undamaged) stiffness and E_f is the final one, where the load carrying capability of the structure is reduced in such a way that the final failure is imminent. This approach is referred to as progressive failure analysis (PFA), where failure will progress in a stable manner until the final structural failure occurs due to reduction of stiffness from E_1, E_2,... E_f, where $E_1 > E_2$, ... $> E_f$.

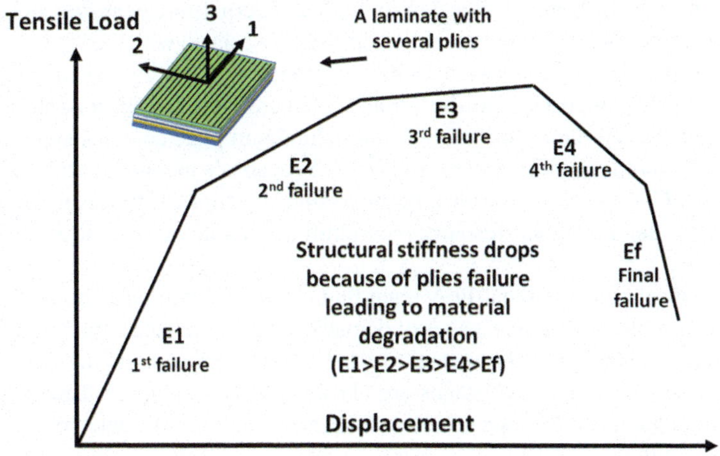

Fig. 8.2 Progressive failure analysis (PFA) concept is shown. It is based on material degradation (decrease in lamina's stiffness, E) as the load magnitude increases and plies' failure occur. Failure of composite structures is a progressive series of events. It often starts as a microcrack in the matrix, cracks between fibers and matrix (the interface), and fiber failure

Failure analysis of composites requires to have not only a comprehensive material property of both laminate and each individual ply but also a reliable failure theory that can evaluate the local and global failures and be able to compare the analytical results with the available material allowables. For example, regarding the maximum strain theory for individual analysis of lamina within the laminate, it requires that the strains developed under load be referred to the lamina principal axes and evaluated against the tensile and compressive allowable for the lamina. Five failure strains are required to have, which are tensile and compressive limits along the fiber's direction, tensile and compressive limits in the transverse to the fiber direction, and an in-plane shear limit. Commercial computer software for analyzing laminated composite structures, for different composite failure analysis theories, are available within aerospace as well as academic institutions. These failure assessment packages, which have the capability to do finite element modeling (FEM), can be customized to allow engineers to input new materials information with allowables available from a reliable source, to select a suitable failure theory, and to conduct failure analysis under the applied load. Results will provide information to engineers whether parts will be able to withstand the service load environment or not.

In general, when materials are subjected to applied stress, their stress analysis can be performed by engineers using different scale levels. These stress analysis tools include atomistic analysis, molecular dynamic analysis, micromechanics analysis, and macromechanics or continuum mechanics analysis. These analysis approaches, that are called multiscale modeling, are discussed in [13]. For example, in the molecular dynamics approach (MD), the displacement of an atom in space in which the atoms are forced into motion, the simulation of the motion can be expressed by the numerical solution of the classical Newtonian dynamic, where each atom is a small mass subjected to applied external force causing atoms to be displaced. Because of computational complexity associated with the large number of atoms needed in the analysis in order to have a meaningful result, the failure assessment of composites will be limited to microscale and macroscale analysis. For example, a 60 mm × 40 mm × 10 mm aluminum plate contains approximately 1.45×10^{24} atoms (multiplying the moles by Avogadro's number to find the number of atoms). Consequently, performing molecular dynamics (MD) analysis on such parts with the computer resources currently available to engineers would be impossible.

The typical computational PFA regarding crack initiation, growth, and the final failure involves five different stages of damage, quantifying the amount of damage at each stage, and relating the damage to the overall behavior of the deteriorating structural stiffness. The five stages are (1) damage initiation, (2) damage growth, (3) accumulation and coalescence of damage propagation, (4) stable propagation up to critical dimensions, and (5) unstable or very rapid propagation to catastrophic failure. The PFA software must include a well-established finite element software by providing stress and strain at each point of the body and using that information to go forward with a meaningful progressive failure analysis based on damage trailing and material property degradation at the microscale of fiber and matrix, where damage and delamination mostly happen. The PFA software should perform multiscale damage tracking from micromechanics to macromechanics. In addition, it must have the capability of using micro and macro interaction methods in the composite structural PFA modules. Microstresses and damages are computed on the constituent level, and the corresponding material degradation is reflected in the macroscopic finite element structural stiffness.

8.4 PFA and the Method of Approach

In this section, an example of progressive failure analysis (PFA) will be provided to the reader to become familiar with damage progression and accumulation in composite material. PFA is based on the material's modulus of elasticity degradation concept using the finite element analysis (FE), together with other tools to monitor the stress distribution in the body of the composite material, but also the gradual material degradation progression under applied load. The structural stiffness will decrease as a result of damage growth, new properties will be assigned to the FE elements, and the process will continue until structural stiffness is depleted, causing the final failure (Fig. 8.2). In addition to PFA analysis for predicting the damage growth and the final failure, linear elastic fracture mechanics (LEFM) is another tool that can be used to study not only the crack growth at the micro level but also the growth of the preexisting damage, such as delamination, up to the final structural failure (Chap. 9). With respect to the structural analysis approach for the advanced composite it is important to have a methodology of how the structure will perform beyond initial local failure. The PFA analysis will be able to eliminate testing as

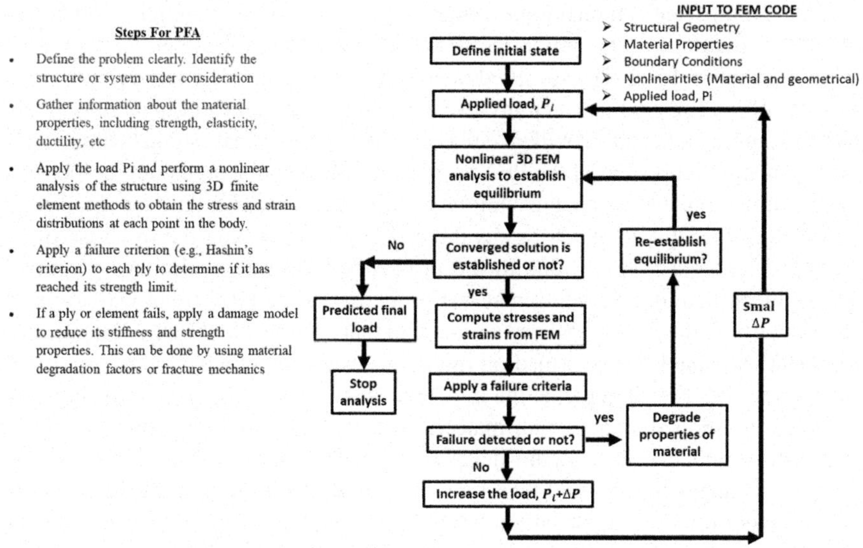

Fig. 8.3 Overview of typical progressive failure analysis (PFA) approach. Input to the FEM, failure theories, and failure modes are shown here

much as possible with the prediction of damage initiation, propagation, and failure mechanisms in composite structures. That is, engineers may be able to replace the costly and time-consuming tests on composite parts with analysis, especially when full-scale tests are required by the customers of using the composite part.

To achieve the goal of predicting the structural behavior at various scales, it is necessary to implement three basic analysis steps. The first step includes the descriptive analysis of the geometrical factors, loading factors, boundary conditions, and material properties (Fig. 8.3). Having those parameters available, the finite element analysis (FEA or simply FE) will be implemented, to provide stress, strain, and displacement fields within the structure with sufficient accuracy. The second step is to capture all possible failure modes within a composite system. Each of the failure modes is independently modeled according to appropriate failure theories. The third step, together with the previous two, provides descriptive equations used to reflect the conditions and the severity of damage through material degradation. Finally, using an iterative process, the designer can achieve a virtual representation of structural behavior over time during the application of service load and the final failure.

The progressive failure analysis (PFA) methodology is a logical chain of all the abovementioned steps. It aims to predict and monitor the entire response of the composite structure, starting from the very first localized damage initiation up to ultimate failure, by employing only a minimum number of coupons tests that are necessary to conduct the PFA analysis.

Figure 8.3 presents an overview of a typical analysis methodology based on the principles of progressive failure analysis approach. The first component is a validated multiscale computational model of the structure, whose aim is to determine stresses and strains at an appropriate scale level for the subsequent analysis at the macromechanics level. The specified operational load is divided into an appropriate number of load steps. Each step involves engaging in a series of numerical calculations, assuming no changes will take place in the material stress–strain behavior. The second major component of the PFA approach deals with the prediction of localized failure initiation, based on the various failure theories. The third component of completing the PFA methodology is the process of describing the response of the damaged material volumes due to the progressive damage evolution and growth within the structure. Degrading material properties is the most usual way to match the response of the failed localized structure to that observed in the experiments. Any degradation should be performed on the same scale level at which local failure has been predicted. For example, at the macromechanics level, when a lamina failure is detected, its properties at certain material axes are modified according to suitable degradation models. Due to the material property reduction, the stiffness of the structure changes locally, and therefore the current stress solution of the computational model does not represent an equilibrium state anymore. The structure equilibrium at the current load step should be established again, taking into account the modified material properties of the failed material volumes, until no additional failures are detected. Only then may the load be further increased to the next loading level and the earlier described main component of PFA method reimplemented iteratively, until catastrophic failure of the structure is detected. The intermediate results of the PFA approach comprise the history of failure patterns and failure modes' evolution within the structure, as well as the damage extent and ultimate load at the final failure.

8.5 Multiscale Computational Model

The analysis method used to calculate the stress, strain, and displacement fields depends on the scale level the analyst will select to perform material property degradation and failure prediction. Furthermore, the analysis level strongly depends on the type of structure under consideration. For example, in the case of composite structures made of conventional unidirectional laminas, stacked in a desired orientation to form a laminate, the macromechanics approach is well established and can lead to results which match the experimental studies. However, more modern composite material systems have been under development in recent times. Fabric composites have reinforcements not only in-plane but also in the third dimension (out of plane) have been developed in order to meet the need for enhanced out-of-plane properties and higher impact resistance. Also, carbon nanotube-reinforced polymeric composites are under investigation, aiming to dramatically increase both stiffness and strength characteristics at minimum weight increase. However, the material system is not yet satisfactorily established, and failure theories or even the damage mechanics of such systems are not fully understood. Therefore, any PFA

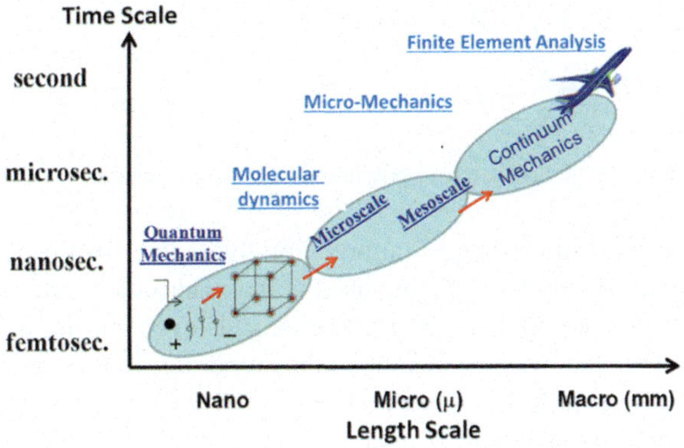

Fig. 8.4 Multiscale modeling plot showing the time and length scale from the bottom-up approach (quantum mechanics, molecular dynamics, micro-mechanics, and continuum regions are shown)

approach of these modern composite systems has to include stress analysis up to a level which goes down to the micro or nanomechanics level (Fig. 8.4). In Fig. 8.4, the simulation should begin at the atomistic level and results must be relayed to the next level microscale and mesoscale. Subsequently, the continuum mechanics will take over, where finite element technique can be used in the analysis. The mesoscale falls between the microscale and macroscale, covering sizes from micrometers to hundreds of micrometers or even millimeters (Fig. 8.4).

The problem in assessing failure of nanocomposites by a computational model is that the analyst needs to include all parameters contributing to the damage process at the nanoscale, which makes it impossible for the entire volume of the structure to be included; that is, there is a need for increasing the computational power to do the damage analysis.

Modeling the structure by FE should ideally be based on three-dimensional finite elements, in order that interlaminar stresses and strains are computed with sufficient accuracy. Interlaminar stress magnitudes are essential for the prediction of delamination, which is primarily a three-dimensional failure mode, as it usually initiates between plies and may propagate in the same ply interface but can also interact with other in-plane failure modes and kink into other plies' interfaces. However, due to the vast computational effort required for three-dimensional analysis, PFA has also been based either on smearing the properties of a certain number of plies within one element through-the-thickness, or on two-dimensional application (using plate and shell finite element models) or using global-local approaches by dividing critical areas (modeled by 3D elements) and noncritical areas (modeled by 2D elements). For more information on this topic, the reader can refer to [14, 15].

8.6 Prediction of Local Failure and the Failure Theories

Whichever the FE computational model is constructed, the outcome is always the same, i.e., nodal displacements, element strains, and element stresses determined by the strains via the material constitutive equations. It is evident that the theories

$$F_1\sigma_x + F_2\sigma_y + F_3\sigma_z + 2F_{12}\sigma_x\sigma_y + 2F_{13}\sigma_x\sigma_z + 2F_{23}\sigma_y\sigma_z$$
$$+ F_{11}\sigma_x^2 + F_{22}\sigma_y^2 + F_{33}\sigma_z^2 + F_{44}\tau_{xy}^2 + F_{55}\tau_{xz}^2 + F_{66}\tau_{yz}^2 = 1$$

Fig. 8.5 Tsai-Wu tensor polynomial criterion equation

applied in predicting failure strongly depend on the scale of the analysis, implying that different criteria are used at the macroscale, microscale, or nanoscale levels.

Concerning the macromechanical level, the initial failure of a lamina can be predicted by applying one of the numerous failure theories available in the literature. It is the purpose of this section to provide guidelines for the selection of failure rules appropriate for PFA analysis, rather than discussing the large number of existing failure criteria that have already been proposed in the literature. Toward this objective and from the PFA perspective, one may categorize the failure theories into two major groups, namely the polynomial and the direct mode determining criteria.

Most of the polynomial failure criteria are polynomials based on curve-fitting built on composite material tests data [16, 17]. They involve interactions between stress and strain components, i.e., they are interactive failure criteria. The most recognized polynomial failure criterion is the Tsai–Wu tensor polynomial criterion [18], which has been widely used due to its simplicity and easy implementation. The quadratic form of the Tsai–Wu failure criterion (Fig. 8.5) may be described as:

Here, the quantities σ_x, σ_y, σ_z, τ_{xy}, τ_{yz}, τ_{xz} are the components of the stress tensor at the principal material directions and the 12 parameters F_1 to F_{66} are experimentally derived material constants. Various other quadratic criteria, different from the Tsai–Wu failure equation, are available and they can be represented by a generalized Tsai–Wu quadratic criterion [18]. The essential disadvantage of these polynomial failure criteria with respect to PFA analysis is that they predict the initiation of failure without providing any information about the lamina failure mode or the damage growth up to the instability. However, when selecting appropriate failure criteria for the PFA analysis, it is important to keep in mind that apart from being accurate and compatible with FE computational models, it is of great importance that these failure criteria be able to distinguish between the different failure modes. It is known that the basic failure modes which can occur in a unidirectional lamina include a fiber mode and a matrix mode which are further subdivided into either tension or compression failure.

In the fiber mode failure, the lamina fails due to fiber breakage in tension or fiber buckling in compression, while in the matrix mode, failure is due to matrix cracking in tension or compression. Furthermore, laminas' interfaces may fail due to the various types of delamination. The mode of failure basically depends upon the load type, stacking sequence, specimen geometry, stress concentrators, initial imperfections, preexisting manufacturing defects, or environmental effects. At a structural component level, the above material failure modes eventually will result in structural failure if dealing with a primary part. A comprehensive PFA analysis of bolted joints subjected to external load is discussed in Sect. 8.8.

X_T, Y_T, Z_T are tensile strengths in the X, Y, Z directions
X_C, Y_C, Z_C are compressive strengths in X, Y, Z directions
S_{XY}, S_{YZ}, S_{XZ} are shear strengths in X, Y, Z directions

Failure Mode	Failure Criterion
Matrix tensile and compressive cracking $\sigma_{yy} > 0$	
Fiber tensile failure $\sigma_{xx} > 0$	$(\frac{\sigma_{xx}}{X_T})^2 + (\frac{\sigma_{xy}}{S_{xy}})^2 + (\frac{\sigma_{xz}}{S_{xz}})^2 \geq 1$
Alternative form of fiber tensile failure criterion (shear terms are excluded)	$(\frac{\sigma_{xx}}{X_T})^2 \geq 1$

Failure Mode	Failure Criterion
Fiber compressive failure $\sigma_{xx} < 0$	$\frac{\sigma_{xx}}{X_C} \geq 1$
Fiber-matrix shear-out $\sigma_{xx} < 0$	$(\frac{\sigma_{xx}}{X_C})^2 + (\frac{\sigma_{xy}}{S_{xy}})^2 + (\frac{\sigma_{xz}}{S_{xz}})^2 \geq 1$
Delamination in tension and compression $\sigma_{zz} < 0$	$(\frac{\sigma_{xx}}{Z_{T,C}})^2 + (\frac{\sigma_{yz}}{S_{xy}})^2 + (\frac{\sigma_{xz}}{S_{xz}})^2 \geq 1$

Fig. 8.6 Hashin's quadratic failure criteria based on material strengths is shown [16, 17]. Note that the tensile and compressive allowable strengths for lamina are denoted by subscripts T and C, and X_T, Y_T, and Z_T denote the allowable tensile strengths in three respective material directions (3D) and X_C, Y_C, Z_C for compression

The direct mode determining failure theories comprise sets of separate polynomial equations based on the material strengths, each one describing one failure mode. From this viewpoint, direct mode determining failure criteria are very useful and necessary in progressive failure analysis. The maximum stress and maximum strain criteria, although representing the simplest type of such direct mode in determining failure criteria, usually fail to provide the required accuracy in complex loading problems, as they are noninteractive. They have no interactions between the stress or strain components, which is sometimes called independent failure criteria. Hashin in 1973 proposed a set of quadratic failure criteria based on material strengths, where each equation predicts one specific failure type [16, 17]. His failure criteria assume that the composite material consists of fibers and the matrix, and each can fail independently. Hashin's criterion considers four different possible failure cases of matrix and fibers: (1) fiber failure under tension, (2) fiber failure under compression, (3) matrix failure under tension, and (4) matrix failure under compression. They are usually applied in the PFA analysis and have been used in all the macro-level PFA applications presented in Sect. 8.8. The mathematical form of Hashin's criteria is presented in Fig. 8.6 shown below, where σ_{ij} are the calculated lamina stress components in the (ij) material direction and the denominators are the ultimate strengths (i.e., allowables) in the corresponding direction. The tensile and compressive allowable strengths for lamina are denoted by subscripts T and C, and X_T, Y_T, Z_T denote the allowable tensile strengths in three respective material directions and X_C, Y_C, Z_C for compression.

As can be seen from Fig. 8.6, Hashin's failure equations can distinguish between matrix tensile and compressive cracking, fiber tensile and compressive failure, fiber–matrix shear-out, and delamination in tension and compression. As mentioned above, the Hashin's failure criteria are based on matrix and fibers failure and it does not account for the effect of the fiber–matrix interface, which can influence the failure behavior of the composite. Finally, *the reader should note that the failure criteria described above are a simplification of failure behavior of composite materials; however, there are parameters in composites that must be taken in consideration which can exhibit variations in failure behavior, such as manufacturing processes, defects, and environmental conditions. Therefore, these failure criteria cannot provide precise estimates of composite failure. It is a good idea to use failure criteria in union with experimental testing and finite element analysis to ensure the safety and reliability of composite structures.*

8.7 Behavior of Damaged Materials

The criteria for detecting any localized failure at certain locations of the composite structure have been discussed. Occurrence of localized material failure implies that the local load carrying capability of the damaged material volumes is not the same as their initial undamaged level. To account for this effect, the local load carrying capability of the structure must be adjusted according to suitable degradation rules. The modeling of damage evolution from the initial to the ultimate failure (the damage growth during the operating load) is much more complicated compared to the prediction of the localized failure. The two main routes toward this goal are based either on damage mechanics, which describes the evolution of damage in terms of material constitutive equations, or on fracture mechanics, which considers damage as being represented by propagating flaws or cracks up to the failure. The degradation of in-plane properties is usually dealt with by damage mechanics, while for the out-of-plane damage (i.e., the delamination) the fracture mechanics approach seems more appropriate where mode I is the opening mode representing the delamination growth. According to damage mechanics, progression of damage in the FE model is accomplished by an appropriate degradation of the material stiffness properties. On the fracture mechanics basis, cracks are assumed at appropriate locations of the structure and modeled by cohesive elements of increased complexity, which are activated to model damage progression [19, 20]. A fracture mechanics approach seems to be more appropriate to predict delamination evolution, despite its generic limitations in describing interaction of delamination with matrix cracks or kinking between different ply interfaces. In both approaches, material properties are needed in order to achieve a meaningful result. In the case of fracture mechanics, for example, the fracture properties must be available for both the matrix and the fibers, as well as the interface between the matrix and fibers. The difficulties associated with both the damage mechanics or fracture mechanics is the anisotropic nature of composite, which has directional mechanical properties and must be accounted for when they are to be used in the analysis.

Finally, from a macromechanics perspective, most of the material property degradation rules existing in the literature may be categorized into two main groups,

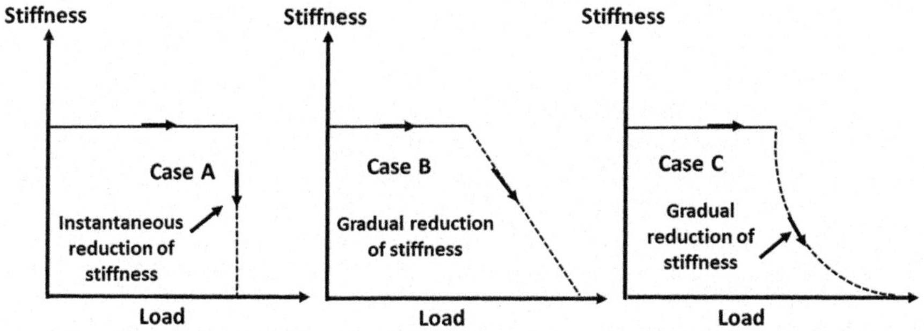

Fig. 8.7 Illustrating the sudden and gradual cases for stiffness property reduction when plies fail upon the application of load

namely sudden degradation (case A) and gradual degradation (cases B & C shown in Fig. 8.7). A third but less common approach is to keep the stress of the failed ply at a constant level [21]. Figure 8.7 schematically illustrates the degradation plots, where the load versus stiffness is plotted for two cases of sudden and gradual degradation. In the gradual property degradation method, the material elastic properties relative to the failure mode are gradually reduced up to the level at which the failure criterion is no longer satisfied. The computational model continues at the same load level, resulting in a partial unloading of elements. Upon increase of load, repeated failures for the same element will occur, representing damage evolution within the element until its complete failure occurs. The unloading can follow a linear (case B) or exponential (case C) curve. The application of gradual property degradation schemes to PFA results in increased computational effort and thus is more complicated compared to the sudden degradation approach (case A).

The sudden degradation method is realized by the instantaneous degradation of the stiffness of the failed material volumes, according to the mode of failure. The most conservative version of sudden degradation is the ply discount method, according to which stiffness properties of a failed ply are set to zero values. A version of the ply-discount method assumes that only selected material properties of a failed ply are reduced, depending on the failure mechanisms responsible for the ply failure. It is generally observed that the ply-discount method underestimates laminate strength and stiffness because it does not account for the residual load-carrying capability that a failed ply has in reality, as well as that in some cases it may lead to the computational problem of not converging.

To overcome these drawbacks, sudden degradation schemes are applied, in which degradation factors different from zero are assigned to the respective stiffness properties, according to the different failure modes. Constant degradation factors of around 10% applied on both elasticity modulus and Poisson's ratio are quite usual, because of their simplicity and the relatively good predictions that they offer of damage propagation [22, 23]. A variation of the simple sudden degradation method assumes that degradation of stiffness properties corresponds to the respective failure mode only, e.g., a fiber failure mode will lead to reduction of elasticity modulus in the fiber direction only; other variations include coupling to other

Mode of failure	**Stiffness *partially reduced***
Matrix tensile cracking	$E_{yy}=0.2E_{yy}$, $G_{xy}=0.2G_{xy}$, $G_{yz}=0.2G_{yz}$
Matrix Compressive cracking	$E_{yy}=0.4E_{yy}$, $G_{xy}=0.4G_{xy}$, $G_{yz}=0.4G_{yz}$
Fibre Tensile failure	$E_{xx}=0.07E_{xx}$
Fibre compressive failure	$E_{xx}=0.14E_{xx}$
Fibre matrix shear-out	$G_{xy} = \nu_{xy} = 0$
Delamination in tension	$E_{zz} = G_{yz} = G_{xz} = \nu_{yz} = \nu_{xz} = 0$
Delamination in compression	$E_{zz} = G_{yz} = G_{xz} = \nu_{yz} = \nu_{xz} = 0$

Fig. 8.8 Sudden material property degradation rules based on modulus and Poisson's ratio degradation factor

stiffness properties, and therefore they also reduce shear elasticity modulus. Variations also exist with respect to degrading Poisson's ratios or not of the failed elements. Of course, degradation factors' values are usually adjusted (deviating from their usual 10% value) with respect to the specific problem under consideration, aiming to achieve the best possible predictions as compared to the corresponding experimental results.

The material property degradation rules are shown in Fig. 8.7 and the degradation factor method, as shown in Fig. 8.8, has been used to represent the response of the damaged material volumes. The sudden material property degradation rules of Fig. 8.8 are a mixed set, comprising the rules set by Camanho and Matthews for matrix tensile and compressive cracking and for fiber tensile and compressive failure [14]. The degradation in Poisson's ratio is restricted by the relationships between the engineering properties of the anisotropic materials. To allow practical application of the degradation rules, the degraded material properties are assumed to have a very small value, instead of zero, in order to avoid numerical problems in the FE solution. The in-plane damage material property degradation method has been modified and extended to the prediction of delamination progression.

8.8 Implementation of PFA in Composite Bolted Joints

The design optimization of composite bolted joints is affected by numerous parameters, such as geometry, combination of material types of the joined parts, stiffness properties, joint configuration, friction, clamping force and loading type, which significantly influence the many interacting and complicated failure mechanisms taking place during the bolted joint operational load. Despite the substantial number of investigations performed in composite bolted joints analysis [24, 25], a PFA approach capable of addressing all damage phenomena and operating in affordable computing times is one of the most suitable approaches for analyzing the problem. Here, a three-dimensional PFA analysis is presented, which can predict damage initiation and progression in a composite bolted joint and assess its residual strength

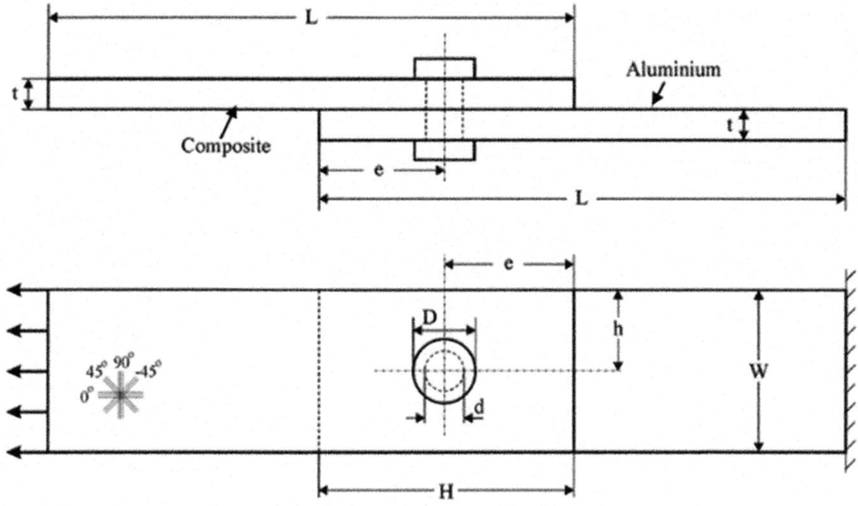

Fig. 8.9 Geometry of the composite-to-metal bolted single-lap joint

under quasi-static loading [26, 27]. The three-dimensional nature of the computational FE model is chosen to address the through-the-thickness phenomena of the laminate, which play an important role in the jointed behavior with several plies through-the-thickness (Fig. 8.9).

The current model comprises the typical PFA components described earlier in Fig. 8.9 and programmed into a parametric Ansys macro-routine using the Ansys finite element analysis software [28, 29]. The computational model of a composite bolted joint by the FE was established. The bolted structure was subjected to incremental in-plane tensile loading. Results of the analysis were compared with the experimental tests. The composite bolted joint consisted of an upper plate made of composite laminate, a lower plate made of aluminum, and a titanium bolt with a protruding head. The joint geometry is schematically shown in Fig. 8.9. Several plies of different orientations were used for the composite plate as shown in Fig. 8.10.

The three-dimensional typical FE mesh of the joint is shown in Fig. 8.10 where finite element model of the bolted single-lap joint (a), detailed mesh of the area around the bolt (b), and model of the protruding head bolt (c) are shown. Perfect fit was assumed between the bolt and the two plates. Modeling of the composite plate was done using the eight-noded Solid46 Ansys layered element, while the metallic plate and bolt were modeled using the Solid45 element type. Eight layered elements stacked together with one element per ply were used through thickness to model the $[0°/90°/\pm45°]_S$ laminate. The 3D type of analysis aims at an accurate representation of the interlaminar stress field developing around the bolt, due to bending effects, clamping force of the washer, and through-thickness pressure in the bolt-hole area. Thus, the out-of-plane effects, such as delamination damage, out-of-plane bending, and stacking sequence effects in the area close to the bolt were taken into account in the model.

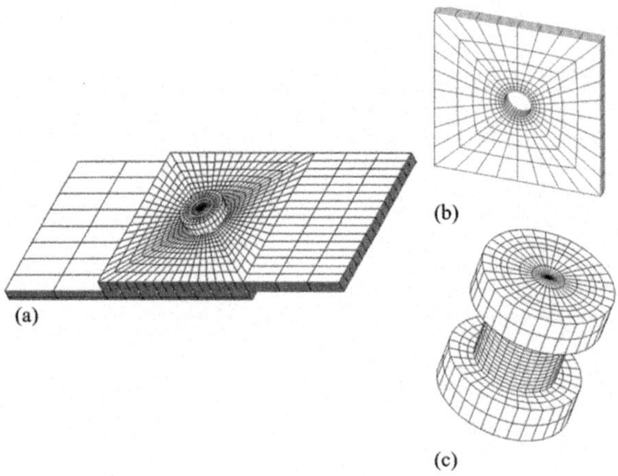

Fig. 8.10 (**a**) Finite element model of the bolted single-lap joint. (**b**) Detailed mesh of the area around the bolt. (**c**) Model of the protruding head bolt

To simulate contact between the two plates, the washer, and the plates, as well as between the bolt and the hole surface, the node-to-surface 3D Contac49 element type was used. The bolt, washer, and nut were considered as one unit to limit the number of contact elements in the model. For all contact surfaces a friction coefficient of 0.3 was assumed.

To reduce the size of the problem, symmetry conditions were implemented. The computational time was reduced by one order of magnitude, when the symmetric model was used for the analysis. The composite plate was loaded with an in-plane uniform tensile pressure at its end, while the aluminum plate was fully built-in. In order to prevent secondary bending of the joint, lateral support was applied in the model. Since the two plates and the bolt were assembled by washers, the clamping torque on the laminate could be simulated by a uniformly distributed pressure. In order to apply the required pressure, at first thermal expansion properties were assigned to the material of the bolt; then an appropriate temperature reduction led to bolt contraction, and hence to the required pressure. Since the procedure was linear, the appropriate temperature reduction could be found easily for each required torque.

Prediction of initial failure for the current bolted joint problem is performed using the Hashin-type stress-based failure criteria shown in Fig. 8.6. Once failure is detected in a ply using Hashin failure criteria, its material properties are degraded by implementing the same sudden property degradation rules used in the compression panel in Fig. 8.7. Using the PFA macro-routine, the bolted joint response is mimicked by gradually increasing the applied load. This procedure continues until excessive damage is reached, which means that failure has propagated to a specimen edge and the applied load can no longer be redistributed in the damaged areas. As a result, the specimen cannot carry any additional load and catastrophic failure is reached. At this point the residual strength is evaluated and the maximum load, which is the last applied load, is determined.

8.8.1 Damage Initiation and Progression Modes Within the Bolted Joint

It has been observed experimentally that there are three basic macroscopic failure modes in composite bolted joints. That is, tension, shear-out, and bearing, as schematically shown in Fig. 8.11. Which failure mode will initiate first and how failure will propagate depends on the bolt position [25]. The geometries examined were chosen such that the three possible macroscopic failure modes could be observed, as shown in Fig. 8.11.

The results presented hereafter refer to basic dimensions of a bolted joint of $L = 150$ mm (5.9 in), $W = 60$ mm (2.36 in), $D = 17.8$ mm (0.7 in), $d = 10$ mm (0.39 in), $e = 30$ mm (1.18 in), $h = 30$ mm (1.18 in), $H = 60$ mm (2.36 in), $t = 4.16$ mm (0.18 in), see Fig. 8.9. The laminated plate has been made of composite material system HTA/6376 (a carbon fiber/epoxy composite material manufactured by Hexcel composites, with designation HTA/6376). The full material properties are shown in Fig. 8.12 [12, 26].

In the initial stage of the different damage modes of a bolted joint geometry that failed at 9.9 kN (2.2 klb) is presented [26]. In the early stages of loading the composite behaves elastically with no sign of damage to the material. As the load increases the first ply failures are detected. Matrix cracking is initiated at the outer 0° ply at the bearing plane in the material located under the washer area due to normal compressive stresses acting on the laminate from the washer, mainly during pre-tension. As the load increases, fiber-matrix shear-out initiates at 45° to the

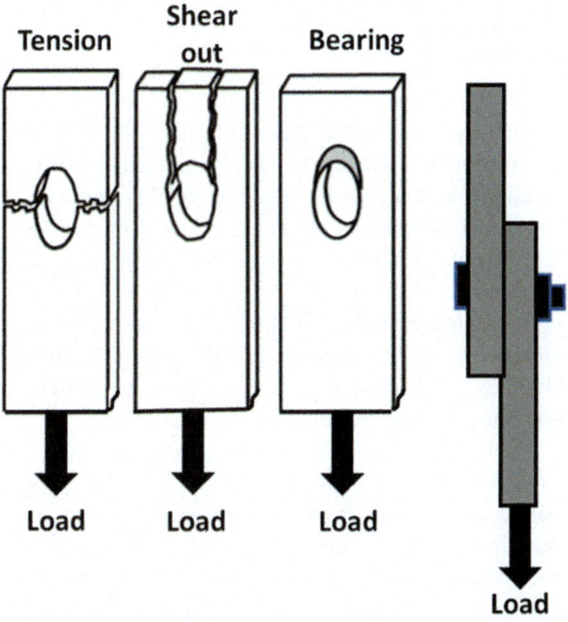

Fig. 8.11 Macroscopic failure modes observed in composite bolted joints

X_T, Y_T, Z_T are tensile strengths in the X, Y, Z directions
X_C, Y_C, Z_C are compressive strengths in X, Y, Z directions
S_{XY}, S_{YZ}, S_{XZ} are shear strengths in X, Y, Z directions

E_{XX}	E_{YY}	E_{ZZ}	G_{XY}	G_{XZ}	G_{YZ}	ν_{XY}	ν_{XZ}	ν_{YZ}
130.4	12.97	12.97	6.38	6.38	4.69	0.3	0.3	0.45
X_T	X_c	Y_T	Y_C	Z_T	Z_C	S_{XY}	S_{YZ}	S_{XZ}
1.38	1.14	0.081	0.189	0.081	0.189	0.069	0.021	0.069

Fig. 8.12 Material properties (stiffness and strength values in GPa) Note: 1 GPa = 1000 MPa, 1 MPa = 0.145 ksi

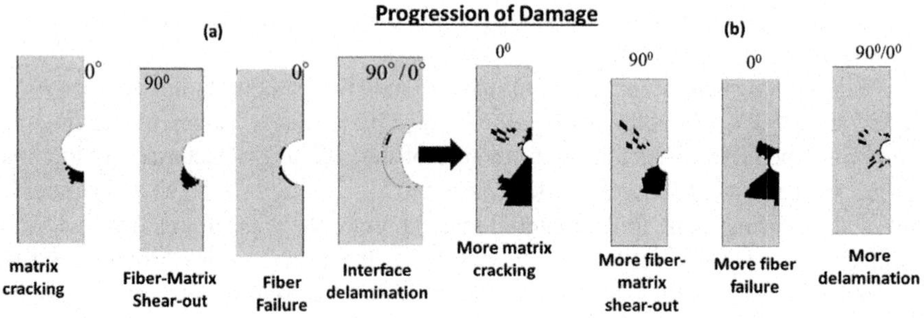

Fig. 8.13 (**a**) Prediction of initial stage of damage for $[0°/90°/\pm45°]_S$. (**b**) Damage prediction at the 0° and 90° plies at the final failure (13470N) of the $[0°/90°/\pm45°]_S$ specimen that failed in bearing (d = 10 mm, e/d = 6 and w/d = 6 – see Fig. 8.9)

load direction along the shear-out plane (90° ply), due to high shear stresses. Fiber failure (compressive and tensile) is initiated at 68% of the failure load across the hole boundary and at the 0° ply due to high normal stresses in the fiber direction. Delamination onset is predicted to occur on the hole edge at the 0°–90° interface at 75% of the failure load, which is in agreement with the predictions and experiments.

The damage prediction at the −45° ply for the fully sheared-out failed joint was observed as shown in Fig. 8.13. Because of the small edge distance *e/d*, a shear-out failure is predicted as expected, at a failure load of 9.98 kN (2.24 klb). In this figure, both the initiation damage (a) and damage near the final failure (b) are shown.

Figure 8.14 shows the damage prediction at the 45° ply at the final failure of the tension-failed joint. The specific type of failure is characterized by the early fiber failure progression from the hole edge toward the specimen's free edge at the direction normal to the loading due to the small *w/d* ratio (Fig. 8.15). Therefore, the residual strength of the joints with such geometry is very small. Figure 8.16 illustrates the prediction of damage at the 0° and 90° plies in the final failure stage of

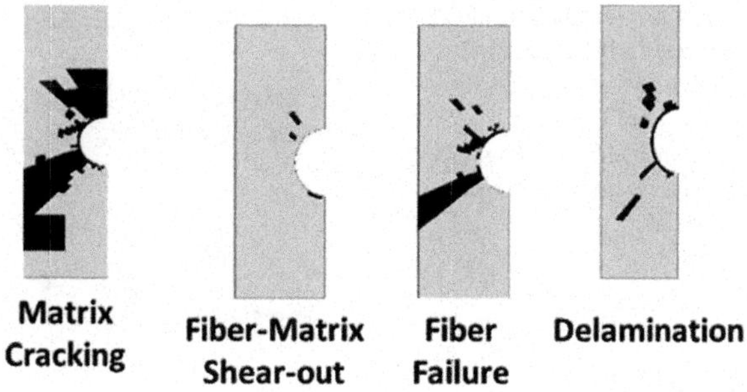

Fig. 8.14 Damage prediction at the –45° ply at the final failure load of the $[0°/90°/±45°]_S$ specimen that failed in shear-out ($d = 10$ mm, $w/d = 3$ and $e/d = 4$)

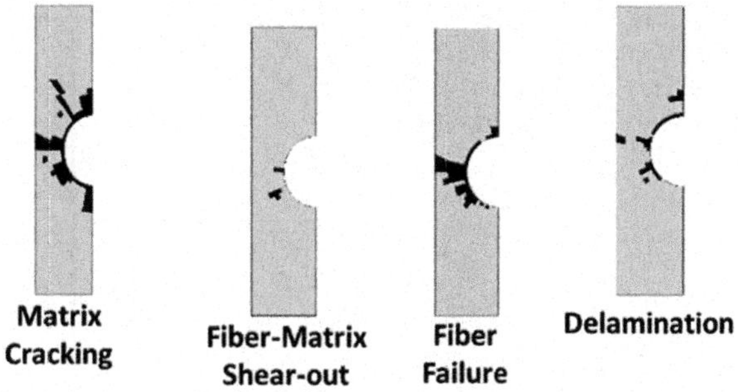

Fig. 8.15 Damage prediction at the 45° ply at the final failure load (6822N) of the [0°/90°/±45°]S specimen that failed in tension (d = 10 mm, w/d = 2 and e/d = 6)

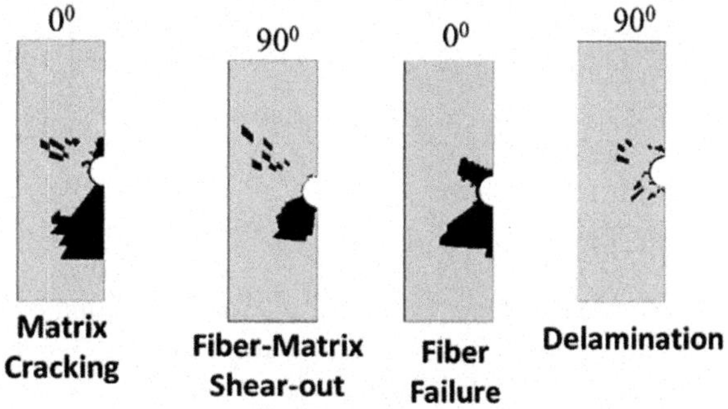

Fig. 8.16 Damage prediction at the 0° and 90° plies at the final failure load (13470N) of the [0°/90°/±45°]S specimen that failed in bearing (d = 10mm, e/d = 6 and w/d = 6)

the specimen that failed in bearing due to the large edge distance ratio *e/d*. Damage was initiated from the contact surface as a result of the compression failure due to a concentrated bearing load.

Occurrence of both fiber-matrix shear-out and delamination damage at the 90° ply in the bearing plane confirms the experimental observations, that the formation of shear cracks and delaminations are the major characteristics of bearing damage [26]. The shear cracks were either propagated toward the free surfaces of the laminate or merged to initiate delamination, resulting in a significant reduction of laminate stiffness and fiber failure. As a result, the joint could no longer sustain any additional load and the failure was catastrophic.

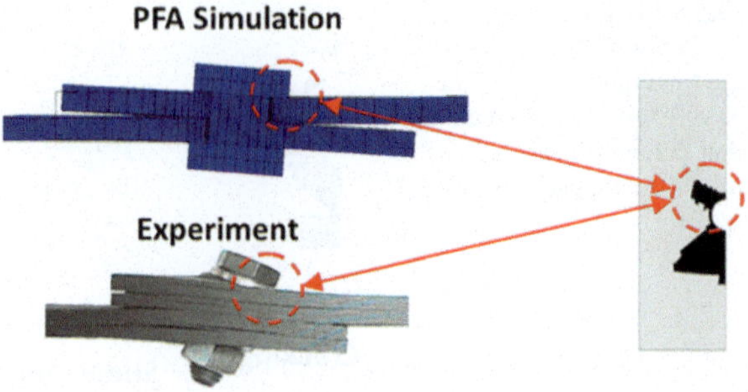

Fig. 8.17 Prediction of damage in the laminated plate due to the penetration of washer. The PFA computer simulation is in good agreement with the experiment

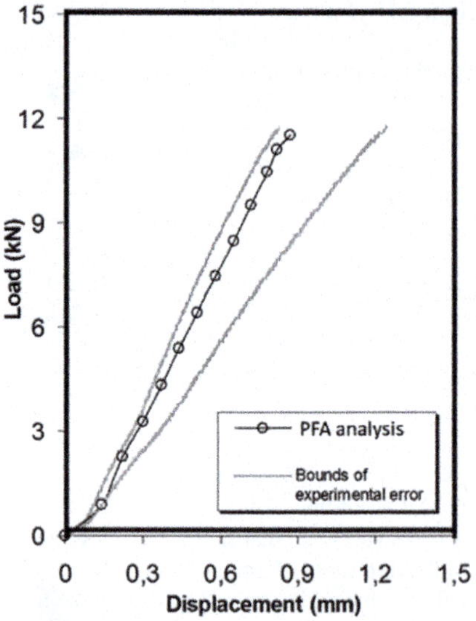

Fig. 8.18 Comparison of load-displacement curve predicted by the PFA analysis to the experimental data [12]

The final failure criteria used for the definition of catastrophic failure and determination of failure load for the cases presented above are based on damage accumulation. For tension and shear failure modes, the criterion is the propagation of fiber failure until the free edge of the specimen at the directions of 90° and −45° with regard to loading direction.

In Fig. 8.17, the predicted deformed shape of the bolted joint, which failed in bearing is compared to the corresponding experimental results. Secondary bending of the area around the bolt leaned the bolt and led to penetration of the washer into the laminated composite plate. As can be seen, this phenomenon is accurately captured by the PFA computer model by predicting both deformation of the joint and damage in the laminated plate resulting from the penetration.

The stiffness and strength of bolted joint configurations for which extensive experimental data existed are predicted and presented hereafter. In Fig. 8.18, the load-displacement curve predicted by PFA analysis is compared to the experimental curve. The stiffness of the joint has been accurately predicted since the numerical curve falls within the bounds of experimental error. The successful strength prediction, which occurs due to fiber failure, indicates the accurate detection by the respective fiber failure criterion implemented in the PFA [12].

8.9 Concluding Remarks

The progressive failure analysis approach was discussed, and its methodology was based on (1) computational modeling based on the 3D finite element, (2) application of a suitable failure theory for the damage initiation, and (3) the implementation of material properties degradation to the final structural failure. That is, the computational modeling of PFA analysis for any structure under applied load shall start with a three-dimensional finite elements modeling, in order to capture sufficient not only the in-plane but also the out-of-plane stresses. The 3D computational modeling includes the geometrical and structural details, capable of solving for the nonlinear behavior. Moreover, in material properties degradation, there are many approaches; however, in the analysis of bolted joints presented in this chapter, the 10% reduction rule for mechanical properties degradation was implemented and the result was in a good agreement with test data. In addition, it provided reduction in the computational time when conducting the PFA analysis. Finally, the combination of fracture mechanics and the PFA can be useful to engineers by assuming the preexisting defects in the structure through the nondestructive inspection enable the application of a damage tolerance design philosophy, in a fashion similar to that currently used for metallic structures.

References

1. R.M. Christensen, A comprehensive theory of yielding and failure for isotropic materials. J. Eng. Mater. Technol. **129**(2), 173–181 (2007)
2. J. Besson, D. Steglich, W. Brocks, Modelling of plain strain ductile rupture. Int. J. Plast. **19**(10), 1517–1541 (2003)
3. Q.M. Li, Strain energy density failure criterion. Int. J. Solids Struct. **38**, 6997–7013 (2001)
4. B. Farahmand, Chapter 6 (FMDM Theory), in *Fatigue and Fracture Mechanics of High-Risk Parts*, (1996)
5. B. Farahmand, Progressive damage assessment (PDA) of composite materials (Chapter 13), in *Application of High and Low Cycle Fatigue on Metallic and Non-Metallic Material (Application of Fatigue for Engineers)*, (Amazon eBook, 2022)
6. S.W. Tsai, E.M. Wu, A general theory of strength for anisotropic materials. J. Compos. Mater. **5**, 58–80 (1971)
7. I.I. Gol'denblat, V.A. Kopnov, Strength of glass reinforced plastic in the complex stress state. Polym Mech **1**, 54–60 (1966)
8. K.K. Kar, Chapter 1, in *Composite Material – Processing, Applications, and Characterization*, (Springer, 2017), p. 70
9. I.M. Daniel, O. Ishai, *Engineering Mechanics of Composite Materials* (Oxford University Press, 1994)
10. S. Alger, M. Mark, *Polymer Science Dictionary*, 2nd edn. (1997)

11. W. Yu, *An Introduction to Micromechanics* (Trans Tech Publications, Switzerland, 2016), pp. 3–24
12. G. Labeas, Prediction of damage propagation and failure of composite structures (without testing), in *Virtual Testing And Predictive Modeling*, ed. by B. Farahmand, (Springer, 2009)
13. G.M. Odegrad, Multiscale modeling of nanocomposite material, in *Virtual Testing and Predictive Modeling*, ed. by B. Farahmand, (Springer, 2009)
14. P.P. Camanho, F.L. Matthews, Delamination onset prediction in mechanically fastened joints in composite laminates. J. Compos. Mater. **33**, 906–927 (1999)
15. T.E. Tay, G. Liu, B.C. Tan, et al., Progressive failure analysis of composites. J. Compos. Mater. **42**, 1921–1966 (2008)
16. Z. Hashin, Failure criteria for unidirectional fiber composites. ASME J. Appl. Mech. **47**, 329–334 (1980)
17. Z. Hashin, A. Rotem, A fatigue failure criterion for fiber reinforced materials. J. Compos. Mater. **7**, 448–464 (1973)
18. S.W. Tsai, E.M.A. Wu, General theory of strength for anisotropic materials. J. Compos. Mater. **5** (1971)
19. L.F. Kawashita, S.R. Hallett, A crack tip tracking algorithm for cohesive interface element analysis of fatigue delamination propagation in composite materials. Int. J. Solids Struct. **49**, 2898–2913 (2012)
20. J. Reeder, S. Kyongchan, P.B. Chunchu, D.R. Ambur, Postbuckling and growth of delaminations in composite plates subjected to axial compression, in *43rd AIAA/ASME/ASCE/AHS/ASC Structures, Structural Dynamics, and Materials Conference*, (2002), p. 10
21. M.N. Nahas, Survey of failure and post-failure theories of laminated fiber-reinforced composites. J. Compos. Technol. Res. **8**, 138–153 (1986)
22. H.T. Hahn, S.W. Tsai, On the behavior of composite laminates after initial failures. Astronaut. Aeronaut. **21**, 58–62 (1983)
23. Y. Murray, L. Schwer, Implementation and verification of fiber-composite damage models, in *Failure Criteria and Analysis in Dynamic Response*, ASME, vol. 107, (1990), pp. 21–30
24. X. Li, Y. Li, F. Li, Z. Huang, H. Chen, Failure analysis and experimental study on bolted composite joints based on continuum damage mechanics. Compos Struct **303**, 1 (2023)
25. A. El-Sisi, A. Hassanian, F. Alsharari, N. Galustanian, H. Salim, Failure behavior of composite bolted joints. CivilEng **3**(4), 1061–1076. MDPI Publisher (2022)
26. K.I. Tserpes, G. Labeas, P. Papanikos, et al., Strength prediction of bolted joints in graphite/epoxy composite laminates. Compos B **33**, 521–529 (2002)
27. K.I. Tserpes, P. Papanikos, T. Kermanidis, A three-dimensional progressive damage model for bolted joints in composite laminates subjected to tensile loading. Fatigue Fract. Engng. Mater. Struct **24**(10), 673–686 (2001)
28. ANSYS User's Manual. Version 5.7. Swanson Analysis Systems, Inc; 2001
29. T. Ireman, Three-dimensional stress analysis of bolted single-lap composite joints. Compos Struct **43**, 195–216 (1999)

Chapter 9
Application of Fracture Mechanics to Composites

9.1 Introduction

Polymer composites may show inherently different behaviors when compared with metallic alloys when subjected to monotonic or cyclic load environments. Metals and their alloys are credited mainly to their attractive properties, such as high toughness, machinability, weldable, castable, and above all, homogeneous and isotropic. As industries transition toward composite materials, there is an increasing need for a deeper understanding of failure mechanisms in PMCs, MMCs, and CMCs when subjected to various service loads. However, the focus of this chapter is primarily on polymer matrix composites, whose application in industries, especially in aerospace and aircraft structures, has notably expanded in recent years.

The conventional low- and high-cycle fatigue approaches to structural life must be distinguished from the linear elastic fracture mechanics (LEFM), which is based on the material being elastic, homogeneous, and isotropic, and the idea that material has inherent defects must be taken into consideration when conducting life analysis. The LEFM concept concentrates on the crack tip stress intensity factor, where the stresses are much higher than the rest of the body when under the far field applied stress [1, 2]. Therefore, an adequate amount of energy must be available at the crack tip to cause the crack to propagate and form two new crack surfaces. In the traditional approach to structural life, when employing the stress to life approach (S-N, representing the high-cycle fatigue) or strain to life approach (ε-N, representing the low-cycle fatigue), it is assumed that the structure is initially free from cracks and after N number of cycles, the crack will initiate in some localized high stressed areas [3]. The total life of the structure will be the sum of crack initiation and stable crack propagation. However, in applying the LEFM, it is assumed that there are always microscopic defects preexisting in the material situated perpendicular to the applied load (i.e., the worst orientation representing the mode I scenario). To evaluate the life of a part, an initial crack size (a_i) in the structure must be assumed, and it will be provided to the analyst using the standard nondestructive inspection method. This

B. Farahmand, *Fundamentals of Composites and Their Methods of Fabrications*,
https://doi.org/10.1007/978-3-031-80201-0_9

flaw size is the largest flaw in the material, but inspection capability is not good enough to accurately determine its size. For example, in the case of penetrant inspection, the maximum initial flaw size that escapes from detection is a surface crack of depth $a = 0.075$ in and a length of $2c = 0.15$ in [4–6]. Other inspection methods will provide slightly different initial flaw sizes for analyzing a structural part based on the LEFM method (see Chap. 14). In addition to the initial flaw size in the material, which its dimensions must be available to analysts for conducting their structural life assessment, there is also a need to have fracture properties, such as fracture toughness and fatigue crack growth data. These data will help analysts to determine the number of cycles from initiation to final failure by using a proper fracture mechanics tool, such as the NASGRO [7] or AFGRO [8] software. NASGRO is a computer program with fracture toughness and fatigue crack growth data for many metallic materials. The code is developed jointly by Southwest Research Institute (SwRI) and the National Aeronautics and Space Administration (NASA). AFGRO was initially established by the Air Force Research Laboratory and is mainly used for crack growth analysis of aerospace components, but it can also be utilized for any type of metallic structure that experiences fatigue cracking. Both NASGRO and AFGRO were developed for isotropic material and their applications to composites may introduce errors when calculating the number of cycles to failure. Also, note that the crack growth in composites can be a combination of both modes I and II; therefore, by assuming the mode I behavior only, the output can be conservative, where the number of cycles to failure will be less. For that reason, it may be acceptable for an analyst to use the mode I LEFM approach to conservatively estimate the number of cycles to failure of their part.

Obtaining material fracture toughness and fatigue crack growth rate data is not an easy task, and these tests are costly, time-consuming, and labor-intensive. Figure 9.1 shows the crack initiation and propagation in metallic and laminated composite material. Metals are considered homogeneous and isotropic, and the assumption of composite being elastic and homogeneous will be implemented in the analysis. In the case of laminated composites, the crack growth analysis in the matrix material can be assessed by applying the LEFM. The above assumptions will be utilized, where both the matrix and fibers are elastic and homogeneous. Another assumption for the cause of delamination is that the delamination initiation can happen between plies due to inadequate bonding (Fig. 9.2). In this figure, the process of delamination growth and the number of cycles to failure (N_f) can be assessed through linear elastic fracture mechanics by considering the delamination is as a result of the adhesive failure between two plies.

As was mentioned in Sect. 7.3 (Fig. 9.8) of Chap. 7, the LEFM is applicable to a lamina of a laminated composites at the onset of instability when the crack tip stress intensity factor, $K_I = K_{Ic}$ or $K_{II} = K_{IIc}$, but not the subsequent stable crack growth. However, applying the Mode I LEFM approach without considering the contribution of Mode II may result in a conservative life assessment analysis.

When the NDI inspection on the part is completed, a comprehensive FEM analysis will follow to obtain the state of stress in the areas where delamination initiation between plies or the case of crack initiation and growth in the matrix,

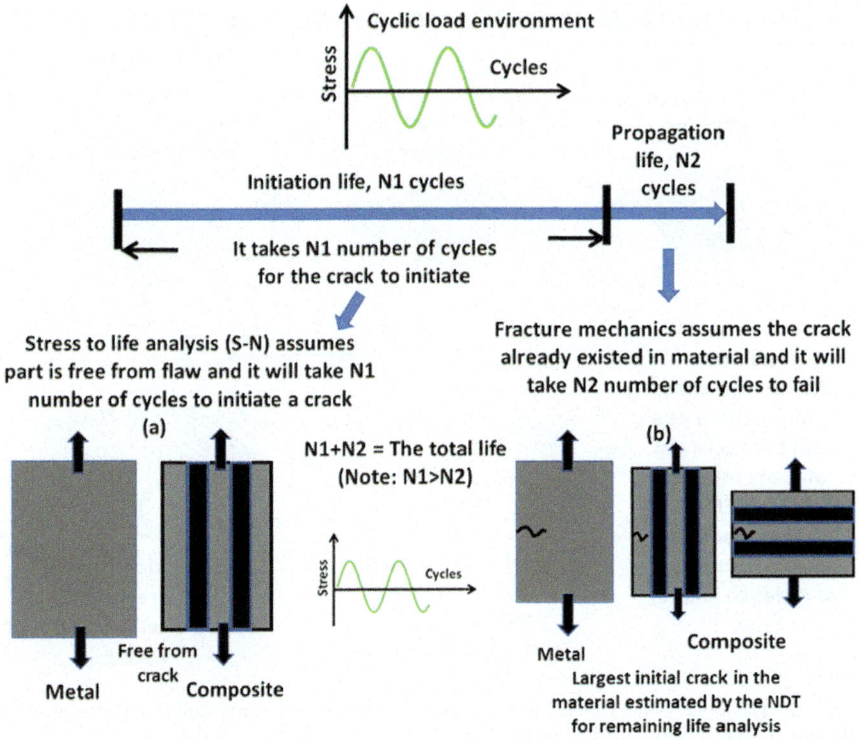

Fig. 9.1 (**a**) Crack initiation life (N_1) can be estimated through the S-N or ε-N methods. (**b**) The propagation life (N_2) to be estimated through fracture mechanics. To estimate the number of cycles, N_2, fracture allowables must be available to analys

interface, and fibers are most probable to occur. In the case of crack initiation from the matrix, the case of a crack perpendicular to the load and parallel to fibers is a strong possibility to be considered and engineers are able to use LEFM for their analysis (Fig. 9.1b).

The case of a crack perpendicular to the load and fibers will also be discussed here. In addition, several fracture mechanics analysis cases will be presented in this chapter.

There are several nondestructive inspection methods that can be used to investigate the microscopic delamination defects within a composite laminate prior to their service usage. These include ultrasonic testing, radiography testing, and thermography, which will be assessed in Chap. 14. The thermography method has the advantage of being more effective, both from time and cost perspectives as compared to other NDI methods. Applying the appropriate NDI method on the composite part before and during its service usage is to determine the size of the initial flaw size, a_i, prior to the part's usage and to monitor the rate of its growth during the operation. If the degree of damage is critical, then the part will be replaced or will be repaired (Chap. 13).

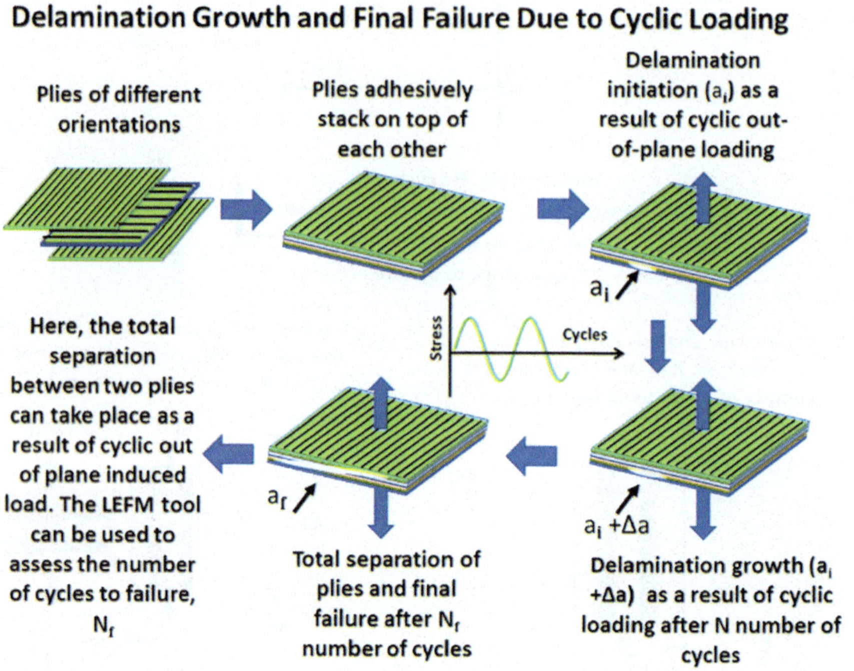

Fig. 9.2 The process of delamination growth and number of cycles to failure (N_f) can be assessed through the linear elastic fracture mechanics, assuming the adhesive material is elastic

When the initial size of delamination inside the layers has been determined by the NDI inspection, the linear elastic fracture mechanics can be used to find out the number of cycles to failure, N_2, as illustrated in Fig. 9.1. Note that, in conducting the LEFM life analysis, the initiation number of cycles (i.e., N_1 as shown in Fig. 9.1) are not considered in the analysis. The whole idea of using the NDI inspection is to look for the maximum initial flaw size (e.g., delamination) that escapes from detection and afterward to conduct a meaningful fracture mechanics analysis of the remaining portion of the life cycles, i.e., N_2. The analysis will help to find out if the part with delamination can survive the load environment or not. If the inspector of the NDI inspection reported that no damage was found in the part prior to its usage, engineers or analysts still must assume a small flaw size preexisted in the material that can be used in the fracture mechanics analysis. The size of the initial flaw size must be slightly below the standard NDI detection capability. By using the assumed initial flaw size in the part, engineers can find out how many load cycles are needed to cause failure, N_f.

9.2 Griffith Energy Balance and Irwin Stress Intensity Factor Approaches

There are two approaches that engineers will use to assess the crack behavior when subjected to an externally applied load. Here both the energy balance approach, G, and the crack tip stress intensity factor, K, will be briefly discussed. The Griffith energy balance approach was established by Griffith [9, 10], and it simply states that for some applied stress, σ, and ideally sharp centered crack of length $2a$ in an infinite elastic plate, the instability will take place when an adequate amount of energy is available to create two new crack surfaces. The energy release rate, G, in terms of crack length and applied stress, σ, can be written as,

$$G = \left(\pi\sigma^2 a\right)/\mathrm{E} \tag{9.1}$$

where E is the tensile modulus of elasticity of the material. The quantity G is not the right parameter to use when assessing the crack tip stresses or fatigue crack growth rate analysis. Instead, the quantity K [10], the stress intensity factor, is more appropriate to use for the crack tip analysis. However, both G and K are related to each other by,

$$G = K^2/E' \text{ where } E' = E \text{ for the plane stress and } E' = E\left(1 - \nu^2\right) \text{ for the plane strain} \tag{9.2}$$

E is the modulus of elasticity, and ν is Poisson's ratio. The critical value of these two parameters is designated as G_c and K_c. Figure 9.3 shows the three modes of failure that are designated as K_I (mode I or the opening mode), K_{II} (mode II or the sliding mode), and K_{III} (mode III or the tearing mode). Accordingly, for the energy release rate, they are shown as G_I, G_{II}, and G_{III}.

9.3 Method of Obtaining Mode I Fracture Properties for Isotropic Material

To continue with the fracture mechanics analysis of a laminated composite part subjected to cyclic loading, the static and fracture properties of laminated material are needed. Figure 9.4 shows the fracture toughness as a function of thickness and fatigue crack growth data for a metallic material, which are necessary to conduct a fracture mechanics analysis in a load-varying environment. After obtaining stresses from finite element modeling analysis, the maximum initial flaw size, which is situated perpendicular to the applied load, will be assumed, and used to assess the delamination growth found in part by the NDI method. The initial delamination length of a_i and the applied far-field stress of σ is used to calculate the stress intensity

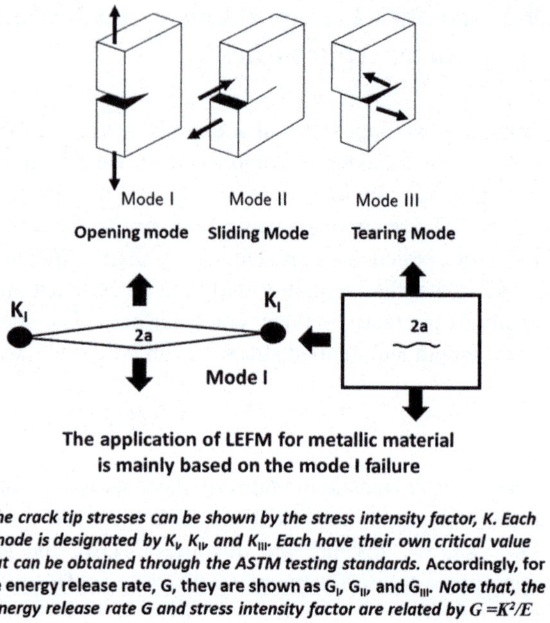

Fig. 9.3 Three modes of failure. Each mode has crack tip stresses that are designated by K_I, K_{II}, and K_{III} called mode I, II, and III stress intensity factors

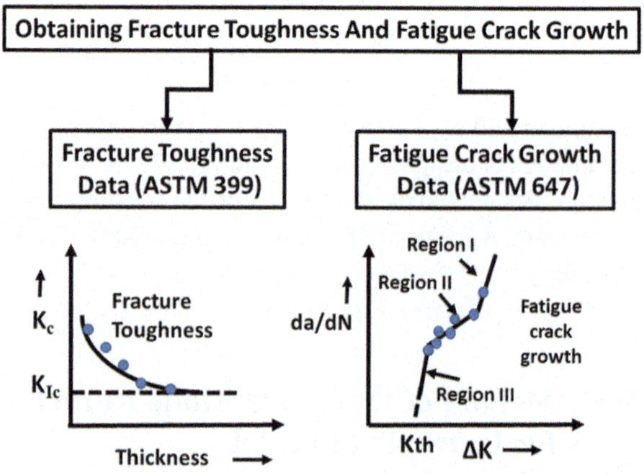

Fig. 9.4 Fracture properties data (both the fracture toughness and the fatigue crack growth data) are needed for a fracture mechanics analysis (i.e., finding number of cycles to failure, *N*)

factor, *K*. The *K* parameter will provide information about the criticality of crack tip stresses and is a function of the applied stress and the crack length, a. That is, $K = \beta\sigma(\pi a)^{0.5}$, where β is the correction factor for the kind of load and crack geometries. When the crack length reaches its critical value, part failure is imminent. That is, when the value of $K = \beta\sigma(\pi a_{cr})^{0.5}$ it increases and becomes large, then the quantity $K = K_c$ (or K_{Ic} under the plane strain condition). Here, the quantity K_c is the mode I fracture toughness of the material. Failure occurs when the crack length is equal to a_{cr}. The Kc value can be obtained through the ASTM testing standards. For brittle material with no plasticity at the crack tip before failure, then its value is independent of the part's thickness, and it is equal to the plane strain fracture toughness, $K = K_c = K_{Ic}$ (Fig. 9.4) and its value can be obtained through the ASTM E-399.

Three cases regarding damage initiation and growth associated with laminated composite will be assessed in this section, and they are outlined in Fig. 9.5. They are,

1. The delamination with an initial estimated NDI crack length of a_i will be assumed between plies or at the edges. The delamination growth will be assessed through fracture mechanics methodology. Both the static and fracture properties of adhesion material, holding the two laminae, must be available for life analysis, as shown in Fig. 9.5 (case 1). Finite element stress analysis is needed to look for the maximum stresses where delamination initiation is most probable.
2. The initial NDI estimated crack preexisted in the matrix of a lamina of a laminated composite. It will grow in the matrix perpendicular to the applied load until final failure occurs. This scenario is a possibility because the matrix material is weaker than the fibers, and if the bond between the fiber and the matrix is weak, a minimum amount of crack arrest will be expected to halt the advancing crack. However, the fiber will be able to bridge the two sides of the crack mouth if fiber failure does not occur (Fig. 9.7, case 2a). Another scenario will be the initiation of matrix cracking perpendicular to the applied load, but parallel to the fibers with no resistance from the interface (case 2b).
3. The third scenario will be the case of the initial estimated NDI crack which will grow in the matrix of a lamina of a laminated composite until it is arrested by the fiber. The interface bond between fibers and the matrix is strong and crack growth will be halted, Fig. 9.5 (case 3). All three scenarios will be discussed later in Sects. 9.5 and 9.6, respectively.

9.4 Obtaining Fracture Allowable Through Testing

In the case of homogeneous, elastic, and isotropic materials, their fracture properties and the ASTM documentation and guidelines for testing are well established. Practically, all metallic alloys currently used for industries, their fracture properties are available, and they are obtainable through the ASTM testing standards (*Annual Book of ASTM Standards*, Volume 03.01). In addition to having fracture properties (i.e., fracture allowables) for life analysis, engineers must realize that the selection of alloys for industrial parts is based on, (1) having adequate ductility, (2) having adequate strength, (3) being machinable without cracking, (4) should maintain their properties at cold temperature, (5) having adequate resistance to corrosive environments, (6) availability of the material and their properties when industries need to utilize them, and (7) Material must be cost-effective. Based on the abovementioned criteria, after engineers or designers have selected their material of choice, then both static and fracture allowables will be established through testing or obtained from a

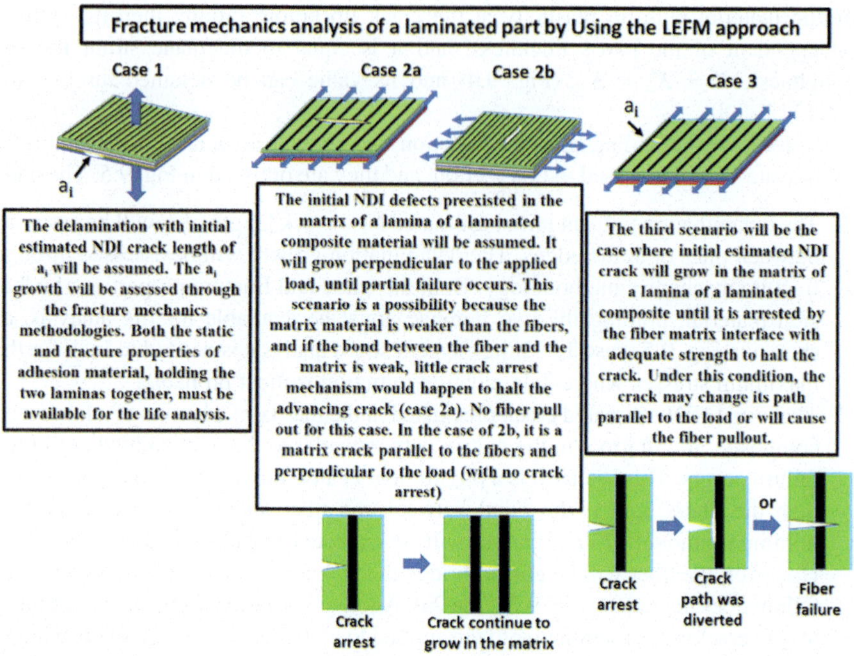

Fig. 9.5 Three crack cases were studied for the fracture mechanics analysis of a laminated composite. All cases are based on initial NDI estimated crack. Here, the crack tip is referred to delamination front

reliable source (if data are already available to the public for use). For the fracture toughness (the material's ability to resist cracking), both the plane strain (K_{Ic}) and the plane stress fracture toughness (K_c) are obtainable through the ASTM-E399 [11] and ASTM-E561 [12] standards, respectively. For the crack growth rate testing, which can determine the rate of crack advancement when subjected to cyclic loading, the ASTM-E647 is used for metallic material [13]. The NASGRO computer code [7] has a significant source of data that contains both static and fracture properties of so many alloys currently being used by industries. NASGRO was initially developed at the NASA Johnson Space Center in the early 1980s to perform fracture analysis for manned space programs. These data will subsequently be utilized for calculating the number of cycles to failure when parts are subjected to a fluctuating load environment. Figure 9.6 shows key fracture allowables that are needed for metallic material. In this figure, the description of all the fracture-related terms is provided to the reader.

When fracture properties are available to engineers, then life assessment analysis tools are used to calculate the number of cycles to failure (N_f) in a fluctuating load environment. The most acceptable equation that can relate the crack growth rate (da/dN) to all the regions of crack growth (regions I, II, and III of Fig. 9.6) is available in NASGRO computer code and it is known as the Forman–Newman–de

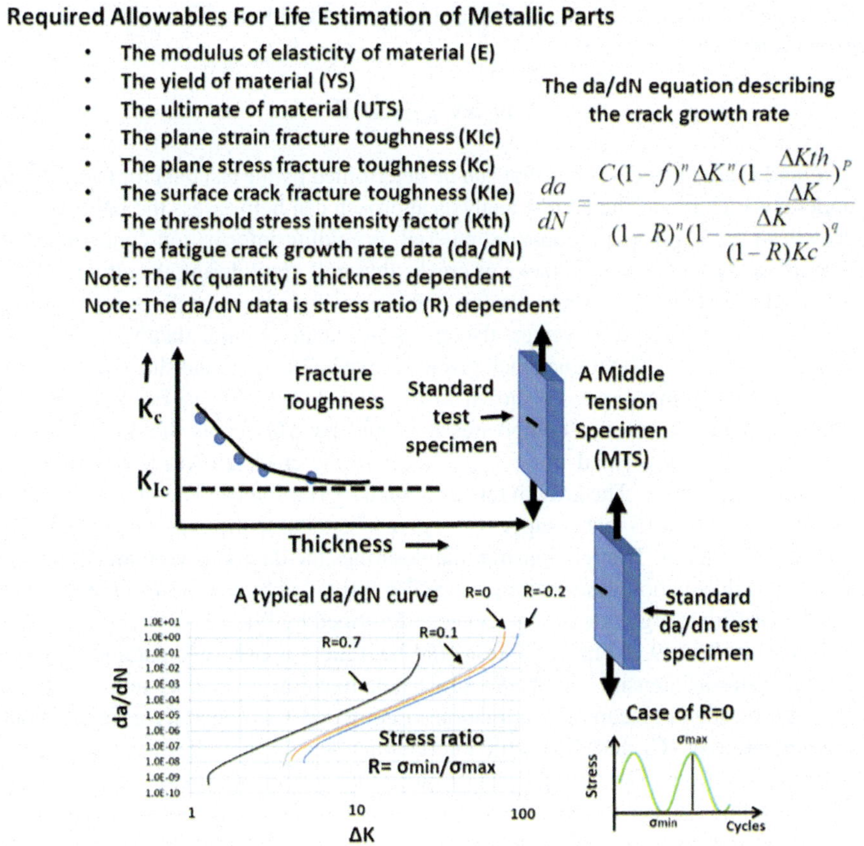

Fig. 9.6 Required allowables for life estimation of metallic parts. Both the fracture toughness and the da/dN specimen are shown. Trends of data for both Kc and da/dN are plotted here

Koning crack growth equation. The crack growth equation for mode I can be written as:

$$\frac{da}{dN} = \frac{C(1-f)^n \Delta K^n \left(1 - \frac{\Delta K\text{th}}{\Delta K}\right)^P}{(1-R)^n \left(1 - \frac{\Delta K}{(1-R)Kc}\right)^q} \tag{9.3}$$

In Eq. 9.3, a = half a crack length, ΔN = the number of load cycles for Δa crack growth, K = the stress intensity factor for mode I, ΔK is the difference between the maximum and minimum value of K during cycles, K_c = critical stress intensity factor for the thickness under consideration, K_{th} = threshold stress intensity factor, where below that crack will not grow and has an infinite life, R = stress ratio ($\sigma_{min}/\sigma_{max}$), and C, n, f, q are constants that are available by the NASGRO computer code for life analysis [7]. The Forman–Newman–de Koning crack growth equation (Eq. 9.3) can be simplified to the Paris equation that is only applicable to the region II of crack growth rate by:

$$da/dN = C(\Delta K)^m \tag{9.4}$$

C and m are Paris constants that can be determined by the test data for the material under consideration. The LEFM analysis approach can help assess the fatigue crack growth in the laminated composite material, providing information concerning the damage growth in terms of cracks or cracks that will grow during the structural life are available. For composites, fatigue crack growth curves are plots of the strain energy release rate, ΔG, versus the crack propagation rate, da/dN, usually on logarithmic axes. Each fatigue crack growth curve has a sigmoidal shape comprising three regions of the curve, as shown in Fig. 9.6. In addition to Eq. 9.3 that was established for metals (for composites it should be ΔG versus the da/dN) a new approach is also presented in Sect. 9.12, where the crack growth is described by Hartman and Schijve. The growth rate presented by Hartman and Schijve is based on the da/dN versus $\Delta\sqrt{G}$ that is supposed to give a better fit to the experimental data for composite material. The relationship that describes the da/dN, is very much like the crack growth equation described by NASGRO provided by Eq. 9.3 (see Sect. 9.12).

The Griffith energy release rate concept described by Eq. 9.1, $G = (\pi\sigma^2 a)/E$, was discussed in Sect. 9.2. The G value would be a more appropriate parameter to use than the stress intensity factor, K, when delamination growth is required to be assessed by the engineers. Delamination in composites is generally a mixed-mode fracture process (G_{I} and G_{II}) with interlaminar tensile and shear stresses at the delamination front. The inter ply out of plane induced tensile stresses give rise to a mode I component (with the associated energy released rate G_{I}), and the interlaminar (in-plane) shear stresses give rise to a mode II component (with energy release rate G_{II}).

Figure 9.7 shows the delamination test for several plies adhesively attached together. The critical value of energy release rate, G, is an important parameter in the failure analysis of composites, and it is designated by G_c. The mode I critical value of energy release rate is G_{Ic}, and for mode II is designated by G_{IIc}. For mode III the critical energy release rate is shown by G_{IIIc}. The method most popularly used to obtain G_{Ic} value is the double cantilever beam (DCB) test [11]. The values of G_{Ic} are consistent within the type of test method employed. They vary depending on the matrix material, fiber orientations and fiber amount, and the nature of the fiber–matrix interface. In Fig. 9.7, the DCB test is shown for the G_{Ic} test based on the ASTM testing standards. There is a wide range of international and national test standards (ASTM, EN, ISO, JIS) developed to test composite materials where,

ASTM (American Society for Testing and Materials)
EN *(European Norms)*
ISO *(International Organization for Standardization)*
JIS *(Japanese Industrial Standards)*

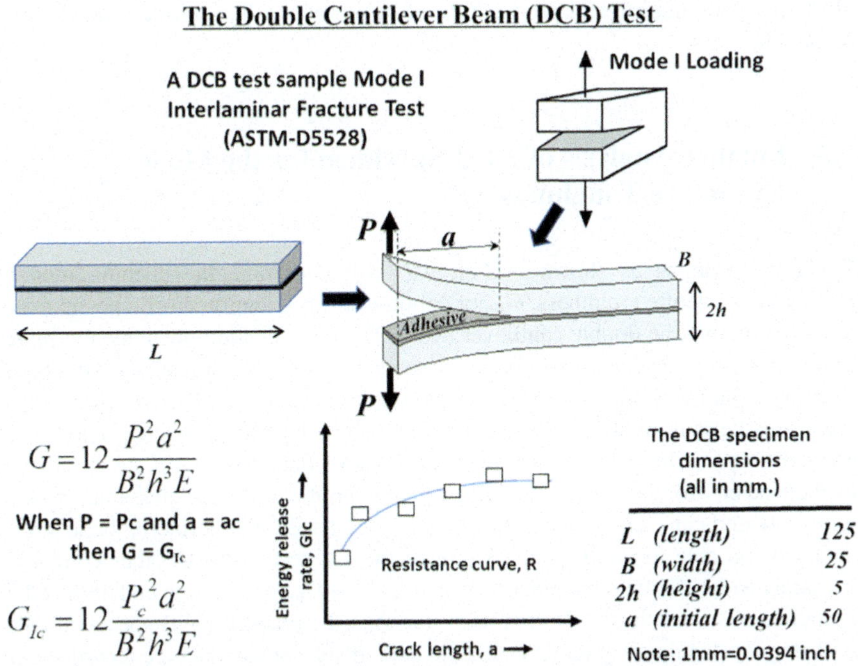

Fig. 9.7 The DCB test specimen geometry for the G_{Ic} test, where the upper and the lower parts are adhesively attached to each other simulating the case of delamination with partial bonding

Several other historical standards, such as those published by SACMA (Suppliers of Advanced Composite Materials Association). This manual contains 27 composite test methods. It covers strength, stiffness, and physical properties, and each method is given an SRM designation (for SACMA Recommended Method). They are numbered 1 through 27. CRAG (Composites Research Advisory Group) is another standard but it is no longer an active organization [12].

The double cantilever beam (DCB) test (ASTM D5528) is the most widely used test method for measuring the Mode I fracture toughness (G_{Ic}) of unidirectional composites. The specimens generally have dimensions of ~125 mm long, 20–25 mm wide, crack length a = 50 mm, and 3–5 mm thick (1 mm = ~0.0393 in), as shown in Fig. 9.7 [11]. The mode II fracture toughness for obtaining the G_{IIc}, and the mixed mode value will be discussed in Sect. 9.9.

In the DCB testing method, at the start of delamination growth when $a = a_i$, the value of mode I energy release rate, G_I, will be calculated. The G_I value will gradually increase upon delamination growth due to the applied load, where the original delamination size, a_i, will advance incrementally to $a = a_i + \Delta a$ and at instability $a = a_{cr}$. A resistance curve (R-curve), like the case of metals, representing the G_{Ic} as a function of delamination length, a, can be generated during the delamination growth. A sample of the R-curve is shown in Fig. 9.7. This curve illustrates the initiation and propagation of delamination in a tensile mode I DCB specimen.

9.5 Further Analysis of DCB Specimen for the Mode I Fracture Toughness, GIc

The test apparatus as shown in Fig. 9.7, will determine the opening mode I interlaminar fracture toughness, G_{Ic}, of continuous fiber-reinforced composite materials by utilizing the double cantilever beam. The delamination phenomenon represents a strong weakness of many advanced laminated composite structures. Therefore, knowing the laminated composite material's resistance to interlaminar fracture is helpful in having a sound product for the industries. The G_{Ic} test allows engineers to conduct a life assessment analysis of their part when subjected to a cyclic load environment and to find out if the NDI initial estimated delamination size will grow and can cause failure during its service usage. Furthermore, the acquired data will be able to compare the value of the G_{Ic} test result to other composite materials with different constituents. In brief, in the G_{Ic} test, (1) one end of the DCB specimen has a notch of ~25 mm deep, (2) then the cracked DCB specimen faces pulled apart (the opening mode I), (3) the load–displacement data are recorded, and (4) the obtained data are used for computation of the energy release rates, G_{Ic}, and to develop delamination failure criteria for composite damage tolerance and durability analyses.

Regarding calculating the G_{Ic} value, the linear elastic behavior is assumed in the calculation of G. This assumption is valid since no nonlinear deformation at the delamination front is expected, which is characteristic of brittle material. Equation 9.5 describes the critical value of G_{Ic}, when the applied load $P = P_{\mathrm{cr}}$ (Fig. 9.7), and the derivation of it, based on the DCB geometry, is shown in Fig. 9.8.

$$\frac{12P^2a^2}{B^2h^3E} \tag{9.5}$$

The DCB testing apparatus is shown in Fig. 9.8a, the cantilever specimen is shown in Fig. 9.8b, the derivation of the energy release rate in Fig. 9.8c, and the method of obtaining the load $P = P_{\mathrm{cr}}$ from the load versus displacement curve that is obtained from the testing is illustrated in the Fig. 9.8d. Finally, it should be noted that the DCB test has a few limitations which influence its ability to accurately measure static fracture toughness, G_{Ic}. For example, fiber bridging can increase resistance to delamination growth in static DCB fracture toughness testing and the result will be higher than the actual value. Moreover, the amount of precracking (i.e., the initial crack length) have influence on the final value of G_{Ic} (see discussions provided in Sect. 9.11).

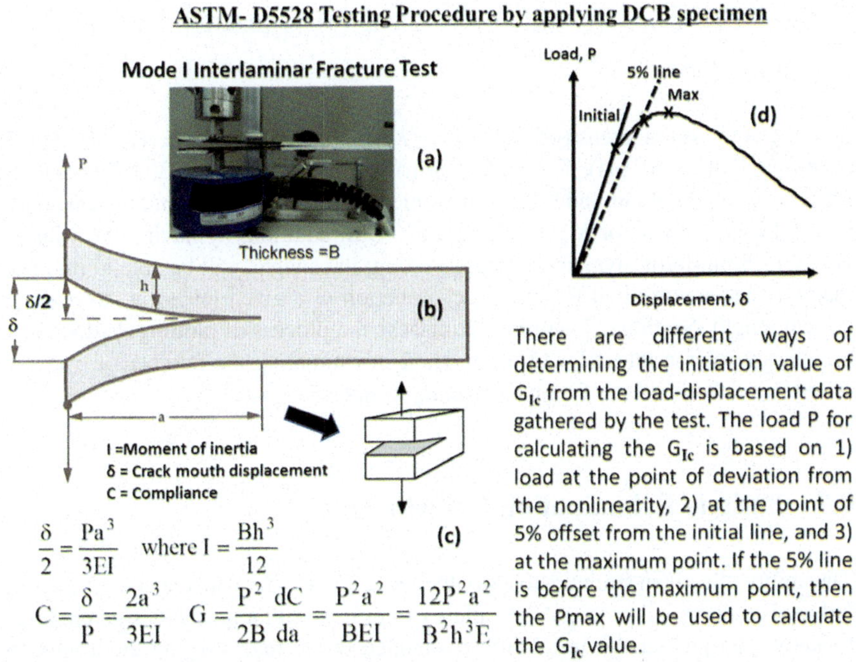

Fig. 9.8 The derivation of energy release rate, *G*, for mode I case using the DCB geometry

9.6 Delamination Growth Analysis (Case 1 of Fig. 9.5)

The initial delamination and subsequent growth associated with it, when subjected to cyclic loading, will be assessed (see case 1 of Fig. 9.5). The assessment of the delamination growth of case 1 requires having fracture properties, as is evident from the example below. The delamination test, based on the ASTM- D5528, requires utilizing the DCB specimen with fibers in the specimen that are unidirectional and parallel to the crack length. Results will provide the critical value of the energy release rate for mode I, G_{Ic}. Having all parameters related to the specimen dimensions (B, $2h$, E, a, and P_c), including the modulus of elasticity available, the critical value of G can be obtained by Eq. 9.5 and shown in Fig. 9.7. For example, let us use specimen dimensions indicated in Fig. 9.7 for a typical resin, where $B = 25$ mm (~1 in), $2h = 5$ mm (~0.2 in or $h = 0.1$), $a = 50$ mm (2 in), and the tensile modulus $E_1 = 25.84$ GPa (~37.45×10^5 psi = ~3745 ksi). If it is assumed that the recorded value of the critical applied load when the DCB specimen fails is $P_c = 30$ lb, then the critical value of the energy release rate (G_{Ic}) will be:

$$G_{\rm Ic} = 12\ (P_{\rm c})^2 (a^2)/(B^2)(h^3)(E_1) = 12(302)(4)/(1)(0.001)(37.45\times10^5)$$
$$= 11.53\ \text{inch-lb/in.}^2$$

In terms of the stress intensity factor, $G_{\rm Ic} = K^2_{\rm Ic}/E$; therefore, $K^2_{\rm Ic} = (G_{\rm Ic}{\rm E}) = (11.53$ in-lb/in^2)(37.45 × 10^5 psi), $K_{\rm Ic} = 6.57$ ksi(in)$^{0.5}$, which means with 30.0 lb. force, the DCB specimen will fail under the monotonically increasing load, and the calculated $G_{\rm Ic} = 11.53$ in-lb/in^2 or $K_{\rm Ic} = 6.57$ ksi(in)$^{0.5}$. In addition, by having the value of $G_{\rm Ic} = 11.53$ in-lb/in^2 available through testing, the analyst will be able to find how many cycles are required for initial delamination of a given length, $a_{\rm i}$, to reach its critical length. Section 9.7 will briefly discuss the process of plotting da/dN versus the energy release rate, G, and the method of estimating the number of cycles to failure for delamination of a given size, $a_{\rm i}$, to fail.

9.7 Fatigue Delamination Growth Test

The ability of a given structure to carry the service load after damage and whether the induced damage will propagate to failure or not is essential for the design engineers to know. To predict the instability of damage under load varying environments, fracture properties of the material must be obtained based on the ASTM or other testing standards. In particular, mode I, II, and the mixed-mode fracture toughness data (the $G_{\rm Ic}$, $G_{\rm IIc}$, and the mixed mode $G_{\rm Ic}/G_{\rm IIc}$ energy release rate) are required to assess the damage criticality. In addition to the fracture toughness, the fatigue crack growth rate data are also an important quantity, which can enable engineers to determine the number of cycles from already existing delamination in the material to failure. Mode II and the mixed-mode fracture toughness will be discussed later in Sect. 9.9.

Fatigue crack growth in metals can be characterized by relating the crack growth rate to the stress intensity factor range, ΔK, during each cycle. On the other hand, for the laminated composite material, the delamination growth rate has been related to the strain energy release rate, G, using a power law format similar to the Paris equation (i.e., da/dN = $C\Delta K^{\rm m}$) to determine the number of cycles to failure, $N_{\rm f}$. For composites, the Paris exponent, m, relating the propagation rate (*da/dN*) to the stress intensity factor, ΔK, is high, especially in mode I. Because of the large exponents used for composites, small uncertainties in the applied loads or when a small error in data is being recorded will lead to large uncertainties in estimating the delamination growth rate. It should be noted that the quantity ΔK is directly related to the applied load magnitude, where $\Delta K = \beta\Delta\sigma(\pi a)^{0.5}$. This makes the derived power-law relationships, using the stress intensity factor, ΔK, not suitable for life assessment analysis of delamination.

For life analysis related to delamination growth and the number of cycles to failure, engineers need to have the variation of da/dN (fatigue crack growth rate) as a function of energy release rate ΔG. To generate the da/dN as a function of ΔG, the

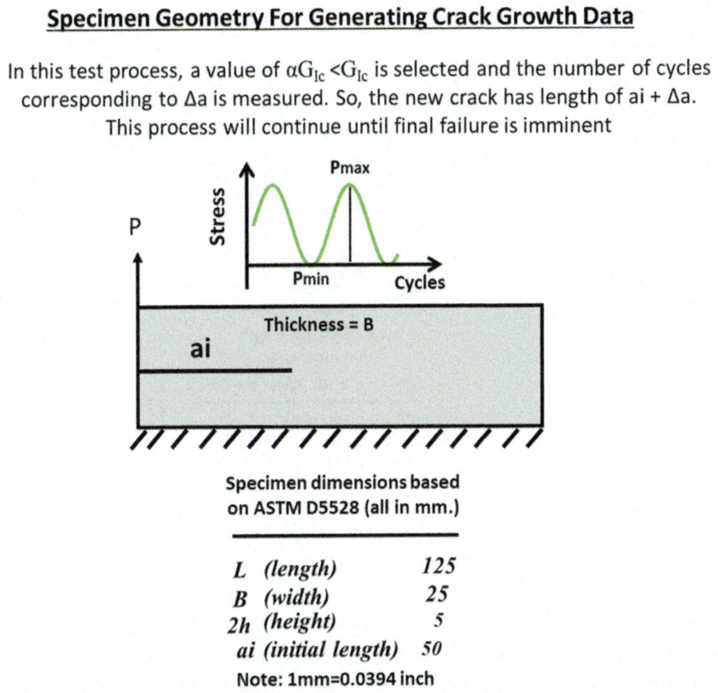

Fig. 9.9 Specimen geometry for generating the crack growth rate by starting a low value of αG_{Ic} (e.g., $\alpha = 0.3$) and measuring the number of cycles, N_i, and amount of crack growth, Δa, up to failure

number of cycles (ΔN) and the corresponding delamination growth (Δa) must be recorded (Fig. 9.9). Later, it can be argued that the use of G_{max} will be more appropriate to use than the ΔG. The quantity G_{max} is the maximum value of G during the load cycle. An optical microscope and a light source can enhance observation and a more accurate measurement of the delamination growth can be obtained. In addition, it will be helpful to mark the test specimen mouth by 0.1 in intervals to make the crack growth reading more feasible. Before starting the test, the delamination onset should be established by inserting a folded aluminum material of 0.005 in thick, during the lay-up of the pre-preg, into the specimen. The presence of aluminum foil will be used as an initial delamination length for the subsequent growth when subjected to cyclic loading. The average fiber volume fraction of the specimens should fall between 60% and 65%, and before testing, the specimens should be vacuum dried for a few hours for the drying cycle to occur. The specimens were allowed to cool to room temperature and stored before testing. Finally, after gathering an adequate amount of data, the strain energy release rate, G_{max}, versus the crack propagation rate, da/dN, is plotted on the logarithmic axes. The idea of using the G_{max} is because this quantity is a stable quantity and does not introduce error into the measurement when compared with $\Delta G = G_{max} - G_{min}$. The error is because of

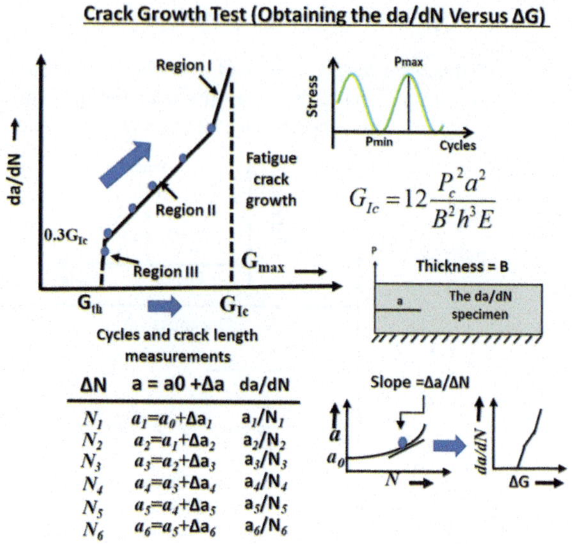

Fig. 9.10 Method of generating the crack growth rate as a function of energy release rate, G. The G value increases as the crack advances toward instability, where $G = G_{\mathrm{Ic}}$. The measurements were recorded starting from $G\sim \alpha G_{\mathrm{Ic}}$. The quantity G should be above the threshold value. Otherwise, crack growth does not happen

the quantity of $G_{\min}$, whose measurement can have errors because the crack mouth is not fully closing, whereas the $G_{\max}$ alone doesn't face that problem.

To continue on the $G_{\max}$ versus the crack growth rate (the da/dN testing), a specimen with an initial delamination length, a_{i}, allowed to undergo a cyclic load with a constant amplitude, P, and with a G value equal to αG_{Ic}, where α is a percentage value of G_{Ic} (e.g., $\alpha = 0.3$). This value must be above the threshold G_{th} (the method of obtaining the G_{th} on the region I will be discussed later in this section). The crack growth and the corresponding number of cycles will be continuously recorded until the calculated value of G is near the failure, corresponding to region III. Delamination is accelerating toward final failure (region III of the curve, as shown in Fig. 9.10). One can determine the constant cyclic load magnitude, P, corresponding to αG_{Ic} by having the G_{Ic} value through the DCB test, specimen dimensions (B, h, and a_{i}), and material modulus E. Note that the quantity $G = \alpha G_{\mathrm{Ic}}$ should be above the threshold value. Otherwise, the crack growth does not happen.

The cyclic testing machine will be cycling with the constant load P and the stress ratio of R (e.g., $R \sim 0$). The maximum cyclic load for $\alpha \sim 0.3$ will be calculated from Eq. (9.5) as:

$$P^2 = (0.3G_{\mathrm{Ic}})(B^2)(h^3)(E)/12a^2$$
$$P = \left[(0.3G_{\mathrm{Ic}})(B^2)(h^3)(E)/12a^2\right]^{0.5}$$

For initial delamination crack length of $a_{\mathrm{i}} = 50$ mm (2.0 in), $B = 25$ mm (~1.0 in), $2h = 5$ mm (~0.2 in or $h = 0.1$), $G_{\mathrm{Ic}} = 11.3$ in-lb/in^2, and $E_1 = 25{,}840$ MPa (3750 ksi), the applied constant fluctuating load based on $0.3G_{\mathrm{Ic}}$ will be, $P_{\max} = [(0.3)(11.3)(1^2)(0.1^3)(37.5 \times 10^5)/(12)(2^{2)}]^{0.5}$, $P_{\max} = 16.3$ lb. The applied fluctuating load of $P_{max} = 16.3$ lb. with $R \sim 0$ is constant throughout the test, but the initial crack length is advancing due to cyclic load. Therefore, the value of G_{max} keeps growing until the crack length reaches a critical value and $G_{max} = G_{Ic} = 11.3$ in-lb/in^2. Note that the stress ratio is $R \sim 0$ and therefore $\Delta G_{max} \sim G_{max}$. The crack advancement (Δa) and the associated number of cycles will continuously be recorded for the applied fluctuating load of $P_{max} = 16.3$ lb. and $P_{min} \sim 0$ to generate the da/dN as a function of ΔG or G_{max}. Figure 9.10 illustrates the steps for generating the da/dN as a function of ΔG. The procedure is like the ASTM-647 [11] standards used for the case of metals, where first, the plot of crack growth versus the number of cycles is plotted (a_0+ Δa versus N). Then the slope of the curve at each point was recorded to generate the da/dN versus the ΔG, where $\Delta G_{max} = G_{max}$, as shown in Fig. 9.10. In this figure, five measurements are shown. The threshold value of strain energy release rate, ΔG_{th}, below which no delamination would occur can be measured separately, but testing results will be complex and time consuming. The maximum applied G for which no delamination growth had occurred until at least 10^6 cycles, can be considered as the threshold strain energy release rate, ΔG_{th}, and it is a function of R. In the case of metal, the value of ΔK_{th} can be approximated by the value of 10% of the K_{Ic} value (i.e., 0.1 K_{Ic}) [14] associated with the stress ratio of $R \sim 0$, as shown in Fig. 9.11 for several metallic alloys used in industries. When data are not available for the ΔG_{th}, the value of βG_{Ic}, where $\beta = 0.1$, can be a reasonable value to be used for the threshold value in terms of G_{Ic}. Having both the G_{th}, G_{Ic}, and also region II of da/dN, it is possible to assess the delamination life of a laminated composite part. For example, the NASA delamination growth study on AS4/PEEK material [15], indicated that for $R \sim 0$, the value of $G_{th} = 2$ in-lb/in^2 with G_{Ic} 14.14 in-lb/in^2, where the value of $\beta = 10\%$ provides $G_{th} = 1.4$ in-lb/in^2. This value is very close to the actual value of $G_{th} = 2$ in-lb/in^2.

$K_{th} = \alpha K_{Ic}$, where $\alpha \approx 0.1$
(For R=0)

Material	KIc	Kth (Test)	Kth (0.1KIc)	Material	KIc	Kth (Test)	Kth (0.1KIc)
2014-T6 (L-T)	27	2.7	2.7	2219-T87 (T-L)	27	2.9	2.7
2020-T651 (L-T)	22.5	2.2	2.3	7005-T6 & T63 (T-L)	40	3.4	4.0
2024-T3 (L-T)	33	2.9	3.3	7010-T73651	31	2.5	3.1
2024-T3 (T-L)	29	2.9	2.9	7050-T736 & T74 (T-L)	24	2.3	2.4
2024-T3 (L-T)	33	2.9	3.3	7050-T76511(T-L)	24	2.0	2.4
2024-T351 (T-L)	29	2.6	2.9	7075-T651 (L-T)	28	3.0	2.8
2024-T62 (L-T)	36	2.9	3.6	7075-T6510 (L-T)	28	3.0	3.0
2024-T62 (T-L)	30	2.9	3.0	7075-T6511(L-T)	28	3.0	2.8
2024-T852	28	2.9	2.8	7075-T73 (L-T)	28	3.0	2.8
2024-T861 (L-T)	23	2.2	2.3	7075-T7351(L-T)	29	3.0	2.9
2048-T851(L-T)	35	2.7	3.5	7075-T73510 (L-T)	31	3.0	3.1
2048-T851(T-L)	30	2.7	2.7	7075-T73511(L-T)	33	3.0	3.3
2124-T851 (L-T)	30	3	3.0	7075-T7352 (L-T)	33	3.0	3.3
2124-T851 (T-L)	26	2.7	2.6	7075-T7651(T-L)	23	2.4	2.3
2219-T62 (-320F)	30	2.9	3.0	7079-T651(L-T)	26	2.0	2.6
2219-T851(L-T)	33	3	3.3	7149-T73511 (L-T)	31	3.0	3.1
2219-T87 (L-T)	30	2.9	3.0	7178-T7651(L-T)	28	3.0	2.8

Fig. 9.11 The estimated threshold value of K (K_{th}) for metallic material in terms of βK_{Ic}, where $\beta \sim 0.1$

The following is an example of the number of cycles to failure of a part subjected to a cyclic load environment that has specified delamination dimensions.

Example (Case 1 of Fig. 9.5)

A laminated composite part is subjected to a fluctuating load environment with an applied load of $P_{max} = 40$ lb, where $R \sim 0$. The critical value of G_{Ic}, based on the DCB test, has been shown to be 15 in-lb/in^2. The initial delamination size provided by the NDI inspection is found to be $a_i = 0.2$ inches. The power law growth rate equation for the adhesive material is given by da/dN $= \eta(G_{Imax})^{\gamma}$, where $\eta = 4.9 \times 10^{-11}$ and $\gamma = 5.8$. That is,

$$\mathrm{da/dN} = 4.9 \times 10^{-11} (G_{Imax})^{5.8}$$

(1) Plot the da/dN versus the G_{Imax} and (2) find out the maximum number of cycles required for the initial delamination length of $a_i = 0.2$ in to reach its final failure. The part dimensions are $B = 1.5$ in and $2h = 0.2$ in. The value of E_1 for this material is 40×10^5 psi.

Solution

To plot the da/dN versus the G_{Imax} in a log-log format, the Paris power law equation will be used. Several G_{Imax} data points are selected and the corresponding da/dN versus G_{Imax} are calculated. For example, based on $G_{Ic} = 15$, values of $G_{Imax} = 3$, 5, 7, 9, 11, and 13 are used and the corresponding da/dN were calculated (Fig. 9.12):

$$\mathrm{da/dN} = 4.9 \times 10^{-11} (G_{Imax})^{5.8}$$

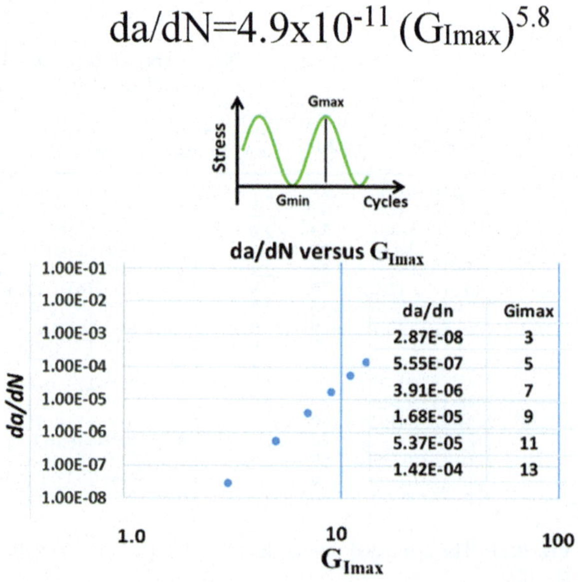

Fig. 9.12 da/dN versus model I energy release rate, G_{Imas}, for the example problem related to case 1

The number of cycles to failure from a small initial delamination of 0.2 in in size to failure can be calculated as follows:

Determine the critical delamination size based on the constant cyclic load of $P_{max} = 40$ lb. and $R \sim 0$. Using the DCB formula with $G_{Ic} = 15$ in-lb/in^2 and by using Eq. (9.5), where the value of $a_{cr} = [(G_{Ic})(B^2)(h^3)E/12P_{max}{}^2]^{0.5} = [(15)(1.5^2)(0.001)(40 \times 10^5)/(12)(40)^2]^{0.5}$. The calculated critical delamination size $a_{cr} \sim 0.7$ in. Therefore, the amount of delamination growth from its original length of 0.2 in to its final length will be a total increase of $\Delta a = 0.5$ in and the average number of cycles to failure will be,

$$\Delta N = (\Delta a)/\left(4.9 \times 10^{-11}\ (15)^{5.8}\right)$$

$$= 1539 \text{ cycles from initial crack length of 0.2 in to 0.7 in.}$$

9.8 Crack Growth in the Matrix (Case 2 of Fig. 9.5)

For case 2a of Fig. 9.5, the initial NDI estimated crack preexisted in the matrix, a_i, will grow in the matrix of a lamina of a laminated composite material perpendicular to both the load and the fibers until partial failure occurs. This scenario is a possibility because the matrix is weaker than the fibers, and if the bond between the fibers and the matrix is weak, the crack will be halted for only a short time and ultimately will continue to advance with no fiber pullout. Figure 9.13 shows the crack growth in the matrix of a laminated composite with applied cyclic load parallel to fibers and perpendicular to the crack length, the mode I condition. Two fracture mechanics analyses will be performed here. First, the analysis will be done for the case of only one ply with fibers parallel to the applied load. The second analysis will be done for one ply with a crack parallel to the fibers but perpendicular to the applied load (case 2b of Fig. 9.5). In case 2a of one ply only, the crack will initiate and grow from the matrix with fibers parallel to the applied tensile load, as shown in Fig. 9.13. Here, the primary assumption is the weak bonds between the matrix and fibers that minimize the crack arrest mechanism, and damage will continue to grow within the matrix. Moreover, it is also assumed that fibers will remain intact during the crack growth at the two sides of the crack mouth, and only the separation between the fiber and matrix will occur during the crack growth process. The unbroken fibers at both sides of the crack mouth will be helpful to prevent the total failure of the ply (Fig. 9.13).

In the first part of the analysis, the LEFM can be used to assess the matrix crack growth behavior. Both the fracture toughness and fatigue crack growth rate properties of the matrix must be available through the ASTM testing standards for the fracture analysis to go forward. Later in Chap. 10, the virtual testing technique will be discussed that can provide the analyst with both the fracture toughness and the fatigue crack growth data without conducting costly and time-consuming tests.

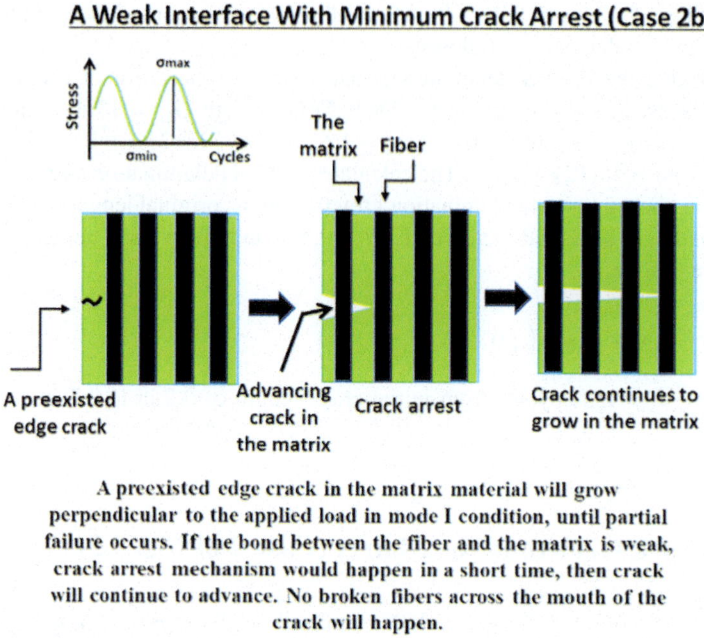

Fig. 9.13 Crack growth in the matrix with minimum crack arrest feature, where the bonds between fibers and matrix are weak (case 2a). The unbroken fibers at both sides of the crack will slow down the crack growth process

Even though, in this case, the mode I crack growth will be in the matrix due to the weak interface between the fiber and the matrix, engineers must realize that the presence of unbroken fibers at both sides of the crack mouth will delay the growth and slow down the rate of advancing crack. The resistance to crack growth will be related to the strength of the interface and the number of unbroken fibers at both sides of the crack (Fig. 9.14). In this situation, the presence of unbroken fibers at both sides of the crack mouth does not allow the crack to fully open. Therefore, the amplitude of fluctuating cycles will reduce and the stress ratio, R, will change. The reduced load cycles amplitude will be less effective, and consequently, they will help extend the life of the composite structural part.

The presence of unbroken fibers on the rate of crack growth will be in such a way that at the tensile portion of the cyclic loading, the resistance from fibers will slow down the growth as shown in Fig. 9.14. Not taking the crack growth resistance into consideration, the result of crack growth analysis will yield a shorter life (i.e., the analysis result will be conservative). Therefore, upon loading during each cycle, the resistance due to unbroken fibers will be effective and during the unloading part of the cycle, there will be no resistance. However, it should be noted during the unloading in a fatigue cycle, the two faces may come in contact with each other, resulting in facial interference. Facial interference can cause rough surfaces due to the presence of debris. These factors can prevent closing the two crack faces during unloading in the fatigue cycle, which can alter the stress ratio, R.

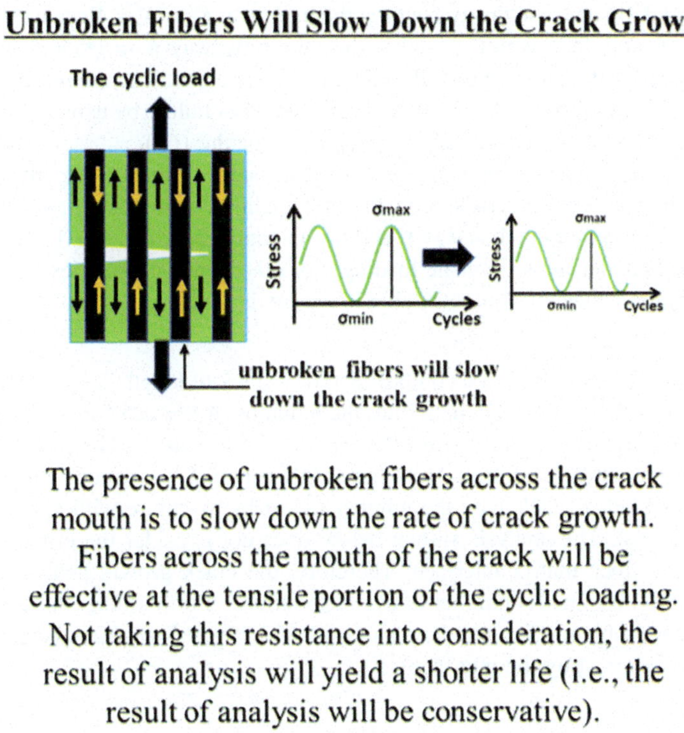

Fig. 9.14 Crack growth in the matrix, where unbroken fibers will slow down the crack growth mechanism

To study the crack growth in the matrix, engineers must realize that the fracture mechanics analysis will yield shorter life by not considering the resistance due to unbroken fibers as was described above. In addition, it is necessary to assess the number of cycles when the crack tip is confronting the fiber that has been stuck with the matrix at the interface. The number of cycles needed to allow the crack to overcome the interface must also be considered in the analysis. The fracture toughness and the crack growth rate properties of the interface are also needed for a meaningful life analysis. However, as was mentioned before by not considering those factors, the results of the analysis will be conservative.

Let us assume it will take N_1 number of cycles for the crack to grow in the matrix until it is stopped by the fiber. Subsequently, it will take another N_2 cycle to break the interface and continue growing further until the crack tip is exposed to another fiber bonded to the matrix. This mechanism will be repeated until the crack length becomes critical and the value of the stress intensity factor or the energy release rate becomes equal to the fracture toughness of the lamina ($K = K_c$ or $G = G_c$).

Furthermore, let us also assume the fracture toughness of the matrix is $(K_c)_m$ and the fracture toughness of the interface is $(K_c)_I$. For a monotonic load, at the interface the K has to increase in order for the crack to go forward. For that to happen, the value of crack tip stress intensity factor at the interface, $K = (K_c)_I$. Because the crack length at the interface is not changing then the only way $K = (K_c)_I$ is when the monotonic applied load increases. Based on the linear elastic fracture mechanics, the K value at the interface can be calculated at the onset of failure by using the equation of the stress intensity factor, $(K_c)_I = \beta\sigma_{cr}(\pi a)^{0.5}$ or the $(G_c)_I = \pi a(\sigma_{cr})^2/E$. Other crack geometries under the opening mode have the same formula except the quantity β must be adjusted for the crack geometry and loading conditions. For example, the value of β for a center crack is $\beta = 1$, and for an edge crack, it is $\beta = 1.12$ when the crack plate is under mode I tensile loading. For other crack geometries and loading conditions, the values of β are available in the literature [16].

Example

During the NDI inspection of a composite part, a crack of length $2a_i = 0.2$ inch was found in the matrix. The resin matrix has the modulus of $E = 1.5$ GPa ~ 214 ksi, and the fracture toughness of $K_{Ic} = 7.8$ MPa $(m)^{0.5}$ ~ 7.1 ksi $(\text{in})^{0.5}$. The Paris constants for the matrix can be calculated from the da/dN versus K_{Imax} data provided by the figure shown below. Calculate the number of cycles to failure with applied stress equal to 5.5 ksi. The analysis shown below does not consider the effect of crack arrest by the fiber–matrix interface. Therefore, the crack growth analysis will be conservative (i.e., smaller number of cycles to failure). Furthermore, the analysis of this example problem will yield the same result as in Fig. 9.5, case 2b (a matrix crack with no crack arrest by the interface).

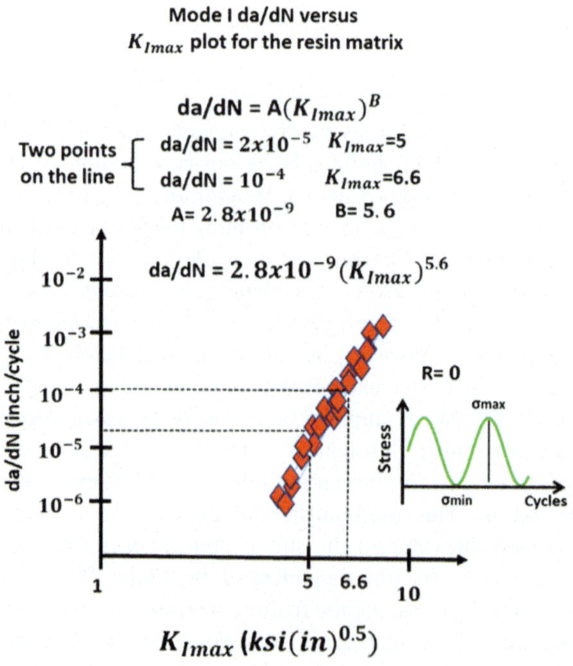

Solution

From the available data, the constants for the Paris region can be calculated. To obtain those data, there is a need to use two data points to solve for the Paris constants, A and B. By selecting two data points, the two values of $A = 2.8 \times 10^{-9}$ and $B = 5.6$ were obtained. Therefore, the equation describing the data for the Paris region can be written as:

$$\mathrm{da/dN} = 2.8 \times 10^{-9}(\mathrm{K_{Imax}})^{5.6}$$

To estimate the number of cycles to failure for the applied stress equal to 5.5 ksi, where for a center crack $K = \beta\sigma\sqrt{\pi a}$, with $\beta = 1$, and half a crack length of $a = 0.1$ inch, the critical value of crack length can be estimated:

$$a_{\mathrm{cr}} = (1/\pi)(K_{\mathrm{Ic}}/\sigma)^2 = 0.32\ (7.1/5.5)^2 = 0.53 \text{ inch (half a crack length)}$$

From the crack growth rate data, da/dN $= 2.8 \times 10^{-9}(K_{\mathrm{Imax}})^{5.5}$, solve for number of cycles, N.

$$\int_0^{N_f} \mathrm{dN} = \int_{a_0}^{a_c} \frac{\mathrm{da}}{2.8 \times 10^{-9}(K_{\mathrm{Imax}})^{5.5}} = 0.36 \times 10^9 \int_{0.1}^{0.53} \frac{\mathrm{da}}{\left[(5.5)(3.14a)^{0.5}\right]^{5.5}}$$

$$= 42 \times 10^3 \int_{0.1}^{0.53} \frac{\mathrm{da}}{a^{2.75}} = 1.42 \times 10^3 \int_{0.1}^{0.53} a^{-2.75}\mathrm{da} = 1.42 \times 10^3 \left[\frac{a^{-1.75}}{-1.75}\right]_{0.1}^{0.53}$$

$$= -0.8 \times 10^3\left[0.53^{-1.75} - 0.1^{-1.75}\right] = -0.8 \times 10^3[3.03 - 56]$$

$$= 4.3 \times 10^4 \text{ cycles}$$

The crack growth will be in the matrix, parallel to the fibers, and perpendicular to the applied load. The crack growth will not encounter crack arrest by the fiber–matrix interface. In the case of 2b mentioned in Fig. 9.5, it is a matrix crack parallel to the fibers and perpendicular to the load (with no crack arrest). Therefore, the analysis of this example problem should yield the same result as in case 2b.

9.9 Mode II Delamination and Related Testing

The fracture mechanics methodology has been expanding not only to describe the brittle material, semi-brittle material with small scale yielding at the crack tip but also adhesion material, where plies in the laminated composites are attached through adhesion and delamination between plies can occur during the service operation. Epoxy adhesives material used in laminated composites has a high strength to a variety of substrates as well as superior resistance to chemicals, moisture, and the environment. However, in many bonding applications, epoxy is considered brittle and can crack, causing delamination initiation and growth due to temperature,

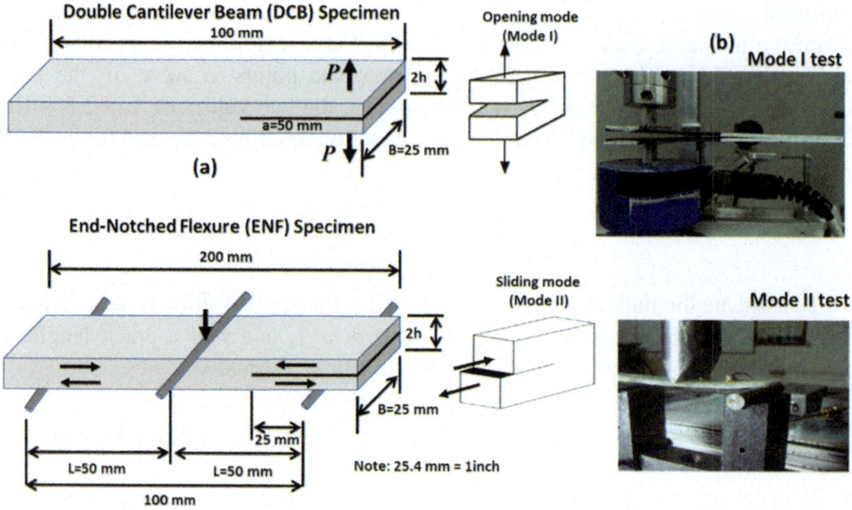

Fig. 9.15 (**a**) Mode I and II test specimen loaded in tension and three-point bending with dimensions shown in the figure. The critical strain energy release rate, G, for both modes, G_{Ic} and G_{IIc}, can be obtained. (**b**) Both mode I and II apparatuses for generating G_{Ic} and G_{IIc} fracture toughness are shown

moisture, load cycles, and mismatch between plies. Because of their brittle nature at room temperature, the LEFM can be used to assess their crack growth behavior when subjected to a cyclic loading environment. Static tests, such as the ASTM double cantilever beam (DCB) and the end-notched flexure (ENF) tests, are available to obtain fracture toughness of both fracture modes, mainly the G_{Ic} and G_{IIc}. The method of obtaining the critical energy release rate, G_{Ic}, and its value was already discussed in Sect. 9.5 in detail. In using the end-notched flexure (ENF) method for mode II, the applied load is by three points bending method, as shown in Fig. 9.15. A typical ENF fracture specimen is essentially a three-point bending beam having a mid-plane initial delamination of a desired length, a_i, at one end of the beam (Fig. 9.15), where the ASTM D 7905–14, "Determination of the Mode II Interlaminar Fracture Toughness of Unidirectional Fiber-Reinforced Polymer Matrix Composites, 2014" can be used. Note that an ASTM standard testing for Mode III is being developed for obtaining the G_{III} values for composites under tearing type loads. In both cases of mode I and mode II fracture toughness tests, the dominant crack propagation is parallel to the fiber direction of the laminated composite. In general, the configuration, specimen thickness, width, and distance between supports are chosen such that large displacements would be avoided and transverse shear effects minimized to maintain the pure sliding mode (mode II) as much as possible. Note that the initial delamination of length, a_i, is situated in the middle of the ENF plate. Therefore, the upper portion and lower portion will be exposed to tensile and compressive forces, respectively, as it is depicted in Fig. 9.15.

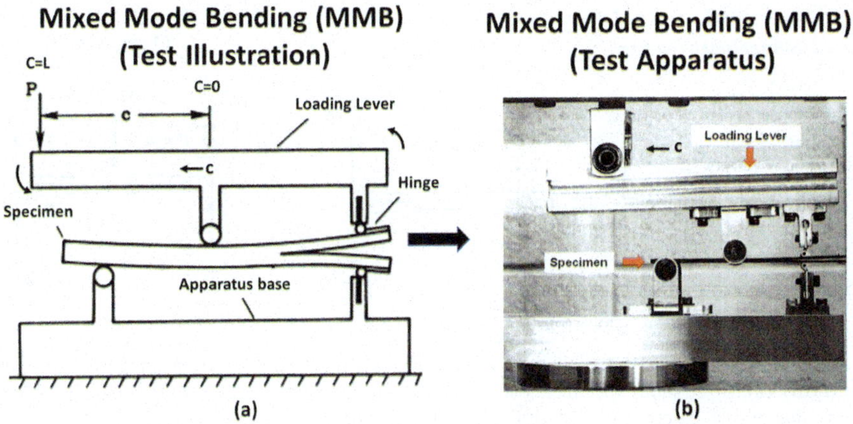

Fig. 9.16 (**a**) MMB test can be used to measure delamination toughness over a wide range of mode I/II ratios as well as pure mode I and mode II. When the load is at $c = 0$, it simulates mode II and when $c =$ L, it is mode I. (**b**) A photograph of the MMB test apparatus is shown

As shown in Fig. 9.16, the DCB is a pure mode I test, whereas the ENF is a pure mode II test, and the mixed-mode beam (MMB) can perform both mixed mode I and mode II tests. Even though it is not yet clear how the combined fracture modes can cause the failure of the structure, is not well established, but an attempt is made to briefly present the MMB in this section (ASTM D6671, 2001). The MMB specimen proposed by Reeder and Crews (1990) [17] is a combination of the double cantilever beam (DCB), and the end notched flexure (ENF) specimen. One specimen can be used for measuring pure mode I and mode II interlaminar fracture toughness. The test illustration that can be used to measure delamination toughness over a wide range of mode I and mode II ratios, as well as pure mode I and mode II, is shown in Fig. 9.16a. In the case of mixed-mode fracture toughness, for G_{Ic} and G_{IIc}, the loading position, c, of Fig. 9.16, determines the relative magnitude of the two resulting loads on the specimen and, therefore, determines the mixed-mode delamination ratio. Pure mode II loading occurs when the applied load is directly above the beam mid-span (where $c = 0$), and for pure mode I, the loading can be achieved by removing the beam and pulling it up on the hinge. The apparatus used for the mixed mode, as well as each individual mode, is shown in Fig. 9.16b.

In Fig. 9.17, the MMB loading is expressed in terms of the applied load P, the loading lever length c, and the specimen half-span L. In this figure, the mode I component of this loading setup can be written as:

$$P_{\mathrm{I}} = \frac{P(3c - L)}{4L}$$

From the beam theory analysis of the DCB specimen (Eq. 9.5), the G_{I} can be written as:

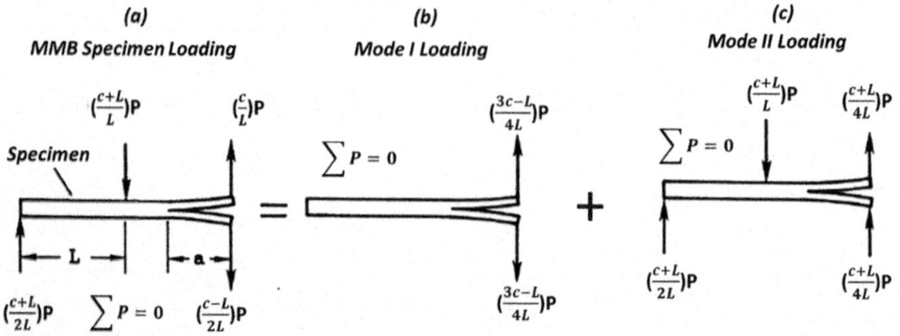

Fig. 9.17 Applying the superposition analysis of mode I and mode II to obtain the mixed mode (MMB). (**a**) Mixed mode specimen loading, (**b**) mode I specimen loading, (**c**) for mode II loading, note that in each case, the sum of all forces is equal to zero

$$G_{\mathrm{I}} = \frac{12a^2 P_I{}^2}{B^2 h^3 E_1} \tag{9.6}$$

By substituting for $P_{\mathrm{I}} = \frac{P(3c-L)}{4L}$ into the G_{I} equation based on the setup shown in Fig. 9.17, the mode I energy release rate, G_{I}, can be written as,

$$G_{\mathrm{I}} = \frac{3a^2 P^2 (3c-L)^2}{4L^2 B^2 h^3 E_1} \tag{9.7}$$

With the same analogy, the equation for the G_{II} can be derived, where the quantity P_{I} will be replaced with P_{II}:

$P_{\mathrm{II}} = \frac{P(c+L)}{L}$ and based on the three-point beam bending analysis, the value of G_{II},

$$G_{\mathrm{II}} = \frac{9a^2 P^2}{16L^2 B^2 h^3 E_1}(c+L)^2 \tag{9.8}$$

By dividing the two quantities G_{I} and G_{II} (i.e., $G_{\mathrm{I}}/G_{\mathrm{II}}$) the ratio for the mixed mode test can be expressed as:

$$G_{\mathrm{I}}/G_{\mathrm{II}} = \frac{4}{3}\left[\frac{(3c-L)}{(c+L)}\right]^2 \tag{9.9}$$

Note that the ratio $G_{\mathrm{I}}/G_{\mathrm{II}}$ is only a function of load position, c, and half span length, L. The $G_{\mathrm{I}}/G_{\mathrm{II}}$ ratio is zero for $c = L/3$.

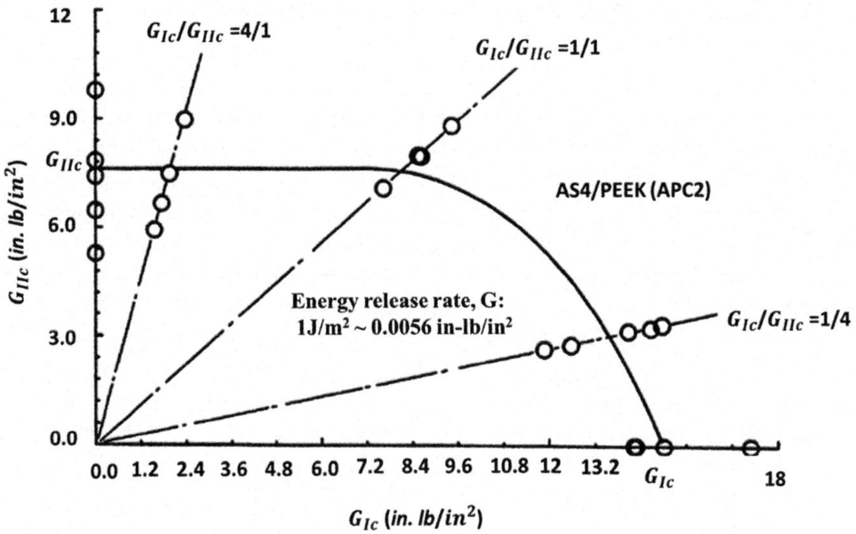

Fig. 9.18 Variation of the mixed mode for different G_{Ic}/G_{IIc} ratios for the AS4/PEEK(APC2) material. Test data is also shown in this figure, which were generated through the MMB testing machine

Finally, the total strain energy release rate, G, for the MMB test is obtained by adding the two equations describing the G_{I} and G_{II}, where for $c = L$, the value of $G = G_{\mathrm{I}}$, and for $c = 0$, $G = G_{\mathrm{II}}$. That is:

$$G = \frac{3a^2P^2}{16L^2B^2h^3E_1}\left[4(3c - L)^2 + 3(c + L)^2\right] \tag{9.10}$$

Figure 9.18 shows mixed-mode data generated for the AS4/PEEK laminated composite with different $G_{\mathrm{Ic}}/G_{\mathrm{IIc}}$ ratio of 4/1, 1/1, and 1/4, including the case of pure G_{Ic} when $G_{\mathrm{IIc}} = 0$ and the case of G_{IIc} when $G_{\mathrm{Ic}} = 0$. In this test, to produce initial delamination, each specimen was precracked to about 30 mm. No significant fiber bridging was observed at this mixed-mode ratio. The $G_{\mathrm{I}}/G_{\mathrm{II}}$ mixed-mode ratios of 4/1, 1/1, and 1/4 were produced by selecting the load positions from Fig. 9.18. Pure mode II tests were conducted by applying the load at the specimen midspan ($c = 0$). For the pure mode I case, the loading lever was removed, and an upward load was applied directly through the upper hinge on the specimen. Each specimen was loaded under displacement control with a rate of 0.5 mm/mm until the delamination started to grow. The maximum load was recorded, and the final delamination size was measured at the specimen edges visually. The edges had been marked to make the delamination growth easier to measure.

9.10 A Final Comment on Mode I and Mode II

In Fig. 9.18, the energy release rate values for mode I and II are shown to be $G_{Ic} \sim 14.5$ in-lb/in^2 and 7.5 in-lb/in^2 for the AS4/PEEK (APC2), respectively. The data obtained from Fig. 9.18 show the value of $G_{Ic} > G_{IIc}$ (higher G value means more resistance to cracking). This contradicts the usual results obtained from tests that provide higher G_{IIc} values when compared with G_{Ic}. Figure 9.19 shows the G_{Ic} and G_{IIc} test results for several fiber-reinforced composites, where in all cases the $G_{IIc} > G_{Ic}$. Additional tests for mode I and II were also conducted separately by NASA Langley Research Center [15] on graphite/epoxy IM6/3501–6. Two panels were manufactured for this investigation and the critical strain energy release rates associated with debonding of the adhesive in graphite/epoxy IM6/3501–6 interlaminar fracture specimens were investigated.

One of the panels did not have the correct lay-ups and was discarded in this study. Double cantilever beam (DCB) specimens were used to determine the opening Mode I interlaminar fracture toughness, G_{Ic}, of unidirectional fiber-reinforced composites. The five specimens were tested and an average value of 946.42 J/m^2 for G_{Ic} (~5.37 in-lb/in^2) was obtained. The critical strain energy release rate, G_{IIc}, for initiation of delamination under in-plane shear loading was explored using the end-notched flexure (ENF) Test. Four specimens were tested for mode II and an average value of 584.9 J/m^2 for G_{IIc} (~3.32 in-lb/in^2) was calculated. Calculations from the DCB and ENF test results show the values of G_{Ic} are greater than the G_{IIc} for the adhesive debonding in the material studied in this investigation. From these tests, one can conclude that a crack in the adhesive requires more energy to grow in the opening

Fig. 9.19 G_{Ic} and G_{IIc} test results for several fiber-reinforced composites, where in all cases $G_{IIc} > G_{Ic}$ (source: Robinson and Hodgkinson 2000, p. 172). The ENF is an end-notched flexure and ELS is an end-loaded split specimen. Note: 1 J/m^2 ~ 0.00568 in-lb/in^2

Composite Fiber/Matrix	Fracture Mode	Fracture Toughness (J/m^2)
T300/6376	Mode I	270
	Mode II (ELS	600
	Mode II (ENF)	650
XAS/913	Mode I	280
	Mode II (ENF)	660
T300/914	Mode I	140
	Mode II (ENF)	720
T800/924	Mode I	220
	Mode II (ELS	440
AS4/PES	Mode I	800
	Mode II (ELS	1230
	Mode II (ENF)	1290
AS4/PEEK	Mode I	1680
	Mode II (ELS	1740
	Mode II (ENF)	1820

mode than in the shearing mode. This is just the opposite of what is typically observed for delamination within the composite. Extensive data and analytical work on crack growth resistance have been found and accumulated for polymer matrix composites (PMCs) related to both modes I and II. In the case of ceramic matrix composites, some works were also conducted and discussed in [17, 18], both the interlaminar tensile (through the thickness loading) and shear strengths (the resistance of the composite to delamination under shear forces parallel to the layers of the laminate) were presented.

There, six different SiC/SiC fiber-reinforced CMCs at room temperature were tested. The CMCs, with 2D woven fabricated and results showed weak interlaminar properties with interlaminar shear strength of 4–7 ksi and interlaminar tensile strength of 1.5–2.9 ksi. The interlaminar shear strength was two to three times greater than the interlaminar tensile strength [16], implying a difference in crack growth resistance between mode I and II loading.

9.11 Precracking DCB Specimen (Dependency of GIc on the Level of Precracking)

When testing for mode I energy release rate, G_{Ic}, to assess the life of a composite part, the initial crack length, and the amount of precracking of the DCB test specimen is important to know. Otherwise, the test result can have unacceptable errors [19–21]. To assess the precracking issue, several DCB specimens were tested, and each specimen had a different level of precracking. Furthermore, each specimen was fatigue tested for a given number of cycles from an initial crack length of $a_p - a_o$, and the crack length versus cycles relationship was monitored and measured. Here, the quantity a_o is the length of the initial film that was originally inserted and a_p is the original crack length + precracking amount ($a_o + \Delta a$). In this process, the method of precracking right after the a_o was performed in such a way that the test stopped for the first specimen, and the crack length was measured for several crack advancements. A new fatigue test is then restarted from this precrack length (i.e., a new $a_p - a_o$). This process is then repeated several times until the final failure of the test specimen occurs. The proposed test process can generate a series of different crack growth curves that are a function of the precrack length, as shown in Fig. 9.20.

In Fig. 9.20, each plot of log da/dN versus log ΔG_I belongs to a different level of precracking in the Paris region depending upon the value of $a_p - a_o$. As for the question of which one of these plots is correct, one may conservatively select the "worst case" curve as it is acceptable and practiced by the NASA-HDBK-5010 [20]. This variation in the da/dN can arise from the degree of fiber-bridging that is present in the DCB test before the fatigue test is run. Thus, as $a_p - a_o$ increases, the fatigue crack growth rate shifts to the right, and growth slows down for a given value of ΔG_I. This leads to a better measured fatigue performance for the carbon fiber-reinforced polymer CFRP. In the case of metals and their alloys, the fatigue crack

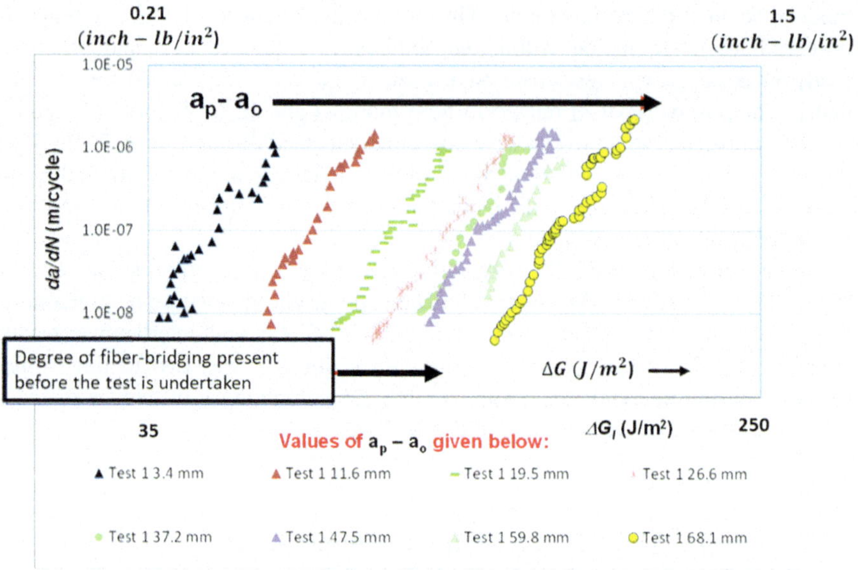

Fig. 9.20 Values of various pre-crack extension lengths, $a_p - a_o$, before the start of each of the fatigue tests is shown in the text box [20, 21]. Test #1, 3.4 mm denotes test specimen #1, where the length of the pre-crack at the start of the test was 3.4 mm, i.e., $(a_p - a_0) = 3.4$ mm. Note: $J/m^2 = 0.0056(inch - lb/in^2)$ and 1mm= 0.0393 inch

growth rate curve is independent of the initial crack length. That is, by testing several specimens with various initial crack lengths, the result of crack growth rate data will be the same, and the simulative law holds. In contrast, in the case of composite, as described above, the simulative concept does not apply.

9.12 A New Method of Plotting Composite Fatigue Crack Growth Results

The question of which parameter should be used when plotting fatigue crack growth results is discussed in this section [22]. One suitable method that can be for the da/dN versus ΔG, where $\Delta G = G_{\mathrm{Imax}} - G_{\mathrm{Imin}}$ for mode I crack growth. This is based on noting that K_I^2 is proportional to EG_I, where E is the tensile modulus and G_I is the mode I energy release rate.

Later in Sect. 9.13, it can be argued that the G_{Imax} can be a more stable parameter to use for mode I when compared with the quantity ΔG_I. However, the following discussion is based on the fact that since K_I^2 is proportional to EG_I, therefore K_I should be related to the quantity $\sqrt{G_I}$. The above fact is the original work of Sih, Paris, and Irwin [21] revealed that the crack tip stress field for a composite was uniquely described by $\sqrt{G_I}$. Hence, the Paris law should be expressed as da/dN versus $\Delta\sqrt{G_I}$, where the quantity $\sqrt{G_I} = \sqrt{G_{Imax}} - \sqrt{G_{Imin}}$. Expressing da/dN in terms of $\Delta\sqrt{G_I}$ will remove incorrect observations about effects, such as the R-ratio, on the measured fatigue crack growth rate data when ΔG_I (or G_{Imax}) is used. The application of G_{Imax} instead of ΔG_I will be described in Sect. 9.13.

Several routes to ensure that the measured fatigue behavior takes into account fiber-bridging effects are being discussed within the ESIS-TC4 Committee. In Reference [23] it is proposed adopting the (mandated) NASA Fracture Control Handbook NASA-HDBK-5010 [20] approach whereby the durability and damage tolerance analysis is performed using the "upper-bound," that is the worst-case, fatigue crack growth rate curve to be sure that the analysis is conservative. Therefore, the upper-bound must have both lower accelerated region (lowest failure point) as well as the lowest possible fatigue threshold value. In such cases, there is no retardation of the da/dN due to fiber-bridging is taken into consideration. This upper-bound fatigue crack growth rate curve must also consider the inherent scatter that is always observed in fatigue tests, e.g., from material inconsistencies or the manufacturing process.

To determine the upper-bound, worst-case, curve for the fatigue crack growth rate which is free from fiber-bridging effects with the threshold value associated with it, it requires to determine the corresponding worst-case fatigue threshold value designated as $\Delta\sqrt{G_{Ithr}}$ below which no fatigue induced delamination (cracking) growth occurs. The notation $\Delta\sqrt{G_{Ithr}}$ is from [19, 24]. The full crack growth equation, similar to the case of Forman–Newman–de Koning crack growth (shown by Eq. 9.3), is provided by Hartman–Schijve as:

$$\frac{\mathrm{da}}{\mathrm{dN}} = D\left[\frac{\Delta\sqrt{G_I} - \Delta\sqrt{G_{Ithr}}}{\sqrt{\left\{1 - \sqrt{G_{Imax}}/\sqrt{A}\right\}}}\right]^n \tag{9.11}$$

In Eq. 9.11, the two quantities D and n are constants and can be obtained by using a few experimental data, and A is the cyclic fracture toughness. Now, by applying the Hartman–Schijve equation to the experimental data, one can calculate a mean value of $\Delta\sqrt{G_{Ithr}}$ and its standard deviation (sd), then to predict an upper-bound (worst-case) fatigue crack growth rate curve [21].

In Fig. 9.21, the upper-bound fatigue crack growth rate curves from Eq. 9.11 for the $R = 0.5$ is plotted and the experimental results for 77 tests, where the length, $a_p - a_o$, of the initial starting crack has been varied for any given set of tests [21]. Here, the longer the starting crack the more fiber-bridging present at the start of the test. How conservative engineers wish to be it depends upon the multiple of the standard deviation that is subtract from the mean value of $\Delta\sqrt{G_{thr}}$ and the quantity A (Fig. 9.21).

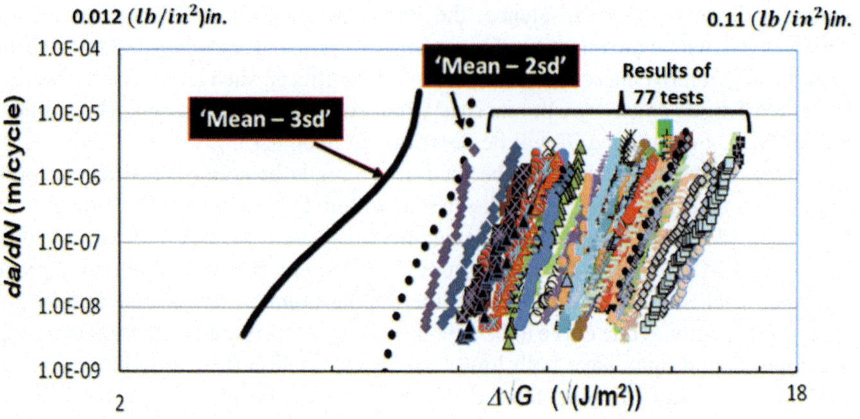

Fig. 9.21 The predicted "upper-bound" curves encompass all the measured results. The level of conservative depends upon your choice of the multiple of the standard deviation that it subtracts from the mean value of $\Delta\sqrt{G_{\text{thr}}}$ and A (cyclic fracture toughness—Eqs. 9–11) [21]

9.13 Importance of GImax Versus ΔG for the da/dN Plot

During the precracking fatigue cycles stage of the DCB specimen, when it undergoes the unloading part of the cycle, the delaminated faces will be meeting each other, resulting in facial interference. Facial interference is a combination of effects including rough surfaces, and the presence of debris. All these factors help in not closing the delamination during unloading properly in each fatigue cycle, when the stress ratio $R \sim 0$. This is very much like the crack closure phenomenon in metallic materials, where the presence of plasticity at the wake of the crack and other factors would not allow the crack to close properly [21, 22]. In the case of both the DCB and ENF precracking and crack growth analysis, facial interference has influence on the G_{min} when the crack mouth approaches zero. Because of the occurrence of closure phenomenon in composite, a better variable that is more reliable to use should be G_{max}, when fatigue crack growth rate test is of interest to engineers. That is, the da/dN versus G_{max} is preferable to use rather than the da/dN versus ΔG (i.e., $G_{\text{max}} - G_{\text{min}}$) because of error arising from using G_{min} quantity [23].

9.14 Summary of Application of Fatigue and Fracture Mechanics in Composites

The number of cycles to failure of composite structural parts with preexisting defects in a high-cycle fatigue load environment was assessed through linear elastic fracture mechanics, where the basic assumptions of the material being elastic and homogeneous were taken into consideration. The presence of defects between plies, resulting from debonding and the analysis of growing defects during the environmental applied load, necessitates the establishment of material allowables. In addition, the growth of preexisting defects in the matrix, because the matrix is weaker than the fibers, was assessed in this chapter. It was mentioned that if the bond between the fiber and the matrix is weak, a minimum amount of crack arrest will be expected to halt the advancing crack, but the fiber will be able to bridge the two sides of the crack mouth if fiber failure does not occur. Another scenario was the initiation of matrix cracking perpendicular to the applied load, but parallel to the fibers with no resistance from the interface was also presented.

Standard tests, such as the ASTM Double Cantilever Beam (DCB) and the End-Notched Flexure (ENF) tests were presented. They are available to obtain the fracture toughness of both fracture modes, mainly G_{Ic} and G_{IIc}. The results from the DCB tests have shown a large degree of scatter in the measurements that are intrinsic to the composite materials was also discussed.

In addition to the fracture toughness of both fracture modes, mainly G_{Ic} and G_{IIc}, a methodology was presented for predicting an upper-bound fatigue crack growth curve from the experimental test data representing laminated composites exhibiting none or very little retardation under fatigue loading. The methodology proposed by the Hartman–Schijve equation was discussed and it was used to assess the upper-bound fatigue crack growth rate curve, which can be thought of as a material-allowable property and was obtained using an A-basis statistical approach. That is, a conservative upper-bound curve can be calculated from the DCB laboratory test data for material allowable characterization that can be used for design and the number of cycles to failure analysis. As mentioned in this chapter, how conservative the crack growth analysis wishes to be, it depends on the multiple of the standard deviation that is subtracted from the mean value of $\Delta\sqrt{G_{thr}}$ and the quantity A, where A is the cyclic fracture toughness.

Defects in composite parts can severely weaken the material properties, resulting in degraded strength and fatigue performance of composite parts, ultimately leading to total structural failure. The remaining life assessment of composite parts with the preexisting manufacturing defects requires material allowables, as discussed in the cases of both the DCB and ENF test specimens for fracture toughness and the da/dN versus Gmax for crack growth. In this chapter, both fracture toughness and fatigue crack growth were discussed, and the method of obtaining their values for estimating the useful life of composite parts was presented.

The useful life assessment approach of critical composite parts during their operations requires starting with:

- A comprehensive 3D finite element analysis to check the high-stress areas in the part is necessary for analyst to have. High-stress areas could act as a damage initiation site during the operation,
- An appropriate, accurate, and proven nondestructive inspection method is needed to identify if the critical part has preexisting defects and to estimate the dimensions associated with the defect,

- Material allowables based on acceptable standards must be available to the analyst,
- An acceptable, accurate, and proven software is required to estimate the number of cycles to failure based on the load environment available to engineers.

Note that, in the case of PMCs, the first-ply failure usually does not influence their residual strength capability and useful life. Damage progression and the failure of more plies are required for total failure to happen. A comprehensive fatigue crack growth analysis approach that can capture mixed-mode failure, and progression of structural damage to a detectable size is required.

9.15 Concluding Remarks

The application of fracture mechanics to composite materials is helpful to better understand both the behavior and performance of these materials in the presence of defects. By utilizing finite element analysis and principles derived from fracture mechanics, researchers and engineers have been able to predict potential failure modes in composite structures, ultimately contributing to the development of safer and more reliable composite parts.

Because of anisotropy nature of composites, there is a need for (1) more research and testing are essential for gaining a robust fracture mechanics theory on composites that will be able to provide better results when compared with the test data, (2) conducting more standard tests to establish material fracture allowables under different loading conditions, (3) more crack growth tests under different loading conditions and accounting for variables such as dependency of final results on the initial crack length, fiber volume, humidity, temperature variation, and different frequency rate. Understanding all the above can play a crucial role in advancing the fracture behavior of composites, paving the way for their widespread use in diverse industries ranging from aerospace to automotive and beyond. As technology evolves, further research and innovation in fracture mechanics will undoubtedly contribute to unlocking new possibilities and optimizing the performance of composite materials in different environments with various applications.

References

1. B. Farahmand, *Fatigue and Fracture Mechanics of High-Risk Parts* (Chapman & Hall, 1996)
2. B. Farahmand, *Fracture Mechanics of Metals, Composites, Welds, and Bolted Joints: Application of LEFM, EPFM, and FMDM Theory* (Springer, 2000)
3. B. Farahmand, *Application of High and Low Cycle Fatigue on Metallic and Non-Metallic Material (Application of Fatigue for Engineers)* (Amazon eBook, 2022)
4. NASA-STD-5009B, Nondestructive Evaluation Requirements For Fracture-Critical Metallic Components, Approved: 2019-05-08 Superseding NASA-STD-5009.

5. B. Farahmand, *Fatigue and Fracture Mechanics of High-Risk Parts* (Chapman & Hall., Chapter 5 (Section 5.3), 1996)
6. B.C. McMaster, Liquid penetrant inspection tests, in *Nondestructive Testing Handbook*, vol. 2, 2nd edn., (American Society for Nondestructive Testing, 1982)
7. Forman, R. G.; Shivakumar, V.; Cardinal, J. W.; Williams, L. C.; McKeighan, P.C. "Fatigue Crack Growth Database for Damage Tolerance Analysis" (2005).
8. A.J. Harter, AFGROW reference manual (version 4.0). Wright-Patterson Air Force Base, AFRL/VASM (2003)
9. A.A. Griffith, The phenomena of rupture and flow in solids. Philos. Trans. R. Soc. Lond. A **221**(582–593), 163–198 (1921)
10. B. Farahmand, *Fatigue and Fracture Mechanics of High-Risk Parts*.," Chapters 3 & 6 (Chapman & Hall, 1996)
11. D 5528 – 01 Standard Test Method for Mode I Interlaminar Fracture Toughness of Unidirectional Fiber-Reinforced Polymer Matrix Composites. Approved 2007
12. P.T. Curtis, CRAG (Composite Research Advisory Group) Test Methods for the Measurement of the Engineering Properties of Fiber Reinforced Plastics (1988)
13. ASTM E647, Standard Test Method for Measurement of Fatigue Crack Growth Rates. Last Updated: Aug 16, 2017
14. B. Farahmand, K. Nikbin, Predicting fracture and fatigue crack growth properties using tensile properties. Eng. Fract. Mech. **75**, 2144–2155 (2008)
15. R.H. Martin, G.B. Murri, Characterization of Mode I and Mode II Delamination Growth and Threshold in Graphite/PEEK Composites, NASA TECHNICAL MEMORANDUM 100577, APRIL 1988
16. D.P. Rooke, D.J. Cartwright, *Compendium of Stress Intensity Factors* (HMSO Ministry of Defense. Procurement Executive, 1976)
17. W.S. Johnson, P.D. Mangalgiri, Influence of the resin on interlaminar mixed-mode fracture, in *Toughened Composites*, ASTM STP 937, ed. by N.J. Johnston, (American Society for Testing and Materials, Philadelphia, 1987), pp. 295–315
18. J. Cresws, J. Reeder, Mixed-mode bending method for delamination testing. AIAA J. **28**(7), 1270–1276 (1990)
19. L. Yao, R. Alderliesten, R. Jones, A.J. Kinloch, Delamination fatigue growth in polymer-matrix fibre composites: A methodology for determining the design and lifing allowables. Compos. Struct. **96**, 8–20 (2018)
20. NASA-HDBK-5010, Fracture Control Handbook For Payloads, Experiments, and Similar Hardware, May 2005, Revalidated 2012.
21. G.C. Sih, P.C. Paris, G.R. Irwin, On cracks in rectilinearly anisotropic bodies. Int. J. Fract. Mech. **1**, 189–203 (1965)
22. A.J. Kinloch, R. Jones, A way forward for industry to determine valid cyclic-fatigue relationships for polymer-matrix fibre composites. Procedia Struct. Integrity **28**(2020), 26–38 (2020)
23. B. Farahmand, *Fatigue and Fracture Mechanics of High-Risk Parts* (Chapman & Hall., Chapter 4 (Section 4.3.1), 1996)
24. R. Jones, A.J. Kinloch, J.G. Michopoulos, A.J. Brunner, N. Phan, Delamination growth in polymer-matrix fibre composites and the use of fracture mechanics data for material characterisation and life prediction. Compos. Struct. **180**, 316–333 (2017)

Chapter 10
Obtaining Fracture Properties Through Virtual Testing Approach

10.1 Introduction

When a critical part is subjected to a load-varying environment, engineers need to know whether the part can survive during its service usage or not. Therefore, both the static and fracture allowables are needed to conduct a meaningful life analysis assessment of the structural parts under consideration. Fracture data for metallic materials are currently obtainable through the ASTM testing standards, which are costly and labor-intensive. Difficulties associated with specimen preparation, precracking the notch, recording and monitoring the data, obtaining the variation of fracture toughness versus part thickness, capturing data in the threshold region of crack growth, repeating the test at least three times, and in some cases, the test must be repeated due to invalid results, they all make the virtual testing technique extremely helpful to overcome the time and cost of testing difficulties. Farahmand [1–3] utilized extended Griffith's theory to obtain a relationship between the applied stress and half a crack length by using the entire stress–strain curve for the metallic material under consideration. The extension of Griffith's theory led to the estimation of material fracture toughness and the construction of an entire fatigue crack growth rate curve for aerospace metallic material. An attempt is used here to apply this methodology to the composite material, and the results will be compared with the existing test data. However, as was mentioned above, the full stress–strain curve for the lamina and its constituents are needed to estimate the energy release rate (or fracture toughness) for the composite material.

B. Farahmand, *Fundamentals of Composites and Their Methods of Fabrications*,
https://doi.org/10.1007/978-3-031-80201-0_10

10.2 The Virtual Testing Approach

It is estimated that the cost of fully characterizing the mechanical properties of a new material that must be used as a part of aerospace, aircraft, or military structures can be substantial. Static, fatigue, and fracture mechanics data must be available to conduct a meaningful stress and damage tolerance analysis of high-risk components. The successful prediction of mechanical properties through the virtual testing technique is the goal of most industries. The proposed technique utilizes the primary data from a simple static tensile test to estimate the fracture mechanics allowables through the extended Griffith theory. The fracture toughness values can be obtained by taking the energy per unit volume under the uniform and nonuniform portion of the stress–strain curve for the metallic material to calculate the energy consumed at the crack tip up to the failure [4].

The fracture toughness value, K_c or G_{Ic}, obtained through this approach will be helpful for the upper portion of the fatigue crack growth rate curve (the region III of the da/dN curve), where the stress intensity factor, K or $G = K^2/E$ will accelerate toward the instability. The upper portion of the da/dN curve is related to the critical value of the stress intensity factor, K_c, therefore, the fracture toughness value estimated through the virtual testing technique can be used to establish the region III of the da/dN versus ΔK data. Other regions of fatigue crack growth rate can be obtained separately and are presented in this section.

Metallic materials used for designing aircraft and aerospace parts exhibit three distinct regions when subjected to static monotonic loading up to failure. The area under the load versus displacement curve (up to the final failure) will represent the degree of ductility that the material possesses. In the case of CMCs and PMCs, both the matrix and fibers usually do not show any ductility at all, and their stress–strain curve represents only the elastic region. Under monotonic loading, purely elastic brittle materials are strictly prohibited from being used in the design of aerospace and aircraft parts, where the vehicle's safety is of great concern. Figure 10.1a illustrates the stress–strain curve for a few polymers and in Fig. 10.1b, the stress–strain curves for several thermoset polymer epoxy composites with no ductility at the onset of failure [5] are shown. Still, when a lamina is subjected to applied load, its fracture toughness is demonstrated through the crack arrest mechanism by the reinforcement and the interface. The crack arrest mechanism is more pronounced in ceramic matrix composites (e.g., SiC/SiC), where fibers or whiskers can stop the crack from growing. This arrest feature of ceramic composites can also occur in laminated composites when the bonds between the fibers and the matrix are adequate to prevent crack growth in the matrix. It will take more energy to overcome the crack arrest for further crack growth, either by resistance through debonding or fiber bridging mechanisms.

As part of the aerospace design requirement, material selection must be limited to those alloys that exhibit elongation larger than 4%. Aluminum and titanium alloys are excellent candidates for aircraft and aerospace parts because of their attractive

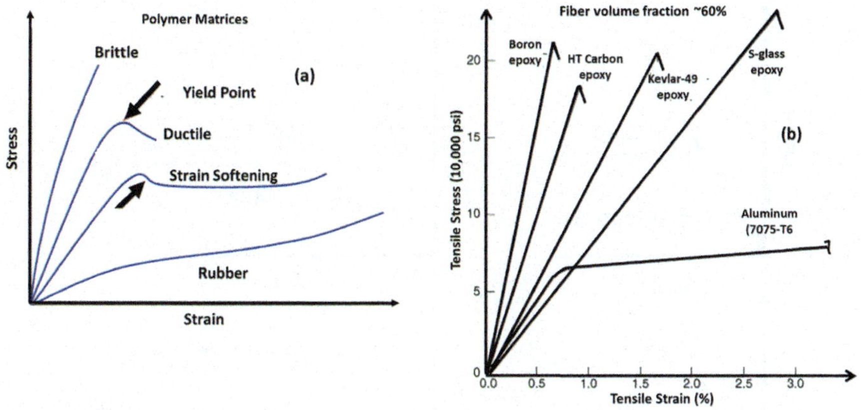

Fig. 10.1 (**a**) Stress–strain curves for several polymer matrixes are illustrated. (**b**) Stress–strain curves for several composite materials with thermoset elastic matrix. The tensile stress–strain curve is linear up to the point of failure. These specimens fail by tensile rupture of fibers. However, in the case of 7075-T6 aluminum, more than 3% ductility is shown

strength-to-weight ratio, as well as their great elongation that almost in all cases exceeds the design requirement of 4%. Since ductility and toughness are directly related to each other, the extended Griffith theory is used to relate the area under the full stress–strain curve to the material fracture toughness via Farahmand's theory [4], which will be briefly described in Sect. 10.3.

10.3 The Extended Griffith Theory and Fracture Toughness

Materials selected by engineers to manufacture fracture critical hardware must exhibit some amount of plastic deformation and stable crack growth at the crack tip region before their final failure. The amount of energy consumed at the crack tip is largely due to material resistance in that region prior to final failure, which is not adequately accounted for in the linear elastic fracture mechanics analysis. Farahmand [4] observed that for tough material, the amount of energy dissipated at the crack tip is of two kinds: local nonuniform straining, U_F, and uniform straining, U_U. Thus, the fracture behavior of ductile material can be characterized by two energy release terms representing plastic deformation at and near the crack tip. Both terms can be derived using the areas associated with uniform and nonuniform plastic deformation under the full uniaxial stress–strain curve for the alloy under consideration, Fig. 10.2. The extended Griffith theory in terms of energy consumed at the crack tip for plastic deformation, U_U and U_F, including the two new crack surfaces term, $2T$, can be written as:

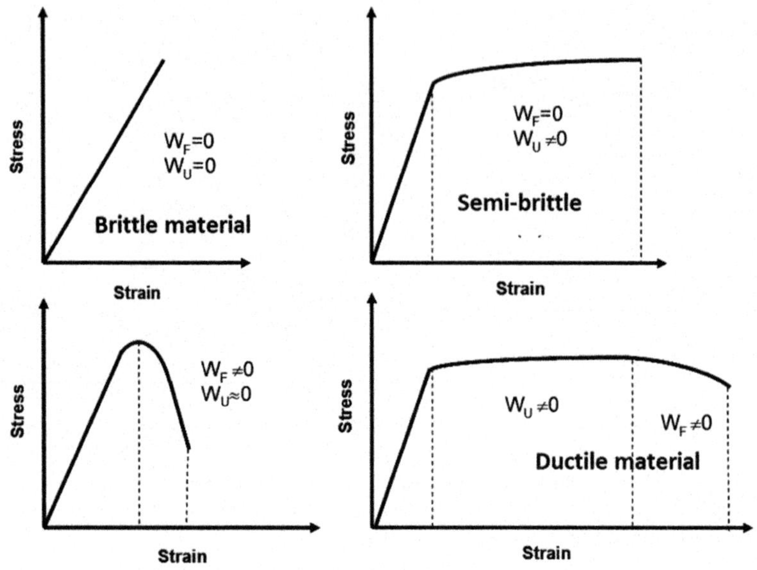

Fig. 10.2 Material with different stress–strain curves. The energy consumed to fail material is related to the area under the stress–strain curve

$$\frac{\partial U_{\mathrm{E}}}{\partial a} = 2T + \frac{\partial U_{\mathrm{U}}}{\partial a} + \frac{\partial U_{\mathrm{F}}}{\partial a} \tag{10.1}$$

All the terms to the right are associated with the energy consumed at the crack tip, where the first term, $2T$, is the energy term needed to create two new crack surfaces, the second term is associated with the uniform straining, and the third term is for the nonuniform straining.

For ductile metals, the terms associated with uniform and nonuniform plastic deformation are much greater than the two new crack surfaces term, $2T$. If the two plastic deformation quantities to the right of the equation are determined, then a relationship between applied stress, σ, and half a crack length, a, at the onset of failure can be established. The energy rate associated with $\frac{\partial U_{\mathrm{U}}}{\partial a}$ term can be derived by assuming that all the crack tip energy, due to nonuniform straining up to the failure, has been consumed by the amount of W_{U} (the area under the stress–strain curve, Fig. 10.2) to create a permanent slip of height h_{U}. That is,

$$\frac{\partial U_{\mathrm{U}}}{\partial a} = W_{\mathrm{U}}\, h_{\mathrm{U}} \tag{10.2}$$

The energy term associated with $\frac{\partial U_{\mathrm{F}}}{\partial a}$ can be derived by assuming all the crack tip energy due to nonuniform straining up to the failure, has been consumed by the amount W_{F} (the area under the nonuniform plastic deformation, Fig. 10.2) to create permanent coarse slip of height h_{F}.

$$\frac{\partial U_F}{\partial a} = W_F h_F \tag{10.3}$$

The final equation describing the extended Griffith theory for ductile metals in terms half a crack length, a, and applies stress, σ, in an infinite can be written as [4]:

$$a = \frac{E}{\pi\sigma^2\mu}\left\{2T + \overline{\sigma}_{UF}\varepsilon_{PN}h_F k + \frac{n}{n-1}\sigma_{TU}\varepsilon_{TU}\left[1 - \left(\frac{\sigma_T}{\sigma_{TU}}\right)^{n+1}\right]h_{min}\left[\frac{\varepsilon_{TF}\varepsilon_{TL}}{\varepsilon_{TU}\varepsilon_T}\right] * \left[\left(\frac{\varepsilon_{TU}}{\varepsilon_{TL}}\right)^{\frac{n-1}{n}} - 1\right]\beta\right\} \tag{10.4}$$

Quantities in this equation are defined as:

$\overline{\sigma}_{UF}$ = *Average stress between ultimate and final fracture*
σ_{TU} = *True stress at ultimate*
σ_{TF} = *True stress at failure*
σ_{TL} = *True stress at limit*
ε_{TU} = *True strain at ultimate*
ε_{TF} = *True strain at failure*
ε_{TL} = *True strain at limit*
ε_{PN} = *Plastic strain at necking*

Thickness parameters describing the plane stress and strain conditions for each term of the above equation are designated by μ, k, and β parameters.

As was mentioned before, by calculating half a crack length, a, for a given applied stress, σ, the critical value of the stress intensity factor, K_c, can easily be calculated (Eq. 10.4) and will be used for region III of the da/dN versus ΔK (detail discussion related to this topic is presented in Sect. 10.4). Fig. 10.3 shows a comparison between the fracture toughness test data and calculated fracture toughness for 2219-T6 and 6061-T6 aluminum alloys through a virtual testing approach. The results are in good agreement with the test data. In Sect. 10.5, the virtual testing technique will be implemented to composite material [6]. As was mentioned in Chap. 7, the application of fracture mechanics is acceptable for composite material if they fall under being homogeneous and elastic/brittle material (Chap. 7, References 16 and 17).

10.4 Fatigue Crack Growth Rate Data Through the Virtual Testing Method

To obtain the da/dN versus ΔK data for all regions of fatigue crack growth rate curve, the virtual testing technique, proposed by Farahmand [6] can be extremely useful to apply when there are budget or time limitations to conduct coupon tests. Farahmand's unique approach is based on establishing each region of the da/dN

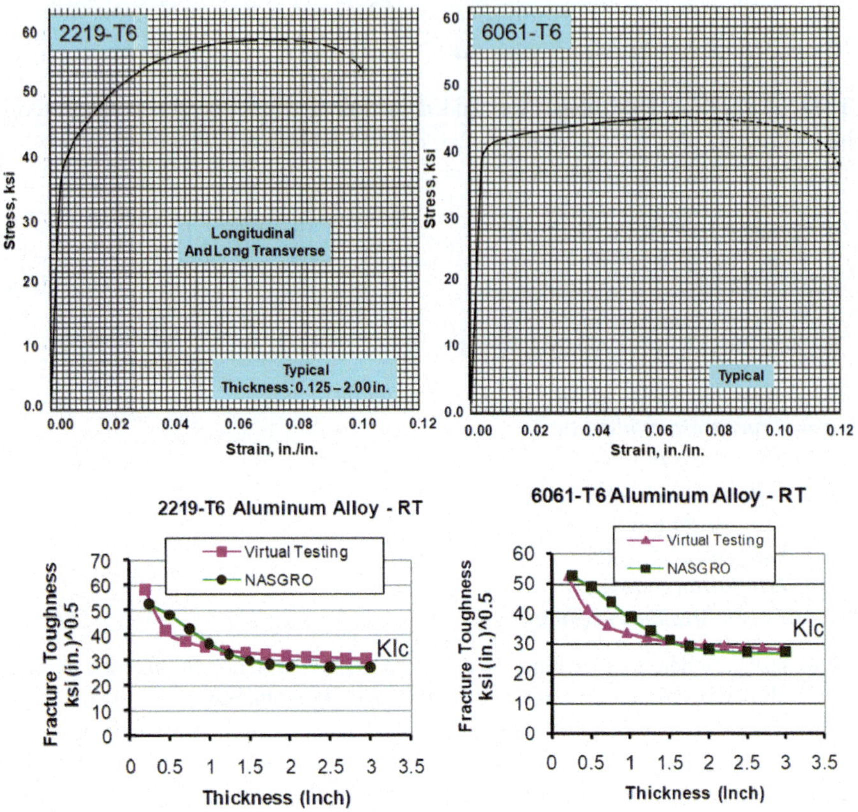

Fig. 10.3 (**a**) The full stress–strain curves for 2219-T6 and 6061-T6 aluminum alloys. (**b**) Virtual testing fracture toughness results versus test data for 2219-T6 and 6061-T6 aluminum

curve separately and relating them through the Forman–Newman fatigue crack growth equation described by Eq. 9.3 of Chap. 9:

C, n, f, p, and q are constants
K = The stress intensity factor
Kc = The fracture toughness
Kth = The threshold stress intensity factor
R = The stress ratio

$$\frac{da}{dN} = \frac{c(1-f)^n \Delta K^n \left(1 - \frac{\Delta K_{th}}{\Delta K}\right)^p}{(1-R)^n \left(1 - \frac{\Delta K}{(1-R)K_c}\right)^q} \tag{9.3}$$

Currently, all fatigue crack growth constants (c, n, K_c, ΔK_{th}), shown in the Eq. (9.3of Chap. 9), will be provided through the ASTM tests which are costly and labor-intensive. However, by using the virtual testing concepts, all constants can be estimated and the da/dN versus ΔK, from Eq. (9.3), can be plotted. Note that in all cases the units for the da/dN versus ΔK are given in the Ksi and the conversion of fracture toughness, K, is: Mpa $(m)^{1/2} \sim 0.91$ ksi $(\text{in.})^{1/2}$.

10.4.1 Estimating the Accelerated Region and Fracture Toughness

The region III of the da/dN versus ΔK curve is related to the fracture toughness, K_c, and it is thickness dependent. As was mentioned previously, the quantity K_c can be estimated via eq. (10.4) for all ranges of material thickness. For the plane strain condition and elastic material with no ductility at the crack tip, the quantity $K_c = K_{Ic}$. Noting that for the brittle material, the fracture toughness is not thickness dependent. In Fig. 10.3, it was shown that the fracture toughness variation with respect to thickness calculated by the virtual testing technique is in good agreement with the NASGRO database generated by test data [6, 7]. Therefore, by using this technique, region III of the da/dN versus ΔK curve can be obtained without conducting tests. At the same time, when fracture toughness, K_c or K_{Ic}, is available, then the energy release rate value, G_{Ic}, can be obtained through the relationship $G_{Ic} = (K_{Ic})^2/E$.

10.4.2 The Region II and Estimating the Paris Constants, C *and* n

A significant number of fatigue crack growth rate data for numerous aluminum alloys were extracted from the NASGRO database and the da/dN versus ΔK for all available alloys were plotted in Fig. 10.4. As a result of this data gathering, two important observations were made that will be helpful to establish the Paris constants (i.e., C and n) of region II of the da/dN curve. The lower point in the Paris region of the fatigue curves (Fig. 10.4) has a material-independent property in such a way that the ratio of the stress intensity factor, K, at the lower part of region II to the material threshold stress intensity factor $\Delta K/K_{th} \sim 1.125$ for stress ration $R = 0$. At this ratio, the crack growth rate per cycle is da/dN ~ 1.0E-7 inch/cycle (~2.54E-6 mm/cycle). An additional observation was also made in the upper region of the da/dN curve at the end of the Paris region shown in Fig. 10.4. The ratio of the upper bound stress intensity factor and its critical value, K_c, ($\Delta K/K_c$ for $R = 0$) is found to be ~0.9 for the da/dN ~0.005 inch/cycle (~0.127 mm/cycle). These two points are helpful to plot the entire region II of the da/dN curve [6, 7]. In Sect. 10.4.3, the method of finding the threshold value, K_{th}, for different stress ratios, in region I, will be discussed.

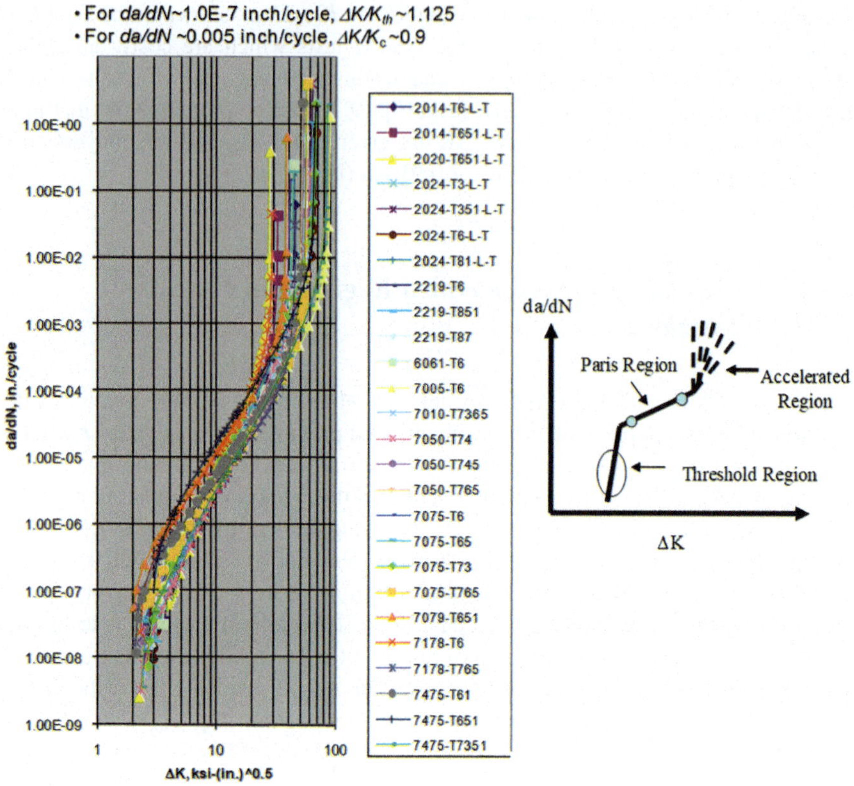

Fig. 10.4 The upper and lower regions of Paris crack growth have special characteristic common among many alloys

10.4.3 Estimating the Threshold Value, Kth *of Region I*

The threshold region of the da/dN versus ΔK curve is difficult to obtain. The complexity is due to the plasticity and surface roughness closure phenomenon that makes measuring the threshold value difficult. The threshold stress intensity factor measurement is time-consuming, and sometimes it may take several days to make a few measurements of ΔK_{th} value. For these reasons, the approximate value of ΔK_{th} is of interest to engineers when conducting life assessments of structural parts. Farahmand [8] was able to establish a reasonable relationship between the threshold stress intensity factor, ΔK_{th}, for the case of stress ratio $R \sim 0$, and the plane strain fracture toughness, K_{Ic}. The result of his observation on the threshold values of more than 100 metallic alloys was in such a way that the quantity ΔK_{th} falls very close to 0.1 K_{Ic} (for $R = 0$). This comparison was already shown previously in Fig. 9.11 of Chap. 9. Farahmand argued that materials with high K_{Ic} value also possess higher K_{th} value, and conversely, the lower K_{th} value belongs to low K_{Ic}. This figure shows the calculated threshold values based on 0.1 K_{Ic} and the corresponding test data extracted from the NASGRO database. It is evident that in most cases, the test values fall very close to the estimated value of $0.1K_{Ic}$.

As was mentioned previously, all regions of da/dN versus ΔK must be estimated separately before being able to plot the fatigue crack growth rate curve by applying Eq. 9.3 of Chap. 9. The da/dN versus ΔK can be plotted by calculating its constants through regions I, II, and III described in Sects. 10.4.1, 10.4.2, and 10.4.3. Figure 10.5, 10.6, 10.7, 10.8 and 10.9 are only a few selected cases for aluminum (2014-

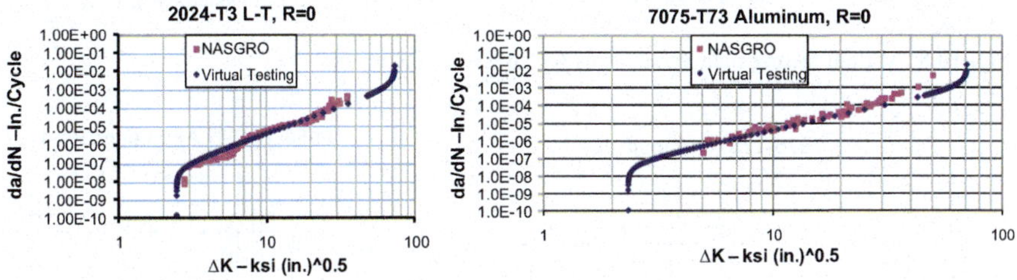

Fig. 10.5 Virtual; testing results versus test data (aluminum alloys)

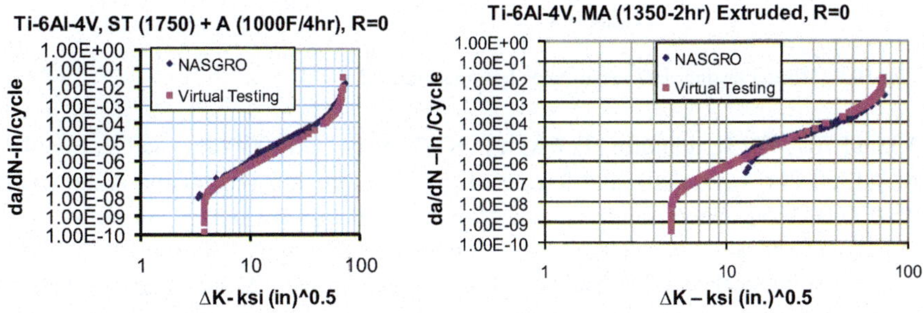

Fig. 10.6 Virtual; testing results versus test data (titanium alloys)

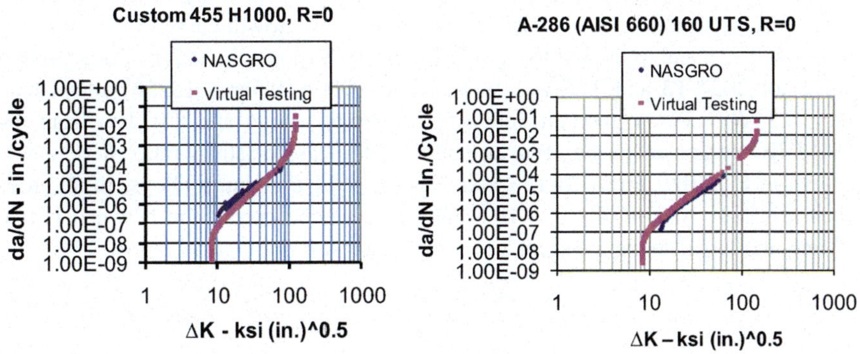

Fig. 10.7 Virtual; testing results versus test data (ferrous alloys)

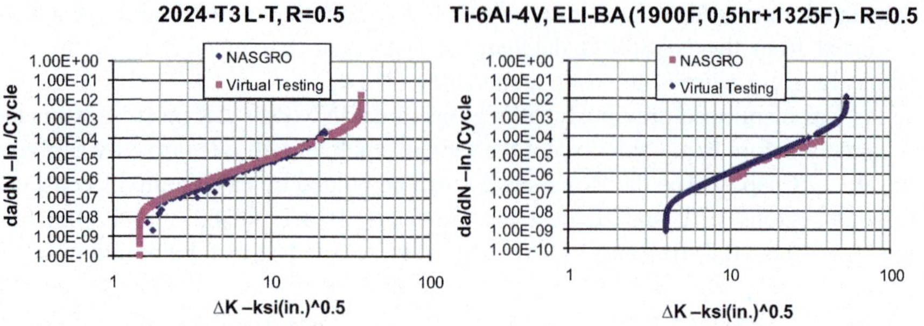

Fig. 10.8 Virtual testing data versus tests for stress ratio $R = 0.5$

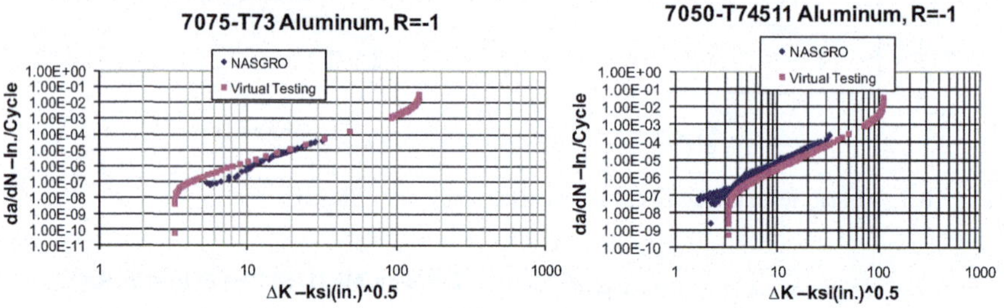

Fig. 10.9 Virtual testing data versus tests for stress ratio $R = -1$

T3 and 7075-T73), titanium (Ti-6Al-4V, ST(1750) + A(1000F-4hr) and Ti-6Al-4V, MA(1350-2hr) Extruded), and ferrous alloys (Custom 455 H1000 and A286 (AISI 660) 160 UTS), where fatigue crack growth rate data were plotted for stress ratio, $R \sim 0$, and checked against available test data provided by NASGRO [9]. Excellent agreement between test data and virtual testing can be seen. More comparisons have been generated and results were compared with test data obtained from the NASGRO database, where good agreements between the virtual and test data were obtained [6]. For other stress ratios, R, the fatigue crack growth rate curve can be established by using Eq. (9.3) of Chap. 9, where the quantity R and f are two parameters that will be able to shift the fatigue crack growth curve of $R = 0$ to the right for $R < 0$ and to the left for $R > 0$. Figure 10.8 shows the fatigue crack growth curve for 2024-T3 and Ti-6Al-4 V, ELI-BA (1900F, 0.5hr + 1325F) alloys with $R = 0.5$ ($R > 0$). A good fit between the virtual testing and test data from the NASGRO database can be seen. In addition, the case of stress ratio, $R = -1$, is shown in Fig. 10.9 for 7075-T73 and 7050-T74511 aluminum alloys. In all cases shown in Fig. 10.5, 10.6, 10.7, 10.8 and 10.9, a good correlation between virtual testing and test data can be seen.

10.5 Estimating Fracture Properties for the Epoxy Matrix by Virtual Testing Method

Just like in the case of metallic materials that were discussed in the previous sections in obtaining both the fracture toughness and fatigue crack growth rate data through the virtual testing technique, an attempt is made to employ the same technique to the composite material. In Chap. 7, it was argued that the crack tip stress intensity factor, K, can be used also for lamina of a laminated composite material, where they fall under elastic, homogeneous, and anisotropic material (References [16 and 17] of Chap. 7).

Virtual testing is desirable for industries to use because the testing approach will be too costly and time-consuming. In the case of composites, a few materials will be examined, and results will be compared with the available test data to check the applicability of the approach. In all cases, the entire stress–strain curve is needed as an input to generate both the fracture toughness and fatigue crack growth data through the virtual testing technique. A simple case of thermoset resin matrix will be examined first, and the method will be extended to the laminated composite. Finally, a computer code has been established to calculate the fracture toughness of epoxy material based on the virtual testing technique described above. The fracture toughness determination (FTD) code, which was used for the metallic material, can also be used for estimating the fracture toughness of the epoxy matrix provided in [10]. The FTD code is based on the energy consumed at the crack tip concept described by the Eq. 10.4. In Fig. 10.10a, the stress–strain curve for the epoxy thermoset resin is shown. In Fig. 10.10b, data were gathered as input into the FTD code [11], as shown in Fig. 10.10c. The fracture toughness estimation for the epoxy of Fig. 10.10a is $K_{Ic} = 0.67$ ksi $(\text{in})^{0.5}$, which is compatible with the typical value obtained through testing [10]. The G_{Ic} value associated with the 0.67 ksi$(\text{in})^{0.5}$ is $G_{Ic} = 1.0$ in-lb/in^2. This is compatible with the value of 1.47 in-lb/in^2 provided by [10].

To conduct the fatigue crack growth analysis of epoxy matrix material shown in Fig. 10.10, the virtual testing method will be implemented for this case. The code used for generating the da/dN data is called FCG (fatigue crack growth) [12]. For the accelerated region, the fracture toughness quantity was already calculated through the FTD code (Fig. 10.10), where $K_c = 0.67$ ksi $(\text{in})^{0.5}$. For the threshold region, it was decided to use the value of $K_{th} = 0.1\, K_c \sim 0.067$ ksi $(\text{in})^{0.5}$. Other quantities to be used as input to the FCG code for obtaining the full crack growth rate data are shown in Fig. 10.11a. In Fig. 10.11b, the da/dN data are generated for the epoxy matrix material.

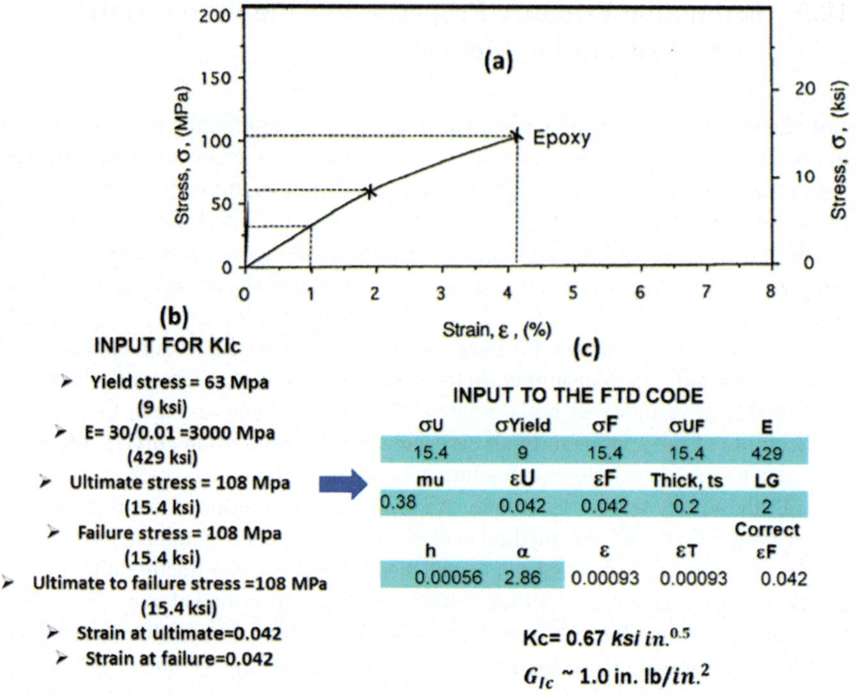

Fig. 10.10 (**a**) Stress–strain curve for the epoxy. (**b**) Data gathering from the stress–strain curve. (**c**) Data from (**b**) was used in the FTD code to calculate fracture toughness

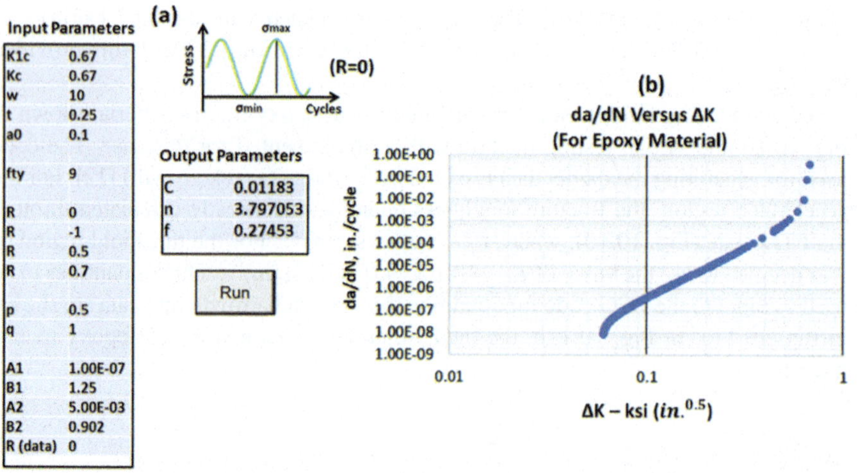

Fig. 10.11 (**a**) The input to the fatigue crack growth (FCG) code is shown for the case of stress ratio $R = 0$. (**b**) The crack growth curve was generated (the da/dN versus ΔK) for the epoxy material

10.6 Fracture Properties of Laminated Composite (Virtual Testing)

In order to increase design reliability, the fracture behavior of adhesively jointed plies under service loading must be carefully studied. For laminated composite, the fracture toughness, as well as the fatigue crack growth data, must be generated for the debonding or delamination process, which is considered an important localized defect that will grow slowly, causing ply separation and final failure. Multiple cracks and debonding may also be distributed throughout the process zone, which can grow and merge with each other to create critical size damage. Note that a laminated composite part is constructed from groups of individual unidirectional plies, which are adhesively bound together with various orientations. Inadequate bonding between these plies can initiate defects and subsequent failure during the part's service life. The mismatch between plies, the compressive applied load during the fluctuating cycles (e.g., when the stress ratio is $R = -1$), and the impact during the manufacturing phase or during the part's service life can introduce a separation between plies. Therefore, the presence of a minor delamination defect will advance through two plies and possibly merge with other existing defects to create a more significant size defect capable of failing the laminated structure. Thus, understanding both the fracture toughness of debonding (i.e., material resistance to debonding, G_{Ic}) and the fatigue crack growth rate da/dN data (how fast the delamination will grow) are important to have for a comprehensive life analysis of a lamina of a laminated composite part.

In applying the virtual testing approach, just like in the case of the resin epoxy matrix discussed in Sect. 10.5, the basic properties of the laminated composite must be available. That is, the full stress–strain curve for the adhesive material holding plies together is needed if the aim of the analysis is to obtain the fracture toughness and fatigue crack growth properties of laminated composite plies during their service usage. It should be noted that in conducting a fatigue test, due to repeated cycling, the load frequency will have an impact on the mechanical properties of the polymer (Fig. 10.12).

The influence of high cycling frequency is much more pronounced in the case of thermoplastic matrices that will undergo softening due to the induced heat during cycling [13]. The temperature will rise to around 150 °F is expected for frequencies above five cycles per second (i.e., 5 Hz). In Fig. 10.12, the variation of time and temperature rise for several frequencies are plotted for the E-glass epoxy. Under five cycles per second (5 Hz), the increase in temperature occurs rapidly, and at 40 Hz, the temperature will reach as high as 110 °F. In Fig. 10.13, the effect of frequency on the fatigue performance is shown for the carbon fiber reinforced PEEK laminate for two laminates with plies of different orientations, $[\pm 45]_{4S}$ and $[-45/0/45/90]_{2S}$ laminates [13, 14]. From this figure, it is clear that the fatigue life for 0.5 Hz is superior to 5 Hz because of the frequency effect causing the degradation of the matrix material due to the heat rise. The lack of thermal conductivity of the matrix is vital for temperature rises in the material. When the conductivity is low, the heat can rise locally, and the polymer properties degrade faster.

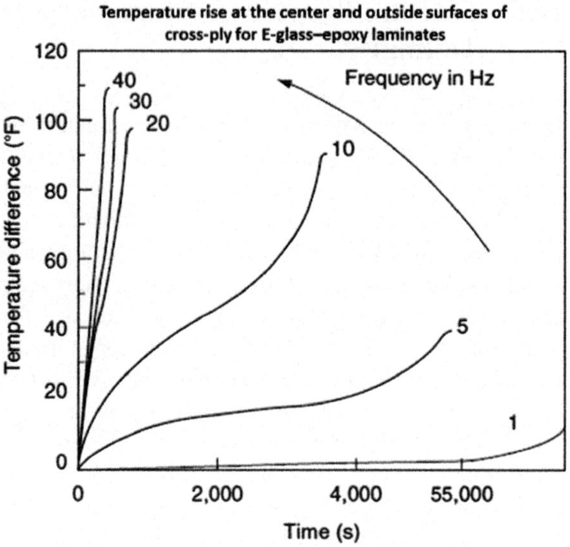

Fig. 10.12 Difference in temperature rise at the center and outside surfaces of cross-ply E-glass–epoxy laminates during fatigue cycling for different frequencies

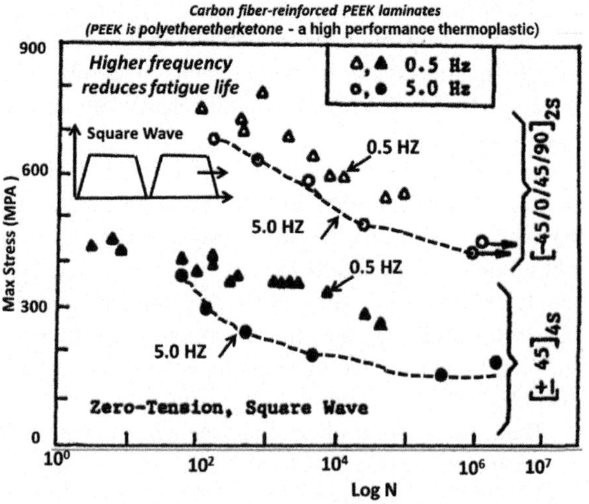

Fig. 10.13 Influence of test frequency on fatigue life for two laminates with different orientations

The fracture toughness of resin epoxy to be used as an adhesive material to attach plies of the composite was already discussed in Sect. 10.5. Another adhesive material that is used extensively in aerospace is the FM-73 and its fracture properties will be assessed here. The stress–strain curve of FM-73 at room temperature is shown in Fig. 10.14a, and it will be used for fracture toughness determination. Note that the adhesive properties change a lot with temperature. One needs to ensure the properties are correct for the temperature of the environment under study. Also note that, the adhesive properties are a function of the test method, adhesive thickness, and temperature, which have already been mentioned before.

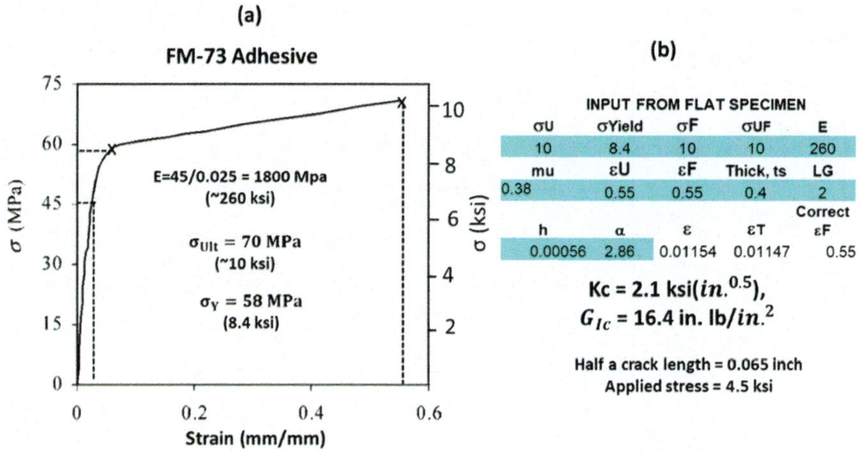

Fig. 10.14 (**a**) Tensile stress–strain curve for adhesive FM-73. (**b**) Fracture toughness estimation through the FTD code

The input data for obtaining the FM-73 fracture toughness were extracted from the stress–strain curve as input to the FTD code. The value of $K_c = 2.11$ ksi $(\text{in})^{0.5}$ was calculated by the code as shown in Fig. 10.14b, and subsequently, the energy release rate was estimated based on $G_{Ic} = (\pi\sigma^2 c)/E = 16.4$ in-lb/in^2. This is compatible with the value of 2000 J/m^2 ~ 12.0 in-lb/in^2 of [15].

The crack growth analysis of FM-73 material will follow by using the FCG code. The input to the code and results are shown in Fig. 10.15. In Fig. 10.15a, the input to the FCG code, and in Fig. 10.15b, the plot of da/dN is shown.

10.7 Fracture Properties of Fibers (Virtual Testing)

In Sect. 10.5, the fracture toughness and fatigue crack growth of epoxy matrix material was presented. However, in this section, both the fracture toughness and fatigue crack growth of fiber material used in polymer composite will be studied through the virtual testing technique. The material under consideration is the Kevlar 49 fiber, and an attempt was made to generate its fracture toughness and crack growth properties through the virtual testing technique using the FTD and the FCG codes. As an input, the full stress–strain curve for the Kevlar is needed for the fracture toughness determination. Fig. 10.16a shows the stress–strain curve for Kevlar 49, where all the necessary information is shown in the figure. Applying those data as input into the FTD code, the fracture toughness can be estimated. In Fig. 10.16b, the estimated fracture toughness is 20 ksi $(\text{in})^{0.5}$ and the $G_{Ic} = 24$ in-lb/in^2. This value is estimated by the code where the final length of crack is calculated to be $2c = 0.62$ in. The applied stress at the onset of failure is calculated to be

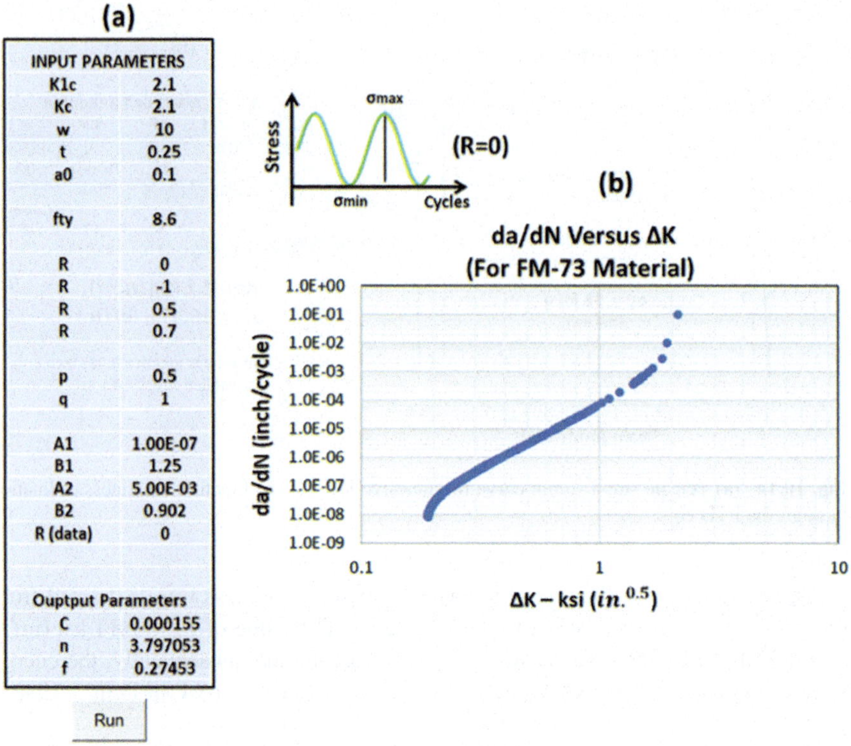

Fig. 10.15 (**a**) The input to the fatigue crack growth (FCG) code is shown for the case of stress ratio $R = 0$. (**b**) The crack growth curve was generated (the da/dN versus ΔK) for the FM-73 material

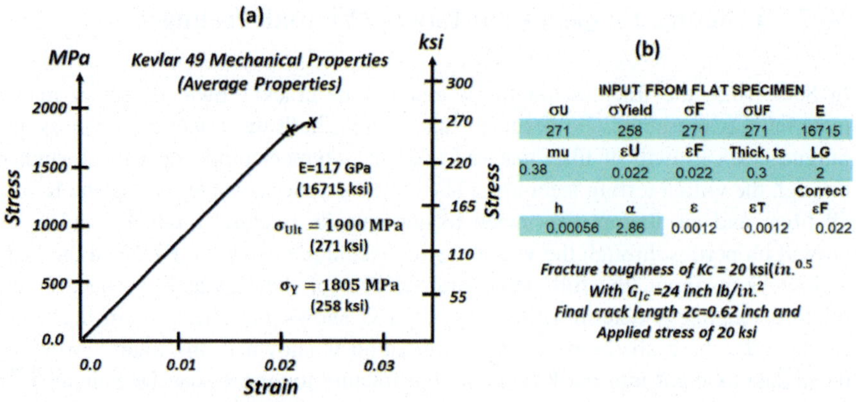

Fig. 10.16 (**a**) The stress–strain curve for Kevlar 49 with average properties. (**b**) The calculated fracture toughness by the FTD virtual testing code

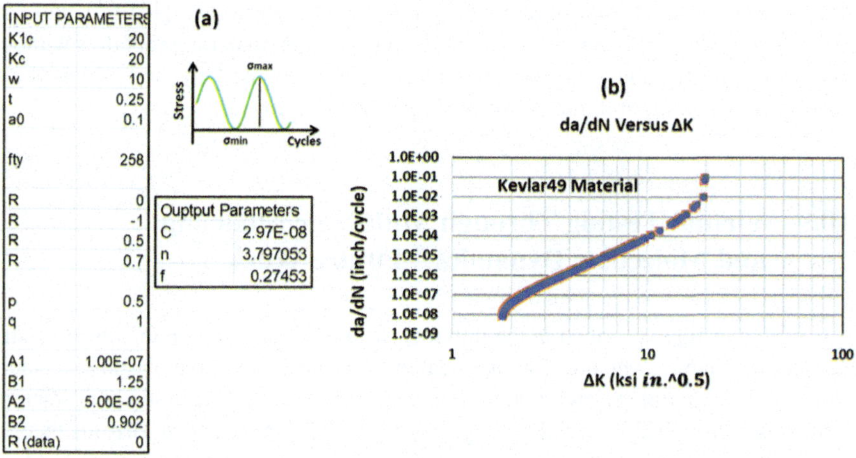

Fig. 10.17 (**a**) The input to the fatigue crack growth (FCG) code is shown for the case of stress ratio $R = 0$. (**b**) The crack growth curve was generated (the da/dN versus ΔK) for the Kevlar 49 material

$\sigma = 20$ ksi. In Fig. 10.17, the da/dN versus ΔK is plotted. The calculated fracture toughness value, shown in Fig. 10.16, was used as an input into the FCG code to generate the crack growth rate data for the Kevlar 49 material.

10.8 Concluding Remarks on Virtual Testing

This chapter has covered the virtual testing technique, offering numerous example problems to illustrate its application. The approach enables engineers to generate the material fracture toughness and the fatigue crack growth rate data without utilizing costly and time-consuming testing. In some cases, industries need to have the fracture properties of a new material where their fracture data are not available to engineers to conduct their life analysis. The purpose of fracture mechanics analysis is to determine if the structural part can withstand the operational load in the presence of defects. If not, the material must be altered, or the design of critical regions with high stresses should be modified to prevent catastrophic failure and possible loss of life. In the case of metallic material, both the FTD and FCG codes proved capable of providing fracture properties with high accuracy when compared with test results. However, in the case of composites, more cases must be examined, and compare the result with the existing test data to be sure that results are compatible. In the abovementioned analysis, the central assumptions were as follows: (1) the material is homogeneous, and (2) the material is elastic and anisotropic. No consideration is given to the nature of the material, which for composite material, as we know, is anisotropic. However, the application of linear elastic fracture mechanics seems to show favorable results when applied to composites that possess brittle behavior and based on the work done by Wu (Reference [16] of Chap. 7). Section 10.9 briefly discusses a novel approach to obtain material properties without conducting costly and time-consuming tests. The proposed method is based on molecular dynamics and multiscale modeling methodology.

10.9 Virtual Testing Through Multiscale Modeling and Molecular Dynamics—Future Work

The virtual testing described in the previous sections was based on the linear elastic fracture mechanics which can be applicable to material that have elastic characteristics. The methodology and assumptions that were used to obtain material fracture toughness through the virtual testing was based on the extended Griffith theory, where the energy release rate at the crack tip was consumed not only for the creation of two new crack surfaces, but also for the formation of plasticity at the crack tip when dealing with ductile metals [1–6]. In the virtual testing analysis described earlier, the full stress–strain curve for the material under consideration must be available to analysts in order to estimate the fracture toughness value. The new proposed method is based on the molecular dynamic approach, which considers all materials are made of atoms and the failure as the result of applied force will initiate when bonds between atoms fail.

The purpose of employing molecular dynamics through computer simulations is to comprehend the properties of materials based on their molecular structure. Ultimately, this understanding can be translated to microscopic and continuum levels, enabling engineers to derive benefits from it. This approach aims to substitute traditional laboratory experiments with analysis and computer simulations. Computer simulations of atoms, when at equilibrium or under external forces, serve as a bridge, connecting atomistic and microscopic scales of length and time to the macroscopic world of the laboratory. Of course, assumptions are necessary when dealing with interactions between atoms to approximate bulk mechanical properties closely. Some of these assumptions stem from limitations due to the lack of computer capability for conducting large simulations. To avoid those limitations, the concept of coarse grain technique will be presented in Sect. 10.10. *This technique is used to simulate a larger number of atoms by grouping atoms or molecules into larger entities or coarse-grained particles [16–21]. Instead of simulating each individual atom, groups of atoms are treated as a single entity, reducing the computational cost while capturing the system's overall behavior.* It should be noted that discussions related to molecular dynamics simulations concerning virtual testing and the ability to obtain material properties are brief and are considered as future work if engineers are interested in pursuing this topic.

Computer simulations serve as a connection between theories established at the molecular time and length scale and experiments conducted in laboratories. The objective of virtual testing lies in predicting material properties based on assumptions made at the atomistic level, then comparing these predictions with experimental results and if results are favorable then it can be of a great help to engineers. Therefore, a direct comparison between simulation and actual measurements on specific materials is important. To achieve a favorable match between testing and simulation, a comprehensive model of molecular interactions, accounting for all relevant parameters in the simulation, becomes essential. However, it is crucial to recognize that the goal of molecular dynamics in predicting material properties is to minimize the need for excessive fitting and guesswork in this process; rather, a model that can cover essential physics can be adequate.

In molecular dynamic simulations, Newton's equations of motion are solved numerically to simulate the behavior of a system of particles, considering forces between them, where the simulation is aimed to analyze the movement of atoms and molecules due to external forces [22–24]. Because the atoms are allowed to interact with each other, therefore any perturbation by outside forces will cause disturbance by all other atoms in the vicinity. The force calculation in MD involves evaluating interatomic forces based on a chosen force field and potential energy function. Calculation of force between atoms that are perturbed by the external forces typically involves several steps, (1) Force field selection where appropriate force field that describes the interactions between atoms or molecules is essential. Force fields contain mathematical expressions for potential energies and forces between particles based on empirical data or quantum mechanical calculations, (2) Potential energy calculation which can calculate the potential energy of the system based on the positions of atoms or molecules, r. Note that a force field is the region where forces exist, whereas potential energy is the energy coupled with the position of an object within that field. They are connected because changes in position within a force field can alter an object's potential energy, leading to corresponding changes in the forces acting on it. Interatomic potentials are used in computational material to describe and predict material properties and for this reason it is a valuable tool to use for the virtual testing concept based on molecular dynamics. Common potential energy functions include the well-known Lennard-Jones potential (Fig. 10.18), Coulombic interactions for charged particles, and bonded interactions like harmonic or Morse potentials for bonds (see Eq. 10.8 describing Morse potential) [25, 26].

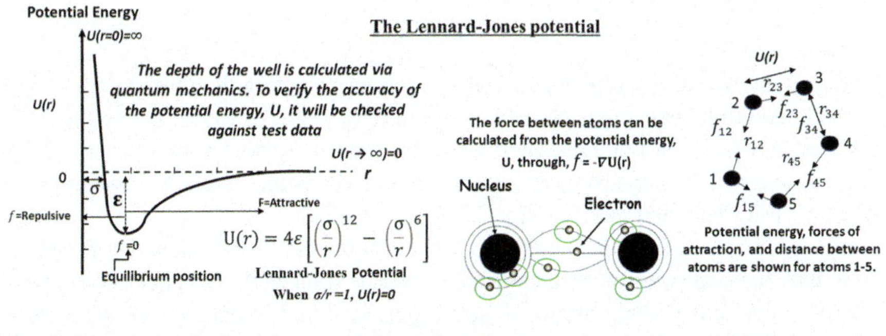

Fig. 10.18 ε is the depth of the potential energy well representing the strength of the interaction between atoms, σ is the distance at which the potential energy between the particles (atoms) is zero, and r is the distance between atoms

Force calculation can be done by computing forces acting on each particle based on the potential energy function and the positions of neighboring particles. Forces are the negative gradients of the potential energy with respect to particle positions. These forces determine how particles move during the simulation (Fig. 10.18). Finally, the reader should realize that in MD analysis, each atom is considered as a mass and furthermore, the presence of electrons must be eliminated and assume that the electrons are part of the total mass (i.e., nuclei) and the interaction between two atoms is simply represented by a potential energy function, $U(r)$.

Integration of equations of motion can be done by using numerical integration methods to update the positions and velocities of particles over small time steps (usually in the range of femtosecond ~ 10^{-15} seconds) based on the calculated force from the potential energy equation. Then repeat the above steps iteratively, advancing the simulation through time to observe the dynamic behavior of the system. Essentially, the molecular dynamics simulation consists of the numerical, step-by-step, solution of the classical equations of Newtonian laws of motion through time, which for a simple atomic system (where nuclei and electrons are viewed as a mass) may be written as:

$$m_i \ddot{r}_i = f_i \tag{10.5}$$

where $f(r)$ is the force between masses that may include the effect of external force, $\ddot{r}$ is the acceleration, and m is the mass of the atom. The force f between atoms can be obtained by taking the derivative of their potential energy expression, $U(r)$, with respect to distance r separating the two atoms (see Fig. 10.18):

$$f_i(r) = -\frac{\partial}{\partial r_i} \mathrm{U}(r) \tag{10.6}$$

The mathematical expression of $U(r)$ for the Lennard-Jones potential can be written as [25–27] (Fig. 10.18):

$$\mathrm{U}(r) = 4\varepsilon\left[\left(\frac{\sigma}{r}\right)^{12} - \left(\frac{\sigma}{r}\right)^{6}\right] \tag{10.7}$$

Here ε is the depth of the potential energy well representing the maximum strength of the interaction between atoms (it is measured by joules or electron volts), σ is the distance at which the potential energy between the particles (atoms) is zero (it has units of length, like angstroms or nanometers), and r is the distance between atoms (Fig. 10.18). The first part of Eq. 10.7 describes the repulsive forces between particles, $(\sigma/r)^{12}$, while the latter part of the equation, $(\sigma/r)^6$, denotes attraction. Note that, the radius, r, for the weak van der Waals force is the distance between two atoms or molecules when attractive and repulsive forces balance each other, typically around the equilibrium position, slightly shifted toward the attraction force. By taking the derivative of the Lennard-Jones potential (Eq. 10.7) with respect to separation distance, r:

$$f(r) = 48\varepsilon\left[\left(\frac{\sigma}{r}\right)^{13} - 0.5\left(\frac{\sigma}{r}\right)^{7}\right] \tag{10.6a}$$

where $f(r)$ is the force between atoms separated by r. The force is negative when the particles are attracted and positive when they are repelled. By letting the force to be equal zero, then the minimum value of $r_{\min} \sim \sigma\ (2)^{1/6}$.

Example

Calculate the intermolecular potential between two Argon atoms separated by $r = 4.0$ Å. The two quantities of $\varepsilon = 0.994$ kJ/mol and $\sigma = 3.41$ Å. To solve for the intermolecular potential, $U(r)$, between the two atoms, Eq. 10.6 can be used:

$$\mathrm{U}(r) = 4\varepsilon\left[(\sigma/r)^{12} - (\sigma/r)^{6}\right]$$

Upon inserting the values of ε =0.994 kJ/mol, σ = 3.41 Å, and the distance of separation, r = 4.0 Å to the above equation describing U(r), and solve for potential energy:

$$\begin{aligned}\mathrm{U}(r) &= 4 \times (0.994)\left[(3.41/4.0)^{12} - (3.41/4.0)^{6}\right]\\ &= 3.988 \times (0.14 - 0.38) = 3.988(-0.24) = -0.96\mathrm{kJ/mol}\end{aligned}$$

Another potential equation similar to the Lennard-Jones potential is Morse potential energy (after physicist Philip Morse). The equation describing the Morse potential energy [25] can be expressed as:

$$\mathrm{U}(r) = D_e\left[e^{-2\alpha(r-r_e)} - 2e^{-\alpha(r-r_e)}\right] \tag{10.8}$$

where r is the distance between the two atoms, D_e = depth of the well, r_e = position of minimum, and α = measure of curvature near minimum. At long range, as $r \to \infty$, the potential energy becomes a constant (since the interaction between the atoms becomes negligible) which is chosen to be almost zero ($U(r) \sim 0$) [25, 26].

As was mentioned above, by having the potential energy, U(r), the force between two atoms as a function of distance (r) can be calculated. Keep in mind that to allow the simulation to go forward, one must consider the following: (1) each atom (i) can be considered as a particle with mass, m_i, (2) from potential energy, the force, f, can be calculated, (3) from force between particles, acceleration, a_i, can be calculated, (4) from acceleration, the velocity, v_i, can be calculated, (5) from velocity, the new separation distance, r_i, can be calculated, (6) finally, the new position of particle at time $t + \Delta t$ can be estimated. Therefore, steps for the trajectories of atoms when atoms are subjected to perturbation are as follows [28–30]:

Step 1: Provide the initial position, $r^{(i=0)}$, and velocity, $v^{(i=0)}$, for the atom. Here, we are dealing with two atoms interactions, where their distance apart is r. By having the distance, velocity, and acceleration (see step 2), then for a time interval, Δt, one can calculate the new position for the atom.

Step 2: Get the force function from the negative gradient of potential energy, where force $f = -\nabla U(r)$. Knowing $a = f/m$, the acceleration, a, can be calculated. The value of acceleration can be obtained from $a = f/m$.

Step 3: for short time Δt, predict next atom position by moving atom and update velocities:

$$\text{Move atom from } r^{(i=0)} \text{ to its updated position, } r^{(p)}, \text{ where } r^{(p)} = r^{(i)} + v^{(i)}\Delta t + \frac{1}{2}a\Delta t^2$$

$$\text{Update velocity : } v^{(p)} = v^{(i)} + a\Delta t$$

Step 4: Adjust atom position based on the new acceleration, a.

$$r^{(i+1)} = r^{(p)} + v^{(i+1)}\Delta t + \frac{1}{2}a\Delta t^2 \text{ and the updated velocity is } v^{(i+1)} = v^{(p)} + a\Delta t$$

Step 5: Calculate the output physical quantities of interest. Note that at each Δt iteration, the distance, r, the potential energy, $U(r)$, the force between atoms, $f(r)$, are now different because they all have a new value of r. Therefore, they all must be calculated for the next position where $t = t + \Delta t$.

Step 6: Move time and iteration step forward with $t = t + \Delta t$, and $i = i + 1$

Step 7: Repeat the iteration as long as needed.

Increase in external forces can affect potential energy in molecular dynamics simulations. For example, in a simple case where you have a collection of atoms interacting within a system (Fig. 10.18), the potential energy is determined by the interactions between these particles. If an external force is applied to this system (for instance, through mechanical forces), it can change the positions of the particles and therefore change the potential energy of the system. When an external mechanical force is applied to the system, it can cause a change in the positions and orientations of these particles, which can change the distances between atoms and the overall structure of molecules [26]. Furthermore, an increase in external mechanical forces can increase potential energy in molecular dynamics by altering the distances and orientations between atoms, which can eventually lead to bond breakage.

When a mechanical force is applied to a material, it creates stress within the system. If the force exceeds the strength of the bonds holding the atoms together, it can damage the bonds. This change occurs because the atoms are pushed farther apart, and if the applied force exceeds a critical value, the bond breakage occurs. Note that the maximum force holding the two atoms together is equal to the value of ε shown in Eq. 10.7.

In a simple case involving Hooke's law where springs are shown by the atomic bond and atoms are represented by mass, the separation due to an external force can be approximated by $U(r) = 0.5k(r - r_0)^2$, where r_0 is the equilibrium distance between atoms and k is the force constant associated with the bond. If the applied external force caused the separation (Δr) from the equilibrium position, then $f_{total} = -dU/dr \approx -k \cdot \Delta r$. These methods provide insights into how bonds respond to external forces and the resulting structural changes. Considering that the external forces have influence on the atomic forces, the total forces are the vector sum of all forces acting upon particles. The total force on atoms can be written as:

$$\text{Total force on atom} = \text{Atomic forces} + \text{External forces}$$

Note that the external force can separate atoms causing failure. The force at the onset of failure can be obtained experimentally using techniques like atomic force microscopy (AFM) [31].

10.9.1 Example of MD for a Small Number of Atoms

A molecular dynamics simulation example is used here based on a small cluster of atoms. The simulation will be on the behavior of a few atoms in a 2D space using a simple Lennard-Jones potential (Eq. 10.7), a common model for interatomic interactions.

Engineers can start with a small number of say Argon atoms randomly positioned in a 2D space. Implement the Lennard-Jones potential between atoms, which models the interatomic forces and then assign initial positions and velocities to the atoms. The velocities are typically randomized based on a certain temperature using Maxwell–Boltzmann distribution. The simulation parameters are the simulation time, and any relevant boundary conditions. Running the simulation and employing numerical integration methods to update the positions and velocities of the atoms are based on Newton's equations of motion already described in Sect. 10.9. Iteratively advance the simulation in small time steps, Δt, allowing the atoms to move according to the forces acting upon them. Record and analyze the trajectories of individual atoms over time to understand their motion, interactions, and structural changes. Compute properties of interest such as displacements, energy distributions, and temperature profiles. Visualize the simulation trajectory and properties using software tools to gain insights into the behavior of the simulated system. Validation and interpretation can be done against experimental data or theoretical predictions if available. In summary, positions (r) and velocities of the atoms (v), the interatomic potential (U), the time step (Δt), and the boundary conditions are all needed for MD simulation to go forward. In addition, a failure theory to indicate the failure of bonds between atoms as a result of external force is needed to go forward for obtaining material failure at the macroscopic level.

If the system of atoms is in equilibrium with no perturbation, atoms will eventually reach their steady state position. However, if there is an external force on the particles, that is increasing in magnitude, it will increase the potential energy over time and cause bonds failure. The amount of force necessary to break the bonds between atoms varies significantly depending on the types of atoms (e.g., covalent, ionic, and others) and the environment in which the bond is being tested, like temperature and pressure. As a rule, breaking atomic bonds involves exceeding the attractive forces that hold the atoms together. Once an atomic bond breaks due to an external force, the potential energy of the system increases and the amount of energy required to break the bond is equal to the depth of the potential well, ε, representing the maximum force of attraction.

Experimental techniques like atomic force microscopy (AFM) [32] or techniques in molecular dynamics simulations can provide understanding into the forces needed to break specific bonds (covalent, metallic, ionic, or others) by applying forces to individual atoms and measuring the response. *Calculating the exact force that is required to break a covalent bond is related to bond energy.* In the case of atoms with a covalent bond, where the bond is formed by sharing electrons between atoms, to break the bond, it requires to supply enough energy to overcome the attraction between the shared electrons. This energy that can break the bond is called the bond energy or bond enthalpy, and it is usually expressed in kilojoules per mole (kJ/mol) of bonds (Note: 1 kJ/mol = 0.2389 BTU/mol). The bond energy is different for different types of covalent bonds, for example, the bond energy for C–H bond is about 413 kJ/mol and in the case of C–C bond, it is 347 kJ/mol. The same is true with ionic bond, where energy is needed to overcome the force of attraction between the oppositely charged ions. This energy is called the lattice energy or lattice enthalpy, and it is also expressed in kilojoules per mole (kJ/mol) of bonds. The enthalpy concept of breaking bonds between atoms is a thermodynamic concept and represents the average energy required to break a particular type of bond in a collection of molecules, whereas the physical energy necessary to break a bond (provided by an external force) refers to the actual energy input required to physically separate two atoms that are bonded together. It involves overcoming the attractive forces holding the atoms together.

To calculate the bond enthalpy energy to break a bond, one can use the following general formula [32–34]:

$$\Delta \mathrm{H} = \sum \mathrm{H}(\text{bonds broken}) - \sum \mathrm{H}(\text{bonds formed})$$

where ΔH is the change in enthalpy or heat of the reaction, $\sum$H(bonds broken) is the sum of the bond energies of the bonds that are broken in the reactants, and $\sum$H(bonds formed) is the sum of the bond energies of the bonds that are formed in the products. For example, to calculate the energy required to break one mole of H–H bonds, one can use the following equation:

$$\Delta \mathrm{H} = \mathrm{H}(\mathrm{H} - \mathrm{H}) - 0 = 432\ \mathrm{kJ/mol}$$

where H(H–H) is the bond energy of one mole of H–H bonds, which is 432 kJ/mol and 0 is the bond energy of the products, which are two atoms of hydrogen that have no bonds. Therefore, you need to supply 432 kJ of energy to break one mole of H–H bonds. In brief, the strength of a bond shows how strongly each atom is bound to another, and how much energy is necessary to break the binding between the two atoms. It is helpful to emphasize the difference between the energy needed to break the bond and the Lennard-Jones potential energy described in Sect. 10.9. While bond energy (or bond enthalpy) aims on the energy required to break a specific bond within a molecule, the Lennard-Jones potential is more general, describing the interaction between particles (atoms or molecules) at a distance, r. The Lennard-Jones potential helps to explain phenomena like intermolecular forces, while bond energy is specific to the breaking and forming of chemical bonds within molecules.

10.9.1.1 MD Simulation of N Atoms

Below is a simple example of molecular dynamics analysis of N atoms interaction. First, the interaction of atom #1 with the surrounding masses (masses #2, 3, 4, and 5) will be assessed (Fig. 10.19). At time t, the distances r_{ij} and hence forces F_{ij} between nearby atoms are calculated. From these forces we can solve the equations of motions and hence, get new positions and velocities. In Sect. 10.9, the Newtonian equation of motion, together with Lennard-Jones potential, can be used to perform MD analysis of N atoms interactions as shown in Fig. 10.19a. The displacement over a time step Δt is denoted with Δr. The Δr must be much smaller than the distance between nearby atoms. Fig. 10.19b shows the detailed MD analysis of body #1 through #5. The time interval, Δt, is usually taken to be around 10^{-15} s (femtosecond). The quantity r in the face centered cubic (FCC) structure is around 2.4A^0, the

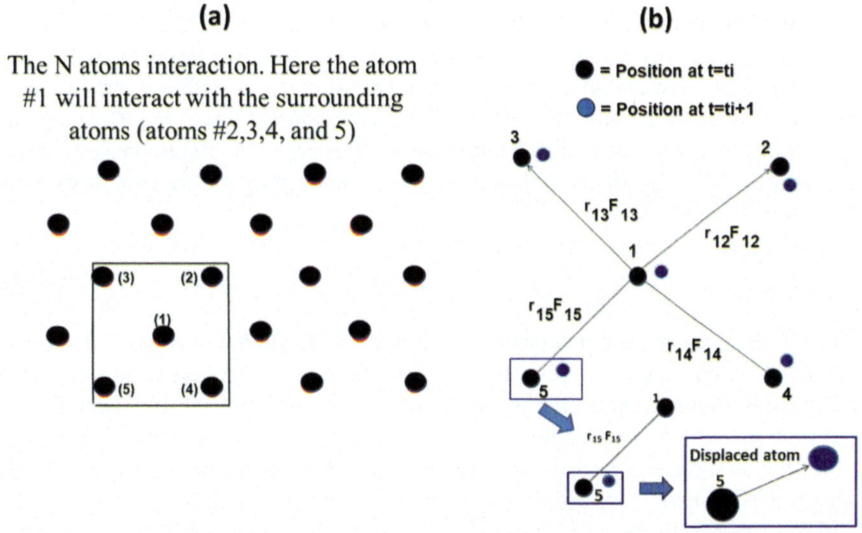

Fig. 10.19 (**a**) Interaction of atom #1 with other atoms in the vicinity. (**b**) The interaction caused the displacement of atoms as indicated in the figure

mass of aluminum atom is (26.98 g/mol)/(6.02 x10^{23}) = 4.48 × 10^{-23} g (Note: 1 kJ/mol = 0.2389BTU/mol), and at the equilibrium position, the force on atoms $f = -\nabla U(r) = 0$. The change in energy between its current position and its position at a small distance away can provide force on the atom. This can be recognized as the derivative of the energy with respect to the change in the atom's position. Knowledge of the atomic forces and masses can then be used to solve for the positions of each atom along a series of extremely small-time steps (in the order of femtoseconds).

10.9.2 External Forces and Bond Failure

When external force F_{ex} applied to atoms, it affects their separation distance r and thus alters the potential energy. The change in potential energy (ΔU) due to the applied force can be estimated using the work-energy formula:

$$\Delta U = -F_{ex} \cdot \Delta r \tag{10.9}$$

Here, the ΔU is the change in potential energy, F_{ex} is the applied external force, Δr is the change in distance between the atoms due to the applied force. This change in potential energy implies the alteration in the interaction energy between the atoms caused by the external force (Eq. 10.9). It should be noted that the Lennard-Jones potential described by Eq. 10.7 represents the interaction energy between two atoms or molecules as a function of their separation distance, r. However, it is not clearly designed to describe the situation where atomic bonds break due to an external force.

When an atomic bond breaks due to an external force, the Lennard-Jones potential ceases to be a suitable model because it is typically formulated to describe the interaction between bonded atoms or neighboring atoms within a certain range. However, once an atomic bond breaks due to an external force, F_{ex}, the potential energy of the system increases and the amount of energy required to break the bond is equal to the depth of the potential well, ε, representing the maximum force of attraction. That is,

$$U_{Failure} = \varepsilon \tag{10.10}$$

Therefore, at failure the potential energy, as a result of external force needed to break the bond, is equal to ε as described by Eq. 10.10. Each broken bond within the structure will contribute to the reduction of the total strength of the material. The question that will be challenging for analysts to assess is how many of these bonds must fail in order to have a total failure upon the application of a monotonic increasing applied load. Additionally, it is crucial for analysts to consider the assumption that the merging of defects can result in a larger defect and subsequent failure of material. Understanding the relationship between the number of broken bonds during the application of external load, and the critical defect size leading to global material failure will also aid analysts in their analysis assessment of virtual testing through molecular dynamics without conducting fracture related tests.

10.10 The Coarse Grain Technique for the MD Analysis

In this section a brief description of the coarse grain technique is provided to the reader when analysts have difficulty dealing with computer simulation of many atoms through the molecular dynamics when properties of material are of interest to acquire.

The coarse grain technique for molecular dynamics is a method of simulating the behavior of atoms in a system by using a simplified representation of a large number of atoms or molecules within the system [16–21]. In this technique, each node (beads or particles are often used in the context of molecular dynamics and is associated with nodes in coarse-grained modeling) represents a group of atoms and the interactions between nodes are described by empirical potentials. That is, just like the potential energy between two atoms that was described in Sect. 10.9.1, in the case of coarse grain approach, the potential energy between nodes that each have many atoms, must be available to be able to conduct the simulation. This reduces the number of degrees of freedom and allows shorter simulation times and larger system sizes. In the area of material science, the coarse grain technique can be of great help to apply because the simulation can be more manageable if it is properly implemented.

Coarse grain (CG) molecular dynamics is different from atomistic molecular dynamics. In the case of atomistic molecular dynamics, it represents only a single atom and potential energy is between two neighboring atoms. However, in the case of node or particle molecular dynamics, the interaction is between nodes that have a large number of atoms, and the overall potential energy between them may be less accurate. On the other hand, atomistic molecular dynamics can provide more accurate information about potential energy, but simulation is limited by the computational cost and great amount of time for obtaining bulk properties.

In conducting the atomistic and coarse grain simulation, it is important to distinguish between two regions of the simulation: (1) The region in which the nodes are in one-to-one correspondence with the atoms, which is called the molecular dynamics (MD) region [16–19], and (2) The region with many atoms per node is called the coarse-grained region (CG) [20–24]. The equation of motion for the atoms in the MD region is just the usual MD equation of motion already described in the previous section. It is based on Newton's laws of motion and the empirical potential describing the force of interaction between atoms. In the coarse-grained region, the equation of motion is the mass, stiffness, and displacement relationship. Figure 10.20 shows the multiscale modeling of a system that contains molecular dynamics (MD) and finite element (FE) region. Also, a narrow region is used to couple the FE and MD regions and is called mesoscale zone. It refers to a level of simulation that lies between the MD scale (where individual atoms and their interactions are

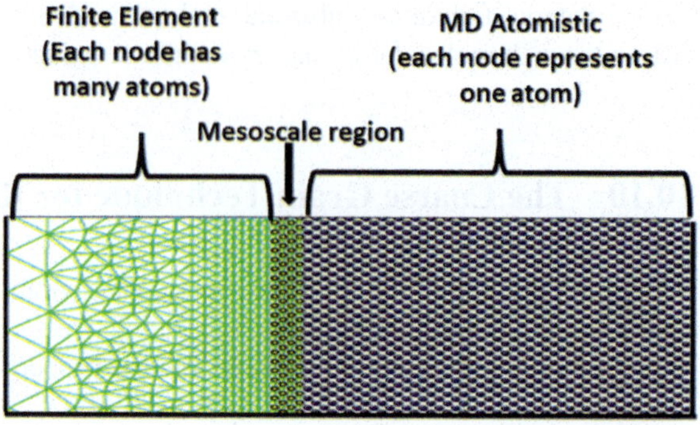

Fig. 10.20 Molecular dynamics and finite element modeling. With the molecular dynamics approach, nodes are in one-to-one correspondence with atoms. However, in the coarse-grain area, each node has a large number of atoms

explicitly modeled) and the macroscale (where bulk properties are observed and can be assessed by finite element, FE). By implementing the coarse grain MD tactic, it can improve the efficiency and better accuracy of the simulation, but at the interface between the two, it requires a coupling method that would provide a smooth transformation of properties between the two regions, Fig. 10.20. In this figure, the critical area is modeled using atoms as the nodes of the mesh and they are in one-to-one correspondence. Then the mesh size increases moving out further into the less critical regions, where each node of the mesh corresponds to many atoms (i.e., bead). It is this reduction in the number of degrees of freedom that increases the efficiency of multiscale simulations. These methods can be employed to generate stress–strain curves for metallic materials that can be used as an input into virtual testing that was described earlier to provide fatigue and fracture data for bulk properties.

Even though coarse-grain molecular dynamics is a useful technique for exploring the properties and functions of complex systems of large numbers of atoms that are beyond the reach of atomistic molecular dynamics simulation, it also has some limitations. That is, the choice of the coarse graining level, the accuracy of the coarse grain potentials, the validation of the coarse grain models, and the interpretation of final computer simulation results are limitations that analysts must take into consideration.

10.10.1 Summary of MD Simulation and the Future Work

Section 10.9 covered the molecular dynamics applications in material science, with a few examples to illustrate its application in obtaining material properties. Molecular dynamics is a computer simulation technique used in physics, chemistry, and material science to capture the behavior and movement of atoms and molecules. It involves tracking the positions and velocities of particles over time by solving Newton's equations of motion, where the force between atoms can be estimated by using the potential energy. This approach allows scientists to understand how molecules interact when subjected to external forces; their structural changes, properties, and provides insights into the dynamics and behaviors of materials at the atomic level. The steps necessary for analyst to take in the molecular dynamics simulation are, (1) Create an initial state of atom, (2) Introduce the interaction potential such as the Lennard-Jones potential, (3) Predict how atoms move as a function of distance r and time Δt, (4) Repeat the iteration as much as needed until to obtain a meaningful simulation result.

The method of obtaining fracture properties for metallic material was presented earlier in this chapter. Virtual testing approach estimates fracture properties of both metallic and to a lesser degree to composites material currently being used in industries [1–5]. As an input it requires data from the full stress–strain curve for the material under consideration. The future work on virtual testing is to use the molecular dynamics approach and multiscale modeling technique to obtain material properties without relying on the ASTM testing. Achieving this goal will eliminate the cost and time of conducting fracture related tests that customers demand from manufacturers of parts. An attempt was made here to provide a road map for engineers to use the bottom-up approach by studying material properties from the atomistic perspective using the molecular dynamics technique in the critical areas (e.g., crack tip), together with finite element method, where each node can have numerous numbers of atoms (Fig. 10.21). For molecular dynamics to work at the bottom of the scale, it requires a valid interatomic potential in order to generate a meaningful simulation.

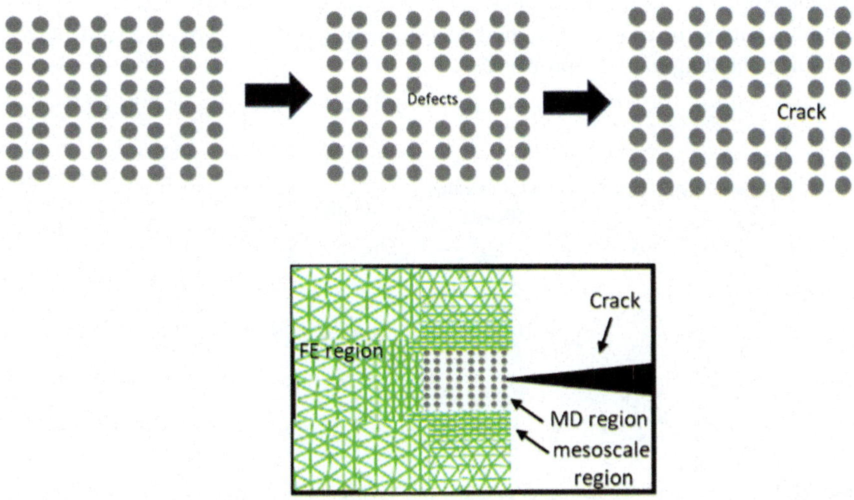

Fig. 10.21 Molecular dynamics modeling of a metallic material with the presence of crack where atoms at the crack tip were used for the simulation. *MD* mesoscale, *FE* regions are shown

Despite today's enhanced computational platforms and sophisticated quantum mechanics techniques, the structures that can be studied are limited at most to a few hundred atoms. A balance must be made between accuracy and feasibility in order to simulate a much larger number of atoms. As was mentioned in Sect. 10.9, the molecular dynamics simulation of billion atoms is much easier to perform by using the effective interatomic potential such as Lennard-Jones or Morse potential. However, the billion atoms simulation still represents a very small solid, only 0.3 microns per side. Another obstacle is the time interval that has been chosen for atomic simulation. As previously mentioned, the time interval, Δt, is usually taken to be around 10^{-15} s (femtosecond), which is close to what it should be. For that reason, molecular dynamics simulation is difficult to perform. Using larger Δt, will be helpful but may not provide for accuracy in the final computational results. With future computers, hopefully, a larger number of atoms and shorter time interval, Δt, will make it possible to simulate material properties through atomistic simulation. Finally, the application of coarse grain technique can be of great help to ease the computer simulation for a large number of atoms. Unlike atomistic simulations that deal with individual atoms, the coarse grain molecular dynamics, and the law of similiarity, simplifies the large system of atoms by grouping atoms into larger particles or beads, reducing the computational complexity while capturing essential characteristics of the system. These beads (aka nodes or grains) might represent multiple atoms or functional groups (Fig. 10.22). By employing the coarse grain molecular dynamics method, researchers can explore phenomena occurring at longer timescales, such as behavior of both metallic and nonmetallic materials. The technique allows for the investigation of larger systems for extended periods, enabling insights into the dynamics and interactions of macromolecular structures. Despite its simplifications, the coarse grain molecular dynamics process remains a valuable tool for understanding material properties.

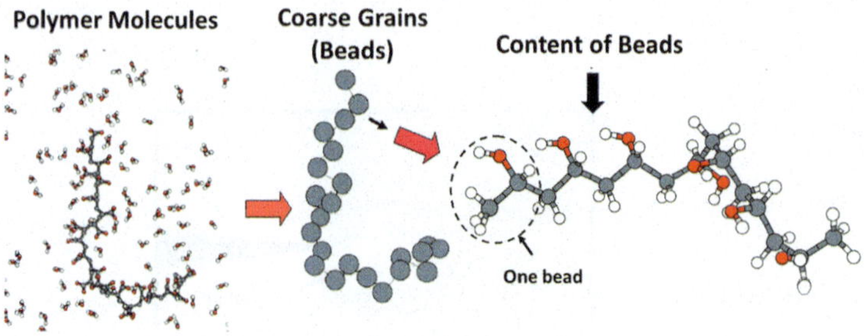

Fig. 10.22 Mapping of the atomistic polymer to a coarse-grained bead-spring model. One bead represents one chemical repeat unit. Bridging the length and time scales (from nano to macro)

10.11 Concluding Remarks

In conclusion, the virtual testing approach has emerged as a powerful and effective method for obtaining fracture properties of various materials used in the industry. By applying a semitheoretical approach based on existing test data or advanced computational simulations and modeling techniques, such as molecular dynamics, researchers and engineers can gain valuable insights into the fracture behavior of materials without the need for expensive and time-consuming physical testing.

The virtual testing approach not only accelerates the understanding of fracture properties (i.e., fracture toughness and fatigue crack growth rate) but also enables engineers and analysts to determine the life of a composite component in a load-varying environment in a short time. As technology in computer simulation continues to advance, the virtual testing approach through the molecular dynamics approach will mature enough to be utilized by analysts to perform their life assessment analysis much faster.

Therefore, the virtual testing technique holds great promise in enhancing our ability to predict and design materials with tailored fracture properties, contributing to the advancement of materials science and engineering.

References

1. George C. Sih, "Multiscale Fatigue Crack Initiation and Propagation of Engineering Materials," Application of virtual testing for obtaining fracture allowables of aerospace and aircraft material, by B. Farahmand, Springer (2008)
2. B. Farahmand, Predicting fracture and fatigue crack growth properties using tensile properties, engineering. J. Fract. (2007)
3. B. Farahmand, D. Xie, F. Abdi, *Estimation of Fatigue and Fracture Allowables for Metallic Materials Under Cyclic Loading* (AIAA, 2007)
4. B. Farahmand, Chapters 6 (Fracture mechanics of ductile metals theory), in *Fatigue and Fracture Mechanics of High-Risk Parts*, (Chapman & Hall, 1996)
5. V.G. Rostiashvili, T. Vilgis, Statistical thermodynamics of polymeric networks, in *Encyclopedia of Polymeric Nanomaterials*, (Springer, 2014)
6. B. Farahmand, Chapter 1 (Virtual testing and its application in aerospace structural parts), in *Virtual Testing and Predictive Modeling*, (Springer, 2009)
7. B. Farahmand, Virtual Testing for Estimating Material Fracture Properties (Reducing Time & Cost of Testing), in 11th International Conference on Fracture (ICF11), Turin Italy, March 2005
8. B. Farahmand, K. Nikbin, Predicting fracture and fatigue crack growth properties using tensile properties. Eng. Fract. Mech. **75**(8), 2144 (2008)
9. NASGRO SOFTWARE OVERVIEW, Software Research Institute (SwRI), Version 8, 2016
10. X. Mi, N. Liang, X. Haifeng, W. Juan, Y. Jiang, B. Nie, D. Zhang, Toughness and its mechanisms in epoxy resins. Prog. Mater. Sci. **130**, 100977 (2022)
11. B. Farahmand, Fracture Toughness Determination (FTD), 2002. Available upon request: bahram1farahmand1@gmail.com
12. B. Farahmand, Fatigue Crack Growth Determination (FCG), 2002. Available upon request: bahram1farahmand1@gmail.com

13. B. Farahmand, Application of High and Low Cycle Fatigue on Metallic and Non-Metallic Material (Application of Fatigue for Engineers), Amazon eBook – Chapter 11 (2022)
14. P.K. Mallick, *Fiber-Reinforced Composites: Materials, Manufacturing, and Design* (CRC Press, 2007)
15. M.D. Aydin, A. Özel, Ş. Temiz, The effect of adherend thickness on the failure of adhesively-bonded single-lap joints. J. Adhes. Sci. Technol. **19**(8), 705–718 (2006)
16. A.Y. Shih, A. Arkhipov, P.L. Freddolino, K. Schulten, Coarse grained protein-lipid model with application to lipoprotein particles. J. Phys. Chem. B **110**, 3674–3684 (2006)
17. P.D. Dans, A. Zeida, M.R. Machado, S. Pantano, A coarse-grained model for atomic-detailed dna simulations with explicit electrostatics. J. Chem. Theory Comput. **6**, 1711–1725 (2010)
18. A. Arkhipov, Y. Yin, K. Schulten, Four-scale description of membrane sculpting by BAR domains. Biophys J **95**, 2806–2821 (2008)
19. S.O. Nielsen, C.F. Lopez, G. Srinivas, M.L. Klein, A coarse grain model for n-alkanes parameterized from surface tension data. J Chem Phys **119**(14), 7043 (2003)
20. R.E. Rudd, J.Q. Broughton, Coarse-grained molecular dynamics and the atomic limit of finite elements. Phys. Rev. B. **58**(10), R5893 (1998)
21. A. Arkhipov, P.L. Freddolino, K. Schulten, Stability and dynamics of virus capsids described by coarse-grained modeling. Structure **14**, 1767–1777 (2006)
22. A.Y. Shih, P.L. Freddolino, A. Arkhipov, K. Schulten, Assembly of lipoprotein particles revealed by coarse-grained molecular dynamics simulations. J. Struct. Biol. **157**, 579–592 (2007)
23. M.P. Allen, *Introduction to Molecular Dynamics Simulation*, John von Neumann Institute for Computing (NIC), NIC Series, vol 23., ISBN 3-00-012641-4 (2004), pp. 1–28
24. S. Chatterjee, B. Gersten, S. Thakur, A. Burin, Molecular dynamics simulation of a single stranded (ss) DNA. Nanotech **2**, 305 (2006)
25. J.E. Jones, On the determination of molecular fields. From the variation of the viscosity of a gas with temperature. Proc. R. Soc. Lond. A, Containing Papers of a Mathematical and Physical Character **106**(738), 441–462 (1924)
26. J.E. Jones, On the determination of molecular fields. —II. From the equation of state of a gas. Proc. R. Soc. Lond. A, Containing Papers of a Mathematical and Physical Character **106**(738), 463–477 (1924)
27. J.E. Lennard-Jones, Cohesion. Proc. Phys. Soc. **43**(5), 461–482 (1931)
28. Plimpton, S., Molecular Dynamics – Parallel Algorithms. sandia.gov
29. W.B. Streett, D.J. Tildesley, G. Saville, Multiple time-step methods in molecular dynamics. Mol. Phys. **35**(3), 639–648 (1978)
30. M.E. Tuckerman, B.J. Berne, G.J. Martyna, Molecular dynamics algorithm for multiple time scales: Systems with long range forces. J. Chem. Phys. **94**(10), 6811–6815 (1991)
31. G. Binning, C.F. Quate, *Atomic Force Microscope* (The American Physical Society, 1986)
32. F. Abdi, M. Garg, Chapter 2: obtaining material properties from the bottom-up approach, in *Characterization of Nanocomposites*, ed. by B. Farahmand,
33. S. Tsuzuki, H.P. Lüthi, Interaction energies of van der Waals and hydrogen bonded systems calculated using density functional theory. J. Chem. Phys. **114**, 3949–3957
34. B.J. Alder, C.E. Hecht, Studies in molecular dynamics. VII. Hard, sphere distribution functions and an augmented van der Waals theory. J. Chem. Phys **50**, 2032 (1969)

Chapter 11
Composites Under High-cycle Fatigue (Industrial Applications)

11.1 Introduction

In the initial stage of designing new structural hardware that must survive during its mission in a load-varying environment, there are several factors that designers and engineers should consider. A comprehensive trade study for different design configurations should be implemented, and based on strength and weight considerations, among others, the most suitable configuration must be chosen. The final selected design must withstand the load environment in question without failure, which can otherwise cause structural damage and loss of life. Therefore, a thorough structural assessment consisting of static, dynamic, fatigue, and fracture mechanics as well as an appropriate amount of coupon and full-scale testing (if necessary) will be conducted to ensure the reliability of the structure during its service usage. However, in this chapter, the emphasis will be placed primarily on the fatigue aspects of the design, necessitating that engineers must have a thorough knowledge of material behavior when it is subjected to both static and dynamic loadings [1]. Attention will be directed to the understanding of fatigue mechanisms of composite parts when they are constantly being exposed to the cyclical loading of different load magnitudes, which can cause crack initiation, followed by slow crack growth, and possibly failure of the structure during its operation.

The core of emphasis in this chapter will be on the laminated composite material. The fatigue behavior in metals is well understood, and there is no need to elaborate on it in this chapter. However, there are still some concepts in fatigue of metallic material that are necessary to discuss because they are common between both materials.

The importance of structural fatigue, namely the formation of cracks from the stress riser sites and subsequent crack growth that can cause eventual failure, cannot be overemphasized in nuclear industries, pressure vessels, aircraft, and aerospace structural parts where the safety of both the structure and loss of life are of the utmost concern. A comprehensive review of catastrophic failures that have occurred in the

B. Farahmand, *Fundamentals of Composites and Their Methods of Fabrications*,
https://doi.org/10.1007/978-3-031-80201-0_11

past century due to structural fatigue and fracture, is compiled in [2], where the emphasis is given to the aircraft, aerospace, ships, and pressurized tanks structures. In reference [2], several catastrophic flight events, leading to the failure of several passenger aircraft due to wings, fuselage, rudders, gas tanks, and landing gear failures, are discussed in detail for both metallic and nonmetallic parts.

This chapter has been written for engineers, designers, stress analysts, as well as material scientists who are concerned with the integrity of the structure under load-varying environments, where crack initiation and growth under the repeated load must be assessed not only from an analysis perspective but also being monitored by a suitable nondestructive inspection method for the status of their growth. It is estimated that in the U.S. economy, the annual cost of structural failure due to fatigue was around $100 billion in 1982 dollars [3, 4]. These costs arise from the fatigue failure of ground vehicles, rail vehicles, aircraft, bridges, cranes, power plant equipment, and other machinery equipment. Poor structural design practices, rough machining, dents, and scratches due to rough handling, lack of adequate inspection before and after their usage, the presence of a corrosive environment, and the presence of tensile as well as compressive residual stresses all can contribute to an accelerated crack growth that may result in premature catastrophic failure and possible loss of life.

Although crack initiation can lead to eventual crack growth and final failure, here, the discussions will focus only on the crack initiation part of the fatigue and the number of cycles associated with it. This is called traditional fatigue, in which engineers assume the structure is free from the crack during its usage, and it will take N number of cycles to initiate a visible crack. In general, the word fatigue in engineering is described as a material failure when subjected to repeated load cycles that could be thermal, mechanical, or sonic. This cyclic loading reduces the material integrity and gives it a finite life, especially when the load magnitude is high. For example, a composite helicopter blade may have a service life of not more than 5000 h and, after that, must go through a comprehensive inspection and possible repair if engineers decide to operate it for an additional number of hours. The topic of crack growth and final failure, which occurs after the crack initiation phase, has already been addressed by numerous authors under fracture mechanics, and the reader can refer to these references for a more detailed understanding of crack growth behavior under cyclic loading [5–7].

Metallic materials are isotropic and homogeneous and are both heat and electrical conductors. On the other hand, composite materials are different from their metallic counterparts. They are becoming increasingly valuable for many industries where weight is an essential factor in designing parts. These materials have nonhomogeneous, anisotropic properties and have problems that engineers must pay attention to when designing parts subjected to repeated impact, bending, and compression loading. This chapter presents the fatigue failure mechanisms of laminated composites with extensive S-N data under different loading conditions. This data can prove beneficial to the reader when distinguishing the fatigue behavior between composites and metallic materials. Later in Sect. 11.8 of this chapter, several fatigue models are briefly presented that can estimate the number of cycles to failure in composite parts when subjected to a load varying environment.

The engineers must remember that the nature of failure for composite materials is distinct from their metallic counterparts when subjected to cyclic loading conditions. Both aerospace and automotive industries are currently utilizing composite materials to design and manufacture their parts and are periodically being challenged to produce superior parts that can withstand environmental loads throughout their service usage. For the user of the composite part, it is vital to know the fatigue data in terms of stress magnitude and the corresponding number of cycles to failure. When the applied maximum cyclic stress is close to the ultimate of composite material, and the number of cycles to failure is still large, then fatigue does not influence the design of the composite structure. This is true especially when the cyclic load is tension–tension, which will be discussed next for several composite materials. Factors that influence the fatigue properties, such as the laminate stacking sequence, fiber and matrix properties, fiber volume fraction, interfacial bonding, and others, are also included in this chapter.

11.2 Composites Under Tension–Tension Cyclic Loading

Primarily, tension dominated fatigue does not pose an issue for laminated composite materials, and the residual strength (strength of material at the present of defects) is not significantly reduced. Under tension–tension fatigue, specimens can achieve as high as 10^6 cycles at load up to 90% of the ultimate strength, indicating that the tension fatigue of composites, even in the presence of holes, is not a major concern [8, 9]. Hence, the stress concentration in composites is not as critical as in metallic materials when subjected to the tension–tension cyclic load, with stress ratio, $R < 0$. In the tension–tension fatigue test provided by the ASTM standard D-3479 test method [10], anomalies are related to internal temperature rise and creep due to the presence cyclic load of mean stress, $\sigma_m > 0$, and the sustained load during the stop time which prolongs the creep effect. The mechanisms of slip and intrusion or extrusion that will lead to the failure of metallic material do not apply to composite materials because of the lack of plasticity, which is the leading cause of the slip mechanism [1].

For most composites, the S-N curve under the tension–tension load environment is almost flat if the specimen is free from preexisting damage. This implies that composites can resist cyclic loading under tension even if the load magnitude is high. Under the $\sigma_m > 0$ environments, such as the one shown in Fig. 11.1, where $R = 0$, it is difficult to determine the value of the endurance limit unless the infinite life is assumed to be above 10^6 cycles. This information is useful for engineers, since it will enable them to compare materials for infinite life. In Fig. 11.1, the S-N curve for high modulus carbon fibers in a polyester matrix with a volume fraction fiber of 40%, under the stress ratio of $R \sim 0$, for the tension fatigue environment is displayed. The S-N data for the case when the cyclic load is parallel to the fiber direction signifies that the curve is almost flat for many cycles [11]. Generally, by testing specimens at various fractions of the ultimate stress or strain, an S-N curve can be plotted, which

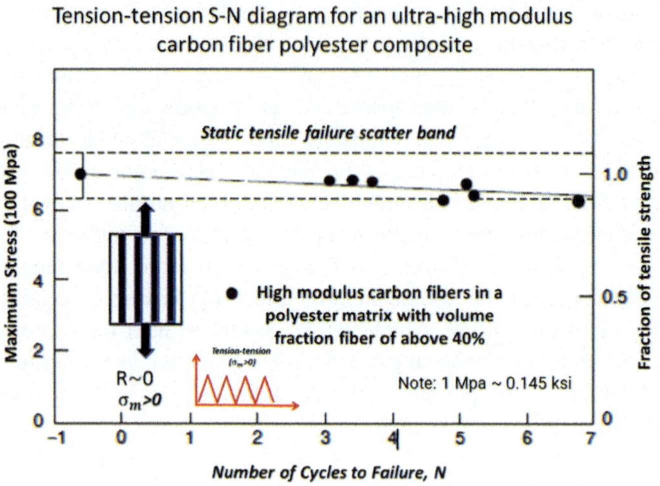

Fig. 11.1 In composites, the S-N curve under tension–tension is almost flat, and it is difficult to determine the value of endurance limit for infinite life. An ultrahigh modulus carbon fiber polyester composite material where *fibers are aligned with the applied load*

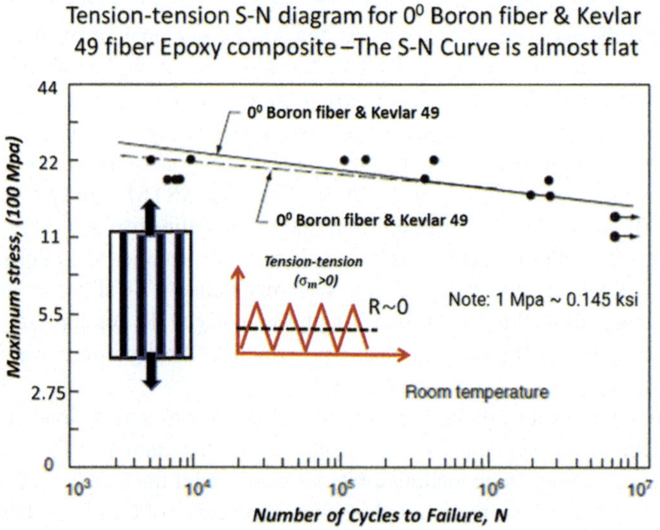

Fig. 11.2 The S-N curve under tension–tension is almost flat for *boron fiber and Kevlar 49 fiber epoxy composite. Here fibers are aligned with the applied load*

can provide information on the number of cycles to failure at a certain stress level before failure occurs. In Fig. 11.2, the S-N curve for Boron fiber and Kevlar 49 epoxy composite under tension–tension cyclic loading is shown. Just like in the case of Fig. 11.1, the S-N curve is almost flat, and it is difficult to determine the value of the endurance limit for infinite life.

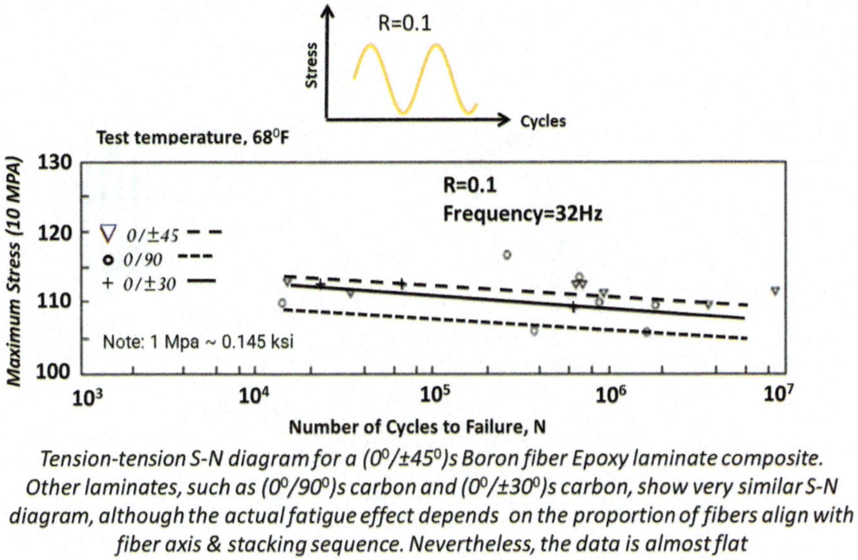

Fig. 11.3 The S-N data for boron fiber epoxy composites with three plies of different orientation under the tension–tension load (*0°/±45°*, *0°/90°*, *0°/±30°*)

In both figures, all the fibers in the composite layers were aligned in the direction of the applied load (0°) with a cycling stress ratio of $R \sim 0$. However, the S-N curves for other cases, where fibers are in different orientations related to the applied load, the S-N curves will manifest almost the same trends under the $\sigma_m > 0$ load environment as in Fig. 11.3 for the boron fiber epoxy laminate composite with (0/±45)s orientations. Other laminates, such as (0/90)s and (0/±30)s carbon epoxies, express almost identical S-N diagrams, all under the stress ratio of $R \sim 0.1$ [6]. The behavior of other stress ratios is not available for comparison. Note that the fatigue effect depends on the proportion of fibers aligning with the fiber axis and stacking sequence. The influence of the fiber volume in a ply is discussed later in this chapter. The situation depicted in Fig. 11.4 is different, and the data clearly states that the trend is unlike the tension–tension cases with almost flat behavior as shown in Figs. 11.1 and 11.2. In this figure, the S-N curve is no longer flat for the E-glass epoxy laminate, with load parallel to the fibers, when compared with carbon epoxy, and furthermore, there is an indication that the endurance limit is present for this material, almost 650 MPa. Again, the data were extracted under the stress ratio of $R \sim 0.1$, and the behavior of other stress ratios are unavailable for comparison [11]. Other loading cases besides the tension–tension cyclic loading were also examined. The fatigue curve for the bending load portrayed that the slope of the flexural S-N curve is greater than that of the tension–tension for high-modulus carbon fibers. The lower fatigue strength in flexure is attributed to the weakness of composites on the compression side of bending that can cause the eventual separation of layers (i.e., delamination) and premature failure of the composite material.

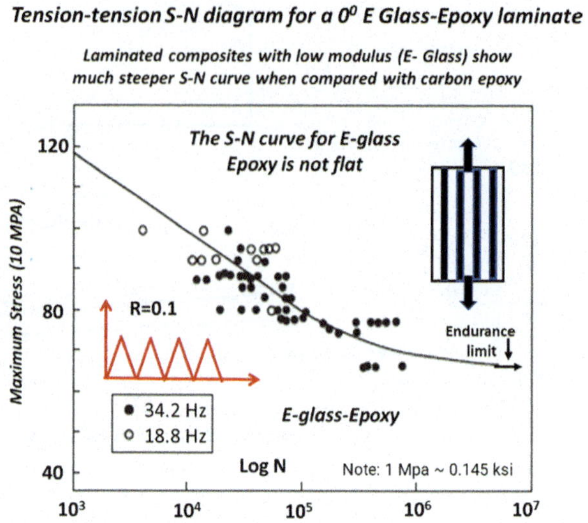

Fig. 11.4 *The S-N diagram for the E-glass-epoxy under tension–tension that behaved differently than the cases shown in* Figs. 11.1 *and* 11.2

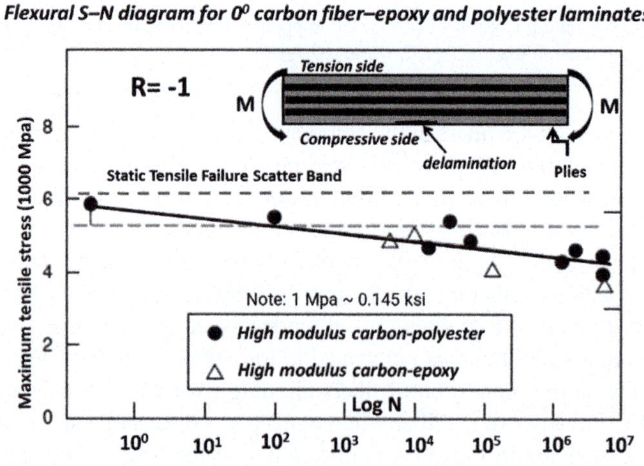

Fig. 11.5 The slope of the flexural S-N curve is greater than that of the tension–tension S-N curve for high-modulus carbon fibers. The lower fatigue strength in flexure is attributed to the weakness of composites on the compression side

Figure 11.5 reveals the Flexural S-N diagram for 0° carbon fiber–epoxy and polyester laminates where the drop in strength can be a consequence of the delamination caused by the compression due to cyclic bending [11]. The delamination mechanism and separation of plies described in Chap. 7 occurred due to impacts, out-of-plane loads, and compression–compression cyclic loading as well as a result of the case of $R = -1$ and bending. Of course, the mismatch between plies as well as the impact load can also contribute to the initiation of delamination.

11.3 Composites Under Compression–Compression Cyclic Loading

For fatigue tests involving compression–compression, the S-N curve is different from the tension–tension case. For composites under compression cyclic loading, the damage will emerge with the debounding and separation of plies leading to eventual material failure. In other words, the damage initiation is because of the delamination caused by compressive cyclic loading. However, for metallic materials, the damage was attributable to the applied tension causing the intrusion and extrusion damage, which resulted in crack initiation, crack growth, and final failure. Moreover, composites with stress concentrations, like a hole in a plate, are also susceptible to damage when under cyclic compressive–compressive load, where $R > 1$. Similar results were obtained for reversed cycle fatigue loading with $R = -1$, where $S_{min} < 0$ and $S_m = 0$, but the damage magnitude is less than compressive–compressive cyclic load.

In Fig. 11.6, three S-N curves with three different stress ratios ($R = -1$, $R = -1.6$, and $R = 10$) are shown [12]. Due to the repeated compressive load, the plies were separated, generating fatigue damage for all stress ratio cases where R is negative or larger than 1, as exhibited in the carbon fiber epoxy composite plate with [0/± 30]6s orientation. The damage initiation, or in this case delamination initiation, resulting from the compressive load is more pronounced in the case of $R = 10$, where both S_{max} and S_{min} are under the compression load (compressive–compressive case). It was concluded that the compressive stress component of a fatigue spectrum is the dominating parameter in damaging the composite part and reducing the corresponding number of cycles to failure, N_f, due to the delamination introduced damage compared to the tension–tension cases, where $R = 0$ or $R < 0$.

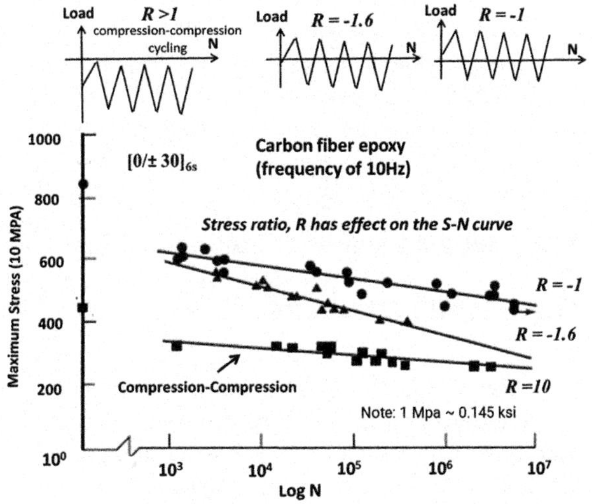

Fig. 11.6 The effect of stress ratio on life for composite material under compressive cyclic load. Delamination initiation and growth for the case of compression–compression cycling reduces the life of the components the most

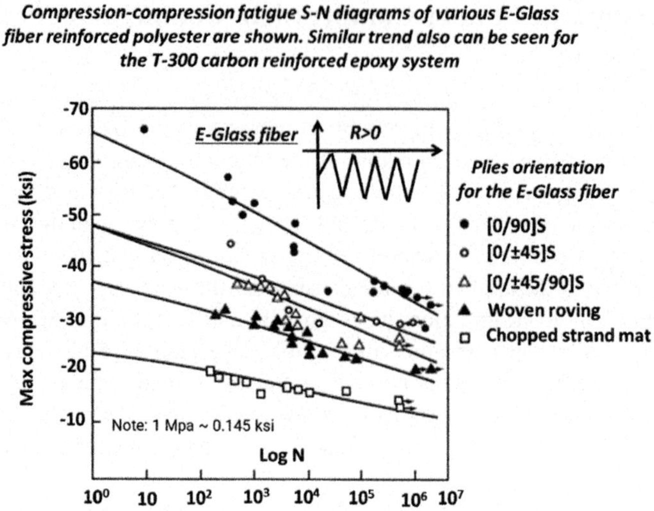

Fig. 11.7 Compression–compression S-N curve for various E-glass composite fibers with different orientations under $R > 0$

Figure 11.7 illustrates another compression–compression fatigue data for the E-glass fiber composites with the S-N data for different fiber orientations. It can be seen that the [0/90]s have a better fatigue property when compared with other orientations under the compressive cyclic load. The same S-N trend was observed on the T-300 carbon reinforced epoxy composites [1, 9]. Note that the 0° and 90° orientations have the orientations of fibers parallel and perpendicular to the applied load and fibers can resist failure. The next orientation that can benefit the composite fatigue properties is the [0/±45]s, where the S-N curve falls below the [0/90]s, as in Fig. 11.7. Moreover, the S-N data for the case of [0/±45/90]s orientation is very close to the [0/±45]s fiber arrangement, as shown in the figure. One would expect that the [0/±45/90]s orientation should be compatible with the case of [0/90]s plies arrangement. One of the explanations describing the data trends between the [0/± 45/90]s and [0/90]s could be the presence of ±45 plies that could cause to shorten the life due to mismatch between plies and possibly more delamination as the result [12].

11.4 Other Factors Affecting Fatigue Life of Composites

11.4.1 Plies Orientation Effects Under Fatigue Loading

The effect of fiber orientation on the fatigue properties of a laminated composite is an important factor when the load environment is cyclic in nature. In Fig. 11.7, the orientation effect under compressive–compressive was observed. In Fig. 11.8a, different fiber orientations were tested for $R = 0$, which indicated that the ±0°

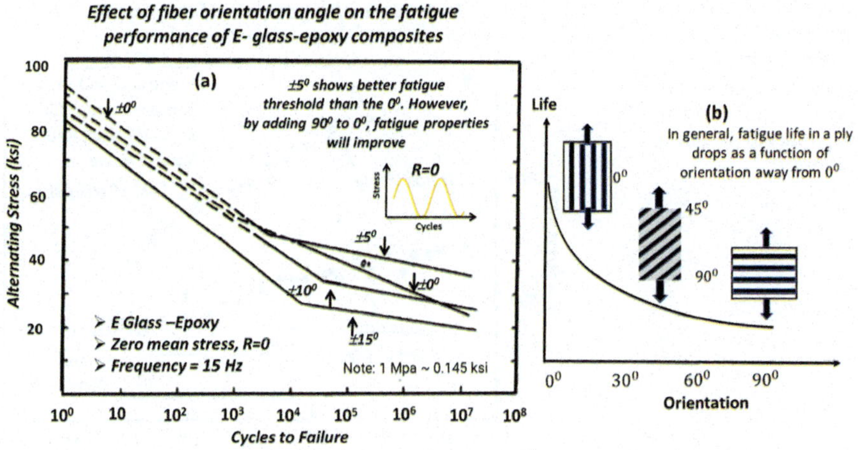

Fig. 11.8 (**a**) Effect of fiber orientation on the fatigue performance of E-glass composites under $R = 0$. (**b**) Effect of fiber orientation on the number of cycles to failure from 0° to 90°

orientation has better fatigue properties, but in the endurance limit region, the ±5° performs better. Furthermore, both ±0° and ± 5° have better fatigue properties than the ±10° and ± 15° orientations. The ±0° orientation is parallel to the fibers and should provide better fatigue curve when compared to all others. The case of ±5° plies orientations, with better endurance limit value, should be subjected to more testing for both orientations. That is, more test results are needed on both ±0° and ± 5° orientations, in the endurance limit region to check for the validity of the result shown in Fig. 11.8a [11]. Note that, near threshold region, the testing is costly and time-consuming, and engineers try to avoid recording data above 10^6 cycles.

In general, the fatigue life will be shorter when the plies' orientation departs from ±0° orientation. This is illustrated in Fig. 11.8b, where the number of cycles to failure as a function of ply orientation is maximum for the case of ±0° and will degrade as the orientation departs from 0° and fatigue life cycles becomes minimum when ply orientation reaches to ±90°.

11.4.2 The Effects of Square Wave Shape Cyclic Loading

Two carbon reinforced epoxy laminates subjected to the $R = 0$ cyclic load environment was compared on their fatigue properties. Higher fatigue strength for [−45/0/+45/90]2s orientations was observed as shown in Fig. 11.9 when compared with the [±45]4s plies orientation. This is due to the presence of 0° fiber in [−45/0/+45/90]2s plies since 0° carbon fiber is parallel to the direction of the applied load [13]. The nature of the applied load in Fig. 11.9 is different and it is termed the square wave shape cyclic load. In this format, the applied load holds a longer duration time per cycle which can cause the matrix to be exposed to the creep phenomenon in the

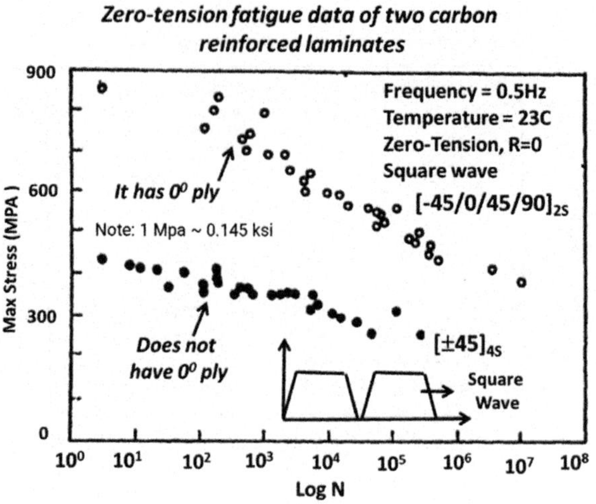

Fig. 11.9 Higher fatigue strength of [−45/0/+45/90] 2s over [±45]4s plies orientation is due to the presence of 0° fiber. 0° fiber is parallel to the direction of load

region if the applied load magnitude is high. The frequency of cycling is only 0.5 Hz, not high enough to create heat and damage to the composites. However, higher frequencies and longer holding time due to the square wave will cause heat to generate leading to lower number of cycles to failure. The effect of high frequency cycling will be discussed in the next section.

11.4.3 The Frequency Effects on Laminated Composite

The high frequency and long duration of the fluctuating load in polymers of a laminated composite can accumulate heat within the material. Due to the material's low thermal conductivity, the heat is not easily dissipated from the plies situated at the center, which, in turn, creates a temperature difference between the center and outer surfaces of the laminate. Variations in temperatures at the center and outside surfaces of the cross-ply in epoxy laminates during fatigue cycling can cause differences in thermal displacement and, hence, bring about the separation of plies (delamination) locally—which can be the source of damage initiation leading to final failure. The frequency effect in terms of heat rise versus the time of cycling is shown in Fig. 11.10a for cross-ply E-glass epoxy laminates and in Fig. 11.10b, the rise in temperature versus number of cycles during cycling at different frequencies for eight-ply ±45 CFRP laminate is shown. It is clear from this figure that a higher frequency, as high as 40 Hz, can raise the temperature of the laminates as high as 110°F. From this figure, high frequencies (20, 30, and 40 Hz) can raise the temperature of the polymer close to 110°F in such a short time (1000 s ~ 17 min). In terms of number of cycles there is not much change at 1 Hz, but the rise in temperature is clearly can be seen for 30 Hz (Fig. 11.10b). This rise in temperature due to high frequencies can induce regional creep in the polymer matrix material when the load

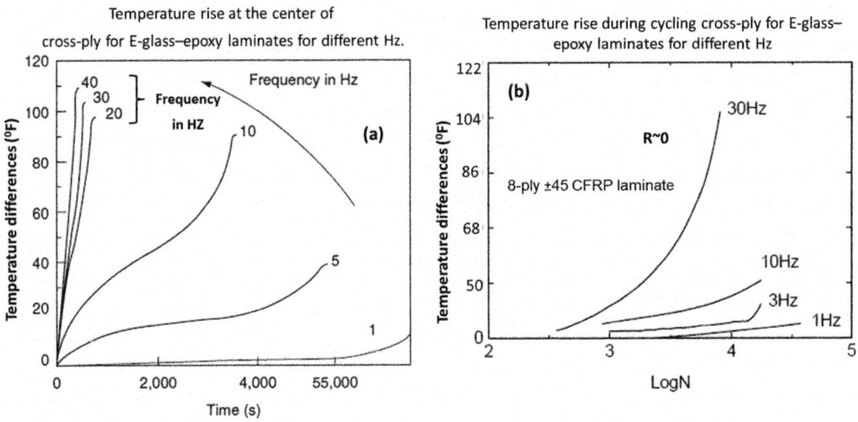

Fig. 11.10 (**a**) Difference in temperature rise at the center of cross-ply E-glass-epoxy laminates during fatigue cycling for different frequencies. (**b**) Rise in temperature during cycling at different frequencies

Fig. 11.11 Influence of test frequency on fatigue life for two laminates with different plies orientations ([± 45]4s versus [−45/0/+45/90]2s)

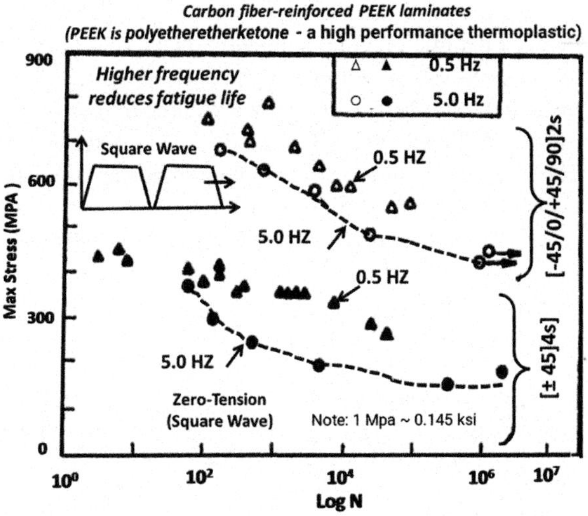

magnitude is high and the mean stress $\sigma_m > 0$. The creep effect will be more for internal plies that cannot dissipate heat to the surroundings and less for those plies that are exposed to the outside temperature. Combined variables, such as load frequency, plies orientation, plies stacking arrangements, and the shape of the wave all have an impact on the fatigue life of composite parts.

In Fig. 11.11, two carbon fiber reinforced laminates (PEEK) with different plies orientations were subjected to 5 and 0.5 Hz frequencies [14, 15]. The one with the ply orientation of [−45/0/+45/90]2s exhibited better fatigue properties than the [± 45]4s plies subjected to fluctuating loads of the same frequencies, 5 and 0.5 Hz.

However, in both cases, the fatigue properties of 0.5 Hz were superior to the 5 Hz, indicating that higher frequencies will reduce the laminated composites' fatigue properties by damaging internal plies that are incapable of dissipating heat and possible creep when subjected to sustain stress associated with high mean stress. However, this effect may not occur if the composite part is cycling at cold temperature environment so the heat would be dissipated to the surrounding.

In summary, the rise in temperature in laminated composites was found to be dependent on the laminate configuration and frequency of the applied cyclic load. For instance, the temperature rises in $[-45/0/45/90]_{2S}$ laminated configuration was only 20 °C above the ambient temperature and up to 150 °C in $[\pm 45]_{2S}$ laminates when both were subjected to the 5 Hz fluctuating load. A higher rise in temperature in the latter was attributed to their matrix dominated lay-up (i.e., orientations and the volume of fibers in the matrix). The fiber volume in the matrix and its influence on fatigue life will be discussed next.

11.4.4 Fiber Volume, V_f, Effect Under Fatigue Loading

Another parameter influencing the S-N curve for the laminated composite material is the amount of fiber volume, V_f, in each ply. The fiber volume fraction is the ratio of the volume of fibers over the total volume of the composite or the percentage of fiber by volume in a lamina. The same is true with respect to the volume fraction of resin. The volume fraction of fibers in a composite laminate is derived from the fiber-to-resin weight ratio. In Fig. 11.12, three S-N curves are presented with different V_f values for the E-glass epoxy laminates all in the 0° parallel to the applied cyclic load. From this figure, it is clear that the case of $V_f = 0.5$ (50%) has much better fatigue properties when compared with $V_f = 0.33$ or 0.16. However, in the endurance limit

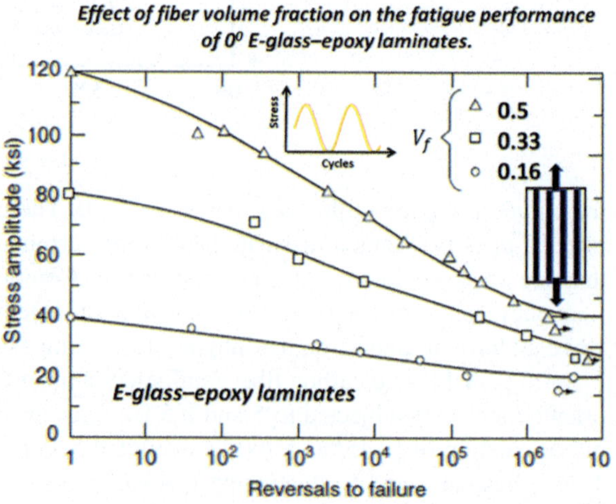

Fig. 11.12 The influence of fiber volume on the S-N curve for E-glass epoxy laminates

region, all three S-N data are closer, and the gaps between the three curves are smaller. The reason for this behavior is because at the low load region of the S-N curve, the influence of fiber volume, V_f, as a load carrying element is not as noticeable when it is compared with the high stress magnitude close to the ultimate of the material. The presence of fibers in the matrix with strong bonding between the fiber and the matrix will be effective in delaying fatigue failure [8, 12].

11.4.5 The Effect of Mean Stress and Constant-Life Diagram

For the fatigue assessment of structural parts and because of the deficiency of fatigue data for cases other than the stress ratio $R = -1$, it is vital to consider the concept of constant life diagrams. The constant life diagrams are a family of curves generated for different fatigue life; it is a design tool that represents the locus of stress states corresponding to a given number of cycles to failure [16, 17].

There are different ways that test data can be represented to engineers, such as the variation of S_{max} versus S_{min}, S_{max} versus S_m, or S_a versus S_m, where S_m and S_a are the mean stress and amplitude stress, respectively. Fatigue test data for the case of $R = -1$ is easily generated by using the Moore rotating machine. The Moore testing machine is designed based on the rotating beam principle. The test specimen acts as a simple beam symmetrically loaded at two points. When the specimen rotates one-half revolution, the stresses in the material initially below the neutral axis are reversed from tension to compression and vice versa, and it simulates the case of $R = -1$ [15]. However, industries need to have fatigue data for different stress ratios, R. The constant life diagram (i.e., the variation of S_a versus the S_m) is a valuable plot that is extremely helpful to engineers. This diagram was first established by J. Goodman [1] that can easily convert data from $R = -1$ with $S_m = 0$ to another format where $R \neq -1$ ($S_m \neq 0$), which will prove beneficial to the analyst. The effect of mean stress, σ_m, and the stress amplitude, σ_a, on the fatigue life of an E-glass reinforced polyester laminate was investigated. The constant-life diagram for this material was established, which is depicted in Fig. 11.13. In this figure, the stress amplitude, σ_a, versus the mean stress, σ_m, is plotted for a different life to failure, where $N_f = 10^3$, 10^4, 10^5, and 10^6 cycles. It must be noted that at each failure point on the constant life diagram curve, there is an assigned value of σ_a and σ_m. To the right of σ_m axis, when $R = 1$ and $\sigma_a = 0$, all the curves will be shifting toward $\sigma_m = \sigma_{ul}$ (ultimate of the material). Moreover, two stress ratio axes are assigned to this diagram: one is for the case of $R = -1$, where $\sigma_m = 0$, and the second is for $R = -4$ with negative mean stresses, as shown in Fig. 11.13. Near $R = -4$, there is a maximum and then a small drop in the stress amplitude value, σ_a, indicating that the presence of compressive mean stress of $\sigma_m < 0$ may have a reduction effect on the constant life diagram of E-glass composite materials. The behavior of other laminated composite materials must undergo an investigation to verify if the same trend is also recorded when $\sigma_m < 0$.

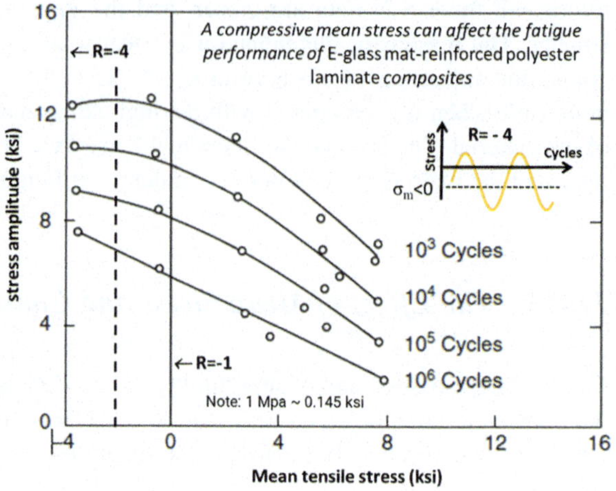

Fig. 11.13 Effect of mean stress on the fatigue stress amplitude of an E-glass mat-reinforced polyester laminate. To the right of σ_m axis, when $R = 1$ and $\sigma_a = 0$, all the curves will be shifting toward $\sigma_m = \sigma_{ul}$ (ultimate of the material). The effect of negative mean stress is shown in this figure. Different cycles to failure for constant life diagram is shown

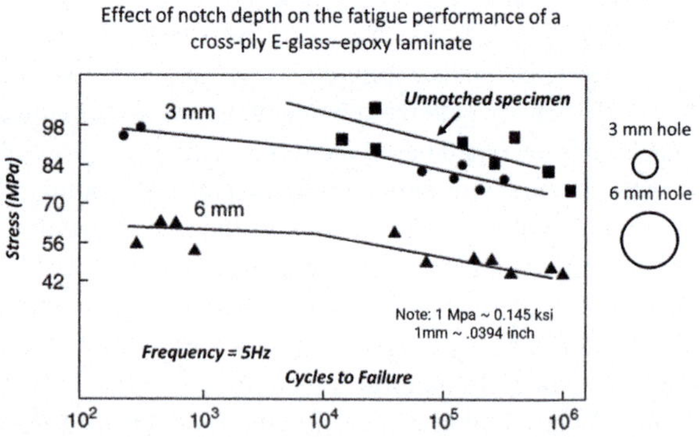

Fig. 11.14 The fatigue strength of a fiber-reinforced polymer decreases with increasing notch depth. Here, the S-N curves of notched laminated composite with 3 mm depth have better fatigue life when compared with 6mm depth, but in both cases, they fall below the unnotched case

11.4.6 Effect of Notch/Hole in the Composite Material Under Fatigue Loading

Other parameters influencing the fatigue life of the laminated composite include the presence of the notch in the material [18, 19]. The fatigue strength of a fiber-reinforced polymer decreases with increasing notch length and depth. Apart from this, its fatigue strength also decreases with increasing notch tip sharpness. The stacking sequence also plays an important role in the notch effect in fiber-reinforced polymers [13]. The effect of notch depth on the fatigue performance of a cross-ply E-glass epoxy laminate can be observed in Fig. 11.14, where S-N curves (unnotched and notched cases) were compared with each other. It is clear from this data that the unnotched S-N data expresses higher fatigue life when compared with the other two notched S-N curves (all under the 5 Hz frequency range). Furthermore, by looking at the two notched fatigue test cases, with 3 mm and 6 mm notched depth, the smaller notch size (3 mm) presents a better fatigue life. In both the notched test specimens, the drop in fatigue strength of the E-glass epoxy laminate up to 10^6 cycles was only around 15% of the original strength.

Finally, the effect of the notch sharpness and geometry on the fatigue performance of a cross-ply E-glass epoxy laminate was investigated for several notch geometries with varied sharpness. The outcomes of the S-N data for three geometries subjected to the 5 Hz frequency are shown in Fig. 11.15. In this figure, a 6 mm hole, a 2 mm hole, a 4 mm slot, and a 6 mm slot were subjected to the same cyclic loading. By observing the data presented in the figure, the fatigue life of the 6 mm slot is lower than the other two cases. Generally, many fiber-reinforced polymers exhibit much less notch sensitivity under a fatigue load environment than conventional metals, but they show weakness under compressive fatigue environments, as was shown in Fig. 11.6.

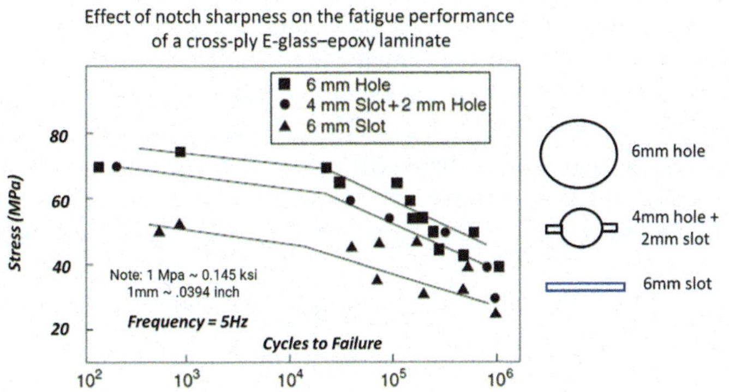

Fig. 11.15 The S-N curve for three different notch geometries were compared (a 6 mm hole, 2 mm hole plus 4 mm slot, and a 6 mm slot) to examine the influence of notch sharpness

Fig. 11.16 Two S-N cases were compared with each other. The long fibers are shown to have better fatigue properties, $R = 0.1$. (Source Harris et al. 1990)

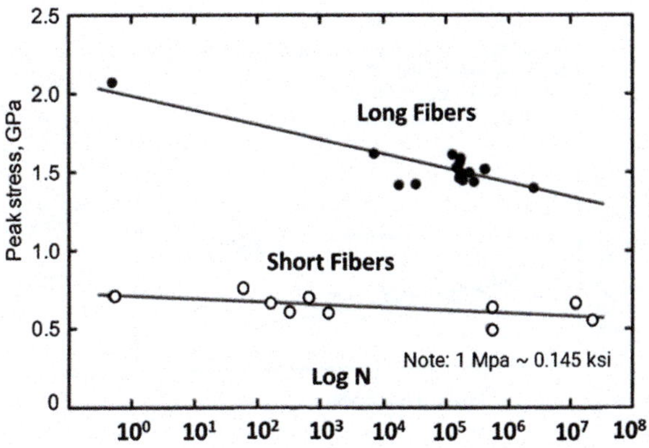

11.4.7 *Effect of Continuous and Discontinuous Fibers*

In Fig. 11.16, two cases of XAS/914 carbon/epoxy laminated composites were compared. One with long continuous fibers and the second one with well-aligned discontinuous fibers that were chopped around 3 mm long. The results indicated that the unidirectional long fibers have superior fatigue properties with both under the stress ratio of $R = 0.1$ as indicated in the figure.

11.5 Formula Describing the S-N Fatigue Data

An attempt is made to describe the S-N data for various fiber reinforced composites by a formula so that engineers will be able to estimate the number of cycles to failure with a simple calculation. Figure 11.17a illustrates the scattered S-N data that were compiled and plotted in a log–log format. All the points along the line represent a failure point due to the fluctuating load with a constant stress ratio. The plot usually shows a straight line (Fig. 11.17a) that can be formulated by an equation of a line like $Y = mx + b$, where Y represents the magnitude of constant amplitude load, x is the number of cycles to failure, m represents the slope of the line, and b is the constant. Because the x and Y axis are in log–log format; therefore, the line can be written as:

$$\begin{aligned} Y &= mx + b \rightarrow \text{or in terms of } S \text{ and } N \\ \text{Log}S &= m\text{Log}N + \text{Log}A \rightarrow \text{where Log}A = b \\ \text{Log}S &= m\text{Log}N + b \end{aligned} \tag{11.1}$$

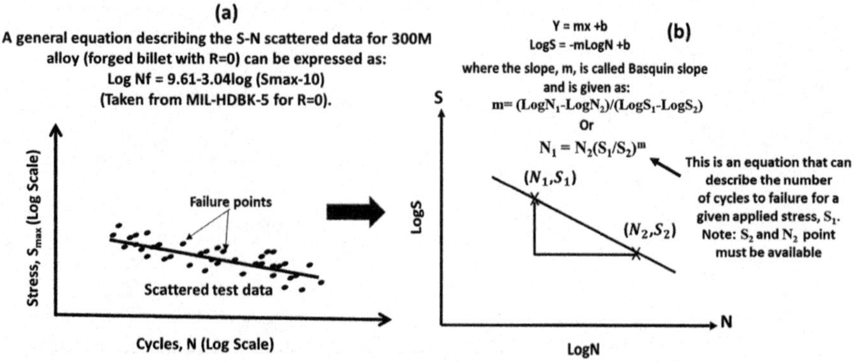

Fig. 11.17 (**a**) The S-N curve in log–log format and the equation that describes the fatigue data for 300M alloy (taken from Mil-HDBK-5 for R = 0). (**b**) The derivation of S-N curve using an equation describing a linear line ($Y = mx + b$)

Having two failure points on the line (N_1, S_1) and (N_2, S_2), the slope of the line in terms of number of cycles to failure and applied stress (Fig. 11.17b) can be written as:

$$m = (\mathrm{Log}N_1 - \mathrm{Log}N_2)/(\mathrm{Log}S_1 - \mathrm{Log}S_2) \tag{11.1a}$$

$$N_1 = N_2(S_1/S_2)^{\mathrm{m}} \tag{11.1b}$$

Equation 11.1b can describe the number of cycles to failure for any given stress S_1 when one data point on the line (the N_2 and S_2 data point) is available. For metallic material, the derived equation in terms of N_1 is often used by engineers when the slope m is available from Eq. 11.1a. For the composite material, the equation of the line describing the S-N data is similar to Eq. 11.1, that is:

$$\mathrm{Log}S = m\mathrm{Log}\,N + \mathrm{Log}\,\sigma_{\mathrm{U}} \quad \text{where } Log\ \sigma_{\mathrm{U}} = b \tag{11.2}$$

S = maximum fatigue stress, N = the number of cycles to failure, σ_{U} = average static strength, m = slope of the line, and b = Log σ_U. Eq. 10.2 has a known point on the failure line that has ultimate stress, σ_{U}, and number of cycles to failure of $N_{\mathrm{f}} = 1$. Figure 11.18 shows the m and b constants of S-N lines for different epoxy materials having different lay-ups. The stress ratio in all cases is approximately 0.1.

For example, in the case of E-glass-ductile epoxy with $R = 0.1$, the S-N constants of m and b can be written as:

$$\mathrm{Log}S = (-\,0.1573\mathrm{Log}N + 1.3743)$$

The S and N plot for E-glass-ductile epoxy is shown in Fig. 11.19a. As another example, the equation that will be able to describe the T-300 carbon–brittle epoxy, can be written (see Fig. 11.19) as:

$$\mathrm{Log}S = (m\mathrm{Log}N + b) = -0.0542\mathrm{Log}N + 1.0420$$

Where the ultimate strength for this material is 270 ksi. Figure 11.19b shows the S-N curve for the T-300 carbon–brittle epoxy material.

Material	*Lay-up*	*R*	*m*	*b*
E-Glass –ductile epoxy	**0^0**	**0.1**	**-0.1573**	**1.3743**
T-300 –ductile epoxy	**0^0**	**0.1**	**-0.0542**	**1.0420**
E-Glass –brittle epoxy	**0^0**	**0.1**	**-0.1110**	**1.0935**
T-300 carbon –brittle epoxy	**0^0**	**0.1**	**-0.0873**	**1.2103**
E-Glass –epoxy	**[0/±45/90]s**	**0.1**	**-0.1201**	**1.1156**
E-Glass –epoxy	**[0/90]s**	**0.05**	**-0.0815**	**0.9340**

Fig. 11.18 The constants describing the S-N curve for different epoxy materials and different lay-up. The stress ratio, R, is approximately is equal to 0.1

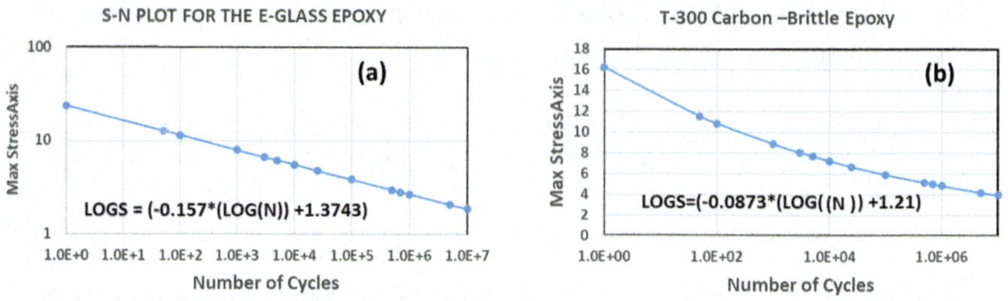

Fig. 11.19 (**a**) S-N curve for the E-glass-ductile epoxy with $R = 0.1$, $m = -0.1573$, and $b = 1.374$, (**b**) S-N curve for the T-300 carbon–brittle epoxy material with $R = 0.1$, $m = -0.0873$, and $b = 1.21$

11.6 Brief Description of Testing Method for the S-N Curve

Fatigue testing involves the application of a constant amplitude cyclic load to several test specimens, where the stress versus the number of cycles to failure is recorded. Unlike monotonic tests in which loading increases until failure, the applied load is cycled between prescribed maximum and minimum levels until a fatigue failure occurs, or until the predetermined number of loading cycles have been reached [1]. If failure does not occur within the specified number of cycles ranging from thousands to millions ($N = 10^6$) the test result is referred to as a run-out. Results from multiple fatigue tests using different cyclic stress levels are plotted as the S_{max}, versus the log of number of cycles to failure, N, and are known as an S-N diagram. All the fatigue test data scattered in one S-N plot are a result of several constant amplitude loading tests having the same stress ratio. That is, all the points on a plot have the same ratio of minimum stress to maximum stress, which is called stress ratio, $R = S_{min}/S_{max}$. For tension–tension fatigue testing, R values will be between 0.1 through 0.7, whereas for reversed tension-compression fatigue tests, $R = -1$. The case of $R = 1$ is assigned to monotonic load to failure where $N = 1$ and the situation where $R = \infty$ is the case with $S_{max} = 0$ and S_{min} is negative. Figure 11.20a shows different plots representing maximum stress versus the number of cycles with various stress ratios, R.

For each stress ratio shown in this figure, engineers can establish one S-N plot, as illustrated in Figs. 11.20b, c, where each S-N curve belongs to a given stress ratio, $R1$, $R2$, and $R3$. Also note that in the case of Fig. 11.20c, for a given stress S, the number of cycles will be $N1 > N2 > N3$ for $R1$, $R2$, and $R3$, respectively.

11.7 Other Parameters Affecting the S-N Diagram

When a composite part is exposed to a given environment, there is a strong possibility that material properties will be affected, and engineers must consider any degradation in properties into consideration. Engineers must study the

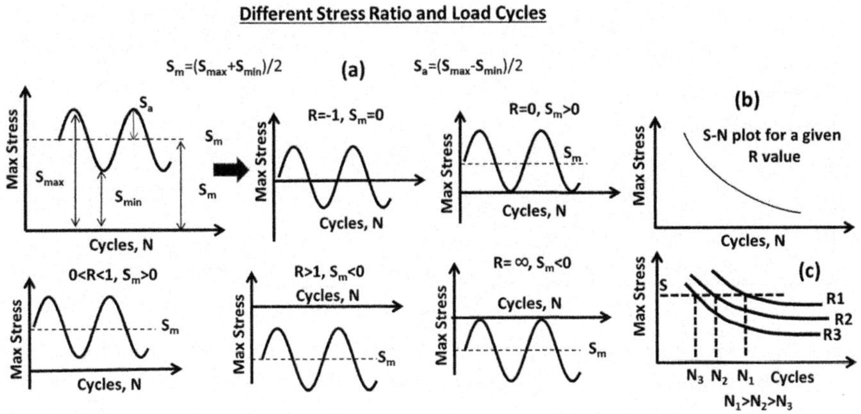

Fig. 11.20 Different stress ratio associated with different load cycles. Here, the cases associated with (**a**) $R = -1$, $R = 0$, $0 < R < 1$, $R > 1$, $R = \infty$, (**b**) shows the S-N curve for a given R, and (c) S-N curves for several R values are shown

environment and the suitability of whether composite parts should be exposed to that environment when designing the part. In addition to parameters influencing fatigue life, which were discussed in Sect. 11.4, the environmental effects such as moisture, temperature fluctuations, any possible radiation exposure, and perhaps the presence of chemicals will be briefly discussed here. These factors should be fully assessed by engineers before fabrication and during the operation of composite parts if the material is going to be exposed for a long time. For example, long exposure to fuel, hydraulic fluids, deicing fluids, paint, and detergents are environmental factors that can significantly change the mechanical properties of a composite material. The matrix, fibers, and the fiber–matrix interface will be all affected, but the degree and severity may be different and time dependent. The combination of humidity and exposure to above room temperature (causing matrix softening) can be detrimental because the matrix and the interface can be affected by the moisture diffusing in the material. The result would be lower properties than what engineers have estimated. These properties are stiffness, strength, fatigue, fracture toughness, and crack growth rate. One way that moisture can penetrate inside the composite part is through voids and porosities present in the composite material. Voids that are present in the material, because of the improper fabrication process, can absorb environmental moisture during parts service usage causing mechanical properties degradation due to their interaction with the matrix and interface.

Figure 11.21a illustrates two S-N curves, where one is exposed to moisture and the other has 0% moisture. The fatigue properties or the number of cycles to failure is reduced, but the degradation is more pronounced in the endurance limit region where the time to failure is higher. That is, at the lower part of the S-N curve, the exposure to the environment is higher because the number of cycles to failure is large.

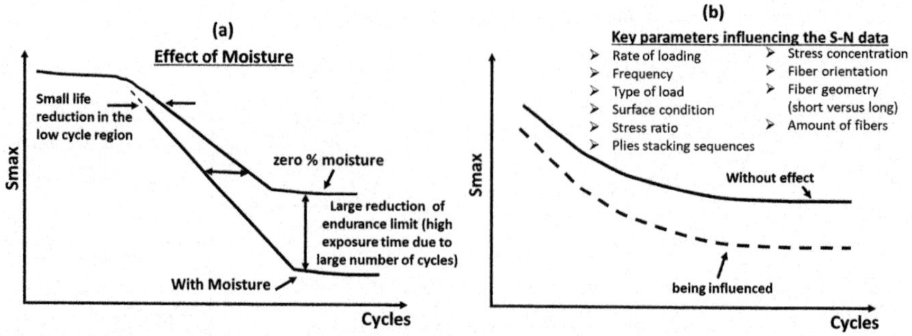

Fig. 11.21 Effect of environmental moisture on the S-N curve. (**a**) The effect is more pronounced in the endurance limit region. (**b**) Other parameters affecting the S-N data

Therefore, in the endurance limit region, the exposure time is high, and the possibility of damaging the composite material is also high.

The effects of moisture at elevated temperatures on the fatigue performance of a ceramic matrix composite (CMC) were investigated by Shankar Mall [20]. Silicon carbide fiber reinforced in silicon carbide matrix SiC/SiC with boron nitride (BN) interphase coating was used for this study. Two sets of tests with a stress ratio of $R \sim 0.1$ (tension–tension fatigue tests) at a frequency of 10 Hz and 750 °C were conducted. The first set of tests was under a dry environment with 0% moisture content and the second was under a humid environment with 60% moisture content. Fatigue life diagrams were developed under these two conditions. Moisture decreased the fatigue life at a prescribed stress level relative to that without moisture (as illustrated in Fig. 11.21a).

In another study [21], the effect of moisture content on the fatigue properties of the composite materials was also investigated. Random samples of woven fibers with polyester resin were used in the study. The test samples were submerged in water for different periods of time. Subsequently, the moisture content was measured. Results showed that the moisture content has a significant effect on the fatigue life of the composite test coupons. The fatigue life decreases as the moisture content increases. A procedure for determining moisture absorption properties through the thickness direction for laminated polymer matrix composite materials is described in ASTM D5229/D5229M-12. It is a standard test method for moisture absorption properties.

In addition to the moisture absorption effect and elevated temperature, a summary of parameters that can influence the fatigue behavior of composite materials are shown in Fig. 11.21b, and they are briefly being summarized here: (1) frequency, where high frequency will cause the softening of the matrix causing shorter life; (2) stress ratio, R, where structural life is shorter when the mean stress is negative because of possible delamination; (3) type of applied load (e.g., tension versus bending or compression); (4) fibers orientations; (5) low fiber volume; (6) fiber geometry (continuous versus short), (7) the surface conditions (the presence of scratches on the surface); (8) the rate of loading; (9) stress concentration factor, which all can influence the S-N data and can shorten the life of the composite part; and (10) changing the stacking sequences of plies (Fig. 11.9). For the metallic parts, stress concentration can reduce the number of cycles to failure much more than the composite material. Nevertheless, in all the abovementioned cases, the affected S-N curve will be shifted below the unaffected S-N curve having shorter life as shown in Fig. 11.21b [1].

11.8 Analytical Models Describing the S-N Data

Cyclic loading of laminated composites can cause damage in the structure that can start from matrix cracking, damaging the fiber–matrix interface, and fiber breakage. Damage can be accumulated from different locations in the composite part that eventually will merge and cause degradation of the mechanical properties, leading to the final failure. In this section, several fatigue modeling is proposed [22, 23] that can be useful to the analyst when performing fatigue analysis of E-glass fiber–epoxy composite. In both fatigue models, for the model to be helpful, it requires conducting a few tests to find constants in the S-N model, which would enable the analyst to obtain the number of cycles to failure for a given stress value. Even though the detailed derivation of the formula is not presented here, the model in its final format can be useful to generate the S-N data for the composite material when subjected to tension–tension cyclic loading.

There are several fatigue models that can provide the life of fiber reinforced composite materials, and they all are categorized into: (1) fatigue life models based on the S-N curves and fatigue failure criteria; (2) the residual stiffness or strength models, where the stiffness or strength will degrade as a function of time or cycles; and (3) the progressive fatigue damage models. In the later case, the mechanical properties of composite part degrade as the failure progresses. Two models will be presented here that are based on the residual stiffness or strength models. Other studies related to fatigue modeling are available for review in literature. More fatigue models based on the reduction of residual strength under cyclic loading will be briefly discussed in Sect. 11.8.3.

11.8.1 Fatigue Damage Modelling of Continuous E-Glass Fiber/Epoxy Composite —1

This model was proposed by R. B. Toumi et al. [22] where both static and fatigue tension tests have been performed on continuous E-glass fiber reinforced epoxy composite for obtaining the constants. These tests were followed by acoustic emission records, microscopic observations, and stiffness monitoring with an extensometer to investigate damage mechanisms occurring under both static and fatigue loadings. These data were used to identify the modes of damage and its effects on the macroscopic behavior of the composite material under cyclic loading. Subsequently, the fatigue model was generated and used to predict the fatigue life of the composite under tension–tension cyclic tests. This model drastically reduces the fatigue tests required to obtain S-N data. Predicated results were in good agreement with the experimental data. Furthermore, the effect of both the frequency and stress ratio are accounted for in this model.

The material under study was an E-glass/epoxy woven fabric composite with a fiber volume fraction of 54%. Mechanical properties of the material were obtained through mechanical tests. Test specimens are cut into the epoxy resin-based 2D glass fiber woven fabric and consist of 10 layers. Each ply had 97% of the glass woven fabric in the warp (longitudinal) direction (0°) while 3% in the weft (perpendicular) direction (90°). The mechanical properties of E-glass/epoxy woven fabric composite were obtained through testing and E_1, E_2, E_3, G_{12}, G_{13}, and G_{23} were obtained.

To model fatigue data, two options were available, the logarithmic equation of Log$S(N)$ = m*log(N) + b and the power law, $S(N) = A{*}N^B$ law, where S is the maximum applied stress, N is the number of cycles to failure and m, b, A, and B are constants. However, the power law models tend to provide improved fit at high cycles, as it was shown in [23, 24]. Based on the deterministic equations developed by Sendeckyj and Herzberg and Manson [25] and recalled by Epaarachchi and Clausen [26], the following equation for tension–tension cyclic load was obtained. Equation 11.3 has only two parameters α, and β that must be determined through experimental data.

$$\left[\frac{\sigma_u}{\sigma_{max}} - 1\right]\left[\frac{\sigma_u}{\sigma_{max}}\right]^{\gamma-1}\left[\frac{1}{(1-R)^{\gamma}}\right] = \alpha\left(N^{\beta} - 1\right)f^{-\beta} \tag{11.3}$$

The quantity γ = 1.6-sinθ, where θ is the smallest angle between the fiber direction and the load direction. For α and β parameters, a trial-and-error method was used. For each experimental data set the left-hand side of the Eq. (11.3) was plotted against the quantity $\alpha(N^{\beta} - 1)f^{-\beta}$ for a trial value of β. Note that all quantities shown to the left of the Eq. (10.3) are known (σ_u, σ_{max}, R, and γ) and to the right the two quantities α and β must be determined through trial-and-error. Figures 11.22 is an example of generating the S-N data for 0° and 90° for E-glass fiber/epoxy composite.

In summary, a semiempirical fatigue model for life prediction was established that includes the effects of stress ratio, R, frequency, f, and maximum applied stress. The model is based on the strength degradation of the composite material according to a power law, $S(N) = A{*}N^B$. The fatigue model predictions were found to be in good agreement with experimental data for tension–tension fatigue behavior as illustrated in Fig. 11.22 for E-glass fiber/epoxy composite. In the above equation, R = stress ratio and σ_u is the material ultimate stress.

An attempt was made to verify the accuracy of Eq. (11.3) shown above by using an independent data related to E-glass fiber/epoxy. From Fig. 11.8, two data points were selected for E-glass fiber with epoxy matrix composite having f = 15 Hz and 0°

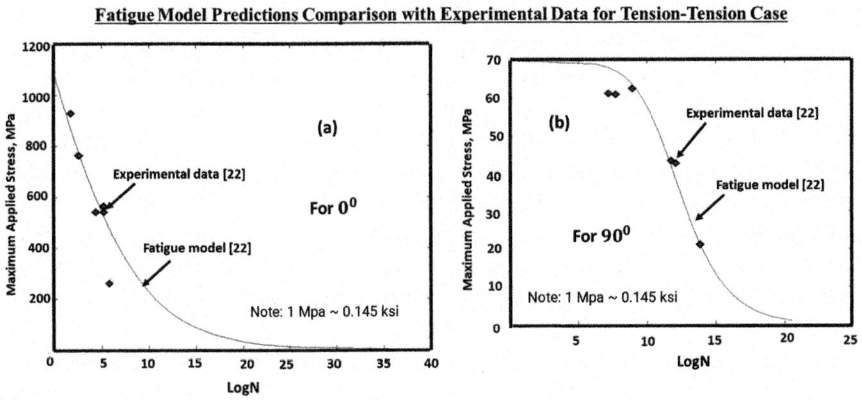

Fig. 11.22 (**a**) The fatigue model prediction plots under tension–tension loading at 0°. (**b**) Fatigue behavior under tension–tension loading at 90°. In both cases good agreement between test data and the model for E-glass fiber/epoxy can be seen

orientation. The first point will be when $N = 100$ with $\sigma \sim 70$ ksi (~483 MPa) and the second point $N = 1000$ with $\sigma = 58$ ksi (~400 MPa) and in both cases $R = 0$. Applying these two points, the two constants α and β of Eq. 11.3 can be obtained ($\alpha = 64$ and $\beta = 0.025$). Therefore, Eq. 11.3 can be written as,

$$\left[\frac{\sigma_u}{\sigma_{max}} - 1\right]\left[\frac{\sigma_u}{\sigma_{max}}\right]^{\gamma - 1}\left[\frac{1}{(1-R)^{\gamma}}\right] = 64.2\left(N^{0.025} - 1\right)f^{-0.025}$$

To check for any other applied stress point on the S-N curve, such as $\sigma \sim 82$ ksi (~566 MPa), the number of cycles was calculated as 34 cycles versus 25 taken from Fig. 11.8.

11.8.2 Fatigue Prediction Modeling of E-Glass Fiber/Epoxy Composite 2

Another empirical model was presented by J. A. Epaarachchi et al. to address the fatigue behavior prediction of glass fiber-reinforced plastic composites. Their model also includes various stress ratios, and cyclic load frequencies, *f*. The fatigue behavior of polymer matrix composites has been shown to be highly dependent on the stress ratio, *R*, and the frequency of applied cyclic load, *f* (see Figs. 11.6 and 11.11). In the previous model formulated by R. B. Toumi et al. [22], described in Sect. 11.8.1, the fatigue life of glass fiber-reinforced plastic composites was modeled that considered the effects of load stress ratios (*R*) and load frequencies (f). The Epaarachchi's model covers several parameters, and it requires several experimental test data to establish an equation that can characterize fatigue curves for the composite material under consideration. Fatigue curves for different stress ratios with $R > 0$ (compressive–compressive cycles) and $R < 0$ (tension–tension) is covered in this model. Available fatigue data from the literature and research laboratories were used to test the validity of the model. Predictions were found to be in good agreement with experimental data where the influence of test frequency as well as stress ratio on the fatigue life of composites were verified. Interesting observations were made by Mandell and Meier [27, 28], where they show the increase of cyclic load frequency, *f*, can result the fatigue crack propagation rate decreases and in doing so there is an increase in the fatigue life of the polymer provided that the increase in polymer temperature is prevented in an air-cooled environment. A good example of this is the estimation of fatigue life of small composite wind turbine blades when they are subject to high frequency loading in an air-cooled surrounding. Otherwise, the rise of temperature, as a result of high load frequency, will have an opposing effect on life because of degradation in the mechanical properties of matrix polymer (softening of matrix), see Fig. 11.11.

With the fatigue model proposed by J. A. Epaarachchi et al., the glass fiber reinforce polymer composites that has both the stress ratio, *R*, and testing frequency, *f*, as independent variables were examined. The new model shows good agreement between predictions and experimental data for a wide range of glass fiber reinforced polymers. The equation describing the degradation of applied stress (σ) as a function of time dσ/dt (where $t = N/f$, e.g., 10 Hz frequency means 10 cycles/second) can be written as:

$$\frac{d\sigma}{dt} = -a\sigma_{\max}\cdot(1-R)^{\gamma}\ t^{-k} \tag{11.4}$$

In Eq. (11.4), the quantities a, σ, t, $\sigma_{\max}$, R, and γ are the applied stress, time of cycling for a given Hz, maximum applied stress, stress ratio, and the set fixed of $\gamma = 1.6$ (derived from assumption on fatigue crack propagation rate), respectively, and a and k are constants. Equation (11.4) suggests that using a set of standard fatigue tests preferably $R = 0$ or 0.1 for tension–tension loading and $R = 10$ for compression–compression loading to be used for obtaining parameters.

11.8.3 Other Fatigue Models

The fatigue life of a composite part is directly related to the fatigue properties of its fiber, matrix, and the fiber–matrix interface. Failure in the composite material is difficult to assess because of parameters involve in laminated composite such as the stacking sequence of plies, the presence of fabrication faults in each ply (such as voids), lack of adhesion between plies, and many others, which they all can degrade the mechanical properties and contribute to the reduction of number of cycles to failure. A lot of research has been done on fatigue life modeling of composite materials over the past few years based on different techniques [29–32]. Two models

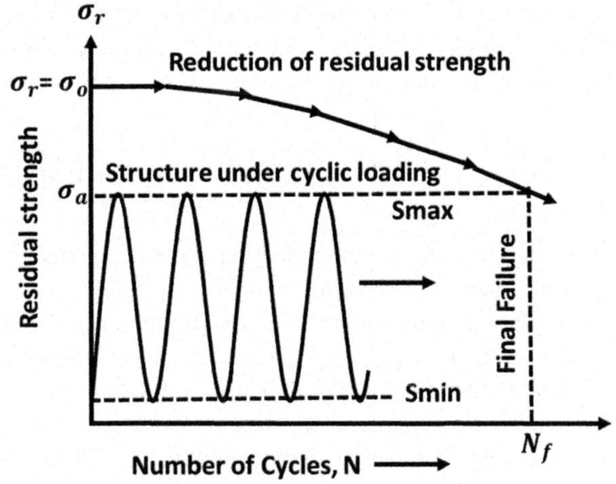

Fig. 11.23 Residual degradation of material as a function of applied cyclic load

were already addressed in previous sections and in this section another fatigue life modeling based on mainly the residual strength degradation philosophy will be presented. The degradation of residual strength under constant amplitude cyclic loading conditions is shown in Fig. 11.23. In this figure the residual strength capability of part before the application of cyclic load is equal to the ultimate strength of material ($\sigma_r = \sigma_o$ when there is no preexisting crack in the material), N_f is the number of cycles to failure, and σ_a is the maximum applied stress. A simple linear model has been proposed by Broutman and Sahu [32] based on the assumption of a linear decrease in residual strength in a fluctuating load environment which is given by:

$$\sigma_r = \sigma_o - (\sigma_o - \sigma_a)\left(\frac{N}{N_f}\right) \tag{11.5}$$

Equation 11.5 describes the reduction of residual strength starting from the ultimate of material σ_o, where $\sigma_r = \sigma_o$ to the point that $\sigma_r = \sigma_a$ in a cyclic load environment when final failure occurs and $N = N_f$. The S-N curve for a given stress ratio, R, is required to estimate the reduction of residual strength for all range of σ_a. For example, from Fig. 11.9, a data point on the S-N data of carbon fiber reinforced epoxy was used with $R = 0$ when $N = 1000$ at applied stress $\sigma_a = 700$ MPa (~101 ksi) and ultimate of $\sigma_o = 900$ MPa (~130 ksi), then the formula for the reduction of residual strength σ_r:

$$\sigma_r = 130 - (130 - 101)\left(\frac{N}{1000}\right) = 130 - 29\left(\frac{N}{1000}\right)$$

For example, for 500 cycles and under 101 ksi applied stress, the residual strength will be 50.5 ksi. Therefore, the reduction of material strength (residual strength before cycling) from $\sigma_o \sim 130$ ksi will reduce to 50.5 ksi after 500 cycles of applied cyclic stress $\sigma_a \sim 101$ ksi.

Equation 11.5 was later modified by Schaff and Davidson [33] by incorporating the strength degradation parameter and assuming failure would occur when residual strength equals the maximum stress. The model is given as follows:

$$\sigma_{r}= \sigma_{o} - (\sigma_{o} - \sigma_{a})\left(\frac{N}{N_{f}}\right)^{\nu} \tag{11.6}$$

where ν is the strength degradation parameter and it depends on initial strength, peak strength and stress ratio, R.

Another fatigue model like the Broutman and Sahu [32] model proposed by Halpin et al. (1973) [34]. The model assumes a monotonical decrease in the residual strength as the function of number of cycles, but its variation can be approximated by the power law equation. The Halpin model include a description of static failure resistance by lamination theory combined with energy-based fracture mechanics criterion for failure and was formulated as:

$$\frac{dR(n)}{dn} = \frac{-A(\sigma)}{m[(R(n)]^{m-1}} \tag{11.7}$$

where $A(\sigma)$ is a function of maximum applied cyclic stress σ, $R(n)$ is the residual strength after n number of loading cycles, and m is a constant. The term to the left is the rate of degradation as a function of number of cycles, n. Another fatigue related model was also proposed by Yang and Liu [35] titled "Life Prediction Methodology for Composite Structures." Their work was based on the residual strength degradation approach for graphite epoxy unnotched laminates. They argued that the residual strength decreases monotonically, and the slope of the residual strength is inversely proportional to the residual strength. Yang and Liu's model in terms of residual strength after n number of loading cycles as:

$$\frac{dR(n)}{dn} = \frac{-f(S, w, \varepsilon)}{cR^{c-1}(n)} \tag{11.8}$$

where S is maximum cyclic stress, w is loading frequency, $\boldsymbol{\varepsilon}$ is stress ratio. The model was developed by making frequency and stress ratio as a constant. The residual strength, (R^{c}), after n number of cycles is given as:

$$R^{c}(n) = R^{c}(0) - k\beta^{c}S^{b}n \tag{11.9}$$

The quantity $R(n)$ in Eq. (11.8) is residual strength after n number of loading cycles and c, b, and k are experimental parameters that have to be determined by test data taken from S-N curve. The failure occurs when the residual strength is equal to the applied stress and the number of cycles to failure, N_{f}, is given by:

$$N_{f} = [R^{c}(0) - S^{c}]/k\beta^{c}S^{b} \tag{11.10}$$

11.9 Concluding Remarks

The understanding of composite structural parts under high-cycle fatigue provides a path for enhancing our knowledge of the material behavior under fluctuating load environment. The complex interaction of several factors such as stress levels, loading frequencies (Hz), fiber volume (V_f), and environmental influences, such as environmental moisture and temperature fluctuation highlights the need for both testing and analytical investigation to improve the durability and performance of composites in various applications.

As engineers being exposed to the above challenges of high-cycle fatigue, continued exploration of innovative materials, advanced testing techniques, and robust modeling methods will be essential for designing reliable and fatigue resistance structures in various engineering applications. The insights gained from such investigations not only contribute to the progression of composite materials, but also hold the potential to shape the future of lightweight, high-strength materials in aerospace, defense and beyond.

References

1. B. Farahmand, *Application of High and Low Cycle Fatigue on Metallic and Non-metallic Material (Application of Fatigue for Engineers)* (Amazon eBook, Seattle, 2022)
2. B. Farahmand, *History of Structural Failures: Failures of Airlines, Fighter Jets, Rockets, Space Shuttles, and Pressurized Tanks* (Amazon eBook, Seattle, 2021)
3. J.J. Duga, W.H. Fisher, R.W. Buxbaum, A.R. Rosenfield, A.R. Burh, E.J. Honton, S.C. McMillan, *The Economic Effects of Fracture in the United States*, NBS special publication 647-2 (U.S. Department of Commerce, Washington, DC, 1983)
4. R.P. Reed et al., Atomic Transition Probabilities. NBS Special Publication 647–1, U.S. Department of Commerce, National Bureau of Standards, (Washington, DC 1983)
5. T.L. Anderson, *Fracture Mechanics: Fundamentals and Applications* (CRC Press/Taylor & Francis, Boca Raton, 2005)
6. D. Broek, *Elementary Engineering Fracture Mechanics* (Kluwer Academic Publisher, Dordrecht, 1986)
7. B. Farahmand, *Fatigue and Fracture Mechanics of High-Risk Parts* (Chapman & Hall, London, 1996)
8. B. Harris (ed.), *Fatigue in Composites: Science and Technology of the Fatigue Response of Fiber-Reinforced Plastics* (Elsevier Science, Amsterdam, 2003)
9. P.K. Mallick, *Fiber-Reinforced Composites: Materials, Manufacturing, and Design*, 3rd edn. (CRC Press, Boca Raton, 2007)
10. ASTM D3479/D3479M-19, *Standard Test Method for Tension-Tension Fatigue of Polymer Matrix Composite Materials* (ASTM International, W. Conshohocken, 2019) (Originally published in 1976)
11. P.K. Mallick, *Composites Engineering Handbook* (CRC Press, Boca Raton, 1997)
12. A. Baumann, J. Hausmann, Compression fatigue testing setups for composites—A review. Adv. Eng. Mater. (2020). https://doi.org/10.1002/adem.202000646
13. M.K.R. Hashim et al., The effect of stacking sequence on fatigue behavior of hybrid pineapple leaf fiber/carbon-fiber-reinforced epoxy composites. Polymers **13**(22), 3936 (2021)

14. A. Bernasconi, R.M. Kulin, Effect of frequency upon fatigue strength of a short glass fiber reinforced polyamide 6: A superposition method based on cyclic creep parameters. Polym. Compos. **30**(2), 154–161 (2008)
15. M. Huidong, B. Xuezong, R. Yawei, W. Xubing, A. Zongwen, *Modeling the Effect of Stress Ratio, Loading Frequency and Fiber Orientation on the Fatigue Response of Composite Materials* (National Library of Medicine, Bethesda, 2022)
16. N.J. Pagano, R.B. Pipes, The influence of stacking sequence on laminate strengths. Int. J. Mech. Sci. **15**, 679 (1973)
17. T. Adam, N. Gathercole, H. Reiter, B. Harris, *Fatigue Life Prediction for Carbon Fiber Composites* (University of Bath, Bath, 1991)
18. X. Mingrui, J. Huang, B. Zeng, Y.F. Dingni He, H. Deng, Effect of notch on static and fatigue properties of T800 fabric reinforced composites. Sci. Eng. Compos. Mater. **27**(1), 335–345 (2020)
19. A. Muc, P. Romanowicz, Effect of notch on static and fatigue performance of multilayered composite structures under tensile loads. Compos. Struct. **178**, 27–36 (2017)
20. S. Mall, Effects of moisture on fatigue behavior of SiC/SiC composite at elevated temperature. Mater. Sci. Eng. **412**(1–2), 165–170 (2005)
21. N. Hassoon, M. Al-Saadi, S.R. Almuradi, An expriemental investigation of effect on fatigue behavior of composite materials. Eur. J. Mech. Eng. Res. **3**(1), 1–18 (2016)
22. R.B. Toumi, J. Renard, M. Monin, P. Nimdum, Fatigue damage modelling of continuous E-glass fibre/epoxy composite. Procedia Eng **66**, 723 (2013)
23. J.A. Epparachchi, P.D. Clausen, A new approach to a fatigue damage model for glass-fiber reinforced plastic composites, in *Proceedings of the Seventh International Conference on Composites Engineering, ICCE/7, Denver Colorado USA 2000*, ed. by D. Hui, (ICCE, 2000)
24. J.F. Mandell, D.D. Samborsky, D.S. Cairns, *Fatigue of Composite Materials and Substructures for Wind Turbine Blades*, Contractor Report SAND2002-0771 (Sandia National Laboratories, Albuquerque, 2002)
25. G.P. Sendeckyj, *Fitting Models to Composite Materials Fatigue Data*, ASTM STP 734 (American Society for Testing and Materials, West Conshohocken, 1981)
26. R.W. Hertzberg, J.A. Manson, *Fatigue of Engineering Plastics* (Academic Press, New York, 1980)
27. J. Mandell, U. Meier, *Effects of Stress Ratio, Frequency, and Loading Time on the Tensile Fatigue of Glass-Reinforced Epoxy* (ASTM, West Conshohocken, 1983)
28. J. Degrieck, W. Van Paepegem, Fatigue damage modelling of fiber-reinforced composite materials: Review. Appl. Mech. Rev. **54**(4), 279–300 (2001)
29. A.P. Vassilopoulos, R.P.L. Nijssen, Fatigue life prediction of composite materials under realistic loading conditions (variable amplitude loading), in *Fatigue Life Prediction of Composites and Composite Structures*, (Elsevier, Amsterdam, 2010), pp. 293–333
30. N. Stojković, R. Folić, H. Pasternak, Mathematical model for the prediction of strength degradation of composites subjected to constant amplitude fatigue. Int. J. Fatigue **103**, 478–487 (2017)
31. J. Degrieck, W. Van Paepegem, Fatigue damage modeling of fibre-reinforced composite materials: Review. Appl. Mech. Rev. **54**, 279–300 (2001)
32. L. Broutman, S. Sahu, A new theory to predict cumulative fatigue damage in fiberglass reinforced plastics, in *Composite Materials: Testing and Design (Second Conference)*, ed. by H. Corten, (ASTM International, West Conshohocken, 1972)
33. J.R. Schaff, B.D. Davidson, "Life prediction methodology for composite structures," Part I—Constant amplitude and two-stress level fatigue. J. Compos. Mater. **31**, 128–157 (1997)
34. J. Halpin, K. Jerina, T. Johnson, Characterization of composites for the purpose of reliability evaluation, in *Analysis of the Test Methods for High Modulus Fibers and Composites*, ed. by J. Whitney, (ASTM International, West Conshohocken, 1973), pp. 5-5–60-5
35. J.N. Yang, M.D. Liu, Residual strength degradation model and theory of periodic proof tests for graphite/epoxy laminates. J. Compos. Mater. **11**, 176–203 (1977)

Chapter 12
Damage Accumulation of Composite Materials Under Fatigue Loading

12.1 Introduction

Failure can be defined as the instance when a part is no longer capable of withstanding the applied load during its service usage. This may result in considerable structural damage or complete structural separation and in many cases, loss of life [1]. The environmental accident during the part operation can cause damage initiation in the material and subsequent stable growth during service operation. In the case of composites, the source of environmental accidents can be lightning strikes, runway impact due to debris, tools drop (impact), impact during handling of the composite part, hail impact, and bird strikes. Upon the impact, the initial damage can be initiated and cause minor delamination and, in some cases, plies separation inside the part which can be seen only by the nondestructive inspections. The initial delamination can extend gradually, especially when the applied service load is compressive (like the case of $R = -1$), causing plies separation and final failure. The extent of the damage, both before or during the part's service usage, can determine whether the component needs to be repaired or replaced.

Under high-cycle fatigue, where the load magnitude is low and the number of cycles to failure is high, structural failure seldom occurs rapidly without some kind of prior indication that the part is failing. In reality, the gradual damage growth that eventually leads to final failure is due to the damage accumulation which starts with damage initiation, damage growth, and final failure. Damage accumulation in a cyclic load environment will require time to grow steadily until it reaches a critical stage causing catastrophic failure. In both metallic and composite materials, the failure is directly related to the magnitude of the applied load and the material's ability to withstand damage. A high load magnitude can shorten the life of the composite structure, and the lack of material fracture toughness will prevent stable crack growth. Lack of toughness and crack arrest mechanism in the material can lead to fast fracture when the critical crack size has been reached. Figure 12.1a clearly shows the initiation, progression, and accumulation of microscopic damages in

B. Farahmand, *Fundamentals of Composites and Their Methods of Fabrications*,
https://doi.org/10.1007/978-3-031-80201-0_12

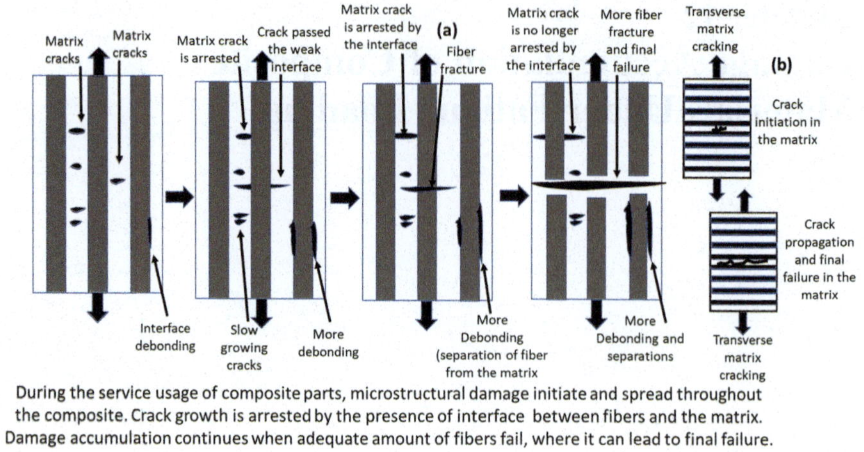

Fig. 12.1 (**a**) Microscopic crack initiation and growth in composites. Damages will accumulate during the structural life causing final failure. (**b**) Transverse crack in the matrix parallel to the fibers and perpendicular to the load is a common phenomenon in matrix reinforced composites

composite material, which started from matrix cracking, debonding, fiber failure, and more matrix and fiber damages that eventually will lead to the final failure of the composite part. The presence of fibers in the matrix brings a fail-safe feature to the structural parts because of the crack arrest behavior of the fibers bonding to the matrix, as illustrated in the figure. The crack arrest phenomenon in composites is more pronounced in ceramic matrix composites (CMCs).

The accumulation of damage can be a result of several load spectrums of different stress magnitudes. The stronger the matrix and fiber constituents of composite material, the longer life is expected. Failure of composite structures is a progressive series of events, where the crack initiation will generally start from the matrix, which is the weakest part of the composite, much weaker than the fiber. Note that carbon fiber material can have up to five times the strength of typical steel while having one-fifth of the weight [2]. Therefore, the initiation must start from the matrix and not the fiber. Moreover, the weak interface between fibers and the matrix can help the initiation and growth of cracks to take place in the matrix during the part usage without much crack arrest to take place. Transverse cracking is another type of matrix cracking that must be given special attention (Fig. 12.1b). In this case, the damage in a laminated composite is usually matrix microcracks [3], which are cracks perpendicular to the load and run parallel to the fibers in that ply (Fig. 12.1b). This usually happens when microcracking occurs in the 90° plies while axial loading is in the 0° direction. Because these microcracks are transverse to the loading direction, they are sometimes called transverse cracks. The terms matrix microcracks, or ply cracks, and transverse cracks, are found often used in composite literature. Microcracks can form in any plies, but they form predominantly in plies off-axis to loading directions. Fibers that are situated unparallel to the applied load will carry less resistance to the load with reduced tensile modulus.

The immediate effect of microcracks is to cause a degradation in the mechanical properties of the laminate, including a reduction in tensile moduli and Poisson ratios. If a given design cannot tolerate microcracks causing a reduction in properties, then the formation of microcracks constitutes a design failure. A secondary negative effect of microcracks is that they can create other forms of damage, such as delamination, which will open the pathways for entry of corrosive liquids. Such damage modes may subsequently lead to laminate failure. The processes by which microcracks form, their effects on laminate properties, and their role in initiating new forms of damage are all important problems in analyzing damage accumulation and failure of composite laminates. In addition to matrix cracking, other weak areas in the laminated composite can be in the interface between plies due to inadequate bonding, which may lead to plies separation (i.e., delamination), leading to final structural failure. Finally, the case of crack initiation, growth, and merging several cracks that are perpendicular to the applied load is another scenario that can happen in the structural parts.

12.2 Damage Accumulation and the Miner's Rule

When the damage in structures is caused by several fluctuating load environments (Fig. 12.2), the total damage will be based on the accumulation of damage from each load environment. Consider the case of an airplane where a section of the fuselage is subjected to several cyclic loadings consisting of the pressurization and depressurization of the fuselage, taxing, takeoff, gust, cruise, and landing. Each load environment can contribute to damage initiation and growth, which eventually adds up and contributes to total structural failure. However, engineers should realize that it is assumed no damage will be accounted for when the stress level is below the fatigue limit of the material. For materials without a recognized fatigue limit, damage can be calculated using the highest N value assigned to the part, which is often assumed to be ten million cycles.

For a constant amplitude load environment, when there is no variation in load magnitude, the damage accumulation can be found directly through the S-N data for the material under consideration. For example, for applied stress, σ_{max}, having a

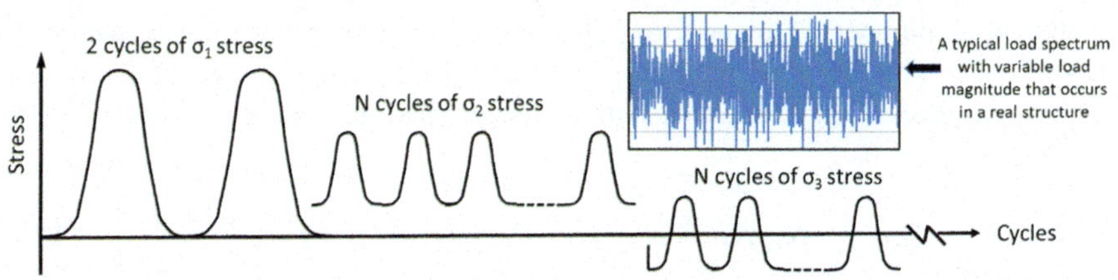

Fig. 12.2 Variable load magnitude that occurs in a real structure. Each has a different load magnitude, σ_i, and stress ratio $R = \sigma_{min}/\sigma_{max}$

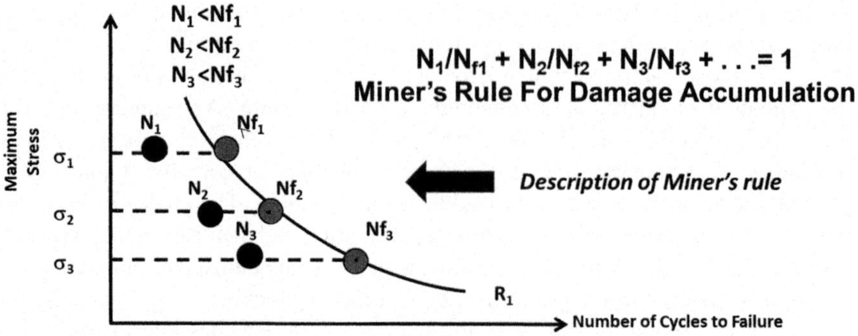

Fig. 12.3 Description of Miner's rule for damage accumulation from several load environment

constant stress amplitude, the S-N data can provide us with the total number of cycles to failure, N_f. Of course, the S-N data for the material must be available to obtain the total number of cycles to failure. However, for variable amplitude load environments, the corresponding total damage accumulation can be estimated via several S-N data with different stress ratios and the well-known Miner's rule (M.A. Miner in 1945). The first step in this process is to reduce and sort the variable amplitude load cycle history to a series of constant amplitude cycles defined by their amplitude and mean value using a cycle counting technique such as the famous rain-flow method. The step-by-step procedures for using a standard rain-flow technique are documented in ASTM E-1049.

An actual structure undergoes different numbers of cycles of varying amplitude loading. An example would be a bridge structure that periodically is subjected to different load magnitudes during its usage. Each load environment contains a different number of cycles of distinct load magnitudes. Figure 12.2 reveals a typical load-varying cyclic environment where each has its own number of cycles and stress magnitude. Miner's rule can be used to estimate the number of cycles to failure of a part in a cyclic environment with various load levels and stress ratios [4]. Miner's rule for damage accumulation is a linear sum of the contribution of each load environment as described in Fig. 12.3, where N_i and N_{fi} represent the contribution of the number of cycles and the total number of cycles to failure for the same load environment, respectively. For example, if the structure sees two cycles of σ_1 stress magnitude, then the contribution of this spectrum with $N = 2$ will be $2/N_f$, where N_f is the total number of cycles to failure corresponding to σ_1 stress magnitude (Fig. 12.3). With the same reasoning, for several load spectrums, each having σ_1, σ_2, σ_3, ..., stress magnitude, the damage accumulation, and Miner's rule can be written as:

$$N_1/N_{f1} + N_2/N_{f2} + N_3/N_{f3} + \ldots = 1 \tag{12.1}$$

Because of several cyclic loadings with different load magnitudes, the linear damage accumulation described by Eq. (12.1), is apparently not sensitive to the way loads are being arranged in the above equation. That is, it doesn't matter how we arrange them, the result will not change the outcome of the damage accumulation. In the case of metallic material, the high stress magnitude cycles will create plastic deformation at the concentration sites and will cause a retardation effect for the following cycles with lower stress values (i.e., slowing down the growth of subsequent cycles) [1, 5]. However, in the case of composites, the situation is different because most matrices and fibers are brittle, and they will not show any plasticity at the crack tip where the stresses are high. Therefore, the retardation effect for the metallic parts will not happen for the parts made of laminated composite.

12.3 The Modulus of the Elasticity Reduction Method for Damage Accumulation

The modulus of the elasticity reduction method in the composite material is employed for estimating the damage accumulation under cyclic loading. As the fatigue damage accumulates and progresses with load cycling of different magnitudes, the dynamic modulus of the material continuously decreases (Fig. 12.4). Dynamic modulus can be measured frequently during fatigue cycling without interrupting the test [6]. It is defined as the material's stiffness under dynamic or varying loads. This cyclic modulus reduction phenomenon [7] causes an increase in strain level if a stress-controlled test is used and a decrease in applied stress level if a strain-controlled test is used, where the gradual degradation of elastic properties, particularly the longitudinal stiffness, will occur. The stress-controlled test means the magnitude of applied stress will not alter during cycling. The variation of drops in dynamic modulus of $[\pm 45]_S$ boron–epoxy laminates are shown in Fig. 12.4 under the

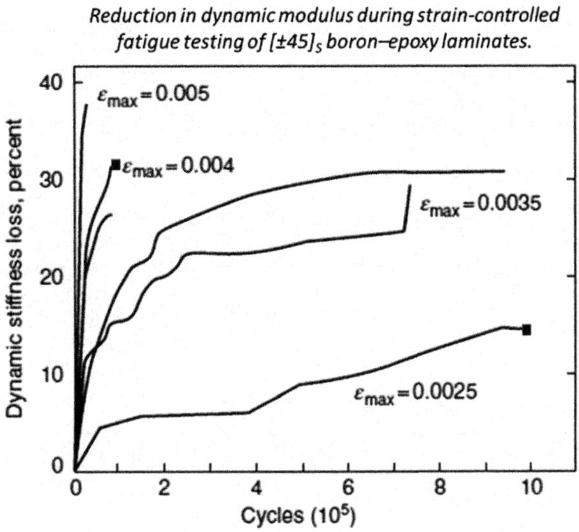

Fig. 12.4 The % dynamic modulus reduction data versus the number of cycles to failure under the strain control condition

strain-controlled condition [8, 9]. That is, the strain level during the test will remain the same. Several strain-controlled levels were considered for the data in Fig. 12.4, which corresponded to strain $\varepsilon_{max} = 0.005, 0.004, 0.0035$, and 0.0025 for $[\pm 45]_S$ boron–epoxy laminates. The variation of the percent dynamic modulus drop versus the number of cycles was recorded under the abovementioned strain-controlled conditions. For example, the number of cycles for the 32% dynamic modulus drop, for a constant strain of $\varepsilon_{max} = 0.004$, is almost 1.0×10^5 cycles, and for the 15% drop associated with a constant strain of $\varepsilon_{max} = 0.0025$ is ~10×10^5 cycles (Fig. 12.4).

The same type of data was generated under the stress-controlled condition, and the variation of number of cycles versus percent dynamic modulus drop for SMC-R65 composite material was plotted and depicted in Fig. 12.5. The sheet molding compound (SMC) has dispersing long strands (usually >1″) of chopped fibers, commonly glass fibers or carbon fibers, soaked on a bath of thermoset resin (typically polyester resin, vinyl ester resin or epoxy resin). Several plots were generated under the flexural fatigue loading for the SMC-R65 composite material. To describe the variation of the number of cycles versus the percent dynamic modulus drop, three data points corresponding to three different stress levels, as a percentage of material ultimate, were considered. These three data points correspond to the polyester matrix with stress levels of $S = \pm 0.56\sigma_{UF}$, $S = \pm 0.34\sigma_{UF}$, and $S = \pm 0.2\sigma_{UF}$ all as a fraction of the ultimate strength of the material, σ_{UF} [1]. An additional two data points with stresses as a fraction of the ultimate, $S = \pm 0.64\sigma_{UF}$, and $S = \pm 0.38\sigma_{UF}$ for the vinyl ester matrix, are also shown in Fig. 12.5. In the case of $S = \pm 0.56\sigma_{UF}$ (point a), the 20% reduction of module took place at around 5000 cycles and at 10% reduction of modulus at $N \sim 800$ cycles. In the case of $S = \pm 0.38\sigma_{UF}$ (point e), the 13% reduction of module took place at $N \sim 10^6$ cycles. In summary, the dynamic modulus reduction can be measured

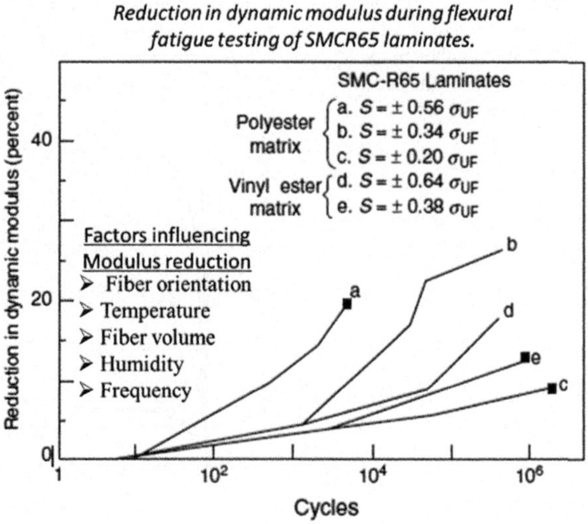

Fig. 12.5 The % dynamic modulus reduction data versus the number of cycles to failure under stress control condition

frequently during fatigue cycling without stopping the test. It is an excellent nondestructive test method and can be used to offer adequate alarm before the part becomes totally broken. In the same figure, factors that can influence the modulus reduction are: (1) fiber orientation, (2) environmental temperature, (3) fiber volume (V_f), (4) humidity, and (5) the frequency associated with the applied cyclic load.

Calculating the residual modulus (the percent modulus left), E_N, after N fatigue cycles in a stress-controlled fatigue test with a maximum stress level, S, by applying the Whitworth equation (Whitworth, H.W., ASME J. Eng. Material Technology [10]). It can be written as:

$$E_N = E_0 - H(E_0 - E_{ff})^a (N/N_f) \tag{12.2}$$

where

E_0 = Initial modulus prior to fatigue cycles,
E_N = Modulus after N number of fatigue cycles, where $E_0 > E_N$.
E_{ff} = Modulus at the time of fatigue failure (final failure). The quantity E_{ff} can be written as $E_0(S/\sigma_{ul})$, where σ_{ul} is the ultimate of the material and S is the maximum stress level. Here $E_0 > E_N > E_{ff}$.
H = Damage parameters determined by fitting the Whitworth equation to the experimental data.
Note: a is in the form of power shown by Eq. 12.2.
N_f = Number of cycles to failure under the maximum stress level S (note $N_f > N$).

Figure 12.6 expresses a good fit between the Whitworth equation and test data, where the dynamic modulus reduction ratio, E_N/E_0, for carbon fiber–epoxy laminates, as a function of fractional life, N/N_f, for $[\pm 35]_{2S}$ is plotted [10].

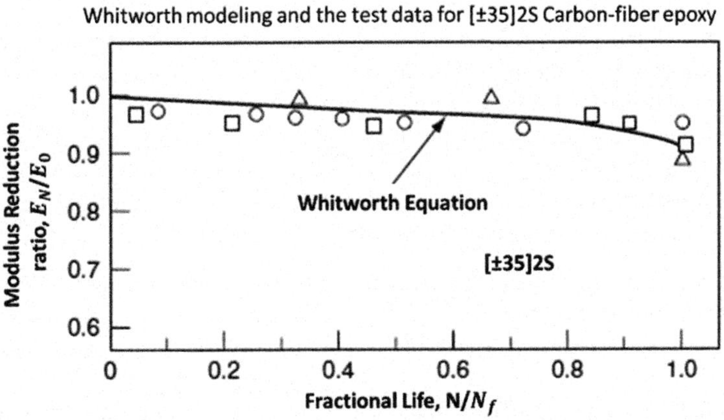

Fig. 12.6 Dynamic modulus reduction ratio as a function of fractional life for $[\pm 35]_{25}$ carbon fiber–epoxy laminates

Furthermore, Whitworth defined the fractional damage accumulation, D_i, in a cyclic load environment for composite laminate after N_i cycles of fatigue loading with a maximum stress level S_i as:

$$D_i = (E_0 - E_{Ni})/(E_0 - E_{ff}) \tag{12.3}$$

where E_{ff} = Modulus at the time of fatigue failure.

In a variable amplitude cyclic loading environment, where there are several load amplitudes, each with a different number of cycles; the failure occurs when the sum of the fractional damages from each load environment will be equal to 1, i.e.:

$$D_1 + D_2 + D_3 + \ldots = \Sigma D_i = 1 \tag{12.4}$$

12.4 Damage Accumulation Assessment Through Fracture Mechanics

The accumulative fatigue damage in composites can also be assessed by developing a damage theory describing the crack progression and accumulation as a function of material degradation in a variable load spectrum environment. An analytical damage approach based on the degradation of the structural stiffness is needed to trace the damage from initiation to the final failure. Such a model should be able to predict the failure of a composite part in terms of the number of load cycles, with the accumulation of damage starting from initiation to final failure. The fracture mechanics approach is one of these tools that can be implemented to assess the growth of microcrack from initiation up to failure [5], when the crack growth will cause structural stiffness to reduce considerably by using either the stress intensity factor, K, or the energy release rate parameter, G. The critical value of the stress intensity factor, K_{Ic}, or the energy release rate, G_{Ic}, can be utilized to measure the incremental growth of, Δa, from the initiation up to the failure ($a_{cr} = a_i + \Delta a$) for each load spectrum and the total number of cycles associated with it, N_{fi}. By using Miner's rule, engineers can find out if the final failure is imminent or not. That is, $N_1/N_{f1} + N_2/N_{f2} + N_3/N_{f3} + \ldots = 1$.

The steps and assumptions toward calculating the total accumulated damage, as a result of several load spectrum, are as follow:

- A comprehensive finite element modeling (FEM) of the structure under study.
- The estimation of maximum tensile stress induced in part through the FEM analysis. The maximum stress will be the initiation site for damage and damage growth.
- Fracture allowables such as fracture toughness and fatigue crack growth data must be available for the material under consideration. They should be obtained from a reliable source.

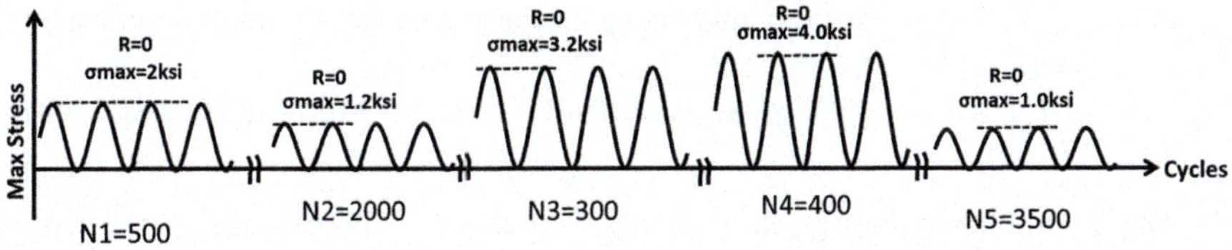

Fig. 12.7 The variable amplitude load spectrum with assigned σmax and the number of cycles for studying the accumulative damage

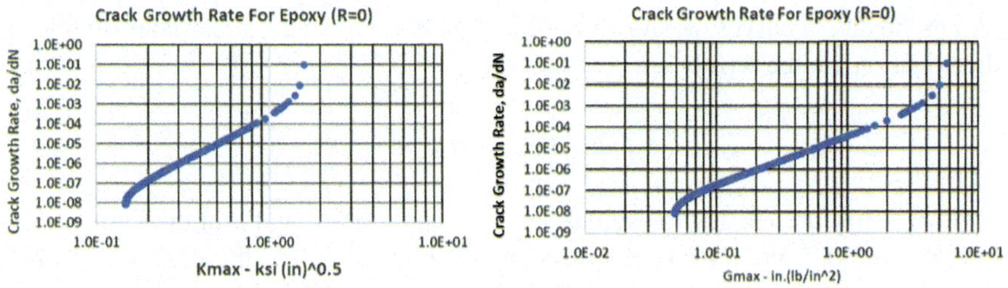

Fig. 12.8 (**a**) The crack growth rate data for the epoxy in terms of Kmax. (**b**) The da/dN plot is in terms of Gmax. Both are in terms of $R = 0$

- The load spectrum that umbrellas all the load events and the number of cycles that the composite part will be exposed to must be available to the analyst. Noting that the load cycles must be sorted based on the stress level (Fig. 12.7).
- Assumptions of an initial flaw size (an intrinsic defect) that can simulate the typical flaw size in the material, a_i, must be available to the analyst. Usually, the flaw size of equal $a_i = 0.005$ in (0.127 mm) can be a reasonable size for the analysis, because it is small enough to simulate natural defects in the material. For metals, this value is chosen as the boundary between small and large cracks [5].
- The analysis in this study is based on a transverse crack in the matrix of a lamina (aka a matrix crack).

Let us assume the epoxy matrix as the material of choice for this study, where the transverse matrix crack will be advancing through the matrix transverse of a lamina by the applied cyclic load. Assumptions to be used for this analysis are as follows:

1. Epoxy material will be utilized as the matrix for matrix cracking.
2. The tensile strength $\sigma_{\mathrm{Ult}} = 13$ ksi, $\sigma_{\mathrm{Yield}} = 10$ ksi, and the tensile modulus $E = 475$ ksi.
3. The stress levels for the load spectrum and the corresponding number of cycles are shown in Fig. 12.7.
4. The G_{Ic} value for the epoxy material used in this analysis is 946 J/m^2 = 5.57 in-lb/in^2 and in terms of $K_{\mathrm{Ic}} = (E\ G_{\mathrm{Ic}})^{0.5} = 1.63$ ksi (in)$^{1/2}$.
5. The crack growth data for the epoxy material is shown in Fig. 12.8. Note that the crack growth rate data is based on the $K_{\max}$ (or $G_{\max}$) versus the *da*/*dN* was generated by a test specimen with a constant load magnitude up to the failure, but

the K_{max} or G_{max} are both increasing because the crack length is constantly increasing.

6. Note that 1 ksi ~ 6.895 MPa, 1 ksi $(in)^{1/2}$ ~ 1.098 Mpa $(m)^{1/2}$, and 1 in-lb/in^2 ~ 176 J/m^2.

The accumulative damage starts by allowing the $a_i = 0.005$ in ($2a_i = 0.01$ in) to grow through the first load spectrum with 500 cycles of stress magnitude $\sigma_{max} = 2$ ksi and $R = 0$. For simplicity of the analysis, all the stress ratio in the variable load spectrum shown in Fig. 12.7 is taken to be $R = 0$.

The Paris constants for both cases can be found by using two points in the linear part of the da/dN curves, as shown in Fig. 12.8. In terms of the stress intensity factor, the equation can be written as:

$$da/dN = \alpha(K_{max})^{\beta}$$

Two points are needed to estimate the two quantities α and β in the Paris region. For $K_{max} = 0.9$, the *da*/*dN* = 1.0E-4, and for $K_{max} = 0.5$, the *da*/*dN* = 1.0E-5. Applying the two above points the Paris equation can be written as:

$$da/dN = 1.5 \times 10^{-4}(K_{max})^{3.91} \tag{12.5}$$

In terms of energy release rate, the two points will be taken from Fig. 12.8, to solve the two quantities α and β. For $G_{max} = 0.2$, the *da*/*dN* = 1.0E-6, and for $G_{max} = 1.9$, the *da*/*dN* = 1.0E-4. The Paris equation in in terms of G_{max} can be written as:

$$da/dN = 2.5 \times 10^{-5}(G_{max})^{2.15} \tag{12.6}$$

To calculate the number of cycles to failure for load spectrum #1 with $\sigma_{max} = 2.0$ ksi, the N_f will be estimated from the crack growth rate data equation by integrating, $da/dN = 1.5 \times 10^{-4}(K_{max})^{3.91.}$

$$\int_0^{N_f} dN = \int_{a_0}^{a_{cr}} \frac{da}{1.5 \times 10^{-4}(K_{max})^{3.91}} = 6.6 \times 10^3 \int_{0.005}^{0.21} \frac{da}{\left[(2.0)(3.14a)^{0.5}\right]^{3.91}},$$

where the critical crack size (based on $\sigma_{max} = 2.0$ ksi), $a_{cr} = 0.21$ in. This value was obtained based on the equation of the stress intensity factor, where $K = K_{Ic} = 1.63$ and $\sigma_{max} = 2.0$ ksi.

The calculated N_f:

$$N_f = 7663 \text{ cycles, and } (n/N_f)_1 = 500/7663 = 0.065$$

To calculate the number of cycles to failure for load spectrum #2, based on $\sigma_{max} = 1.2$ ksi, first, the amount of crack growth after 500 cycles of load spectrum #1 must be calculated. This value will be used as an initial flaw size for load spectrum #2.

From the crack growth rate data, $da/dN = 1.5 \times 10^{-4}(K_{max})^{3.91}$

$$\int_0^{500} dN = \int_{a_i}^{a} \frac{da}{1.5 \times 10^{-4}(K_{max})^{3.91}} = 6.6 \times 10^3 \int_{0.005}^{a} \frac{da}{\left[(2.0)(3.14a)^{0.5}\right]^{3.91}}$$

Where $a = 0.007$ inch after 500 cycles with $\sigma_{max} = 2$ ksi

For the next load spectrum, #2, where $\sigma_{max} = 1.2$ ksi, the total number of cycles to failure will be:

$$\int_0^{N_f} dN = \int_{a_0}^{a_{cr}} \frac{da}{1.5 \times 10^{-4}(K_{max})^{3.91}} = 6.6 \times 10^3 \int_{0.007}^{0.587} \frac{da}{\left[(1.2)(3.14a)^{0.5}\right]^{3.91}}$$

where the critical crack size based on $\sigma_{max} = 1.2$ ksi is 0.587 in.

The number of cycles to failure from $a = 0.007$ to $a = 0.587$ based on $\sigma_{max} = 1.2$ ksi is:

$$N_f = 39864 \text{ cycles and } (n/N_f)_2 = 2000/39,864 = 0.05$$

For the load spectrum #3, where $\sigma_{max} = 3.2$ ksi, the total number of cycles to failure will be:

$$\int_0^{N_f} dN = \int_{a_i}^{a} \frac{da}{1.5 \times 10^{-4}(K_{max})^{3.91}} = 6.6 \times 10^3 \int_{0.008}^{0.08} \frac{da}{\left[(3.2)(3.14a)^{0.5}\right]^{3.91}}$$

Where $a = 0.008$ in after 2000 cycles from $a_i = 0.007$ with $\sigma_{max} = 1.2$ ksi. The critical crack size based on $\sigma_{max} = 3.2$ ksi for the load spectrum #3 is 0.08 in.

The number of cycles to failure from $a = 0.008$ to $a = 0.08$ in based on $\sigma_{max} = 3.2$ ksi is:

$$N_f = 686 \text{ cycles and } (n/N_f)_3 = 300/686 = 0.44$$

Where $a_i = 0.008$ in will grow after 300 cycles to a = 0.015 in with $\sigma_{max} = 3.2$ ksi.

The critical crack size based on $\sigma_{max} = 4.0$ ksi for the load spectrum #4 is 0.015 in.

For the next load spectrum #4, where $\sigma_{max} = 4.0$ ksi, the total number of cycles to failure will be $N_f = 121.6$ and $(n/N_f)_4 = 100/126.6 = 0.822$.

Finally, for the last load spectrum (#5), the number of cycles to failure must also be calculated. The critical crack length for this case with $\sigma_{max} = 1.0$ ksi is 0.946 in., and the initial crack length is the final crack length based on load spectrum #4,

$a = 0.015$ in. Based on $\sigma_{max} = 1.0$ ksi, the total number of cycles to failure will be $N_f = 39{,}310$ and $(n/N_f)_5 = 3500/39{,}310 = 0.089$.

Based on the Miner's rule of damage accumulations (Eq. 12.1):

$$N_1/N_{f1} + N_2/N_{f2} + N_3/N_{f3} + N_4/N_{f4} + N_5/N_{f5} = 1 \tag{12.1}$$

$$0.065 + 0.05 + 0.822 + 0.089 = 1.03 > 1$$

In the above analysis, the part under the load spectrum shown in Fig. 12.8 caused failure to occur. Note: the accumulation of all damages came out to be slightly larger than $\sum[N_i/N_{fi}] = 1$.

In this section, the same technique will be taken for the analysis of damage accumulation by using the energy release rate approach, G. From Eq. 12.6, the Paris equation in this case can be written as:

$$da/dN = 2.5 \times 10^{-5}\ (G_{max})^{2.15}$$

To calculate the number of cycles to failure for load spectrum #1 with $\sigma_{max} = 2.0$ ksi, the N_f will be estimated from the crack growth rate data equation by integrating the $da/dN = 2.5 \times 10^{-5}(G_{max})^{2.15}$. In this analysis, only the first load spectrum (#1) will be assessed, and results will be compared with the previous analysis performed through the stress intensity factor approach.

$$\int_0^{N_f} dN = \int_{a_0}^{a_{cr}} \frac{da}{2.5 \times 10^{-5}(G_{max})^{2.15})} = 2.5 \times 10^4 \int_{0.005}^{0.21} \frac{da}{\left[(2.0)(3.14a)^{0.5}\right]^{2.15}},$$

where the critical crack size (based on $\sigma_{max} = 2.0$ ksi) $a_{cr} = 0.21$ in. This value was obtained based on the equation of energy release rate, where $G = G_{Ic} = 5.57$ in (lb/in^2) and $\sigma_{max} = 2.0$ ksi.

The calculated N_f:

$$N_f = 7790 \text{ cycles, and } (n/N_f)_1 = 500/7790 = 0.064$$

Comparing the G_{max} approach versus the K_{max}, it was concluded that the two approaches yield almost the same result. That is,

$$N_f = 7790 \text{ cycles, and } (n/N_f)_1 = 500/7790 = 0.064 \text{ versus}$$
$$N_f = 7663 \text{ cycles, and } (n/N_f)_1 = 500/7663 = 0.065$$

12.5 Concluding Remarks

The study on the damage accumulation *of composite materials* under fatigue loading has illuminated the intricate mechanisms governing the structural degradation of these advanced materials. It has been shown that their behavior under cyclic loading can be distinctly different from metallic alloys. Through a comprehensive assessment of fatigue-induced damage and the accumulation of damage after N number of cycles, analysts can predict the possibility of failure by employing Miner's rule. Therefore, a better understanding of how repeated loading cycles impact the integrity and performance of composite structures, specifically through the degradation of composite material properties, should be explored.

Due to the anisotropic nature of composite materials, the prediction of final failure using the Miner's rule methodology may introduce errors in the results. The concept of damage accumulation and the Miner's rule methodology is currently applied in many industries for isotropic materials, where parts are subjected to cyclic loading. Further studies and tests are needed to determine if this approach can be successfully implemented for composite materials. The study must consider the influence of material properties, loading conditions, and environmental factors in understanding fatigue damage and predicting final failure. The observed damage mechanisms and damage progression start from matrix cracking, crack arrest due to interface, and fiber breakage, all of which have significant implications for the design, maintenance, and safety of composite components.

Furthermore, this chapter has highlighted the necessity for predictive models that can accurately anticipate the fatigue life of composite materials, enabling engineers to proactively address potential failure points and optimize component design. The application of fracture mechanics in damage growth and the accumulation of damage has been explored in this chapter with example problems that the reader can follow. However, the methodology must be extended to anisotropic materials, where their behavior under cyclic loading can differ from traditional metallic parts.

Finally, it is evident that research on the failure of composite materials is essential for advancing the durability of composite structures, contributing to the continued growth of industries dependent on these materials. The insights gained from this study provide a foundation for future attempts aimed at enhancing the fatigue performance of composite materials.

References

1. B. Farahmand, *Application of High and Low Cycle Fatigue on Metallic and Non-metallic Material (Application of Fatigue for Engineers)* (Amazon eBook, Seattle, 2022)
2. R.M. Jones, *Mechanics of Composite Materials* (McGraw Hills Book Company, New York, 1976) Chapter 1
3. S.L. Ogin et al., *Modeling Damage, Fatigue and Failure of Composite Materials* (Woodhead Pub, Cambridge, 2016)

4. M.A. Miner, Cumulative damage in fatigue. J. Appl. Mech. **12**, 149–164 (1945)
5. B. Farahmand, *Fatigue and Fracture Mechanics of High-risk Parts* (Chapman & Hall, New York. Chapter 4, 1996)
6. R.H.W. Brook, J.S.C. Parry, Cumulative Damage in Fatigue: A Step towards Its Understanding. J. Mech. Eng. Sci. **1–23**, 1959–1982 (1969)
7. The Open University (UK), *T838 Design and Manufacture with Polymers: Solid Properties and Design* (The Open University (UK), Milton Keynes, 2000), p. 30
8. T. Kevin O'Brien, K.L. Reifsnider, Fatigue damage evaluation through stiffness measurements in boron-epoxy laminates. J. Compos. Mater. (2016). https://doi.org/10.1177/002199838101500105
9. J.N. Yang, D.L. Jones, S.H. Yang, A. Meskini, A stiffness degradation model for graphite/epoxy laminates. J. Compos. Mater. (2016). https://doi.org/10.1177/002199839002400705
10. H.A. Whitworth, Cumulative damage in composites. ASME J. Eng. Mater. Technol. **112**(3), 358–361 (1990)

Chapter 13
Composite Repairs (Industrial Applications and Analysis)

13.1 Introduction

Composite laminates are currently utilized for a wide range of applications in industries, and recently they have been used in critical components of aircraft structures, aerospace, as well as automobiles. The term critical components are often used in industries, and it is referred to as primary structural parts. When critical parts get damaged during service usage, repairing may be needed to assure structural safety and prevent loss of life. Unintentional service damages are different from damages resulting from fabrication anomalies, such as the presence of porosity, microcracking, and delaminations. In both cases, they will shorten the life of parts and must be prevented as much as possible. However, in this chapter, attention will be given to the structural damage introduced to composite parts during their service usage and the possibility of being repaired to continue their operational life. For example, the fuselage of a passenger aircraft is a critical component of an aircraft structure because its failure will cause a catastrophic failure that can harm passengers during flight. Damages in composite parts may occur before their operations, which should be checked and monitored before and after usage through proper interval inspection. For example, during assembling parts, grinding, or through machining, there are always possibilities of introducing damage in the composite parts [1]. Figure 13.1 illustrates a damaged area that could happen in a laminated composite part before its usage by an impact or when under compressive loads that must be repaired. Note that composite patches can be bonded to various substrates, such as metals, composites, and plastics to repair damage in different structures. In Sect. 13.7, repairing metallic substrate with composite patch will be discussed in detail. Several example problems will be presented that will discuss both the strength margin of safety and life analysis of the repaired area.

Induced damage to composite components is not always visible to the naked eye, and the extent of damage is best can be determined by applying a suitable nondestructive inspection (NDI) method that must be conducted by the technical team that

B. Farahmand, *Fundamentals of Composites and Their Methods of Fabrications*,
https://doi.org/10.1007/978-3-031-80201-0_13

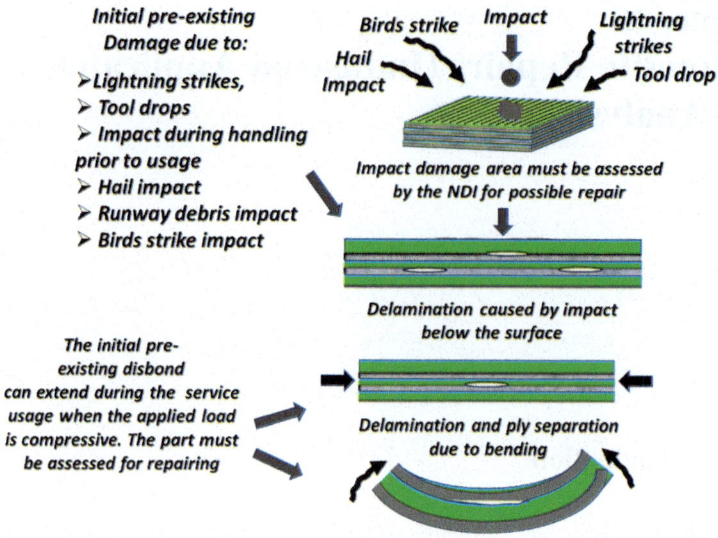

Fig. 13.1 Preexisting damage can occur prior to the part's operation. They can grow during the service and engineers must decide if the part needs repairing or not

has been certified by a trustworthy organization, such as the Federal Aviation Administration, FAA. The FAA organization is the largest transportation agency in the U.S. that regulates all facets of civil aviation in the country, as well as overseas (see Chap. 14 for a brief discussion on the NDI methods).

Repairing sizable damage is heavily dependent on the details of the structure, and therefore, fixing a damaged area has been chosen based on a specific case. If repairing the damaged part is expensive and time-consuming, replacing the part can be a choice. In aerospace and aircraft environments, a plan must be implemented for critical parts to avoid damage happening before service usage, and it is called a fracture control plan (FCP).

A comprehensive fracture control plan, FCP, written for space flight hardware is provided in [2]. It is defined as the rigorous application of engineering, manufacturing, and operational disciplines to all phases of structural hardware development, from design and analysis to customer delivery, to prevent structural failure caused by flaws and/or cracks during fabrication, testing, handling, and service life. The FCP document specifies comprehensive guidelines for the design, analysis, inspection, testing, and handling. The implementation of a fracture control plan requires to have a Fracture Control Board (FCB) that will be able to monitor all fracture critical parts before and after their operations to avoid catastrophic damage associated with the presence of damage. The monitoring process must also consider the status of damage growth during their service life. In general, the fracture control plan will ask the following questions when dealing with the possible failure of critical parts during their operations:

1. What are the functions of the structural part during the operation?
2. How critical components may fail during the operation?
3. What are the probabilities of parts failure during the operation?
4. How can these failures be categorized?
5. How will the failure be detected before and during the operation?
6. How can structural failure be avoided?
7. What are the design options to eliminate failure?
8. If partial failure was detected, then what appropriate actions (i.e., repairing) should be taken to avoid total structural failure?

13.2 Important Steps Regarding the Composite Repair

The purpose of any composite repair is to restore the structural load capability in the damaged area so that it is adequate for the overall composite structure to perform safely under maximum design load during the service operation without failure. To achieve that, there are a few fundamental steps regarding composite repair that must be followed and are listed below. These steps are general to most repairs and industries and are helpful to know about them [3]:

- Assess the extent and degree of damage. Engineers must decide whether to go forward with the repair or not. The cost and time of repairing should be taken into consideration.
- If necessary, remove damaged material without damaging the surrounding areas.
- Clean and prepare the damaged area for repair.
- Complete the repair without damaging the adjacent areas.
- Conduct the repair process in an environment without humidity at room temperature.
- Inspect the repaired area for quality assurance and the possibility of any delaminations. Proper nondestructive inspection (NDI) methods should be required to ensure the repaired part is free from damage, especially in the jointed areas. Inspection methods will be presented in Chap. 14.
- The evaluation of the repaired part must include an assessment of the probability of occurrence of damages due to fatigue, corrosion, or accidental damage and the location of the damage. First, conduct the static analyses supported by the test. Previous service experience and knowledge must also be incorporated into the evaluation. Because of the possibility of widespread fatigue damage, engineers must be sure that this type of damage will not occur within the design service goal of the structural parts. This can be achieved by applying proper NDI, fatigue, and fracture mechanics analysis tools. Alternatively, engineers may design to allow the stress level in the repaired area to be low enough to fall under the infinite life category.
- Finally, by merging the repair composite parts with the rest of the parts, vacuum-bagging can be used. The assembly is placed inside an air oven for curing or over

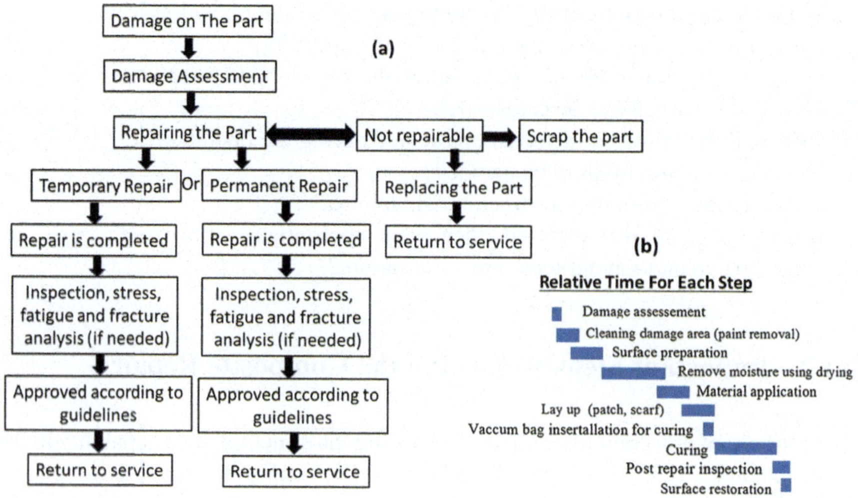

Fig. 13.2 (**a**) This chart can guide engineers to decide on the status of the damaged part on the repairing and the repair type. Subsequent inspection, testing, and analysis maybe needed. (**b**) The chart indicates the usual repair process and the time required to complete the activity [5]

a heater mat in field conditions. Autoclave processing is used to repair high-quality structural components, where the processing conditions of vacuum, pressure, heat-up rate, and cure temperature are controlled.

The residual strength evaluation must also show that with any damage left behind, the remaining damaged structure can withstand the applied load. The residual strength is defined as the remaining strength left in the presence of a defect, if any. When there are no defects presence in part, the residual strength is equal to the ultimate strength of the material. Figure 13.2 shows the flowcharts that can be utilized as a guideline for the repairing of a damaged composite part [4]. If the chart shown in Fig. 13.2 indicates that the repair is necessary, then before starting with the repair, the area should be assessed carefully. Some damage to composites is apparent and quickly evaluated. Still, in many cases, the damage may first look small without needing repair, although the actual damage could be much vaster (Fig. 13.1). Impact damage during the manufacturing phase, or impact by a foreign object during the part's service usage, can appear as a small dent when viewed by the naked eye on the reinforced composite surface, but the underlying damage can be much more extensive. The engineering decision to repair or scrap the damaged part is determined by considering whether the repair will be able to maintain the part's original functionality or not. Other considerations are the cost of repairing, the position and accessibility to the damaged area, and the availability of materials to be used for the repaired section. Detailed guidelines for assessing the damaged area of a composite part (repair or no repair) are shown in Fig. 13.2a. In Fig. 13.2b, the chart indicates the usual repair steps and the relative time required to complete the repairing process for each step [5–7]. For example, if the damage assessment takes 4 h, then the surface cleaning will require at least 12 h. After the damage assessment and cleaning of the damaged area are completed, then surface preparation will begin. Removing moisture and other subsequent steps, with their relative amount of time, are shown in Fig. 13.2b. Note that, in some cases, the surface preparation time for scarf repair may be time consuming when using high skills and experienced technician. However, in focusing on automating technique, by applying advance tools and avoiding the typical surface preparations that includes blasting and sanding, can be helpful to expedite the repair process in much less time.

Repairs are either temporary or permanent. They are temporary (aka time limited) will be implemented typically because of low margin of safety in critical areas such as boundary areas, faulty materials with lower properties, existing damage due to improper repair, and the inability to inspect the damaged area that can be critical to the structure. If, for some reason, a part must be repaired, and it is classified as a temporary repair where the repaired part will meet the strength requirements, it eventually should be replaced by a solid, satisfactory, and permanent one. The time limited category repairs should be taken care of as soon as practicable, and engineers must limit the number of cycles imposed on the repaired part. In some cases, a patch can be used for a temporary repair. It is adhesively attached to the damaged area to ensure the structure's safety until the component can be repaired correctly later before failure can occur. Finally, because the temporary repairs are time limited, they are not intended to last for the entire service life of the component and may need to go through interval inspections, if possible, to be sure the repaired area is safe and free from any sizable defects that initiated after repair and during service operation.

If fatigue analysis is required to be conducted on the permanently repaired area, then engineers should identify the areas of concern having high stresses and conduct structural and fatigue analysis. Usually, the areas with high stress magnitude, or the weak spots such as the interfaces between fibers and matrix, the possibility of delamination due to mismatch or inadequate adhesion between plies, and the matrix cracking, where the load is perpendicular to the fibers, need to pay attention to. After stress analysis, if damage tolerance analysis is also requested by the engineers, the initial cracking must be assumed based on the appropriate NDI inspection or applying a sound progressive failure analysis by using proper PFA software code. In either case, material allowables should be available to analysts.

Another concern is the stiffness difference between the repaired and parent material areas. The stiffness of the repaired part (E_R) should match as close as possible to the stiffness of the parent structure (E_P). However, this may not happen likely, but it should be very close to the actual structure. A repair stiffer than the parent structure ($E_R > E_P$) would attract more load to the repaired region, possibly causing it to exceed its allowable strength. On the other hand, a repair with a lower stiffness than the parent structure ($E_R < E_P$) would divert the load from the repair, possibly causing it to overstress the parent structure. The case of $E_R \sim E_P$ is difficult to achieve unless the same materials and processes are used. In most cases, a different repair material is used, but it must be sure the stiffnesses match each other. In all cases, the stress analysis should show a positive margin of safety based on the ultimate strength achieved. Another point to remember is the repair operating temperature which should not exceed the maximum operating temperature of the resin system. Otherwise, both the matrix and fibers may be exposed to damage. Hence shorter life and structural failure may occur. Humidity while repairing the damaged part can influence in degrading the mechanical properties. Finally, the most crucial element in designing a repair is the adhesive joint. Repair engineers should ensure that the adhesive does not become a weak link causing joint failure. That is, the final failure must occur due to bulk failure of the part, not the joint between the repair and the parent material.

Several types of repair techniques are prominent in industries, and these are mainly referred to as external bonded repair and scarf repair [8–11]. In all cases, either adhesives or fasteners merge the repaired section to the parent material. The scarf bonded repair requires removing material through and around the damaged area. For that reason, it is used for thick parts. On the other hand, the external patch repair is used for thin parts, where the process is simple and faster when compared with scarf repair. One of its applications is in the aircraft where a speedy repair is required to bring back the airplane in operation. The presence of patch on the damaged area will bring stresses low enough to slow down the crack initiation and growth (see Sect. 13.7).

The quality of the adhesive region, bonding parent, and repair material together, is largely dependent on the cure process and pressure settings. In addition to adhesively merging the repair section with the parent material, the fastener joint is also used widely for metallic repairs. However, the case of adhesive method is preferable and is acceptable practices in repairing damaged parts. Adhesive bonding eliminates stress concentrations when repairing is done by fasteners and it will be free from crack initiation sites caused by neighboring fastener holes and the possibility of multisite cracking. Figure 13.3a shows the external bonded composite patch method for repairing the FedEx fuselage and in Fig. 13.3b, the typical DC-10 fuselage skin damage repair by conventional riveted metallic patch, that was later repaired by composite patches.

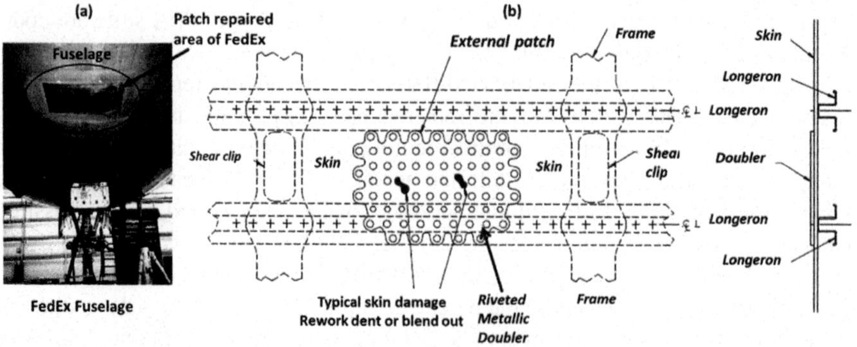

Fig. 13.3 (**a**) The composite bonded patch method of repairing the damaged area of aircraft is shown. Installation of bonded repair for the FedEx fuselage. (**b**) Typical DC-10 fuselage skin damage repair by conventional riveted metallic patch. It was later repaired by composite doubler [27]

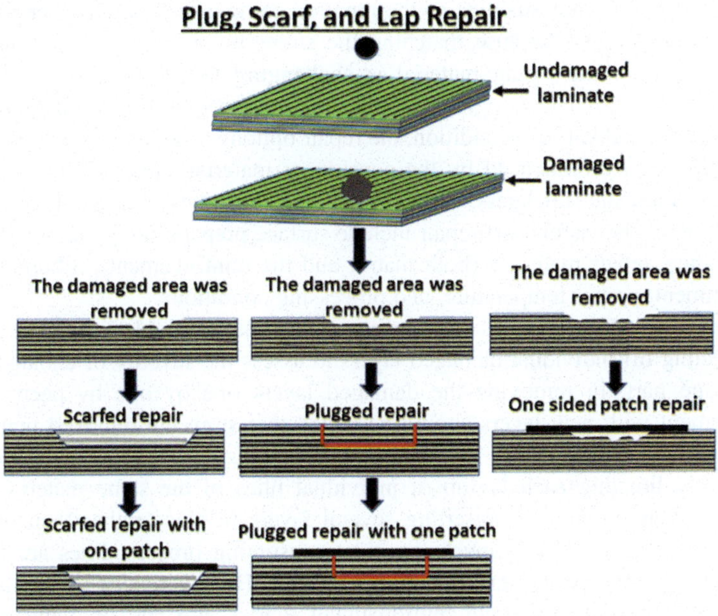

Fig. 13.4 Repair of a damaged laminated composite by scarf, plug, and lap (patch) repair. The damage was repaired

In the scarf technique, the repair material is plugged by prepared tapered composite material with an angle into the laminate in place of the material removed due to the damage (Fig. 13.4). In the lapping technique, the patch material is placed either on one or on both sides of the laminate over the damaged area (depending on the depth of the damaged area). The plug technique will be the same as the scarf approach method, except it is not tapered. In some cases, in addition to scarf repairing, an additional patch is used at the top of repaired area, as shown in Fig. 13.4 to reduce the stress in the damaged area. The prepared patch can be cured directly on the repaired laminate by applying heat and pressure on the laminate using a good-quality adhesive.

13.3 Scarf Repairing Process

The tapered scarf joint repair has long been recognized as the best method for repairing laminated composite parts because of its uniform shear properties through the jointed area [12–14]. In this repairing technique, the already prepared repair material is incorporated into the laminate in place of the material removed due to the damage. Repairing a damaged composite part, to make it to be exactly like the original part with the same mechanical properties, is very difficult. As was mentioned before, the qualities of the repaired composite generally cannot be the same as the original because matching the strength, stiffness, and weight of the repaired area to the parent material is challenging to achieve. For example, if attempting to match the original strength of a composite, the resulting repair is often heavier and stiffer. In addition, the repair options, whether it is a scarf or a lap joint, are heavily influenced by the constituent materials, fiber orientations, core material, laminate thickness, and the number of plies. Essential factors for employing a successful scarf repair include surface preparation, good quality adhesive choice, repair materials (both matrix and the reinforcement), a humidity-free environment, proper temperature, and processing conditions.

Steps for scarf repairing a laminated composite with several layers can be done by first peeling off individual damaged layers to assess the severity of damage on the composite part. In removing the damaged layers one by one by peeling, their replacement with new layers should be added to restore the removed material of laminated composite. After machining and cleaning the surface, it is ready for adhesively bonding each lay-up of individual plies of the same materials. Note that the repair is done by inserting several layers in place rather than repairing layer by layer, which is a much less time-consuming process. Plies are fitted to match the original axial orientation of the structure. However, occasionally layers of composite material, resin and reinforcement fabrics, are laid onto the prepared surface in a way that matches the shape of the scarf. This involves impregnating the reinforcement fabric with resin and carefully applying it layer by layer to rebuild the damaged area with the original orientation.

The scarf repairing method is the preferred method of repairing composite structures, especially when other methods with externally bonded patches can no longer meet the stiffness, strength, and flushness requirements, which cannot offer aerodynamically smooth surface.

The scarf repair process starts with the removal of the top layer and then using a hand-held pneumatic grinder to machine the damaged portion of the laminate into a circular shape, Fig. 13.5. The sides of the circular cutaway must then be machined to a taper at the required angle. The scarf angle is typically defined as a distance-to-thickness ratio, which is between 20 to 1 and 50 to 1, all the way to 100 to 1, or simply using a nominal ply thickness of 0.01 in at a 50:1 ratio, which would give $0.01 \times 50 = 0.5$ in distance per ply. For the case of 4-ply laminate, the scarf distance would be $4 \times 0.5 = 2.0$ in [15]. As the scarf distance increases, more of the original structure is removed, and that can weaken the parent structure. To avoid that, engineers may use a double-sided scarf repair to minimize the amount removed from the original structure (Fig. 13.5). Note that the double-sided scarf is used when the damage is deep and serious. After the machining process and cleaning are completed, the prepared composite material is inserted, and the surface is prepared for vacuum and curing process. Lay-up of an adhesive layer, and each ply of the same materials are positioned to match the original axial orientation of the structure (Fig. 13.5a). In some cases, a doubler is used to make the damaged area safer. In the case of double-sided scarf repair, doubler on both sides may be necessary (Fig. 13.5a).

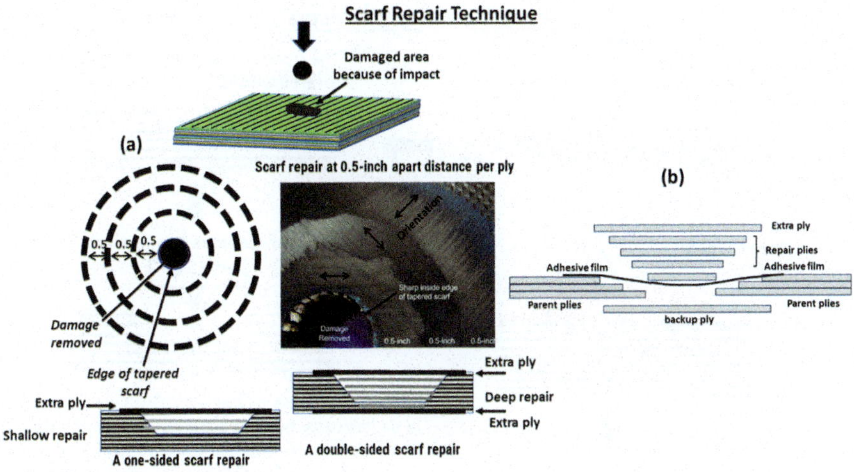

Fig. 13.5 (**a**) Scarf tapered repair. Hand-machined scarf through three plies at 0.5-inch (nominal) distance per ply. A double-sided scarf repair can be used to minimize the amount removed from the original structure. (**b**) Step repair technique (like scarf-tapered repair shown in (**a**))

In some cases, an additional ply plate is used to achieve uniform compaction to the repaired area, resulting in a highly compacted patch and a near-flush surface. The repair patch is then vacuumed bag and cured if applicable. After a final cleaning, the process provides a near-flush surface on the part.

Another composite repairing technique, like the scarf patch repair, is called stepping repair, that can be an alternative repair method for removing material in preparation for repairing the damaged area. In stepping repair, the overall angle is achieved by cutting or machining steps around the damaged area of parent material. Available repair plies are inserted and adhesively attached in the damaged area with additional plies on the front and a backup ply on the back surface. A step repair is considered more difficult to implement and it leaves abrupt edges in each repaired ply, that can create stress concentrations and weaken the joint area (Fig. 13.5b).

13.4 Plug Repairing Process by Adhesive Bonded Joint

The purpose of any composite repair is to reestablish the structural loads capability through or around the damaged area in such a way that it is adequate for the overall composite structure to be able to perform under maximum service loads without failure. Here, the adhesive bonded joint repair technique using the plug approach will be briefly discussed.

The plug repairing process is a rapid method for repairing composite structures, especially when externally bonded patches can no longer meet the stiffness, strength, and flushness requirements. Figure 13.5 illustrates the method of repairing the

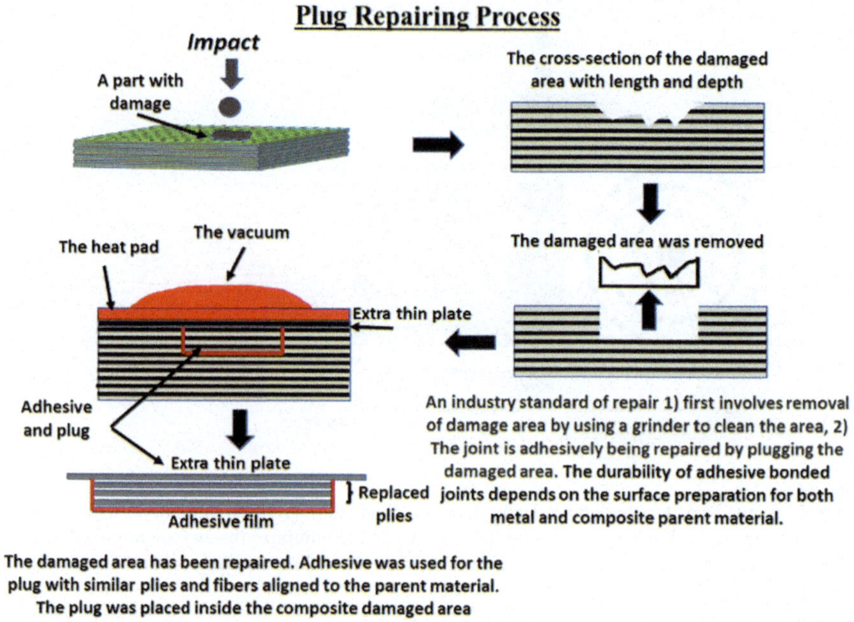

Fig. 13.6 Repair of a damaged laminated composite by using adhesive, heat pad, and vacuum. The durability of adhesive bonded joints depends on the surface preparation of composite, plugged, and parent materials

damaged area and the steps for a good repair. Under low energy impact, damage can be much deeper than the top surface when viewed by the naked eye. The NDI inspection must check beneath the face sheet surface for de-bounding and possible core damage when dealing with sandwich structure. The extent of damage after the NDI inspection determines whether the component needs to be repaired or replaced (see the flowchart shown in Fig. 13.2).

To ensure the repair's effectiveness and to maintain structural integrity, the following considerations must be preserved. That is, the stiffness, strength, stability, durability, and aerodynamic smoothness; all must be maintained. First, it is required to remove the damaged and some amount of undamaged material to initiate the repair. The surfaces of removed material (interior) must be cleaned and ready for bonding together the repair material to parent material adhesively. Figure 13.6 shows a plug repair with the adhesive joint to a composite parent part that has been damaged on the surface and through the thickness. Here, the damaged area has been removed, and adhesive was used for the replacement plug to be placed inside the composite part. Note that, in the plugged repair technique, the removed area need not to be like the scarf repair with angle as shown in Fig. 13.5. The heat pad and the vacuum were used to make the joint compatible with the surrounding material. The durability of adhesively bonded joints depends on the surface preparation and the steps that must be followed for curing the adhesive for the joint areas to have properties close to the parent material. Finally, the repaired part must be inspected before the service operation. The final inspection is for quality assurance and to check if the curing process in the jointed area was done properly. There are a few points regarding adhesive bond failures within a joint that can occur in several ways: (1) poor surface preparation between adhesive and adherend; (2) poor adhesion at the interface (nothing to do with surface preparation); (3) humidity can degrade the bond between adhesive and adherend; (4) attack of the adhesive itself to the repaired area as well as parent material at the interface region; and (5) the environmental temperature during and after the repair can degrade the mechanical properties of the repaired part. For example, the temperature and moisture levels that can affect an aircraft structure vary throughout the flight from one location to another—that is, from takeoff to the landing, which could both deteriorate the structural properties and lead to premature failure of the structure.

13.5 Lap Joint and Patch Repairing Process

Composite patches are increasingly being used to repair metallic and nonmetallic structural parts. There are different types of composite repair using bonded joints and the most applied techniques in the industries are external adhesively bonded patch repair and scarf repair. The scarf repair was already discussed in the previous section. However, the two mentioned repair techniques differ in terms of manufacturing and application point of view. Scarf repair joints require special equipment to remove a considerable amount from the parent material, and its application is for thick laminate composite parts by tapering the parent material with a required angle. The scarf angle is typically defined as a distance-to-thickness ratio, which is usually between 20 and 1 all the way to 100 and 1. On the other hand, external patch repair is relatively simple and faster, and there is no need for tapering and removing the parent material. Hence, it is widely used in aircraft for damaged parts to keep an airplane in operational condition [8]. The main challenge is how to guarantee that the repaired joints can survive for extended operations throughout their service life. The gradual degradation of joint strength is due to exposure to unpredicted and uncontrollable environmental conditions, primarily moisture and temperature variations during the service operation as was discussed before. To ensure that the adhesively bonded parts have an acceptable life, there is a need to have mechanical testing under moisture and temperature variations, damage tolerance analysis, and NDI inspections. Regarding the NDI inspection, special attention should be given to the quality of the repaired area, specifically the interface between the original part and the repaired area as well as peeling/separation at the edges that can separate the patch during the operation.

There are two kinds of damage that engineers must consider. The first one is an impact as a result of small mass and small altitude (height). Damage in this case looks small on the surface and may not have much depth. On the other hand, impact because of larger mass and larger altitude, can create considerable damage both on

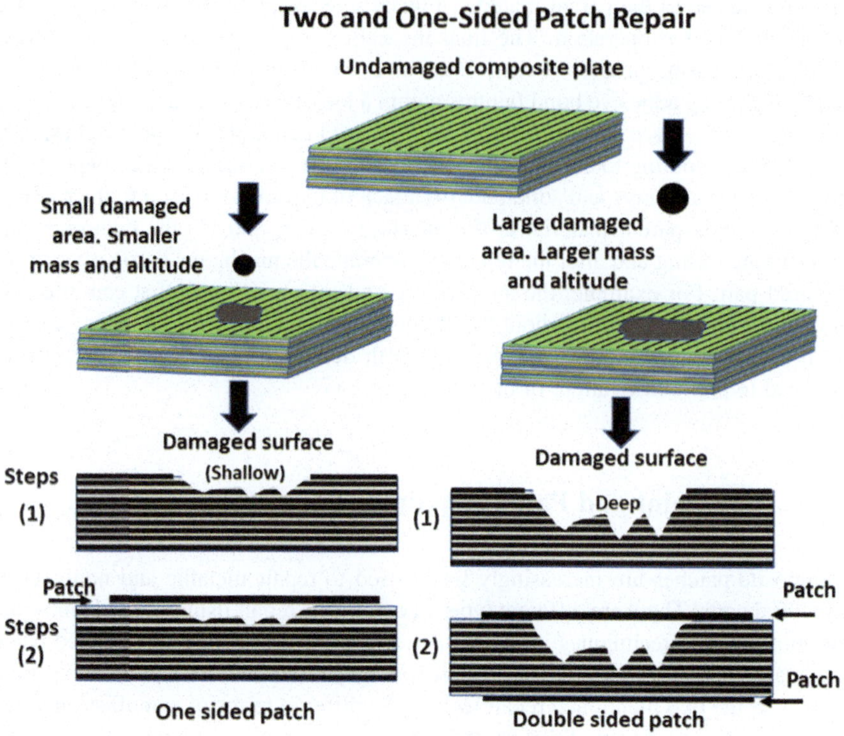

Fig. 13.7 Repair of a damaged laminated composite by using two-sided and one-sided repair (steps 1–2). The durability of adhesive bonded joints depends on the surface preparation of composite, integrity of repair, quality of adhesive, and the patch material

the surface as well as the interior of the part. In both cases, a thorough inspection is required to find out the intensity of damage. Figure 13.7 shows a patch repair technique, where the patch lap material is placed on the damaged area. First, the damaged area was cleaned, and subsequently, the repair patch was adhesively positioned on the top, as shown in the figure. Next, the heater mat and vacuum bag are used to provide a high-quality lap repair. The heater mat is used for the purpose of curing the adhesive and the vacuum is for applying pressure on the patch in order to provide a good quality joint.

In Fig. 13.7, there are two kinds of repairs shown for the patch repair process to be completed. In the first case, the damage is not deep, and the NDI inspection can provide information about the depth of damage. In this case, the damaged area will be cleaned (step 1), then a composite patch selected by engineers will be placed on the top of the damaged surface. Finally, the curing and vacuuming processes will be followed (Fig. 13.7). For the situation where the damage is deep, another patch will be placed on the other side of the surface to have a sound repair. This is called a

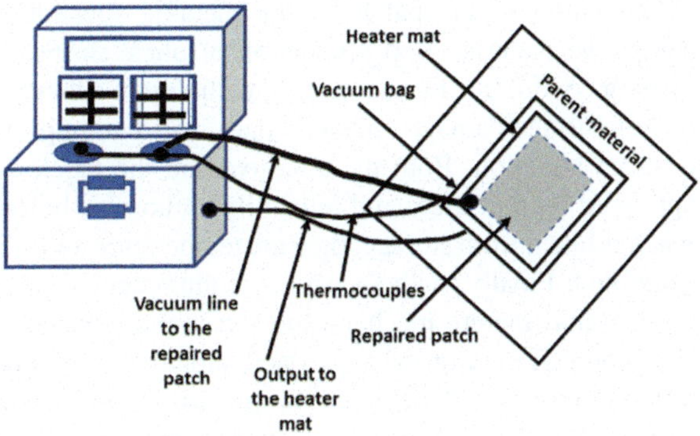

Fig. 13.8 A typical patch repair with portable unit that can apply vacuum, heat up the repair area, and constantly monitor and regulate temperature

double-sided patch repair. The presence of patch will be helpful to reduce the stresses under the patch and will delay the possibility of damage growth.

In a typical patch repair, the patch is applied under high enough temperature and vacuum for the curing of the adhesive applied for the repair. A portable device can be used for the situation where the structure is outside the lab in the field and repair must be done (Fig. 13.8). Of course, when the part can be repaired inside, where engineers have more access to proper equipment, a higher quality device, such as an autoclave should be used.

13.6 Application of PFA in Composite External Bonded Repairs

Knowing the fact that damage during service life of any structures is a possibility, performing repairs to the damaged area is necessary because it can be helpful to restore the fatigue life of the structure. This is true with regard to the aircraft, helicopters, automotive, and marine industries, where a needed repair can increase the operational life of the structural part. In this section, the bulk of discussion is on the progressive failure analysis of an external composite bounded repair on a thin metallic sheet that has been damaged during the operation or before operation due to rough handling.

Repair of damaged parts can be performed by applying the bonded or bolted methods, depending on the type and extent of damage. For metallic substrate that has damage, adhesively externally bonded repairs offer many advantages over mechanically fastened doublers, knowing that the weight of the metallic structure does not increase much with the presence of composite patch. Not only the metallic part would be operational in a cyclic load environment after adhesively repaired by composite patch, but also the presence of patch will be helpful to prevent the metallic part when exposed to a corrosive environment. The most attractive aspect of composite patch installation is that there is no need to drill holes and for that reason the parent material does not get weaker during the service operation. The effectiveness of the patch on the metallic substrate with crack depends on the composite materials and orientation of plies, shape of the patch, dimensions, and loading conditions. Furthermore, the quality of adhesive used to patch the composite material to the damaged metallic part is also an important factor that must be given especial attention. Both the adhesive and the patch composite material should be approved by the safety and reliability authorities before considering it for repair. If patch is taken from storage, then the temperature and humidity of stored area must be examined. Finally, without using a comprehensive NDI inspection and a thorough fatigue and fracture mechanics analysis, the repairing of composite part cannot be implemented on a primary part and should be confined with only secondary structural components. To expand the repairing application from secondary to primary structural parts, a detailed knowledge of their mechanical performance under several loading conditions that may occur during service usage is required analytically with limited amount of testing and the proper NDI inspection is essential.

Most of the computational modeling existed in the literature for calculating the stress field around the repaired area and optimizing the patch geometry regarding the reduction of the crack tip stress intensity factor, K, do not consider the failure analysis of the repaired part adequately. The main failure modes that might occur in an externally bonded repair of a cracked fuselage metallic skin are patch debonding or the possibility of fatigue failure in the parent material outside the patch [16] (Fig. 13.9). Patch debonding is a critical failure mode since its existence reduces the effective area of the patch causing the possibility of crack growth. Therefore, both the crack growth and the patch debonding must be given special attention. Patch debonding occurs when adhesive does not stick to the substrate material properly, which in this case is a composite patch bonding to the metallic

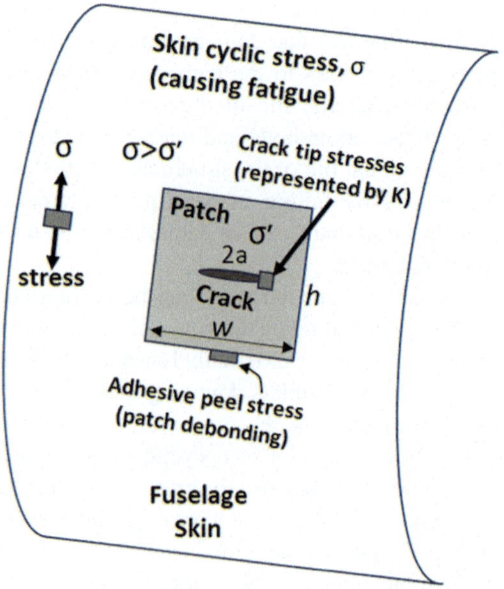

Fig. 13.9 Failure modes that might occur in a bonded repair of a crack in the fuselage skin. The width and height of the composite patch are shown as w and h, respectively

plate (Fig. 13.9). Patch debonding can occur when high shear stresses or peel stresses exist at the interface between the adhesive and the substrate.

In [17, 18], the PFA methodology has been implemented for assessing the mechanical behavior of composite bonded repairs in a metallic sheet having a crack. The crack tip stress intensity factor reduction, with the presence of composite patch, was observed when compared with unpatched configuration. The effects of composite patch material, geometry, thickness, as well as the initiation and progression of patch debonding, induced by mechanical loads, were studied. Based on the developed Ansys PFA macro routine, described in [19], a 3D computational model for the repair patch configuration has been built. The model had several parameters regarding repair geometry, material properties and mesh density. Only one quarter of the repair has been modeled due to symmetry of geometry and loading. A very fine mesh has been adopted in the area around the crack tip to achieve better accuracy in the calculation of the *K* value. Away from the crack tip, a coarser mesh has been adopted in order to reduce the total number of elements and therefore the time of computation. The metallic plate and adhesive material have been modeled as different bodies. For the patch, eight layered elements have been used through the thickness of the composite patch, keeping the rule of using one element per layer (i.e., ply). For the adhesive and the metallic plate, eight and two elements have been used, respectively. The use of more elements through the thickness will result better accuracy for obtaining the through thickness stresses. Therefore, by knowing the through thickness stresses it can be helpful on the prediction of patch debonding stresses, as well as other failures associated with the adhesive and patch, and debonding failure analysis. Note that, the stress, σ, outside the patch is larger than the stress inside the patch, σ', that has direct relationship with the quantity *K*.

The interactive Hashin-type failure criteria (see Fig. 13.8 of Chap. 8) can be applied for detection of the composite patch failure modes. For the adhesive and the adhesive–metal or adhesive–patch interface failure assessment, three major failure modes are considered, namely: (1) adhesive shearing, (2) adhesive to metal peeling, and (3) adhesive/patch peeling. The failure criterion used for the detection of adhesive shearing is given by Tresca [20] as,

$$\frac{\sigma_1 - \sigma_3}{2} \geq \sigma_s$$

where σ_1 and σ_3 are the maximum and minimum principal stresses in the adhesive and σ_s is the shear strength of the adhesive. Adhesive/metal peeling is predicted when $\sigma_{za} > \sigma_{am}$, while adhesive/composite patch peeling is predicted when $\sigma_{za} > \sigma_{ap}$, where σ_{za} is the stress in the normal direction (z), σ_{am} and σ_{ap} are the peeling strengths (allowables) of the adhesive–metal and adhesive–patch interfaces, respectively [17]. To account for the shearing failure of the two interfaces, σ_s can be considered as the minimum of the adhesive shear strength and the interfaces shear strength values. From the modeling point of view, this consideration seems satisfactory since the breakage of a single element of the adhesive is considered equivalent to the debonding of the either of the interfaces in this element.

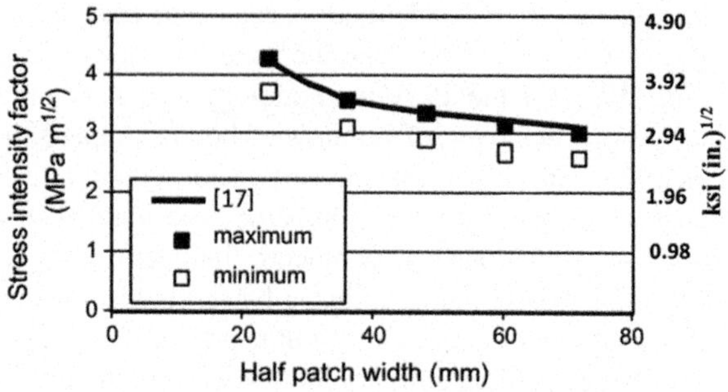

Fig. 13.10 The variation of stress intensity factor, K, versus the patch width (w) [17]

When a mode of failure is detected in the composite patch, the degradation rules of Fig. 8.6 of Chap. 8 are applied to disable the failed material region from carrying a specific load. For the adhesive material, the degradation is always achieved by setting the Young's modulus close to zero value, to avoid numerical problems that would be caused if zero values were used. The iterative PFA procedure runs until the crack tip stress intensity factor in metallic substrate ($\sigma\sqrt{\pi a}$) reaches its critical value, that is $\sigma\sqrt{\pi a} = K_c$. More details about the failure criteria used to predict debonding can be found in [17].

In this repair configuration shown in Fig. 13.9, the repair was considered as a metallic sheet with a central through-thickness crack loaded in tension and the repair was a rectangular carbon fiber-reinforced plastic (CFRP) patch with tapered edges. The dimensions of the patch plate were 240 mm × 240 mm × 3 mm (~9.5 × 9.5 × 0.12 in). The metallic plate is made of an aluminum alloy with $E = 71{,}709$ MPa (10,392 ksi) and $\nu = 0.33$. The thickness of the patch is 2.85 mm (0.11 in without adhesive). The patch height 2 h and width 2w, as well as the crack length 2a are parameters of study. The thickness of the adhesive is 0.15 mm (0.006 in). Therefore, the total patch thickness is 3 mm as was shown above.

The variation of the stress intensity factor with regard to the patch width w (parallel to crack line) was studies, as obtained by the finite element modeling and compared to the results of analysis presented in [17], where the results refer to half crack length of $a = 12$ mm (0.39 in). It was observed that the stress intensity factor variations decreased with the patch width, as it is expected, because the stress intensity factor reduction is a function of patch dimensions (Fig. 13.10).

13.6.1 *The Effect of Patch Lay-Up on the Composite Bonded Repairs*

To assess the effect of patch lay-up on the repair effectiveness, four different lay-ups with eight layers have been considered. The results, shown in Fig. 13.11, suggest that the stress intensity factor variation with the applied load is linear in all cases,

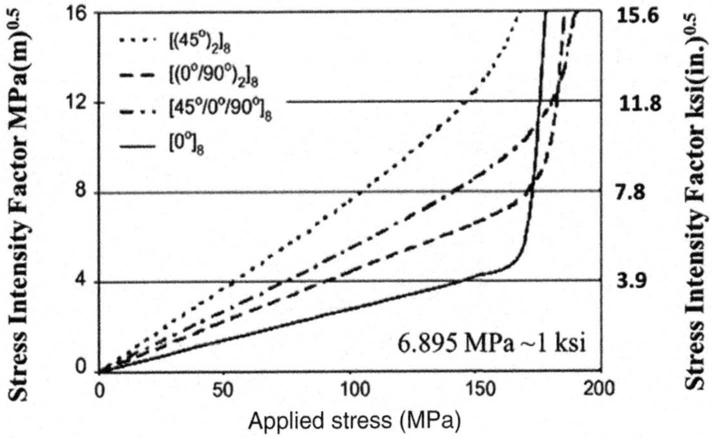

Fig. 13.11 Effect of patch lay-up on the stress intensity factor through the failure analysis as a function of applied stress [17]. The lay-up $[\pm 45°]_S$ is less effective and other two cases $[0°/90°]_S$ and $[45°/0°/90°]_S$ falling in between. Finally, the most effective lay-up is $[0°]_8$, where the 0° layers coincide with the direction of the load application

until the load at which patch debonding causes sudden increase of the K. Since the panel of Fig. 13.9, in which the bonded composite patch is implemented, is subjected to fatigue loading conditions with the K threshold of 4.4 $MPa(m)^{0.5}$ ~ 4.2 $ksi(in.)^{0.5}$ for the aluminum material. The crack tip K value will increase when $K > K_{th}$ value. According to these conditions, the most effective lay-up is the $[0°]_8$, where the 0° layers the fibers coincide to the direction of the load application. Moreover, the analysis shows that the performance of the patch lay-ups with regard to the crack tip stress intensity factor is less effective for the lay-up $[\pm 45°]_S$. Other two cases $[0°/90°]_S$ and $[45°/0°/90°]_S$ falling in between as shown in Fig. 13.11. By investigating the failure progression within the repair patch structure, as predicted by the PFA model [17], it is observed that debonding initiation at the crack tip region is due to crack opening, but it is not catastrophic for the repair in the sense that the applied load can be further increased significantly. In Fig. 13.12, the variation of the amount of debonding area regarding the applied load is presented. Debonding initiated (patch separation) at about 220 MPa (31.9 ksi) and it increases up to 350 MPa (50.7 ksi) where the separation of the upper patch edge will propagate very suddenly leading to complete dissociation of the patch from the metallic sheet.

In Fig. 13.12, the debonding progression for the adhesively attached patch is illustrated based on the percentage of separation. The debonding at the upper patch edge propagates very fast after the applied load reaches 350 MPa, leading to the catastrophic failure of the repair.

A fracture mechanics analysis was conducted to study the number of cycles to failure for the patch repair shown in Fig. 13.9. The finite element analysis of aluminum plate (7050-T76) of Fig. 13.9 was shown to have 16 ksi stress outside the patch area, and the stress magnitude inside the plate with the presence of patch

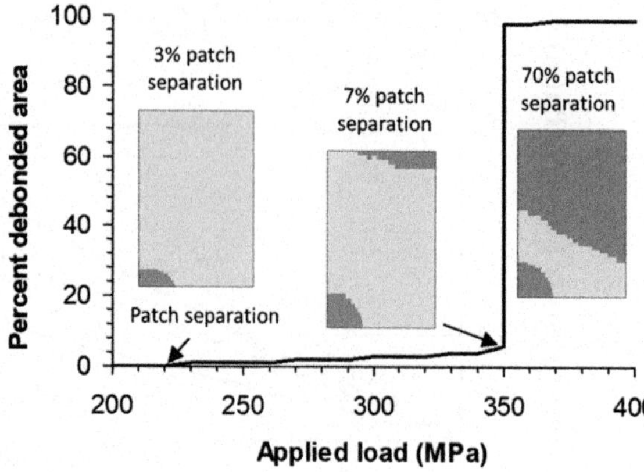

Fig. 13.12 Increase of the debonding area (patch separation) with the applied load in the composite repair patch [17]. Note: 6.9 MPa ~ 1 ksi

was estimated to be 8 ksi. The life analysis conducted by the NASGRO computer code shows the number of cycles to failure for the crack inside the patch area is $N_f = 113{,}217$ cycles with $R \sim 0$. Note that the same analysis was conducted for the crack plate without the patch and the same original half a crack length of 0.39 inch was employed in the life analysis. The number of cycles to failure was shown to be much smaller with $N_f = 16{,}131$ cycles.

13.7 Fracture Analysis of Metallic Plate with Composite Patch

Following the development of analytical work presented in [21–25], it was concluded that the stress intensity factor for a patched crack approached a constant value as the crack length increased. This observation was based on the assumption that, for a long crack in a structure, which is subjected to a remote uniform stress field, the central region of the patch (under the patch region) behaves like an overlap joint. Subsequently, it follows that the stress distribution in this central region should be independent of crack length. This assumption of uniform stress field under the patch was later confirmed experimentally using thermal emission techniques [26]. Here, it was also shown that crack growth rate (da/dN) as a function of number of cycles was a constant. The following is a brief description of important points leading to the damage tolerance assessment of metallic plate with composite patch with several examples that can be beneficial to the reader.

An approximate closed form solution for the repair of a center notch panel is presented here [23]. After patching the composite patch on the surface of metallic plate, the stress intensity factor for a crack in the thin metallic skin, under a remote uniaxial stress σ, approaches an asymptotic value of K_∞ which is independent of the crack length. Under this condition where the stress intensity factor is independent of the size of crack under the patch, the value of the stress intensity factor, K_∞, (the value of the stress intensity factor for the case when there is no disbond), can be approximated by the formula:

$$K_\infty = \sigma_o \sqrt{\pi\lambda} \tag{13.1}$$

Where the quantity $\pi\lambda$ is the crack length independent, but it is the plate, adhesive thickness, and modulus dependent [23]. It is given by:

$$\pi\lambda = \sqrt{E_p t_p/\beta\left(1 + t_p E_p/E_r t_r\right)} \tag{13.2}$$

$$\sigma_o = \sigma E_p t_p/\left(E_p t_p + E_r t_r\right) \tag{13.3}$$

The quantity σ_o is the far field stress in the skin and inside the patch where the patch adhesively attached to the metallic plate ($\sigma_o < \sigma$). The applied stress outside the patch and in the plate, σ, is shown in Eq. 13.3. The quantity β shown in Eq. 13.2 is given by [23]:

$$\beta = \left(t_a/G_a + t_r/3G_r + t_p/3G_p\right)/\left(t_a/G_a + 3t_r/8G_r + 3t_p/8G_p\right)^2 \tag{13.4}$$

where G is the shear modulus and subscripts a, p and r stand for the adhesive, plate, and patch respectively. Quantities t_a, t_p, and t_r are the thicknesses of the adhesive, metallic plate or parent material, and patch. Furthermore, E_p shown in Eq. 13.2 is the modulus of the plate. Also, the quantity λ is dependent on the mechanical properties of both the patch and the plate as well as their thickness. Additionally, λ is strongly dependent on adhesive thickness and mechanical properties through β, as shown in Eq. 13.4.

The crack tip stress intensity factor value, K_∞, thus obtained should be small due to the presence of composite patch, which is dependent on the value of far field stress in the metallic skin, σ, and inside the patch, σ_o. The stress intensity factor should be less than the threshold stress intensity factor for fatigue crack growth, K_{th}, indicating there is no crack growth is expected below this value (i.e., $K < K_{th}$).

If the patch has disbonded over the entire length of the crack the resultant energy release rate, G, can be estimated [23] from the formula:

$$G = (K_\infty)^2/E + \sigma^2 L\left(t_p/t_r\right)/\left(2E_r\left(1 + t_r E_r/E_p t_p\right)\right) \tag{13.5}$$

where K_∞ is the value of the stress intensity factor for the case when there is no disbond and in the case of disbond, the total length of disbond perpendicular to the crack length is 2 L. To assist in meeting damage tolerant considerations, it is desirable that for a disbond of 2 L = 3 mm (0.118 in.) on either side of the crack, the resultant value of G is less than the threshold value for fatigue crack growth.

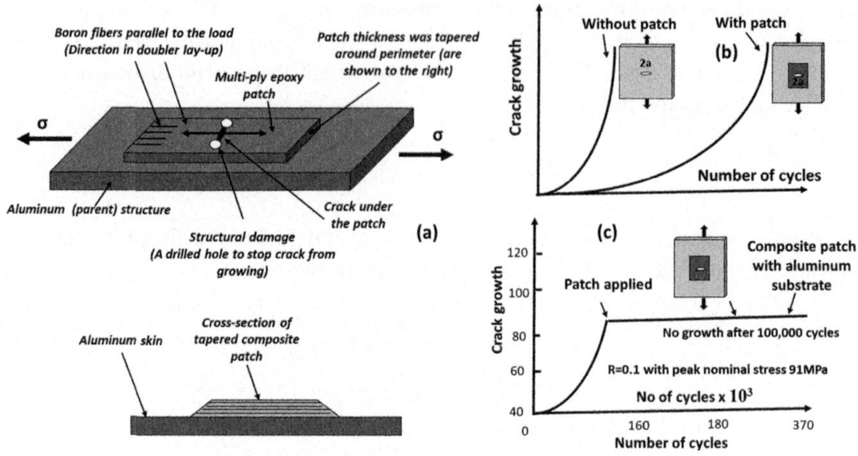

Fig. 13.13 (**a**) Bonded composite doubler installation on an aluminum skin with crack. Note that the patch is tapered through the thickness and around the perimeter (are shown on the right). Moreover, at both crack tips, a small hole is drilled to eliminate the crack tip singularities and reduce the crack tip stress concentration. (**b**) Crack growth versus number of cycles with and without patch. (**c**) Experimental data when patch was able to halt the crack growth [23]

Moreover, the length of the patch perpendicular to the crack length should be greater than $6d$, and if possible greater than $8d$, where $d = \sqrt{[E_p E_r t_p t_r t_a/(G_a(E_p t_p + E_r t_r))]}$. The term d is related directly to the load transfer length λ and it is adhesive, plate, and patch thickness and mechanical properties related. The patch configuration should be such that the edges be tapered with one ply dropped off every 3 mm (0.118 in). Figure 13.13a illustrates a bonded composite doubler installation on an aluminum skin [27]. The patch has been tapered through its thickness and around the perimeter (as it is shown in the figure). Moreover, at both crack tips, a small hole is drilled to eliminate the crack tip singularities and reduce the crack tip stress concentration.

For repairs to thin metallic skins of ~3 mm thick, the minimum patch width should be such that the patch covers as much of the crack as possible, and the patch should preferably extend at least 25 mm (~1 in) ahead of the crack tip at both ends. If this is not possible a minimum distance of 12.0 mm (~ 0.47 in) is acceptable. The presence of composite patch will be able to halt the crack from advancing when subjected to cyclic load. In Fig. 13.13b, the crack growth with the presence of patch and without patch is shown. In Fig. 13.13c, the test results on aluminum substrate with the presence of crack and subjected to constant amplitude cyclic load are shown. The crack growth as a function of number of cycles is shown with patch applied after some number of cycles, where patch essentially was able to stop the crack growth [23].

13.7.1 Comparison with Experimental Results

To illustrate the accuracy of those above formulas, the following three example cases will be considered:

Case 1 A centrally located crack, 38 mm long (~1.5 in), in a rectangular sheet of aluminum alloy with dimensions given as 300 mm × 320 mm × 2.29 mm (~11.8 in × 12.6 in × 0.09 in). The crack is patched with a unidirectional boron epoxy laminate with dimensions 160 mm × 160 mm × 0.762 mm (~6.3 in × 6.3 in × 0.03 in) and bending is prohibited. The adhesive is 0.1 mm thick (~0.004 in) and has a shear modulus, $(G_{12})_a$ of 700 MPa (~101.5 ksi). The aluminum alloy has a Youngs modulus E of 69 GPa (~10,000 ksi) and a Poisson's ratio of 0.3, while the moduli of the boron epoxy laminate are taken as:

$$E_{11}=208.3\ \text{GPa}\ (\sim 30,188\ \text{ksi}), E_{22}=24.5\ \text{GPa}\ (\sim 3550\ \text{ksi}), \nu_{12}=0.1667, G_{12}=7.24\ \text{GPa}\ (\sim 1049\ \text{ksi})$$

substituting these values into Eqs. 13.1, 13.2, 13.3 and 13.4, the quantity $\beta=0.94$, $\pi\lambda=0.71$, it gives the ratio of the stress intensity factors to be $K\infty/K_{up} \sim 0.22$, where K_{up} is the unpatched value of the stress intensity factor ($K_{up}=\sigma\sqrt{\pi a}$), which compares favorably with the value of 0.19 given in [24]. Based on the analysis provided above, where $K\infty/K_{up} \sim 0.22$, the presence of composite patch with adhesive properties shown above can reduce the crack tip stresses by a factor of 0.19 provided by [24]. For example, if the applied stress is 10ksi and crack length 2a = 1.0 inch, then $K=K_{up}=\sigma\sqrt{\pi a}=12.52$. Therefore, with $K\infty/K_{up} \sim 0.22$, the quantity of $K\infty$=2.75 ksi$(\text{in})^{0.5}$, which is below the threshold value of $K_{th} \sim 3.0$ ksi $(\text{in})^{0.5}$ for aluminums.

Case 2 This case is similar to case 1, but with the patch containing a rectangular disbond, over the entire length of the crack, with dimensions 25.4 mm × 38 mm (~1.0 in × 1.5 in). That is, the disbond is along and perpendicular to the length of the crack, where $L=12.7$ mm (~0.5 in) and 2a = 1.0 in. Substituting these values into Eq. 13.5 it gives the stress intensity factor K, where the energy release rate $G=K^2/E$ and:

$$K=0.447\ K_{up} \tag{13.13}$$

which compares favorably with the value of $K=0.381K_{up}$ given in [24]. Here, we have used the result of case 1 that without a disbond $K_{\infty}=0.19\ K_{up}$. Because of the rectangular disbond over the entire length of the crack (as was described above), the K_{∞}/K_{up} ratio is ~0.447 (slightly higher than 0.19).

Case 3 In case 1, the crack was patched on one side of the metallic plate only and the unidirectional boron epoxy laminate patch has the thickness of 0.889 mm (~0.035 in). The direction of fibers in boron epoxy are perpendicular to the crack to resist the crack propagation (as illustrated in Fig. 13.13). The adhesive is 0.1651 mm (~0.0065 in.) thick, has a Young's modulus $E_a = 189$ MPa (~27.4 ksi), and a shear modulus of $G_a = 70$ MPa (~10.2 ksi). The moduli of the boron epoxy laminate are taken the same as shown in case 1. In this case bending is allowed. To analyze this problem, a fully three-dimensional finite element modeling and symmetry are used. The aluminum sheet is modeled by 41 20-noded isoparametric bricks and thirteen of the 15-noded isoparametric elements while the composite patch is represented by 21 of the 20-noded isoparametric bricks and thirteen of the 15-noded isoparametric elements.

The elements at the crack tip are triangular form and have the midpoint nodes moved to the quarter points in order to simulate the $r^{-0.5}$ singularity at the crack tip. With these geometric and material constants, Eq. (13.1) gives the stress intensity factor value of K_p (plate) of 10.2 MPa $\sqrt{m}$ (~10 ksi$\sqrt{in.}$). The value for the Bending Correction factor (BC) is computed to be 1.876 [23]. This gives the value of the mean stress intensity factor to be 1.876 $K_p = 19.1$ MPa$\sqrt{m}$ (~18.7 ksi $\sqrt{in.}$), which compares favorably with the 3D finite element value of ~22.5 MPa $\sqrt{m}$ (~22 ksi $\sqrt{in.}$).

13.7.2 Other Repair Related Example Problems

In the previous section, three cases were studied using the approach that was provided in [23] by using Eqs. 13.1, 13.2, and 13.3. In this section, a few case studies related to patch repair will be presented that will be useful to the readers.

Example 1 (1″ Diameter Repair—Stress Analysis)

Figure 13.14 shows a 1-inch hole plate that was repaired by a composite patch placed on the plate adhesively. The hole was originally a damaged area that was first

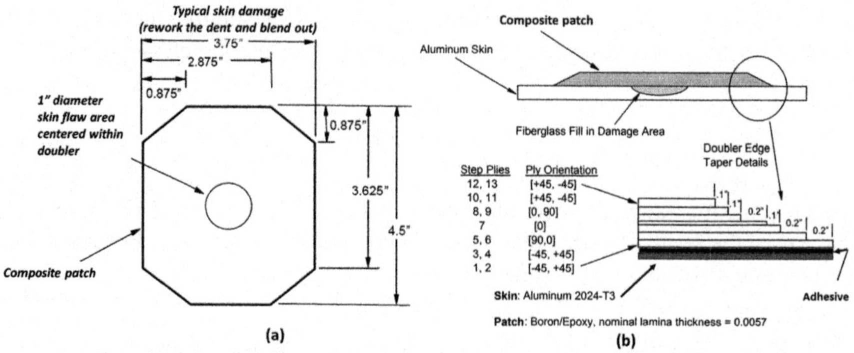

Fig. 13.14 Repair for 1″ diameter damage (rework the dent and blend out). (**a**) The composite patch and dimensions are shown. In (**b**), the patch is adhesively positioned on top of the 1 inch diameter hole. The number of plies, thickness, and orientations of plies are shown [17]

Table 13.1 Boron epoxy and adhesive properties

Material	F_{tu}	F_{ty}	F_{su}	Strain allowable
Boron Lamina	185 KSI	185 KSI	14.7 KSI	4340×10^{-6} in/in
Adhesive	N/A	N/A	5.6 KSI	1750×10^{-6} in/in
The aluminum skin is 2024-T3 clad sheet and material properties are from Mil-handbook 5G Sect. 3.2.3.				
2024-T3 Properties				
Material	F_{tu} (LT)	F_{ty} (LT)	F_{su}	
2024-T3 Bare sheet	63 KSI	42 KSI	39 KSI	
2024-T3 Clad sheet	61 KSI	40 KSI	38 KSI	

Where F_{tu} = tension ultimate stress, F_{ty} = tension yield stress, and F_{su} = shear ultimate stress

cleaned, and the final geometry prior to placing the patch is a hole as shown in Fig. 13.14. Patch geometry and typical dimensions are shown in Fig. 13.14a. The composite patch had 13 plies, quasi-isotropic, with symmetrical lay-ups and ply thickness are also shown in Fig. 13.14b. The maximum stresses provided by the engineers through the finite element analysis for plate, skin, and adhesive during the worst-case DC-10/MD-11 were obtained and the margin of safety for each were analyzed. Aluminum parent material, adhesive, and boron-epoxy properties used in the composite patch design analysis all are shown in Table 13.1.

For the aluminum skin region, the margin of safety is:

$$\sigma_1 = 41.9 \text{ ksi } \rightarrow \left(\sigma_{\text{yield}} = 47 - 50 \text{ ksi and } \sigma_{tu} = 61.0 \text{ ksi}\right)$$
$$\sigma_2 = -\ 8 \text{ ksi and } \tau_{max} = 22 \text{ ksi} \rightarrow \left(\tau_{\text{allowable}} = 39 \text{ ksi}\right)$$

Margin of safety in aluminum skin for tension is calculated as follow:

$$\text{M.S.} = \frac{F_{tu}}{\sigma_1} - 1 = \frac{61000}{41900} - 1 = +0.46$$

And the Margin of safety for shear:

$$\text{M.S.} = \frac{F_{su}}{\tau_{max}} - 1 = \frac{38000}{22000} - 1 = +0.73$$

Maximum Stress in Adhesive

$$\tau_{ZX} = 4.8 \text{ ksi } \rightarrow \left(\tau_{\text{allowable}} = 5.6 \text{ ksi}\right)$$
$$\tau_{ZY} = 5.0 \text{ ksi} \rightarrow \left(\tau_{\text{allowable}} = 5.6 \text{ ksi}\right)$$

Margin of Safety in Adhesive

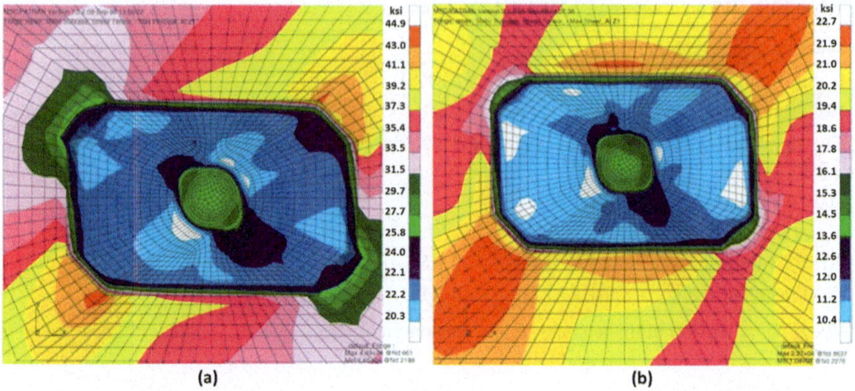

Fig. 13.15 (**a**) Maximum principal tensile and (**b**) shear stress in composite patch and aluminum skin for 5″ diameter repair design and worse-case load spectrum. Stresses were obtained through the finite element analysis provided by [27]

$$\text{M.S.} = \frac{F_{su}}{\tau_{max}} - 1 = \frac{5600}{5000} - 1 = +0.12$$

Maximum stress in composite patch by applying the maximum strain failure criteria, the laminate critical load can be determined from the transverse ply strain (ε_z) in the $-45°$ ply. Moreover, the margin of safety is calculated based on the largest strain value found in the FEM. That is:

$$\text{M.S.} = \frac{\varepsilon_2(\text{Allowable})}{\varepsilon_2(\text{critical location in} - 45° \text{ ply})} - 1 = \frac{4340x10^{-6}}{3735x10^{-6}} - 1 = +0.16$$

Example 2 (5″ Diameter Repair—Stress Analysis)

The analysis of a damage that was cleaned and the final geometry is in the form of 5 in. diameter hole. A composite patch is used to prevent any crack initiation and growth from the periphery of hole when subjected to operating stress. Figure 13.15a, b shows the maximum tensile and shear principal stresses in the composite patch and aluminum skin for 5 in diameter repair design and worse case load spectrum. The margin of safety for skin periphery, adhesive, and the composite patch are shown below.

Margin of safety analysis for the 5 in hole with the composite patch are shown here. The maximum stress from the finite element analysis for the aluminum skin (outside the patch area shown with yellow color) are:

$$\sigma_1 = 44.9 \text{ ksi } \left(\sigma_{\text{yield}} = 47 - 50 \text{ ksi and } \sigma_{\text{tu}} = 61 \text{ ksi}\right)$$
$$\tau_{\text{max}} = 22.7 \text{ ksi } \left(\tau_{\text{allowable}} = 38 \text{ ksi}\right)$$

Margin of safety for the aluminum skin in tension (Fig. 13.15):

$$\text{M.S.} = \frac{F_{tu}}{\sigma_1} - 1 = \frac{61.0}{44.9} - 1 = +0.12$$

Margin of safety for the shear (Fig. 13.15):

$$\text{M.S.} = \frac{F_{su}}{\tau_{max}} - 1 = \frac{38.0}{22.7} - 1 = +0.36$$

The maximum stress from the finite element analysis for the adhesive are:
τ_{ZX} = component has lower magnitude and is less critical than τ_{ZY}:
$\tau_{ZY} = 4.1$ ksi (τallowable = 5.6 ksi). The margin of safety for the adhesive:

$$\text{M.S.} = \frac{F_{su}}{\tau_{max}} - 1 = \frac{5.6}{4.1} - 1 = +0.37$$

Maximum stress in composite patch is based on maximum strain failure criteria. The laminate critical load is determined from the transverse ply strain (ε_z) in the – 45° ply at the second tapered step up from the patch edge (base). The margin of safety is calculated based on the largest strain value found in the finite element model:

$$\text{M.S.} = \frac{\varepsilon_2(\text{Allowable})}{\varepsilon_2(\text{critical location in} - 45^\circ\ ply)} - 1 = \frac{4340x10^{-6}}{3735x10^{-6}} - 1 = +0.16$$

13.7.3 Damage Tolerance and Crack Growth Analysis

The damage tolerance analysis of skin, patch, and adhesive were studied. The mode I stress intensity factor (K_I) for the repaired and unrepaired configurations was determined using a computer program for the aircraft repair analysis provided by [9, 29]. Table 13.2 compares the stress intensity factor for the three composite patch sizes with different initial flaw sizes. The crack is assumed to be initially the same size as the damaged area (i.e., 1, 3, and 5 in), and the applied stress level is 20.68 ksi (the hoop stress in the skin is considered as the applied stress). Values of the stress intensity factor for crack growth with incremental growth of 0.5 in are recorded in Table 13.2. Note that the material has no residual strength when fracture toughness is $K_I = Kc = 77$ for 2024-T3 aluminum skin (t ~ 0.080 in). In the case of a 5 in diameter hole plate with a patch size of 21 × 13 in, the stress intensity factor for unrepaired case exceeds the material fracture toughness when the crack size reaches 9.0 in. in length.

In Table 13.2, the calculated stress intensity factor for repaired case, where crack in the metallic skin is under the composite patch, is a constant for a given patch size. Even though the level of stress intensity factor for repaired patch is a constant, but

Table 13.2 Stress intensity factors for unrepaired and repaired skin with composite patch

Patch Size 4.5" x 3.75"		
Crack Length	K_I Repaired	K_I Unrepaired
1.0"	9.32	25.94
1.5"	9.32	31.77
2.0"	9.32	36.69
2.5"	9.32	41.02
3.0"	9.32	44.94

1" Diameter

Patch Size 13" x 9"		
Crack Length	K_I Repaired	K_I Unrepaired
3.0"	9.06	44.94
3.5"	9.06	48.54
4.0"	9.06	51.89
4.5"	9.06	55.03
5.0"	9.06	58.01
6.0"	9.06	63.55
7.0"	9.06	68.64

3" Diameter

Patch Size 21" x 13"		
Crack Length	K_I Repaired	K_I Unrepaired
5.0"	8.85	58.01
5.5"	8.85	60.84
6.0"	8.85	63.55
6.5"	8.85	66.14
7.0"	8.85	68.64
8.0"	8.85	73.38
9.0"	8.85	77.83
10.0"	8.85	82.04

5" Diameter

A comparison for the three composite patch sizes. The crack is assumed to be initially the same size as the damaged area. The stress level is 20.68 ksi. For unrepaired case where the crack length of 2a>9.0", the calculated $K=K_C=77$ (Skin: AL 2024-T3 and t=0.080 inch).

they are slightly different (i.e., $K_I = 9.32 > K_I = 9.06 > K_I = 8.85$ ksi(in)$^{0.5}$), which can be related to having different patch size used for 1.0, 3.0, and 5.0 in diameter hole (Table 13.2). A simple fracture mechanics analysis of a crack with 2c = 5.0 in subjected to applied stress of 20.68 ksi, and the stress ratio of $R = 0$, provides the number of cycles to failure of $N_f = 70$ cycles. The analysis was performed by the NASGRO computer code using fatigue crack growth equation described in Chap. 9 (Eq. 9.3) for 2024-T3 aluminum skin ($t \sim 0.080$ in). In the cases of 2c = 3 and 1.0 inches crack length, the number of cycles to failure will be $N_f = 495$ and 3406 cycles, respectively. These analyses were conducted for metallic plates with crack lengths of 2c = 5.0, 3.0, and 1.0 in all situated outside the patch. The same analysis will be conducted for the case when cracks are under the composite patch.

The fracture mechanics analysis and the number of cycles to failure for the three above mentioned cases (2c = 5.0, 3.0, and 1.0 in) when cracks are under the patch will be performed by first calculating the magnitude of the applies stress under the patch with $R = 0$. By dividing the stress intensity factor K_I for the repaired and unrepaired cases (Table 13.2), and by considering the magnitude of the stress for unrepaired cases is 20.68 ksi, then the stress magnitude for the repaired can be calculated. That is,

$$(K_1)\text{unrepaired}/(K_1)\text{repaired} = 25.94/9.32 = 2.78 \text{ (for 2c} = 1.0 \text{ inches)}$$

$$(\sigma)\text{unrepaired}/(\sigma)\text{repaired} = 2.78$$

$$(\sigma)\text{repaired} = 20.68/2.78 = 7.43 \text{ ksi}$$

With the same analogy, the values of applied stress inside the patch for the cases of 2c = 3 and 5 in will be 4.16 and 3.15 ksi, respectively.

By using Eq. 9.3 of Chap. 9 and the NASGRO computer code, the number of cycles to failure $N_f = 487410$, 397484, and 171187 cycles for 2c = 1.0, 3.0, and 5.0 in, respectively. The calculated number of cycles to failure for the repaired cases clearly indicates that when applying the composite patch on the damaged area, it will enhance the life of the part and will slow down the crack from advancing. That is, the ratio of number of cycles to failure for repaired versus unrepaired cases will be at least 50 times larger for all cases:

$$(N_f)_{\text{repaired}}/(N_f)_{\text{unrepaired}} > 50$$

13.8 Parameters That Must Be Considered When Conducting Patch Repair

There are instances that an adhesive patch bonded repair may not be appropriate or not feasible to implement because of its effect on the metallic substrate. The following are brief descriptions of cases that engineers found it difficult to go forward with using composite patch repairing:

- Accessibility to the repaired area. Must have enough access to conduct the repair. That is, the ability to do surface preparation to allow the patch repair to be performed.
- The ability to access for inspecting the repair after installation to check the presence of any mistakes.
- Geometric limitations to repair the part. The presence of protruding fasteners, changes in section, tight curvatures make, they all make the repairing less effective.
- Repairing should consider the influence of the environmental hazard. The presence of surface contaminants reduces the quality of the bond. Typical environmental conditions that can be detrimental are high humidity, oils, and a large temperature fluctuation.
- The application of surface preparation prior to the actual repair must not damage the parent material. If surface preparation involves heating, the reduction of strength of parent material must be taken into consideration.

13.9 Summary of Composite Repair

Composites are gaining acceptance in major industries within the USA and their usage is mainly in wind energy, aircraft, aerospace, military equipment, boats, bridges, sporting goods, automotive parts, pressure vessels, and others. Composite parts will eventually fatigue during their service usage, and the degree of damage due to fatigue will determine if they need to be repaired or replaced. The flowchart shown in Fig. 13.2 will help engineers decide whether repairing the part is necessary.

In some cases, a temporary repair may be adequate, but later the temporarily repaired part should go for replacement or a permanent repair. Regarding the analysis of repaired parts, a 3D finite element modeling and subsequent progressive damage failure for bonded composite repairs was proposed. In the case of aircraft adhesively bonded repair, the analytical model shown in Sect. 13.7 has been successfully applied to the cracked metallic plate of fuselage having a through crack. The model can assess the crack tip stress intensity factor, K, patch debonding and damage accumulation in the composite patch as functions of the applied load. For the patch to be effective, it should be placed over the damaged area after proper surface preparation to ensure adequate adhesion between the two surfaces has been implemented. Note that, a major worry with any bonded repair is the disbond that can occur at installation of the composite patch at any time during the service life of the part.

Finally, by following these procedures, engineers can be confident enough to go ahead with their repair projects. The key steps for a successful repair are as follows:

1. *Inspection and damage assessment*: Engineers should inspect the damage thoroughly, mark the damage, identify what material the original part was made from, and decide if the repair makes sense based on Fig. 13.2.
2. *Removal and Preparation*: The number one area that dictates the success of a composite repair is the removal of damaged materials and proper preparation of the bonding surface.
3. *Laminate the repair*: Make the structural repair ply by ply, have your materials precut, and use the appropriate resin. For cosmetic repairs, have the appropriate materials on hand, and do not do more work than is necessary which may cause more damage to the part.
4. *Inspect the Repair and finish*: After repair is completed, be sure that you inspect the repaired part before putting it back into service. After engineers have made a cosmetic finish, spend the extra time to ensure it is flawless.

Finally, it is important to note that the effect of temperature on the strength of adhesive joints is an essential factor to consider when designing the adhesive joints. The mechanical properties of adhesive joints vary as a function of temperature, humidity, and it is also dependent on the coefficient of thermal expansion of both the adhesive and adherend. Generally, adhesive joints strength improves at lower temperatures and reduces at higher temperatures [28, 29].

13.10 Concluding Remarks

The study of composite repair and its industrial applications underscores the crucial role that repairing parts play in maintaining the structural integrity of components in various industries. The presented overview of repair methodologies, materials, and case studies emphasizes the adaptability and effectiveness of composite repair techniques in addressing various types of damage. This not only pertains to repairing composite parts but also extends to the repair of damaged metallic substrates.

Currently, engineers are actively promoting the use of lightweight and high-performance composite materials in designing critical parts. The ease of their contributions in repairing damaged parts, rather than replacing them, proves to be economically beneficial for industries. The ability to extend the service life of components through repairs, along with the potential time and cost savings, is crucial for industries to continue performing their functions.

Looking ahead, composite repair of industrially damaged parts is economically useful, and companies must place more emphasis on research and study to provide more acceptable and robust repair techniques, ensuring widespread repair adoption across industries. In summary, the application of composite repair for industrial use is marked by achievements, lessons learned, and a clear trajectory toward greater efficiency and reliability. Engineers must also bear in mind the importance of achieving an acceptable repair to the customers must come with a comprehensive damage tolerance analysis and a complete nondestructive inspection to be sure the damaged repair will survive the operating environment.

References

1. M. Hoke, *Advanced Composite Structural Repair During Manufacturing* (Aerospace Manufacturing and Design, 2013)
2. National Aeronautics and Space Administration Lyndon B. *Johnson Space Center Houston, Fracture Control Plan for JSC Space Flight Hardware*, By Mohammed Shoeb (2018)
3. Composite World (CW), *Composites Repair*, Published 2014
4. C. Kassapoglou, K. Rangelov, S. Rangelov, Repair of composites: Design choices leading to lower life-cycle cost. Appl. Compos. Mater. **24**(4), 945–963 (2017)
5. A.A. Baker, L.R.F. Rose, R. Jones, *Advances in the Bonded Composite Repair of Metallic Aircraft Structure* (Elsevier, 2003)
6. M. Hoke, Composite Repair (Aerospace Manufacturing and Design, June 2011)
7. S.H. Myhre, J.D. Labor, Repair of advanced composite structures. J. Aircr. **18**(7), 546–552 (1981)
8. S.H. Ahn, G.S. Springer, *Repair of Composite Laminates* (Department of Aeronautics and Astronautics Stanford University, 2000)
9. Fiberglass, *The Ultimate Fiberglass Repair Guide*, The Ultimate Fiberglass Repair Guide
10. S. Mohammadi, M. Yousefi, M. Khazae, A review on composite patch repairs and the most important parameters affecting its efficiency and durability. J. Reinf. Plast. Compos. (2020)
11. K.B. Armstrong, W. Cole, W.G. Bevan, *Care and Repair of Advanced Composites* (SAE International, London, 2005)
12. C.H. Wang, C.N. Duong., Design of scarf and doubler-scarf joints, in Bonded Joints and Repairs to Composite Airframe Structures (2016)
13. Boeing 777 Structural Repair Manual. Boeing Commercial Aircraft Co., 1995
14. E.F. Chesmar, *Metal Bond and Composite Repairs: Similarities and Differences*, in Composites '96 Manufacturing and Tooling Conference, Society of Manufacturing Engineers (1996), pp. 281–290
15. E. David Reedy Jr., *The Mechanics of Adhision, Adhesion Science and Engineering*, 2002

16. R. Fredell, C. Guijt, J. Mazza, An integrated bonded repair system: A reliable means of giving new life to aging airframes. Appl. Compos. Mater. **6**, 269–277 (1999)
17. P. Papanikos, K.I. Tserpes, G. Labeas, et al., Progressive damage modelling of bonded composite repairs. Theor. Appl. Fract. Mech. **63**, 219–230 (2004)
18. P. Papanikos, K.I. Tserpes, Initiation and progression of composite patch debonding in adhesively repaired cracked metallic sheets. Compos. Struct. **81**, 303–311 (2007)
19. ANSYS User's Manual. Version 5.7. Swanson Analysis Systems, Inc, 2001
20. H. Tresca, Mémoire sur l'écoulement des corps solides soumis à de fortes pressions. C. R. Acad. Sci. Paris **59**, 754 (1864)
21. R. Jones, W.K. Chiu, S. Hanna, Potential failure mechanisms of bonded composite repairs for metal and concrete. Theor. Appl. Fract. Mech. **21**, 107–119 (1994)
22. Baker, R. Jones, *Bonded Repair of Aircraft Structures* (Martinus Nijhoff, 1988)
23. R.F. Rose, A cracked plate repaired with bonded reinforcements. Int. J. Fract. **18**, 135–144 (1982)
24. Ratwani, *Development of Bonded Composite Repairs for Cracked Metal Structure*, in Proc. International Workshop on Defence Applications of Repair Technology, NRL, Washington D.C., 22–24 Jul 1981 (1981), pp. 307–403
25. R. Jones, R.J. Callinan, A design study in crack patching. J. Struct. Mech. **1**(7), 107–130 (1979)
26. R. Jones, M. Heller, D. Lombardo, S. Dunn, J. Paul, D. Saunders, Thermoelastic assessment of damage growth in composites. Compos. Struct. **12**, 291–314 (1990)
27. D. Roach, K. Rackow, *Development and Validation of Bonded Composite Doubler Repairs for Commercial Aircraft*, SAND2007-4088, Unlimited Release Printed July 2007
28. J.M. Sousa, J.R. Correia, J.P. Firmo, S. Cabral-Fonseca, J. Gonilha, Effects of thermal cycles on adhesively bonded joints between pultruded GFRP adherends. Compos. Struct. (2018)
29. Y. Zhang, A.P. Vassilopoulos, T. Keller, Effects of low and high temperatures on tensile behavior of adhesively-bonded GFRP joints. Compos. Struct. **92**(7), 1631–1639 (2010)

Chapter 14
Application of Nondestructive Inspection for Composite Materials

14.1 Introduction

In recent years composite structures have been used extensively in aviation, space, defense, aircraft, automotive, and many others. The implementation of nondestructive inspection testing (NDT) of metallic and recently composite structures is necessary after the manufacturing phase, and before their operations, to assure the quality of products. Their application for in-service damage detection has provoked the development and adaptation of several inspection methods and techniques over the years [1–4]. In this chapter, various nondestructive testing methods (also referred to as nondestructive inspection, NDI) will be discussed, which describes the method of inspecting composite parts that will be utilized by the users of composites. Different NDI methods are currently available to detect flaws in the composite structures before and during their service usage. If the structure went through repair, then the damaged areas should also be inspected after the completion of the repair, in order to be sure of the integrity of the repaired area. In some cases, additional interval inspections and testing can be requested by the engineers to check for any stable damage growth during the part's usage. Therefore, it becomes vital for the industry to apply the NDI methodsto detect defects in composite parts after the manufacturing phase, during their usage, and if the repairing process was performed on the part, a limited amount of testing and inspection might be necessary to implement. The strict quality assurance policy of some organizations enforces that critical components made of composites should be inspected for the presence of defects by utilizing the proper nondestructive inspection techniques. These inspections can help detect any preexisting delamination or defects and to determine how fast the rate of their growth has been during the part's usage. Specific acceptance criteria will be helpful for engineers to know to determine whether a composite material meets the ASTM acceptance criteria or not. For example, the ASTM-E2533 has requirements that are intended to control the quality of NDT data regarding the size and location of defects, but not to establish acceptance criteria for the aircraft and aerospace service

B. Farahmand, *Fundamentals of Composites and Their Methods of Fabrications*,
https://doi.org/10.1007/978-3-031-80201-0_14

[5]. The ASTM-E2533 provides information to help engineers select appropriate nondestructive testing (NDT) methods to characterize aerospace polymer matrix composites (PMCs). Another standard and NDI acceptance criterion is the SAE ARP5605 for aerospace and aircraft (established by the Society of Automotive Engineers that addresses the inspection requirements for composite materials). The ARP (Aerospace Recommended Practice) is to describe the design and production of solid composite laminate calibration standards to be used in ultrasonic, resonant, and tap test NDI equipment calibration for accomplishment of damage assessment and post-repair inspections.

A list of several NDIs currently being used by industries is provided here. Due to the large variety of composite materials and the type of NDI techniques, it is not feasible to address all aspects of the NDI applications in composites here. There are other inspection methods available besides those mentioned here that currently are being used in industries, but only A-scan, B-scan, C-scan, tap hammer device, and the mechanical impedance analysis will be briefly described here. Other NDI methods are X-ray, and thermography inspection, which can be used to detect hidden damage, but will not be discussed in this chapter. For more information, the reader can refer to [6, 7]. A brief description of each NDI method will be described below. These methods are:

- Through-transmission ultrasonic (TTU) A-scan, B-scan, and C-scan.
- Manual tap test (tap hammer device).
- Mechanical impedance analysis method (MIA).

14.2 Through-Transmission Ultrasonic (TTU)

The TTU is a nondestructive inspection technique that can be utilized for composites in which a short pulse of ultrasonic energy is incident on a composite specimen to detect the presence of defects in the material. This NDT method (also known as pulse echo ultrasonic) is a highly effective testing tool for corrosion mapping and flaw detection. It is important for the reader to know the role of the probe and transducer when dealing with NDI [8–10]. The ultrasonic probe or a transducer is a device that converts the pulse of electric energy generated from the equipment to mechanical energy. That is, high-frequency sound waves from the probe travel through the tested material to check for anomalies in the material (Fig. 14.1). These sound waves, after bouncing back from the test subject, are captured, and converted by the probes into an electrical energy pulse that can be assessed for flaw detection.

In brief, the transmitter emits ultrasonic pulses and analyzes the signal transmitted back for the inspection of the component. There can be many formats for collecting and displaying the ultrasonic data such as A-scan, B-scan, and C-scan which will be discussed later in this section. All techniques use sound wave energy with a frequency above the audible range, usually several MHz. Most UT inspection applications use short pulse waves with frequencies ranging from 0.1 to 15 MHz,

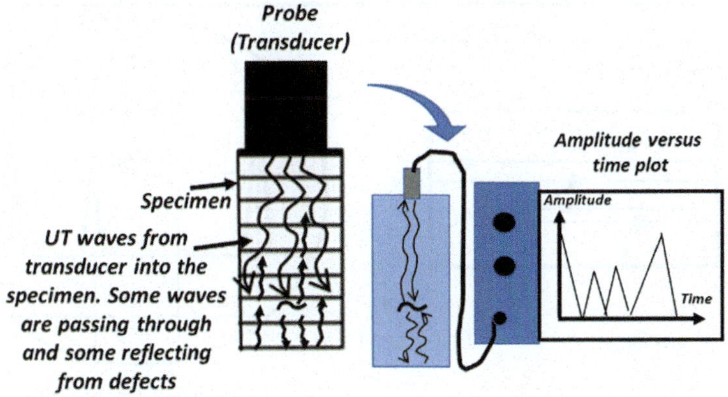

Fig. 14.1 These high-frequency sound waves are generated by small ultrasonic probes also known as ultrasonic transducers. These probes are an essential part of any ultrasonic testing device and come in a wide range of frequencies, sizes, and case styles for different inspection purposes including flaw detection in composite parts

although frequencies up to 50 MHz can be used. Therefore, a high-frequency sound wave is launched into the part, which can be directed to travel normally to the part surface (through the thickness), along the surface of the part, or at some programmed angle to the part surface. It is necessary that the operator tries different directions, besides the direction parallel to the part thickness, to locate the flaw in the structure, if any.

The measurement of the transmitted pulse indicates the degree of attenuation or reduction of the incident pulse. The reduction of the pulse is because of voids, delaminations, the state of resin cure, the fiber volume fraction, the condition of the fiber–matrix interface, and any foreign inclusions present. When there is a discontinuity, such as a crack or delamination along the wave path, part of the energy will be reflected from the flaw surface. Yet some waves will go through and reflect from the back surface. The reflected wave signal is transformed into an electrical signal by the transducer and is displayed on a screen (Fig. 14.2). The transducer can convert the reflected signal from one form to another. In this case, the electrical signals are converted to physical quantities, such as energy. The conversion of one form of energy to another is known as transduction. In the case of TTU, the ultrasonic wave is used for locating defects in a material by passing acoustic energy (sound waves) through the part. The couplant agents, such as water or gel are employed to allow the maximum amount of wave to transmit to the part. It enables the transmission of sound energy between the transducer and the test specimen.

As can be seen from Fig. 14.2, the ultrasonic waves are transmitted through the thickness of the inspected structure [1]. In the case of A-scan (short for amplitude scan), which will be discussed later, one transducer is used which will act as a transmitter and act as a receiver of the sound energy waves when they are reflected. A reduction or a loss of the signal indicates the existence of a flaw, where an image is generated on the oscilloscope screen as the part is being inspected.

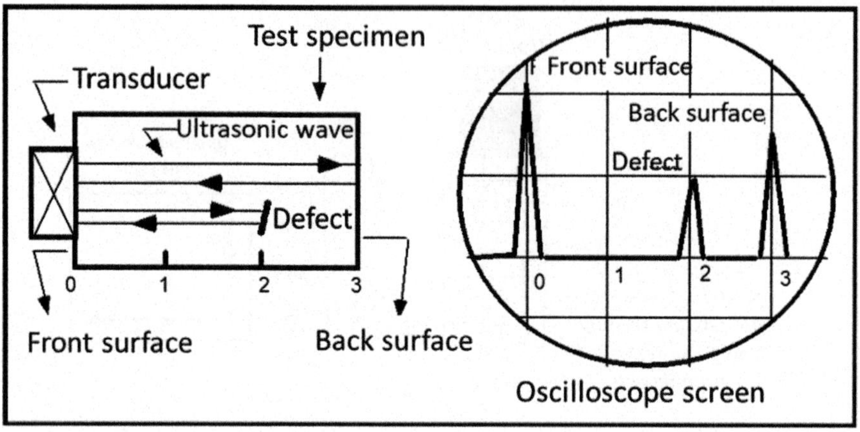

Fig. 14.2 When there is a discontinuity (such as a crack) in the wave path, part of the energy will be reflected from the flaw surface. The reflected wave signal is transformed into an electrical signal by the transducer and is displayed on the oscilloscope screen. The reflected wave from flaw surface and the one from the back surface will be compared

The instrument shown in Fig. 14.2 (like Fig. 14.1) illustrates the loss as a percent of the original signal strength, where the signal loss is compared to a reference standard. The reference original signal strength is established and utilized not only to calibrate the ultrasonic equipment but also to measure the actual thickness without the presence of defects. Areas with a greater loss than the reference standard indicate the presence of defects.

It should be mentioned that the path of the ultrasonic beam is not a straight line but follows a complex path (the beam is reflected one or more times before reaching the receiver). It is also worth mentioning that the media has the tendency to resist the transmission of ultrasound waves by virtue of its density. That is, the denser the medium, the slower will be the velocity of sound through it.

There are three types of TTU inspections currently being used in the industry, and each will be discussed as part of Sect. 14.2.

14.2.1 Through-Transmission Ultrasonic—A-Scan

Through transmission, ultrasonic A-scan (or amplitude scan) was used in the early days of ultrasonic testing and is still used to probe flaws in the part [11, 12]. It is a point-to-point ultrasound technique that can be time consuming when large areas must be inspected. A probe on one side of the composite part transmits an ultrasonic pulse to the specimen, as shown in Fig. 14.3, where both the reflected and transmitted signals are plotted as amplitude versus time or depth. The absence of a full pulse coming to the receiver indicates a defect. That is, the presence of defects shows the reduction of the signal, which means the signal was reflected sooner as the result of

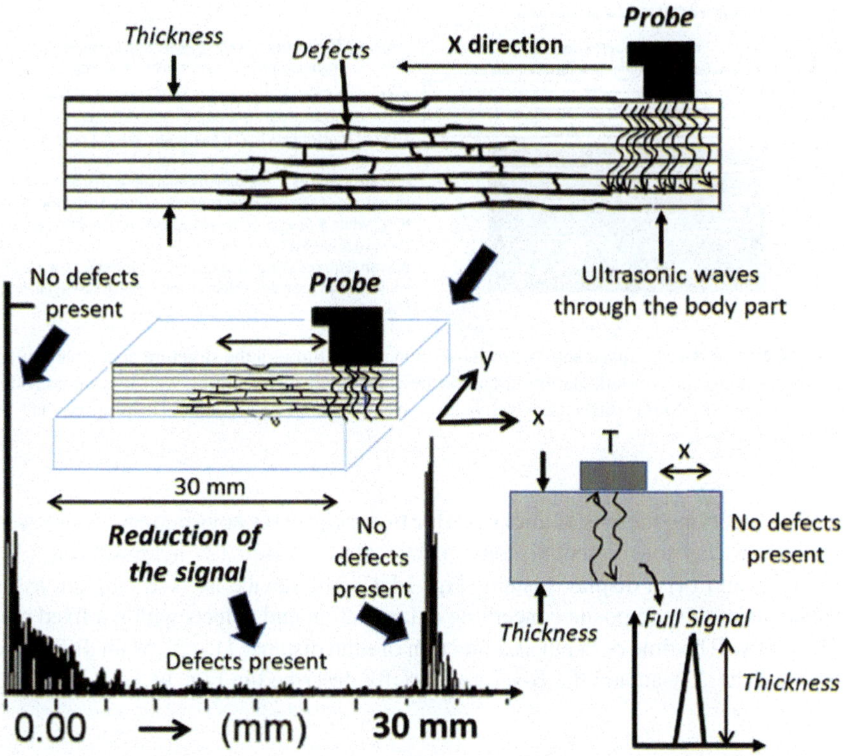

Fig. 14.3 This is the standard TTU (scan A) inspection method and relies on the operator to move the ultrasound probe over the surface of the structure in the X direction. At each measurement location, the operator checks the reading on the flaw detector

the defect's presence rather than going all the way to the backwall. In all cases, the signal will be compared to a standard signal prepared by the inspector (Fig. 14.3). Several ultrasonic systems using imaging techniques to find defects have been developed. However, the enhanced capabilities come with greater complexity, higher training requirements, and higher costs.

14.2.2 Through-Transmission Ultrasonic—B-Scan

A B-scan concept is taken from the A-scan, where it is constructed from a series of A-scan signals. In the case of the A-scan, the operator moves the incident ultrasound wave, emitting from the probe, over the surface of the structure to capture defects in the thickness direction at each location, as was shown in Fig. 14.3 for the A-scan. The B-scan image shows a cross-sectional view through the structure and is particularly useful for analyzing ply orientation in carbon fiber laminates, or in the case of

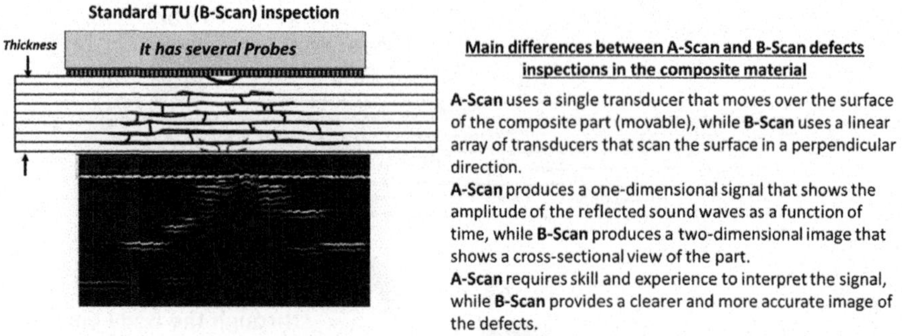

Fig. 14.4 The B-scan image shows a cross-sectional view through the structure and is particularly useful for analyzing ply orientation in carbon fiber laminates or recording wall loss on corroding steel structures. Main differences between A-scan and B-scan defect inspections is also described here

metals, recording the loss of thickness due to the corrosive environment. Any B-scan equipment has four essential components: (1) a pulser, (2) a transducer, (3) a receiver, and (4) a display system (Fig. 14.4). The device is fixed and can assess the status of the composite concerning delamination and defects within a fixed area. The plot will be time or depth as a function of scan distance [1, 13]. Main differences between the A-scan and the B-scan are briefly described in Fig. 14.4.

14.2.3 Through-Transmission Ultrasonic—C-Scan

A color scan image, also known as a C-scan, is constructed by moving the ultrasound beam over the structure's surface to capture any defects in the material. C-scan allows for top-view reflector area (plain view) and depth estimation using the time-of-flight information to capture the defects in composite material. At each measurement position, the depth and amplitude of diagnostic echoes within the scan signal are recorded. The C-scan image then shows a clear plain view of the recorded data revealing the size, orientation, and corrosion spots, as well as visualization of cracks and inclusions in the component. In the case of the A-scan and B-scan, which both can provide a 2D image of defects; however, the C-scan ultrasonic testing displays the data alongside the depth, also known as the time-of-flight diffraction (TOFD), which measures the time of flight of a pulse as it travels from a transmitting probe to a receiving probe [14–16]. In this method, a pair of ultrasonic probes (transmitter and receiver) sit on opposite sides of the test sample under consideration. The transmitters emit an ultrasonic pulse received by the probe on the other side. In the undamaged part with no defects, the signals picked up by the receiver probe are from two waves, one that travels along the surface and one that reflects off the backwall. When a defect exists in the part, there is a diffraction of the ultrasonic wave from both tips of the defect, as illustrated in Fig. 14.5. Using the measured time of flight of

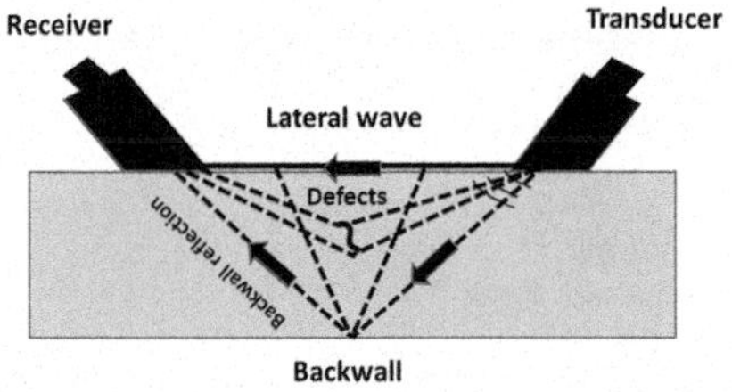

Main differences between B-Scan and C-Scan defects inspections in the composite material

C-Scan uses a water tank to couple the sound waves between the transducer and the part, while B-Scan uses a linear array of transducers that scan the surface in a perpendicular direction.
C-Scan produces a two-dimensional image of the part, showing the defects as different colors or shades, while B-Scan produces a two-dimensional image that shows a cross-sectional view of the part, with the defects appearing as bright spots or lines.
C-Scan is more suitable for inspecting composite parts, the part thickness, or the amount of corrosion on a metallic sheet. However, the B-Scan is more suitable for both solid laminate as well as sandwich structures for detecting defects in the core and skin

Fig. 14.5 In this setup, the receiver sees four signals: one from the lateral wave that has traveled along the top surface, one from the wave that has reflected off the back surface, and two from the defect in the test specimen. The main differences between C-scan and B-scan are described here

the pulse, the object's depth (defect or crack) can be calculated automatically by simple trigonometry. Engineers who are interested in accurately sizing the height of defects should be using the C-scan technique. Main differences between the C-scan and B-scan are briefly described in Fig. 14.5.

Finally, ultrasonic inspection has some limitations in that it requires a skilled specialist to set up the gauges and be able to interpret the results correctly. That said, the benefits of ultrasonic inspection far outweigh the potential disadvantages. That is the reason why the UT inspection method is an essential NDT method for a wide variety of materials, not just composites, across many different industries.

14.3 Manual Tap Test (Tap Hammer Device)

The manual tap inspection test, sometimes called the audio or sonic tap inspection test, utilizes frequencies in the audible range between 10 and 20 Hz [17–19]. Amazingly, the method is accurate if the experienced personnel is familiar with the method of approach. As can be seen from Fig. 14.6, tap testing is perhaps the most common, simple, and inexpensive inspection technique used for the detection of delamination or ply separation. Flaws can be detected by the operator when a change in the audible response occurs as the structure is tapped. That is, the operator will tap the composite part with a lightweight hammer-like device in different locations of the composite part, and the audible response is monitored by the human ear as illustrated in Fig. 14.6. While listening, a clear, sharp, ringing sound is suggestive of a well-bonded solid structure, while a dull-like sound is indicative of an anomaly in that area, which is indicative of a flaw or delamination. The dull-like sound means the incoming sound wave was absorbed by the defect in the material. The whole idea is the possibility that changes within the internal elements of the structure might

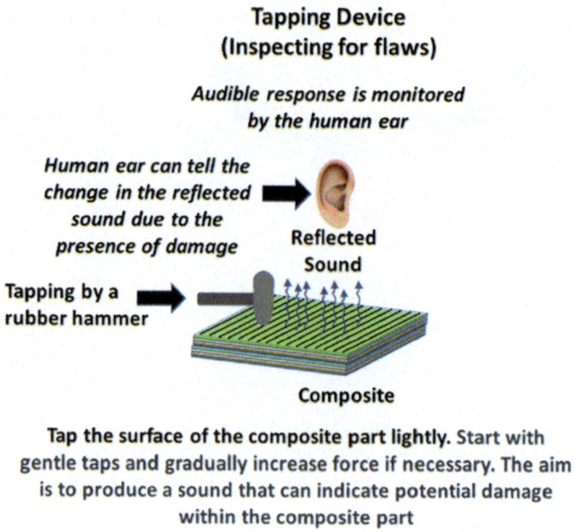

Fig. 14.6 Flaws can be detected when a change in the audible response occurs as the composite structure is tapped

produce pitch changes that are interpreted as defects. This tap inspection method should be accomplished in a quiet area if possible and to be conducted by experienced personnel familiar with the part's internal configuration. The tap hammer device process is also called rapid damage detection device or simply RD.

14.3.1 Automated Tap Test

Automated Tap Test is like the manual tap test except that a solenoid is used instead of a hammer for tapping the composite material. The solenoid produces multiple impacts in a single area, where the tip of the impactor has a transducer that records the force versus time signal of the impactor. A solenoid is a device that converts electrical energy to mechanical energy, using an electromagnet formed from a coil of wire.

The magnitude of the force depends on the impactor, the energy of impact, and the mechanical properties of the composite structure. Therefore, the sound signal from an unflawed region is used for calibration, and any change from this unflawed signal indicates the existence of damage. This can be done by computer software that displays a color-coded map of the part, showing the defects as different colors. Automated tap testing provides faster, more accurate, and more consistent results than manual tap testing. Below is the list of the automated tap testing being helpful to industries:

- Detection of disbond in composite materials, tiles, coatings, etc.
- Detection of disbond of joints in composite structures.
- Location of beams/ribs/spar in bonded structures.

- Detection of boundaries between different materials.
- Detection of dryness/aging/hardening of bonding agents, etc.
- Comparison and evaluation of surface hardness (elasticity) of various plates.
- Detection of the unfilled area within foamed sandwich panels.

14.4 Mechanical Impedance Analysis for the NDI Method

The mechanical impedance analysis (MIA) method is a well-established NDI method for detecting voids or a separation (disbond) between plies because of the lack of adhesive in laminated structures up to a depth of a few millimeters (~6 mm). Mechanical impedance is a measure of how much a composite structure resists the incoming propagating wave when subjected to a harmonic force and is a function of the frequency of the applied force, which can vary significantly over the frequency of the wave. In this NDI technique, the mechanical impedance (or stiffness) measurements are aimed at evaluating the damage-induced changes of structural parameters reflected in the frequency response, where the frequency response of a system is the quantitative measure of the magnitude and phase of the output as a function of input frequency (Fig. 14.7) [20, 21]. In the case of impact on the surface of a composite part, the depth of damage can be assessed through the MIA method. If there appears to be no disbond indication, then this would suggest no damage was done on the surface of composite due to the impact. However, it is a good idea to conduct the MIA inspection on the opposing surface of the impacted area to be sure no ply separation was done on the composite.

The MIA method is a fast, simple, and reliable NDI method. It requires no liquid couplant, whereas in the case of the TTU ultrasound method a couplant is needed. Furthermore, the MIA method can be used to inspect flaws on irregular or curved surfaces, while other methods may have difficulty inspecting for defects embedded in curved surfaces. But one of the limitations of MIA is that it is sensitive to surface

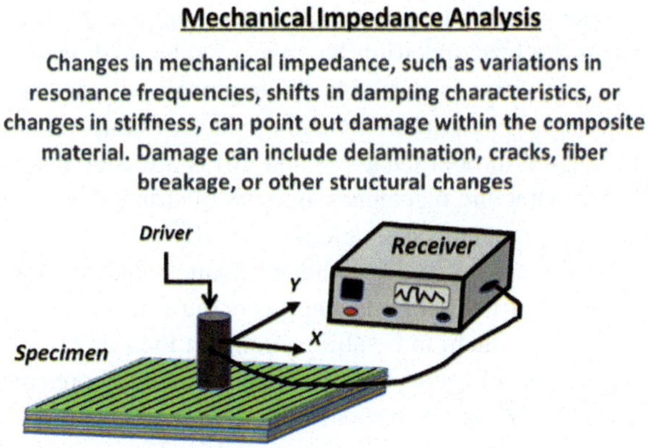

Fig. 14.7 The driver generates the sonic vibrations into the composite test part, and the receiver converts the sonic response to an electrical signal

roughness. It also has a limited penetration depth which may not be able to detect flaws in thick materials. Therefore, the operator should examine the back surface for the presence of any damage as the result of impact.

Summary of the mechanical impedance analysis (MIA) method:

- Sonic vibrations are input into the structure using a probe with a driver and receiver crystal in series (Fig. 14.7).
- The driver generates the sonic vibrations, and the receiver converts the sonic response to an electrical signal that is analyzed by the instrument (Fig. 14.7).
- This device measures voids, stiffness, and disbonds of a composite structure.
- The returned signal is proportional to the stiffness of the structure. An unflawed region is used to calibrate the instrument.
- A loss of or decrease in signal from this calibration corresponds to a loss in stiffness, therefore, the existence of a flaw.

14.5 Application of NDI in Fatigue and Fracture Mechanics

The application of nondestructive inspections in aircraft, aerospace, military, and many other industries is important because of the possibility of finding a sizable surface or embedded cracks, that if undetected would grow in the structure and cause catastrophic failure. In industries, when dealing with a critical part, it is a requirement to conduct a comprehensive NDI before structural usage and, if necessary, periodic inspection during their operation. If NDIs reveal that the part has no defects, then the analyst will perform their fatigue analysis by assuming the critical parts are free from defects and based on the damage accumulation concept and using Miner's rule, the possibility of failure will be estimated. However, in some other industries, such as aerospace, the fracture mechanics approach is more acceptable and preferable for determining the remaining life of the critical components, that is, to estimate the number of cycles for a preexisting small defect (or initial defect) to evolve into failure by using the linear elastic fracture mechanics methodology. In both cases of fatigue analysis or fracture mechanics approach, analytical tools must be available when the load spectrum has many cycles with different stress ratios. Proper computer software is available for engineers to use, which makes the analysis much simpler and less time-consuming. As an example, the NASGRO computer software (discussed in both Chaps. 9 and 10) is a trustable source for conducting a comprehensive fracture mechanics analysis of structural parts based on the minimum initial flaw size that has been detected by NDIs.

In the case of the fracture mechanics analysis approach to number of cycles to failure of structural parts, engineers must have some estimate of the initial flaw size available to them to be able to conduct their crack growth analysis. In other words, the number of cycles that it takes for the initial preexisting defects to grow to failure is important for engineers to have because it can be compared with the number of load cycles that part will be exposed to. The concept of initial flaw size in structural parts has a lot to do with the limitation of the NDI that will be used on the structure for detecting flaws. The minimum flaw size that will escape the detection can be used as the initial flaw size for the fracture mechanics analysis. Each NDI has a limitation regarding the size of defects that can be detected. Furthermore, with each NDI measurement, there is a probability of detection and a confidence level associated with the measurement that arises from deviation in the measurements. When conducting the NDI measurement, there are four possible results that the operator will encounter: (1) a true positive, which means the flaw exists and is identified; (2) a false positive, which means the flaw does not exist but is identified; (3) a false negative: it means flaw exists but is not identified; and (4) a true negative, which means flaw does not exist and is not identified.

Even though several NDI methods were discussed in Sect. 14.1, several other NDI methods are currently being used in industries for detecting flaws in the material. In this section, all methods will briefly be discussed from the flaw size detection limitations perspective that they have. They are: (1) standard penetrant inspection, (2) standard eddy current inspection, (3) standard magnetic particle inspection, (4) standard ultrasonic inspection, and (5) standard radiographic or X-ray inspection. Note that each inspection method has its own variations like the A-scan, B-scan, and C-scan as part of the ultrasonic inspection, but no need to discuss them in this book. Figures 14.8, 14.9, 14.10, and 14.11 show the minimum initial flaw size capability of each NDI method for metallic parts, which is helpful for industries when fracture mechanics analysis is requested by customers. For example, if a given aerospace industry is asked by its customers to conduct a life analysis assessment of a critical part, the minimum initial flaw size provided by Figs. 14.8, 14.9, 14.10, and 14.11 should be helpful to go forward with the analysis. Finally, not all the abovementioned NDIs could apply to composite materials, but an attempt is made to discuss them in this section for the metallic parts. These initial flaw sizes provided for different NDI methods, may will have different flaw sizes associated with them when dealing with composite parts. That is, the minimum initial flaw size that can escape detection can differ between metals and composites due to their varying material properties and behavior under stress and inspection methods used. In summary, the different composition, anisotropic properties, and structural complexities of composites make ultrasonic waves behave differently when compared to the more uniform and isotropic nature of metals. Understanding these differences is crucial for effective NDI testing and evaluation in both types of materials. More discussions on this topic will be presented in Sect. 14.6.

14.5.1 Penetrant Inspection Method

In Fig. 14.8, the standard flaw size for the penetrant inspection is shown [1, 22, 23]. In brief, liquid penetrant inspection is used for the detection of all types of surface cracks. This method is mainly used for metallic material after the machining process is completed. The existence of any surface crack can be detected on the

Penetrant Inspection

Minimum initial flaw size for fracture mechanics analysis

Note: 1inch=25.4 mm

NDI method	Area	Part thickness (in.)	Crack type	Aspect ratio $(a/2c)$	Crack depth (in.)	Crack length (in.)
PT standard	Open surface	$t < 0.075$	TC		t	0.200
			PTC	0.1		0.250
				0.2		0.200
				0.5	(0.075 max.)	0.175
		$t \geqslant 0.075$	TC		t	0.150
			PTC	0.1		0.250
				0.2		0.200
				0.5		0.150
	Surface edges	$t < 0.075$	TC		t	0.100
		$t \geqslant 0.075$	CC		0.075	0.100

PT = Penetrant Inspection
PTC = Part through crack
CC = Corner crack
TC = Through crack

t = Specimen thickness
a = Crack depth
2c = Crack length

Fig. 14.8 Initial flaw size detectability assumptions for standard penetrant inspection method. Initial flaw size for different crack geometries is shown here. This flaw size is associated with the minimum flaw size that escaped from detections

Fig. 14.9 Initial flaw size detectability assumptions for standard magnetic particle inspection method. Initial flaw size for different crack geometries is shown here. This flaw size is associated with the minimum flaw size that escaped from detections

Magnetic Particle Inspection

Minimum initial flaw size for fracture mechanics analysis

Note: 1inch=25.4 mm

Part thickness (in.)	Crack type	Aspect ratio $(a/2c)$	Crack depth (in.)	Crack length (in.)
$t \geqslant 0.070$	TC		t	0.250
	PTC	0.1		0.375
		0.2		0.275
		0.5		0.250
$t \geqslant 0.070$	TC		t	0.250
	CC		0.070	0.250

MT = Magnetic Particle Inspection
PTC = Part through crack
CC = Corner crack
TC = Through crack

t = Specimen thickness
a = Crack depth
2c = Crack length

Eddy Current Inspection

Minimum initial flaw size for fracture mechanics analysis

NDI method	Area	Part thickness (in.)	Crack type	Aspect ratio $(a/2c)$	Crack depth (in.)	Crack length (in.)
		Note: 1inch=25.4 mm				
ET standard	Open surface	$t \geqslant 0.020$	TC		t	0.100
			PTC	0.1		0.100
				0.2		0.100
				0.5		0.100
	Surface edges	$t \geqslant 0.020$	TC		t	0.050
			CC		0.020	0.050

ET = Eddy Current Inspection
PTC = Part through crack
CC = Corner crack
TC = Through crack

t = Specimen thickness
a = Crack depth
2c = Crack length

Fig. 14.10 Initial flaw size detectability assumptions for standard radiographic inspection method. Initial flaw size for different crack geometries is shown here This flaw size is associated with the minimum flaw size that escaped from detections

Radiographic or X-Ray Inspection

Minimum initial flaw size for fracture mechanics analysis

NDI method	Area	Part thickness (in.)	Crack type	Aspect ratio $(a/2c)$	Crack depth (in.)	Crack length (in.)
		Note: 1inch=25.4 mm				
RT standard	Raw stock or machined	$t \geqslant 0.050$	Embedded	Ellipse	0.7t	t (0.150 min.)
			PTC, TC, CC	1/2 Ellipse	0.7t	t (0.150 min.)
		$t < 0.050$	Embedded	Ellipse	0.7t (0.025 min.)	0.150
			PTC, TC, CC	1/2 Ellipse	0.7t (0.025 min.)	0.150

t = Specimen thickness

RT = Radiographic Inspection
PTC = Part through crack
CC = Corner crack
TC = Through crack

t = Specimen thickness
a = Crack depth
2c = Crack length

Fig. 14.11 Initial flaw size detectability assumptions for standard radiographic inspection method. Initial flaw size for different crack geometries is shown here. This flaw size is associated with the minimum flaw size that escaped from detections

surface of the part by the fluorescent light due to the capillary action of dye trapped by the presence of surface cracks. Penetrant inspection can also be performed on various types of material, and it is not limited to metallic parts only. Ceramics, organic, and composite parts can also be inspected by penetrant inspection for the detection of any surface defects. However, the minimum flaw size for other materials, such as composites, can be different.

14.5.2 Magnetic Particle Inspection Method

Magnetic particle inspection is a nondestructive inspection method that can be used to detect cracks at the surface of ferromagnetic materials such as steel and nickel-based alloys. The magnetic inspection process is seldom used for composite materials because composite parts don't have magnetization properties. In the case, there are magnetic fibers present in the matrix, or the matrix is made of magnetic material, then this inspection may be applicable. For the ferritic material, the inspection begins with the magnetization of the component. The surface is then covered with small magnetic particles. The particles will be interrupted when defects are present on the surface of the material. This method is not as popular as other NDI methods and is not much employed in industries anymore for inspections. The minimum initial flaw size associated with the magnetic particle inspection is shown in Fig. 14.9 [22].

14.5.3 Eddy Current Inspection Method

Another nondestructive inspection technique that industries are using to detect surface and near-surface flaws is the eddy current inspection method. This method is based on induced electric currents in a conductive material, that in turn generate their own magnetic field. The presence of defects causes great disruptions in eddy current flow, and it results in impedance and voltage variations. The calibration is performed on a known defect size in the material. Figure 14.10 shows the standard flaw size generated for the eddy current inspections and they are used extensively in the aircraft industry for finding corner cracks emanating from the corner of fasteners [23]. Eddy currents are the desired method for the inspection of surface and near-surface flaws in metallic parts and metallic joints, such as welds and riveted joints. The application of eddy current on the composite material is possible because carbon fiber reinforced polymer exhibits some low-level electric conductivity due to the intrinsic conductivity of the carbon. This low-level amount may allow eddy current to be utilized for the NDI of fiber-reinforced composites. However, the minimum standard flaw size for other materials, such as composites, can be different and must be established by a reliable source before conducting fracture mechanics analysis.

14.5.4 Ultrasonic Inspection Method

The ultrasonic inspection method was extensively discussed in Sect. 14.2. Regarding flaw detection, ultrasonic evaluation methods are most used to detect subsurface flaws in composite parts. Not only this inspection method is used for detecting surfaces before and after service usage, but also can be utilized to check the quality of raw material before machining. In terms of initial flaw size estimation for fracture mechanics analysis, the size of the minimum flaw that can escape detection by ultrasonic inspection is shown in Fig. 14.10 for metallic parts [23].

14.5.5 Radiographic or X-Ray Inspection Method

Radiography or X-ray NDI technique is another method that is used extensively in industries to examine the internal structure of manufactured composite or metallic components, detecting any preexisting subsurface flaws prior to structural usage. The method relies on the differential absorption of X-ray or gamma-ray energy as the incoming rays propagate through the material. In Radiography inspection the part to be examined is placed between the incoming radiation source and the film (or detector) situated on the opposite side. That is, the X-rays that penetrate the material are imaged on the photographic film, where each part under evaluation may have differences in density, thickness, shapes, and sizes, and therefore absorbing different amounts of radiation. The unabsorbed radiation that goes through the part is recorded on film. The minimum flaw size that escapes detection by using the radiographic NDI technique is shown in Fig. 14.11 and they can be used to conduct fracture mechanics analysis [22]. The analysis will determine whether the part can survive the cyclic load environment or not. This NDI method can be also used for the composite material; however, the minimum standard flaw size for composite materials can be different and must be established by a reliable source before conducting fracture mechanics analysis.

14.5.6 Special NDIs

For some critical design applications, the maximum initial flaw sizes by employing the standard NDI techniques, are considered too large because the result of life assessment yields inadequate life for the structural part. Analysts like to have smaller flaw sizes than those shown in Figs. 14.8, 14.9, 14.10, and 14.11 to assure the parts have adequate safe life throughout their service operation. In a situation like this, a smaller flaw size will be helpful to use, because the fracture analysis will show the longer number of cycles to failure. For that to happen, it requires implementing a special NDI. To perform special level NDI, an inspector must first undergo NDI reliability testing in that particular NDI discipline (for example, the ultrasonic inspection). The results of this testing must verify that the inspector can detect smaller flaws than the standards shown in Figs. 14.8, 14.9, 14.10, and 14.11 with a 90% probability of detection and 95% confidence level. As an example, if the analyst is using standard penetrant inspection with initial flaw size $a = c = 0.075$ in (from Fig. 14.8) and the results of the analysis show the part cannot survive during its service usage, then it is necessary to use a special NDI with smaller flaw size. That is, the analyst will determine through fracture mechanics analysis that the initial flaw size should be smaller, for example, $a = c = 0.05$ inch, then the inspector should show that with a 90% probability of detection and 95% confidence level, it is possible to find the quantity $a = c = 0.05$ inch flaw [1]. The same methodology can be applicable to composite material as well. That is, if the analyst is requesting to have a smaller flaw size to show the composite part can have adequate life, then based on the 90% probability of detection and 95% confidence level the smaller flaw size should be available to use.

14.6 NDI Initial Flaw Size for Composites

In Sect. 14.5, various NDI methods for metallic parts were discussed. The minimum initial flaw size necessary to conduct a meaningful damage tolerance analysis could be obtained based on the type of NDI method applied to the structural parts. Inspection techniques such as magnetic particle, dye penetrant, ultrasonic, and radiographic inspections are commonly used for detecting flaws in metals. The maximum flaw size that can escape detection in metals can vary depending on the type of flaw (a surface crack, embedded, or a through crack) as well as the inspection method used on the part. From inspection perspective, several differences exist between metals and composites due to their distinct material properties and their behavior under stress, as well as the inspection methods used. However, the rapidly increasing use of composites in primary structural parts necessitates reliable NDI methods capable of inspecting damage and determining the maximum initial flaw size that can escape detection, both before and after structural usage are needed. Having the maximum initial flaw size, a damage tolerance analysis can be initiated by the analyst. That is, the number of cycles to failure can be estimated and determine whether the part can withstand the load environment or not.

Accurate nondestructive inspection (NDI) methods are essential to ensure structural reliability, and in the case of passenger aircraft, safety is always the number one priority from airlines perspective. Traditionally, tap tests and a few ultrasonic-based inspection methods have been used to inspect delamination in composite structures, as already discussed in Sect. 10.5. However, recent developments, including new and more advanced NDI techniques, have increased the number of available inspection options [32] for inspecting anomalies in composite material and damage tolerance assessment of structural parts.

Over the last several years, a series of experiments have been designed and implemented to assess how well both conventional and advanced NDI techniques can detect voids, disbonds, delaminations, cracks in the matrix, and impact damage on adhesively bonded composite structures. The test program, called Composite Flaw Detection Experiments, was undertaken at the Federal Aviation Administration (FAA) Airworthiness Assurance NDI Validation Center (AANC), operated, and staffed by Sandia National Laboratories (Albuquerque, New Mexico) since 1991. Results of these studies are available in [32].

Metals often demonstrate more predictable failure mechanisms, typically exhibiting signs of deformation, cracking, or yielding before catastrophic failure occurs. On the other hand, composites display a more complex material behavior because of having constituents like fibers and resins, which they manifest as different failure modes such as delamination, fiber breakage, or matrix cracking, which might not always be evident on the surface. Detecting flaws in composites frequently demands advanced inspection techniques like ultrasonic C-scan, thermography, acoustic emission testing, or computed tomography (CT) scanning. Furthermore, the maximum flaw size escaping detection in composites can hinge on factors like the composite's layup, fiber orientation, and the capabilities of the specific inspection method. For instance, the maximum initial flaw size overlooked in composites during ultrasonic inspections relies on several factors, including the composite material type, its structure, and the equipment's inspection capabilities in use. Moreover, most composite structures contain plies with fibers of different orientations, various stacking sequences, or fibers that are chopped and randomly situated in the matrix. In addition, flaws in composites aren't solely crack-like; they can manifest as delaminations, manufacturing voids, fiber breakage, or matrix cracks. Detecting these flaws through ultrasonic inspection or other aforementioned NDI techniques can be challenging due to the anisotropic and heterogeneous nature of composites. Finally, it should be noted that when inspecting materials using methods like ultrasonic examination, the waves transmitted through different materials might encounter different responses when it is exposed to defects. Both the incoming waves as well as reflected ones will show different responses for different microstructures (e.g., metals vs. composites). Consequently, the initial flaw size expected in homogeneous and isotropic metals should differ from that in composites. Using the initial flaw size established for metals, to assess damage tolerance of composite critical parts, may yield unconservative results.

14.7 Concluding Remarks on NDI Techniques for Laminated Composite Materials

Industries continue to expand their use of composite materials from secondary to primary structures. This expanded application is due to having a better knowledge of the damage tolerance analysis of composites and the implementation of accurate nondestructive inspection (NDI) methods executed before and after structural usage.

In the previous sections of this chapter, a few ultrasonic-based inspection methods were discussed for the inspection of laminate composite parts. However, recent advancements in NDI techniques have produced several new inspection methods for better detecting defects in composites. Many of these methods have the capability of producing the presence of 2D flaw maps in the material remarkably. An experiment has been developed to assess the ability of both conventional and advanced NDI techniques to detect voids, disbonds, delaminations, and impact damage in adhesively bonded composite aircraft structures. A series of solid laminate carbon composite specimens with statistically relevant flaw profiles are being inspected using traditional, hand-held pulse echo UT, as well as new NDI methods that have recently been introduced to improve the sensitivity and repeatability of inspections [24]. The principal factors affecting flaw detection in laminates are included in this study, and they are: (1) material type, (2) flaw profiles, (3) presence of complex geometries, and (4) environmental conditions. One aspect of this effort [24] is to utilize technical personnel to study the Probability of Detection (POD) in the field and to formulate improvements to existing inspection techniques. In addition, advanced NDI methods for laminate inspections, such as thermography, shearography, laser ultrasonics, microwave, and phased/linear array UT [25–30], were used to quantify the improvements achievable using more sophisticated NDI. A comprehensive study related to the POD results for advanced NDI with comparisons to results achieved by airline technical inspectors using traditional UT methods is presented in [24].

Future NDI efforts and related data for flaw detection will be helpful to assess the damage tolerance analysis of critical parts. The recommended focus is on defining reliable initial damage size, which can be related to residual strength and damage tolerance life assessment under repeated loads. More reliable NDI procedures for maintenance and a better understanding of the results can help to move toward an advanced approach to damage tolerance for composite laminated structures while reducing the conservatism in the associated design criteria [31].

14.8 Concluding Remarks

The application of nondestructive inspection techniques for composite materials offers many benefits across various industries, including aerospace, automotive, marine, and construction. These nondestructive techniques are used for the detection of internal defects, damage, or irregularities without causing harm to the structure itself. By applying early inspection to critical composite parts, it can be helpful to enhance safety, reduce maintenance costs, and prolong the service life of composite parts. In many industries, when dealing with critical parts, comprehensive nondestructive inspection is mandatory before the part's usage. Additionally, interval inspection must be extended to parts during their applications. Several nondestructive inspection techniques were presented in this chapter, and their advantages and disadvantages were outlined.

The application of nondestructive inspection on critical parts for finding defects throughout their service operation is essential. However, their applications in finding the minimum initial flaw size escaping detection for life analysis of structural components are also essential for analysts. As inspection techniques continue to advance, further research and innovation in nondestructive inspection may be needed to provide a smaller initial flaw size that can lead to a larger number of cycles to failure when conducting life analysis.

References

1. B. Farahmand, *Fatigue and Fracture Mechanics of High-Risk Parts Nondestructive Inspection*. Chapter 5 (Chapman & Hall, 1996)
2. D.K. Hsu, *Nondestructive Inspection of Composite Structures: Methods and Practice*, in 17th World Conference on Nondestructive Testing, 25–28, Shanghai, China (2008)
3. K.B. Armstrong, R.T. Barrett, *Care and Repair of Advanced Composites* (Society of automotive Engineers, Warrendale, 1998)
4. C.F. Buynak, T.J. Moran, Characterization of impact damage in composites, in *Review of Progress in Quantitative NDE*, ed. by D.O. Thompson, D.E. Chimenti, vol. 6B, (Plenum Press, New York, 1987), pp. 1203–1211
5. ASTM E2533, *Standard Guide for Nondestructive Examination of Polymer Matrix Composites Used in Aerospace Applications*, 1 Feb 2021
6. I. Jandejsek, J. Jakubek, M. Jakubek, P. Prucha, F. Krejci, P. Soukup, D. Turecek, D. Vavrik, J. Zemlicka, X-ray inspection of composite materials for aircraft structures using detectors of Medipix type. J. Instrum. (2014)
7. J. Kastner et al., *Defect and Porosity Determination of Fiber Reinforced Polymers by X-ray Computed Tomography*, in Proceedings: 2nd International Symposium on NDT in Aerospace (2010)
8. By Ravi Kumar at Trinity NDT (Engineering Quality), Composites Ultrasonic Testing UT For Aerospace at Trinity NDT
9. D. Richard, *How Ultrasonic Testing Can Find Flaws in Composite Materials* (Composite World (CW), 2021)
10. I. Papa, V. Lopresto, A. Langella, Ultrasonic inspection of composites materials: Application to detect impact damage. Int. J. Lightw. Mater. Manuf. **4**, 37–42 (2021)
11. J. Escobar, *Non-destructive Testing for Composites: Different Inspection Methods*. By AVIATION PROS, 31 Mar 2003
12. T. Nelligan, D. Kass, *Ultrasonic Testing of Fiberglass and Carbon Fiber Composites*, 11 Aug 2014
13. Z. Zhang, Q. Li, M. Liu, W. Yang, Y. Ang, Through transmission ultrasonic inspection of fiber waviness for thickness-tapered composites using ultrasound non-reciprocity: Simulation and experiment. Ultrasonics **123**, 106716 (2022)
14. TECSCAN, *Ultrasonic Testing: Defect-Sizing in Composite Structures*, November 2015
15. T. Hasiotis, E. Badogiannis, N.G. Tsouvalis, *Application of Ultrasonic C-Scan Techniques for Tracing Defects in Laminated Composite Materials*, in 4th International Conference on NDT, October 2007
16. A. Fahr, Ultrasonic C-scan inspection of composite materials. Eng. J. Qatar Univ. **5**, 201–222 (1992)
17. J. Escobar, *Non-destructive Testing for Composites: Different Inspection Methods* (AviationPros, 2003)

18. D. Hagemaier, R. Fassbender, Non-destructive testing of adhesive bonded structures. SAMPE Q. **9**, 36 (1978)
19. G.E. Georgeson, S. Lea, J. Hansen, *Electronic Tap Hammer for Composite Damage Assessment*, in SPIE Proceedings, vol. 2945, Nondestructive Evaluation of Aging Aircrafi, Airports and Aerospace Hardware, ed. by R.D. Rempt, A.L. Broz (1996), pp. 328–338
20. Y.V. Lange, The mechanical impedance analysis method of nondestructive testing. Nondestr. Test. Eval. **11** (1994)
21. P. Cawley, The impedance method of non-destructive inspection. NDT Int. **17**(2), 59–65 (1984)
22. MSFC STD-1249, *Nondestructive Evaluation Guidelines for Fracture Control Plan* (Marshal Space Flight Center, 11 Sept 1985)
23. NASA-STD-5009, *Nondestructive evaluation requirements for fracture-critical metallic components* (National Aeronautics and Space Administration, 2008)
24. R. Dennis, P. Rice, M. Thomas, *A Quantitative Assessment of Advanced NDI Techniques for Detecting Flaws in Composite Laminate Aircraft Structures* (U.S. Department of Energy (Office of Scientific and Technical Information), 2014)
25. L.D. Favro, X. Han, Thermal wave materials characterization and thermal wave imaging, in *Sensing for Materials Characterization, Processing and Manufacturing, ASNT TONES*, ed. by G. Birnbaum, B.A. Auld, vol. 1, (American Society for Nondestructive Testing, 1998), pp. 399–415. ISBN 978-1571170675
26. Y.T. Hung, Shearography: A new optical method for strain measurement and nondestructive testing. Opt. Eng. **21**, 391–395 (1982)
27. C.B. Scruby, L.E. Drain, *Laser Ultrasonics* (Adam Hilger, Bristol, 1990)
28. Synthetic aperture radar-based techniques and reconfigurable antenna design for microwave imaging of layered structures. Retrieved 2014-05-07
29. Microwave Testing (µT): An Overview Johann Hinken, FI Test- und Messtechnik GmbH Magdeburg, Germany, July 2016
30. ASTM, *E2491 Nondestructive Testing vol 3.03, Evaluating Phased Array Characteristics of Phased Array Ultrasonic Testing Instruments and Systems* (American Society for Testing of Materials, Conshohocken, 2012). pp. 1358–75
31. P. Shyprykevich, J. Tomblin, L. Ilcewicz, A.J. Vizzini, T.E. Lacy, Y. Hwang, Guidelines for Analysis, Testing, and Nondestructive Inspection of Impact Damaged Composite Sandwich Structures, DOT/FAA/AR-02/121, FAA Final Report, Mar 2003
32. D. Roach, *Assessing Conventional and Advanced NDI for Composite Aircraft* (CW CompositesWorld, Published 6/30/2008)

Chapter 15
3D Printing and Its Applications for Composites (Additive Manufacturing)

15.1 Introduction to 3D Printing

The 3D printing process, also known as Additive Manufacturing (AM), is a relatively new method that creates physical parts from digital models by adding material layer by layer until the desired thickness and shape are achieved [1–6]. The term "3D printing" is more commonly used among consumers, while "additive manufacturing" is a more technical and academic term. It differs from subtractive manufacturing techniques, which involve cutting away material from a larger block to construct objects with the desired shape (like conventional machining). The first additive manufacturing part was processed in 1984 by Chuck Hull, who founded 3D printing technology, and since then, many new and complex 3D AM parts have been created. It is important to note that during the manufacturing of parts through 3D printing, it is necessary, in almost all cases, to protect the part from falling, distortion, and cracking, as the part cannot be printed without support. In Sect. 15.6, a brief discussion on the support structure is presented.

Figure 15.1 illustrates the two manufacturing processes: (a) the traditional approach where the original larger block is machined to its final shape, and (b) the additive manufacturing method where material is added layer by layer, line by line, in a horizontal plane to build the part. In Fig. 15.1c, a complex geometry was generated through additive manufacturing (a bicycle frame) using metal powder as the additive and in Fig. 15.1d, steps for creating an object through the 3D printing are—demonstrated. Another simple example, resembling the additive manufacturing process, is the creation of a sculpture. In this process, material is continually added to the piece until the final shape is achieved. The AM creation process is based on fusing materials through heating to generate items, typically layer by layer, piece by piece, based on a digital model that must accurately represent the desired structure [3]. Digital modeling involves using computer software to construct a virtual representation of the part's geometry, which is then fed to the 3D printer to generate

B. Farahmand, *Fundamentals of Composites and Their Methods of Fabrications*,
https://doi.org/10.1007/978-3-031-80201-0_15

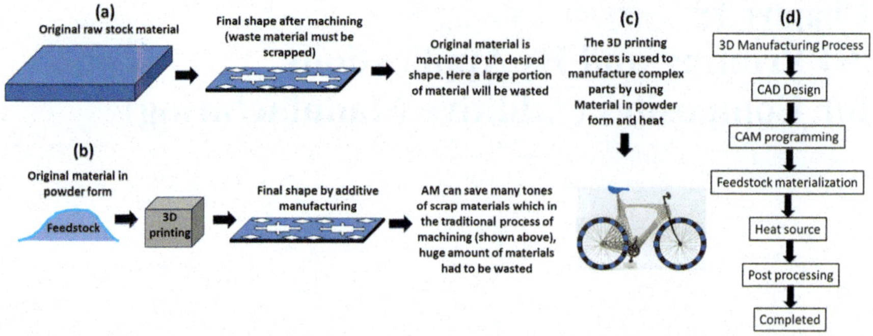

Fig. 15.1 (**a**) The traditional manufacturing parts starting with a big block of material to create the final shape, where a huge portion of material will be wasted. (**b**) Creating the same part by using additive manufacturing method with no material waste. (**c**) A bicycle with a complex geometry was created through the additive manufacturing using metallic powder (*METALS 3D PRINTING—WIRE AND ARC ADDITIVE MANUFACTURING (WAAM)—YouTube*). (**d**) Steps for completing a 3D manufacturing process

the part with minimal human interaction during the manufacturing process (Fig. 15.1). In other words, a complex geometry is generated in the digital world and can be created and utilized in the physical world.

The 3D printing process is capable of creating structures with extremely intricate and complex designs using various materials, primarily plastics, ceramics, and metals either in powder or wire form. Therefore, AM is a technique that lets engineers and designers produce unique parts with high precision in low volumes and because of no material waste (an additive not subtractive process), it is considered as a friendly manufacturing process for the environment with no leftover material to be wasted. However, in many cases, post-processing operations are required by customers after additive manufacturing (AM) has been completed. These operations can be time-consuming and labor-intensive, often resulting in a small amount of material waste that can be recycled/reusable. For AM of metal parts, these post-processing steps may include heat treatment, surface material grinding, and finish machining to bring the fabricated part to its final state and meet customer expectations. In the case of polymer parts, the post-processing requirements are generally less extensive and may involve some material removal and finishing [7].

Overall, the extent of post-processing depends on the customer's demands and expectations for the final AM product. This includes considerations related to its shape, tolerance, surface finish, heat treatment to reduce residual stresses, nondestructive inspection (NDI) to detect sizable flaws, and limited amount of testing to determine the mechanical properties of the part.

15.2 AM Applications for Different Industries

There are several additive manufacturing methods being used for 3D printing that each differ by the materials that may be utilized, and the way that layers are added to build parts. The AM processes are created either by softening or melting the materials that are referred to as feedstocks, like powder bed, wires, filament, or sheet. They will be used as an input material for the additive manufacturing process [7–11].

When a laser beam is used as the source of energy to melt and fuse materials together, it is called laser additive manufacturing (LAM). There are two LAM techniques that are currently used to create thick parts for many industrial applications. They are powder bed fusion (PBF) and direct energy deposition (DED). The PBF and DED AM process had held revenue market shares of 81.5% and 8.5%, respectively in the market for metal in 2019 where it is estimated that the DED technologies' revenue share will rise to 11.1% or more by 2024 [12, 13].

Plastics now dominate the AM market in terms of materials for the 3D printing process, but the metal AM process also is rapidly expanding. AM has been increasingly popular in industries like aircraft, aerospace, and motorsport where large weight and fuel savings are of prime concern. From the total of 8.1 billion euros generated in 2019, the amount of 5.3 billion euros (i.e., 65%) came from polymer AM and 2.8 billion euros (35%) came from metal AM [14]. However, there is a lot of work needed to improve the reliability of AM processes to generate parts that can be used not only for secondary, but also primary parts of structures that could handle the load environment during their service operation. Another concern that needs to be paid attention to, when dealing with AM, is the time needed to complete the part, which customers in many cases require the manufacturer to deliver in a short time.

In recent years, the metal additive manufacturing (AM) market has expanded more rapidly than other material sectors. It is anticipated that the combination of AM-based repair technologies for damaged parts will soon become operational in the automobile and aerospace industries, potentially saving these sectors from having to replace entire components [15]. Another industry that can benefit from 3D printing is in the area of medicine, such as hearing aids, stethoscopes (an instrument for listening to patient's heart), surgical guides, and robots. As another example, the AM can create prosthetic limbs and orthotic or braces devices that are lightweight, comfortable, and functional. Additive manufacturing is able to customize the appearance, size, and shape of the prosthetics to be in accordance with the patient's requirements and demand. A 3D printed prosthetic limb, or a 3D printed leg can help the prosthetist quickly determine if the size and position of the prosthesis are correct during both standing and walking. Finally, in the area of medical devices, AM can also create models of some human organs, tissues, and structures that can be used for surgical planning, simulation, and education [16].

15.3 Several Additive Manufacturing Methods

There are several additive manufacturing methods being used for 3D printing that each differ by the materials that will be utilized and the heat source that layers are added to build parts (Fig. 15.2). There are two main techniques that fall under the laser additive manufacturing (LAM) process that can create fully thick metallic parts for many industrial applications. These two techniques, which briefly were mentioned in the previous section, are powder bed fusion (PBF) and the direct energy deposition (DED) [17–20]. As a heat source, most industries are using YAG lasers, diode lasers, fiber lasers, and CO_2 lasers for their 3D printing laser base process. The typical laser power is 200–400 Watt. In brief, the PBF processing utilizes thin layers of powders of metals and their alloys, ceramics, plastics, and biological materials as raw material and will be melted and fused together layer by layer with a laser beam (or other heat sources) according to the digital shape of the part. Note that the digital file is a necessity part of 3D printing and by not having a digital file, a 3D printer will not have the instructions to build the part needed by the users. The melted powder material will fuse, solidify, and subsequently will form a layer, and then a new layer of powder is spread over the previous layer and the process will be repeated until the part is finalized. PBF processes can produce parts with high resolution, accuracy, and surface quality, but it needs a controlled environment (e.g., at the presence of inert gases) to avoid oxidation if metallic powders are utilized. In this AM technique, a minimum amount of post-processing work may be required based on the customer's demand. Regarding composites, PBF techniques can be used with composite powders that consist of a combination of materials such as polymers, ceramics, or metals as the matrix with reinforcing fibers like carbon fiber, fiberglass, or other additive materials. The challenge with composites lies in ensuring proper spreading

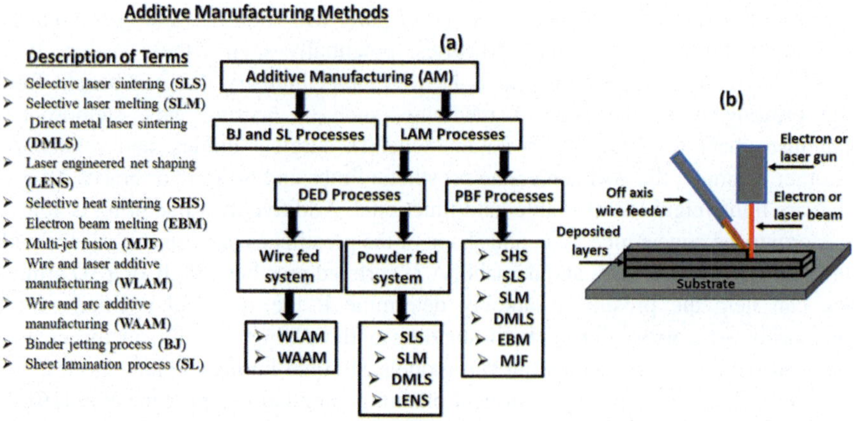

Fig. 15.2 (**a**) Different additive manufacturing methods. Subclasses of DED and PBF are shown here. (**b**) Both the laser and electron beam as the heat source are illustrated with off-axis wire feeder creating layers on substrate

and alignment of the reinforcing fibers or particles within the powder mixture to achieve adequate mechanical properties in the final fabricated printing process. However, advancements in printing technology continue to expand the capabilities of PBF for composite materials, enabling the production of components with tailored properties and improved performance.

For the case of metal powders, aluminum, cobalt, copper, Inconel, and stainless steel are utilized in the 3D manufacturing parts. Nylon, glass-filled nylon, and polypropylene are examples of polymers and finally, for manufacturing ceramic parts, silica, zirconia, and alumina powders are used for the PBF manufacturing process to create 3D geometry [21].

Regarding the direct energy deposition, DED, it is a process that uses mainly either a metal wire or metal powder as the feedstock material and feeds it into the focal point of a laser beam, where it melts and deposits on the substrate or the previous layer. Both the laser beam and the material feed nozzle move along a predefined path dictated by the digital model (or the CAD data), creating layers with finite width and thickness. This process can also be applied to polymer materials, but it is usually used in the form of pellets or filaments. The polymer is melted and extruded onto a substrate to build up the desired 3D shape. The DE*D* process can also be applied to ceramics, a high-energy heat source is used to melt and fuse ceramic powder layer by layer to create ceramic components. The advancements in the DED process hopefully will enable engineers to fabricate 3D composite parts, with enhanced performance characteristics suitable for various industries such as aerospace, automotive, and manufacturing.

DED can produce parts with high deposition rate, high efficiency, and size, but it has lower resolution, accuracy, and surface quality than powder bed fusion. Just like the PBF process, the DED requires post-processing steps like machining, nondestructive inspection and testing to ensure good quality parts and performance. The amount of post-processing work on the final manufactured part has a lot to do with the customer's demand and request.

15.3.1 Laser Versus the Electron Beam-Based DED Systems

It has been discussed previously regarding the heat source utilized in the 3D manufacturing process. In the case of LAM technique, both lasers and electron beams can be used to fabricate parts. When considering the laser as the heat source, there are limitations in terms of the materials they can handle. Some materials have high melting temperatures, for which the level of heat provided by the laser is not adequate to create a 3D object through the fusing process. On the other hand, electron beam-based systems can penetrate various materials and create components with higher melting temperatures, but they are costly and require specific knowledge. The following are some key differences between electron beam and laser-based DED systems, when dealing with 3D printing objects.

Material In using the DED technique for creating the 3D parts made of metals, polymers, and composites, both laser and electron beam as the heat source, can be used. Because ceramics need to have a higher heat energy source for melting, they need to utilize the electron beam that has a higher energy to melt.

Precision In both cases, a high degree of accuracy and precision can be obtained.

Adaptability In both cases, they are highly adaptable.

Melting Temperature In the case of a laser as a heat source, it is difficult to produce 3D DED parts when material has high melting temperatures. However, the electron beam is able to produce parts with higher melting temperature materials.

Material Penetration The laser beam has Difficulty to penetrate materials that are highly reflective or transparent materials. Whereas the electron beam can penetrate a wider range of materials.

Cost Utilizing the laser beam, it is more cost effective. In the case of electron beams, it is more expensive.

Maintenance The laser beam is easier to maintain. However, the electron beam is more difficult to maintain and may require more frequent calibration.

15.3.2 Several Subclasses of AM Processes

This section provides a brief description of several subclasses used in additive manufacturing techniques, with attention given to laser additive manufacturing, LAM, which has the ability to fuse polymers and solid metals, or sinter powdered materials into solid structures. Figure 15.2 illustrates subclasses of LAM process, which DED is one of them. The DED also has several subclasses, which fall under the wire fed and powder fed systems. They are selective laser sintering (SLS), selective laser melting (SLM), direct metal laser sintering (DMLS), laser engineered net shaping (LENS), wire laser additive manufacturing (WLAM), and wire additive arc manufacturing (WAAM). Note that in the case of the WAAM process, the primary heat source is an electric arc that is generated between an electrode and the wire being fed, but it is falling under the DED manufacturing process. The WLAM process, however, can achieve higher precision and surface quality of the manufactured object than the WAAM process [7, 22, 23].

Another subclass of LAM is the PBF process that can be utilized with different sources of energy besides the laser, such as the electron beam or electric arc to deposit powder material layer by layer on the substrate (Fig. 15.2). Brief descriptions of both DED and PBF subclasses and others mentioned in Fig. 15.2 are given by the ASTM F2792 standard, that divides AM technologies into several categories which are briefly defined below:

- Selective laser sintering (SLS) is a process that uses a high-power laser to sinter small size particles of polymer powder, such as polyamide, into a solid structure after cooling down based on a 3D model.
- Selective laser melting (SLM) is a process that uses a high-power laser to melt metal powder particles into a solid structure after cooling down based on a 3D model. This process can be also utilized for fabricating PMC, MMC, and CMC composites.
- Direct metal laser sintering (DMLS) is a process that uses laser power to reach near melting point to partially sinter metal powder particles. It is like the SLM process.
- Laser engineered net shaping (LENS) process uses a high-power laser to melt metal powder particles that are fed through nozzles onto the workpiece. LENS has been adapted to work with composite materials by combining different constituents such as polymers, ceramics, or even metal matrix composites.
- Wire and arc additive manufacturing (WAAM). In this process, the primary heat source is an electric arc that is generated between an electrode and the wire being fed. Even though this process is used primarily for metallic materials, the variations of WAAM might emerge to handle composite materials.
- Wire and laser additive manufacturing (WLAM). Here the heat source is the laser and can achieve higher precision and surface quality than WAAM process. Therefore, the difference between WLAM and WAAM is the source of heat wherein the former uses a laser beam and the latter an electric arc.

A brief description of PBF subclasses is given below. A few of these processes fall under both DED and PBF AM, as shown in Fig. 15.2, and there is no need to repeat SLS, SLM, and DMLS processes.

- Selective heat sintering (SHS) is a process that uses a thermal print head to selectively sinter polymer powder layer by layer.
- Electron beam melting (EBM) is a process that uses an electron beam to fully melt metal powder layer by layer in a vacuum chamber to avoid oxidation.
- Multijet fusion (MJF) is a process that uses an inkjet array to apply fusing and detailing agents to polymer powder, which are then fused by heating elements into a solid layer.

Due to LAM's ability to manufacture complicated shapes via digital modeling, there is a great possibility to change the invention of industrial items and fabricate less heavy, more efficient parts that allow for faster, more affordable production. That is, by implementing the finite element tool it can be helpful to engineers to pinpoint areas that have much less stress. These areas can be removed by the engineers through the digital code. Therefore, the final AM parts can be lighter than traditional production techniques since the AM can remove unnecessary materials and thus lower both weight and costs. Furthermore, in the additive manufacturing (AM) process, there is a reduced need for assembling multiple parts, eliminating the necessity for drilling holes to use fasteners. This can significantly expedite the product development cycle. Laser Additive Manufacturing (LAM) has garnered substantial interest from the aerospace and medical sectors, profoundly influencing research in the utilization of four main material types: steels, nickel-based alloys, and titanium-based alloys [24–27]. However, it is crucial to remember that producing sound (defect-free) parts using LAM technology from a variety of metals and alloys at a reasonable cost and performance level is still difficult to achieve [28]. Besides metals used in the LAM process, other materials can also be utilized to manufacture parts, such as polymers and ceramics.

The laser additive manufacturing method (LAM) is attractive from design flexibility, material saving, and with practically no material waste. However, the LAM technique has parameters involved during the manufacturing process that have an impact on the quality of the parts. These parameters are the laser power, the scanning power, the distance between two adjacent scanning tracks, and layer by layer thickness. Not enough heat may cause porosity in metallic material. In processes like selective laser melting (SLM) or direct metal laser sintering (DMLS), porosity can occur, and insufficient heat can cause incomplete fusion in polymers. On the other hand, too much input heat can cause residual stresses after the cooling process. These stresses often result from the rapid heating and subsequent cooling cycles inherent in the additive manufacturing process. In brief, laser power is the amount of energy furnished to the material by the laser beam and must have the right amount of energy. It contributes to the melting temperature, depth and width of the molten pool, the cooling rate, and thermal gradient. Higher laser power can increase the strength, density, and hardness of the parts. It also causes more residual stress, cracking, and distortion.

15.4 Commonality and Differences Between PBF and DED Processes

Regarding their commonality, both PBF and DED are falling under the LAM process, where the two techniques both utilize a focused energy source such as laser or an electron beam as a heat source to melt and fuse metal wires or powders together layer by layer. Regarding their differences, the powder bed technique is used when dealing with the PBF process, where a thin layer of metal powder is spread over a heated plate and melted by the laser or electron beam to fuse the powders together. The second powder bed will be spread, and the process of heating and fusing powder is repeated until the part is completed. On the other hand, the DED technique utilizes a powder or wire as the feedstock and is fed through a nozzle or other deposition device. The material (feed) from the nozzle is melted by the laser, or the electron beam, or other similar heat sources until the layer is completed, and the process will repeat itself until the desired thickness is achieved. The DED process can be implemented and create parts on existing substrates. One important advantage of direct energy deposition is its ability to create complex 3D structures with high precision and in some cases possible mending of portions of parts that may considered undesirable by the engineers. On the other hand, problems such as controlling the heat input during the manufacturing process are challenging. The possibility of introducing too much heat that can cause distortion or warping of the part being created always exists in the DED process. The high heat input, because of the applied laser beam or electron beam, can cause residual stresses on the part upon cooling that must be taken into consideration by engineers. To avoid that, post-processing may be needed by the annealing process to reduce or eliminate the residual stresses [29, 30]. Finally, designers need to consider the presence of support structures to hold the actual 3D part during the additive manufacturing (AM) process if it is needed. Additionally, the removal of these support structures is a post-processing task that engineers involved in manufacturing 3D parts must carefully plan for.

15.5 Other AM Manufacturing Methods

In addition to DED and PBF processes and their subclasses shown in Fig. 15.2, there are two more AM methods that are available to manufacture parts through 3D technology. They are binder jetting (BJ), and sheet lamination (SL) methods, where binder jetting utilizes an inkjet printer to deposit a liquid binder onto metal powder layers (Fig. 15.3). In this process, the liquid binding agent is injected onto a powdered material bed layer by layer. The binding agent effectively glues the layers together until the desired thickness and shape is achieved. Finally, the part is then cured in an oven or furnace to strengthen the bonds.

In the BJ manufacturing method, the process works with a variety of materials, including metals, ceramics, and polymers, if they are in powdered form. The powdered materials in the binder jetting process can be arranged to suit the desired properties of the final object requested by engineers. This is one of the advantages of the BJ technique, which can work with a wide range of powdered materials, as was mentioned above. *Here, the bind*ing agents used in the 3D printing process are adhesive in nature. It essentially acts as an adhesive, holding ceramic, metallic, composites or any other types of powder in place to create the desired shape. The steps toward manufacturing parts through the BJ technique are as follows*:* (1) a thin layer of powder of material is spread over the build platform by a recoating blade and a roller; (2) a printhead with inkjet nozzles deposits the binding adhesive agent onto the powder material based on the parts' geometry; (3) the built platform will lower and a new bed layer of powder of material is spread over the previous layer; (4) the process will continue until the part is completed and the desire thickness has been achieved; (5) the part is removed from the machine and the loose powder will be brushed away or blown away. The final part may go through the post processing operation, such as sintering, to improve its strength and density (Fig. 15.3).

Several advantages of the binder jetting method are: (1) the speed of the process, (2) its suitability for producing intricate geometries, (3) ability to produce parts with low cost and high speed. It is important to emphasize that the BJ process requires post-processing steps such as sintering or infiltration to increase the density and

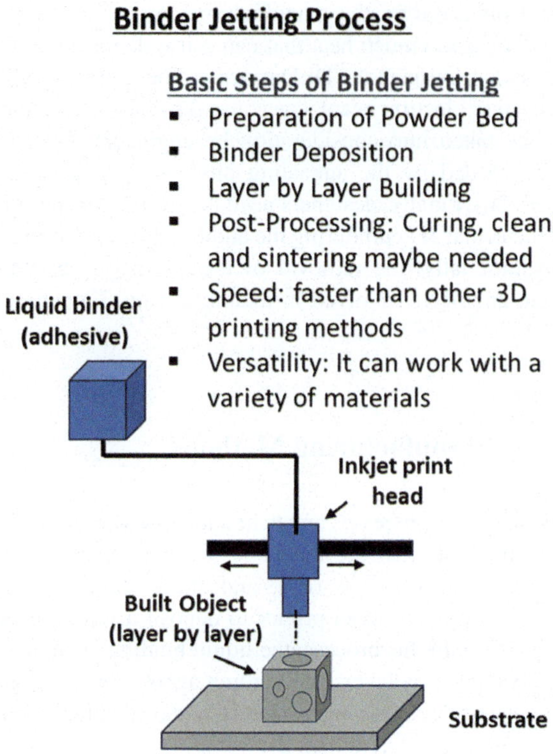

Fig. 15.3 Illustrating the binding jet process where adhesive is used to bind powders together layer by layer

strength of the parts [31, 32]. Finally, just like any other AM processes mentioned in this chapter, the binder jetting 3D printing process needs to have both CAD and CAM software to create the object that engineers have in mind. In brief, binder jetting 3D printing requires having both CAD and CAM software. CAD software is used to create a digital model of the part you want to print, and CAM software is used to slice the model into layers and generate the instructions for the printer.

On the other hand, the *sheet lamination* process, SL, is an additive manufacturing technique that creates objects by layering and bonding sheets or foils of material together. These sheets can be metal, plastic, or other materials. The process involves the following steps: (1) it begins with a stack of thin sheets of material (usually plastic, or metal foil) adhered together on a platform. Each sheet is typically between 0.1 and 0.5 mm thick; (2) a computer-controlled laser or blade system precisely cuts the outline of the object''s cross section onto a single sheet from the stack according to the 3D model''s specifications. The laser or blade cuts through one layer of material while leaving the other layers intact; (3) once the shape is cut out, an adhesive or bonding agent is applied selectively onto the exposed area of the topmost layer; (4) the platform holding the stack of sheets moves down by one layer thickness, and the next sheet is brought into position. The process repeats as the laser or blade cuts the shape onto the new sheet. The cut shape bonds to the layer below due to the adhesive application; (5) this process repeats, layer by layer, cutting the shapes and bonding them together until the entire 3D object is formed. The excess material around the object remains as a support structure for stability during the building process; (6) once the object is fully built, excess material is removed, leaving the finished 3D object. Note that the final shape of the object is achieved by cutting the excess material with a laser or a CNC machine. The formation of parts through the SL process is either to cut the sheet material to shape first and then bonded to the previous layer to create a 3D object. Or the sheet material layers are bonded together before cutting them into the desired shape. Finally, the post-processing steps, such as surface finishing or additional treatments, may be applied depending on the desired specifications.

The SL process can produce parts with minimum energy consumption and has multi-material capability. However, the weaknesses of this process can include high surface roughness in the final part and weak interlayer bonding, which may compromise the mechanical properties of SL 3D fabricated parts if the process is not followed step by step as instructed.

The SL manufacturing process can be used with a wide variety of sheet materials (similar and dissimilar material), depending on the specific process and application. Typical materials include plastic and metal sheets. The advantage of the SL process is its minimum amount of waste because the sheets of material are precut and ready to be processed. The technique can work with varying materials and is suitable for large-scale parts. The application of this technique can be in aerospace for large structural components, packaging, and other areas where layering and bonding sheets are preferable. As was shown in Fig. 15.2, both the BJ and SL techniques are falling under the additive manufacturing process defined by ISO/ASTM 52900-2015.

15.6 Support Structures in the 3D Printing Process

Support structures play an important role in the fabrication process of additive manufacturing parts. The support structure is a temporary structure that is created alongside the actual part being manufactured (Fig. 15.4). It mainly provides stability and prevents distortion or collapse of the part during the 3D printing or deposition process. In addition, they can also be helpful to dissipate heat and reduce the residual stresses in the part during the fabrication process [33–35].

Important parameters related to support structure in additive manufacturing process are: (1) stability of support structure during the process, (2) support structural material, (3) 3D CAD generation of support, (4) *support structure* removal, (5) design consideration of support structure, (6) types of supports, (7) automatic vs. manual generation of support structure, and (8) post-processing operation after removal of support structure. A brief description of each above-mentioned items is provided below:

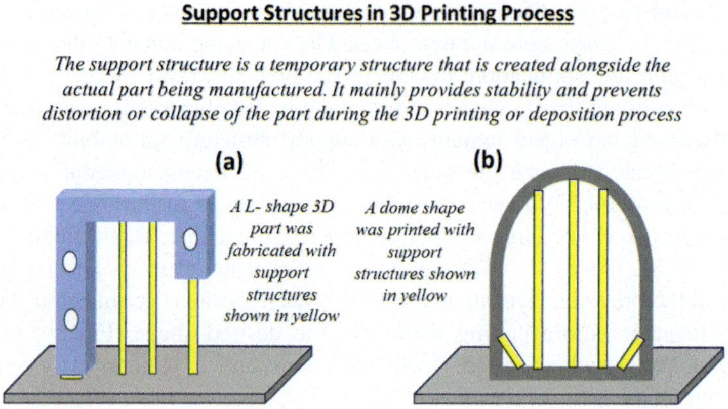

Fig. 15.4 An L-shaped rectangular beam and a dome-shaped geometry were fabricated through additive manufacturing process with support structures. In both (**a**) and (**b**), supports are shown with yellow colors

Stability of the AM support structure: Support structures are designed to create stability of overhanging or cantilevered features of a 3D-printed object when layers are created vertically layer by layer. Without support, these features may bend or *collapse during the printing process.* Therefore, support structures need to be stable to be able to support the main structure.

Support structure and material: Supports in most cases are made from the same material as the part will be printed or the selected support material will be in such a way that can be easily removed after completion of the AM process. The choice depends on the specific AM technology and the part's geometry.

3D CAD generation of support: Support structures are usually generated by slicing software, which takes the 3D CAD model of the part and generates instructions for the 3D printer. The 3D CAD software also generates the necessary support structures based on the design and orientation of the part, but the whole idea is to provide support to the main structure.

Support structure removal: After the 3D printing process is completed, the supports need to be removed from the actual part. The ease of support removal can vary depending on the material and printing technology used. Some supports can be snapped off, dissolved in a solvent, or mechanically removed using tools and in some cases could be time-consuming and labor-intensive.

Design consideration of support structure: Engineers need to consider the placement and necessity of support structures for the stability of the part when designing parts for additive manufacturing. It is a good idea to minimize the need for support because it can reduce material usage, printing time, and post-processing efforts.

Types of Support: The types of support can vary based on the 3D printing technology. For example, in fused deposition modeling or material extrusion, supports are often lattice-like structures. In stereolithography and digital light processing, they can be tree-like or solid.

Automatic vs. *manual generation of support structure*: Some slicing software can automatically generate support structures, while others allow users to manually place and customize supports for more control over the process.

Post-Processing after removal of support structure: After removing support structure, additional post-processing steps like sanding, polishing, or surface finishing may be required to achieve the desired surface quality and accuracy of the final part.

In summary, support structures are a critical part of the 3D printing process, and their proper design and removal are essential to producing high-quality, functional parts with complex geometries. The choice of support material and the efficiency of support generation and removal can impact the overall cost and time required for additive manufacturing projects. Figure 15.4 illustrates two examples of support structures where the 3D printing process is needed to stabilize the fabricated part. In some cases, the support structure can be utilized to obtain mechanical properties of the main structure through testing, provided that there is a sufficient amount of support material available for testing.

15.7 Feeding and Additive Delivery System of AM Process

It does not matter what method will be used for manufacturing parts through AM technology, they all require a feeding system. One of the mechanisms that additives are delivered as an input into the systems is the distribution of powder or solid material to the working area and it is considered an important part of the AM process. The feeding system is essential for creating uniform and consistent layers, which affect the quality and properties of the final printed object. There are different types of feeding systems, depending on the type of feed, the location of the feed supply container, the feeding direction, and the techniques used. Some of the common feed systems are filament feeding method, pellet feeding, wire feeding, and powder feeding. In this section, all these feeding systems will be briefly discussed.

15.7.1 Filament Feeding Mechanism

In the case of a filament feeding mechanism, the process involves using a spool of filament made of thermoplastic polymer, which is fed into a heated extruder by a motor-driven gear. The extruder melts the thermoplastic filament and extrudes it through a nozzle onto the build platform (as shown in Fig. 15.5). As a result, the filament material, which is a thin strand of plastic material, is pushed into the hot end of the 3D printer, is melted, and extruded to create the layers of the object that engineers have in mind [36–38].

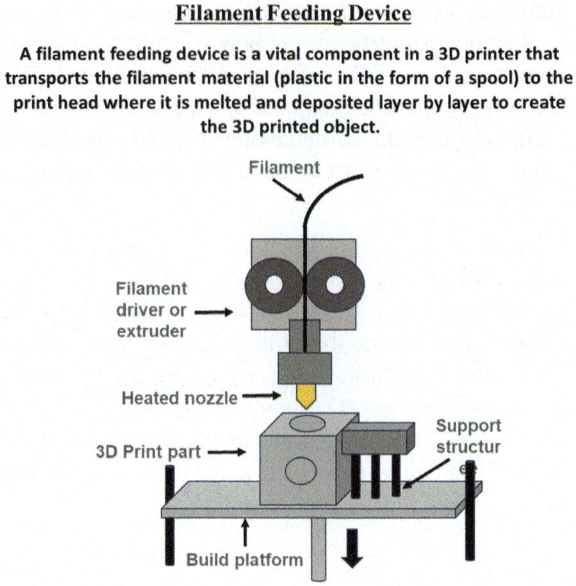

Fig. 15.5 Filament feeding device for manufacturing 3D printing part. Here a spool of plastic filament is fed into a heated extruder by a motor-driven gear

The filament feeding process requires an understanding of the behavior of the polymer filaments as they undergo feed rate. The feeding rate, the heating temperature, and the deposition speed affect the shape and the strength of the printed part. The feed rate determines how fast the filament is pushed into the heater that can affect the quality and accuracy of the printed part. The heating temperature affects the viscosity and flow of the melted filament, and it should be adjusted according to the type and diameter of the filament heating, deposition, and cooling in a highly controlled manner. The deposition speed and the erection affect the shape and strength of the 3D printed part and they should be coordinated with the movement of the nozzle. The cooling rate affects the solidification and shrinkage of the printed part, and it should be optimized to avoid warping and cracking.

Factors that can cause problems with the filament feeding process can be clogging, jams, damage, or low quality of the filament or other elements involved in the process. For example, the problem associated with filament not coming out of the nozzle, or the extruder not extruding, and the poor quality of printing are issues that must be considered seriously and taken care of.

15.7.2 Pellet Feeding Mechanism

The pellet feeding mechanism utilizes small pieces of plastic known as pellets, which can be fed into a heated barrel through a hopper. Unlike the filament feeding mechanism, which uses plastic filament spools, this system is designed to handle plastic pellets. The pellets are pushed toward a nozzle, where they are melted and deposited onto the substrate. This method offers several advantages when compared to filament feeding, including lower material costs, faster manufacturing speeds, and a broader range of material choices. The pellet feeding system is a variant of extrusion additive manufacturing (EAM) that utilizes pellets as feedstock instead of filament [39, 40]. To employ pellets as feedstock, a specialized extruder with a screw rotating counterclockwise and barrel system is required to melt and push the pellets toward a nozzle (Fig. 15.6). In essence, the screw rotates inside the barrels to build up pressure, and friction causes the pellets to heat up and melt. Initially, pellets are delivered from a dryer (moisture free) to a hopper where they are gravity-fed to a screw that runs through a heated barrel. Moisture can be collected on the outside of any plastic pellet which can go into an extruder having a significant impact on the quality of the final product. It will affect both the appearance and the mechanical properties of the 3D printing part.

As the screw rotates, pellets are pushed through a barrel with several heating zones, aided by the friction created by the screw. This gradual process allows the material to heat up and melt progressively. Subsequently, the molten material is extruded from the nozzle, forming a continuous fiber that is deposited either on the build platform or the previous layer (as shown in Fig. 15.6). The advantages associated with pellet feeding include cost-effectiveness, a wider variety of available materials, and the potential for higher production rates.

To ensure a consistent and predictable printing flow, there are several key process variables that must be controlled. They include: (1) pellet size and geometry, which impact the way materials enter and flow through the barrel, (2) maintaining the proper moisture content of the material fed into the extrusion unit, (3) regulating the

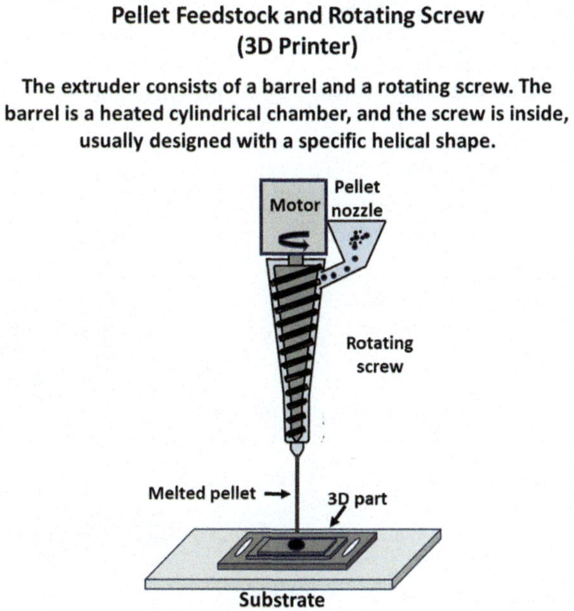

Fig. 15.6 To employ pellets as feedstock, a specialized extruder with a screw and barrel system is required to melt and push the pellets toward a nozzle for the 3D printing parts

sequential heating zones in the extruder and print head to achieve a safe and consistent material flow, (4) the need to adjusting the screw speed to achieve the correct deposition rate, (5) implementing an accurate motion control system to maintain precision, and (6) precisely synchronizing the extruder output with print head motion to ensure a consistent deposition pellet size and shape.

15.7.3 Wire Feeding Mechanism

Wire feeding mechanisms are typically used in a 3D printing process called Wire Arc Additive Manufacturing (WAAM). It is a metal 3D printing technique that uses an electric arc welding process to build up metal parts layer by layer using a wire as the feedstock material. The WAAM technology is like the DED process, except that it does not use expensive laser devices as the heat source. Note that the WAAM process is more cost-effective and has a higher deposition rate than the DED process, but there is more post processing work that must be given attention to, such as reducing the residual stresses through annealing or to apply a milling and grinding process for surface finishing [7, 41, 42].

This process is known for its ability to build large metal components quickly and cost-effectively. Here is how the wire feeding mechanism works in WAAM (Fig. 15.7). The source of heat for the wire feeding mechanism in 3D printing, specifically in wire feeding mechanisms, is through an electric arc welding process.

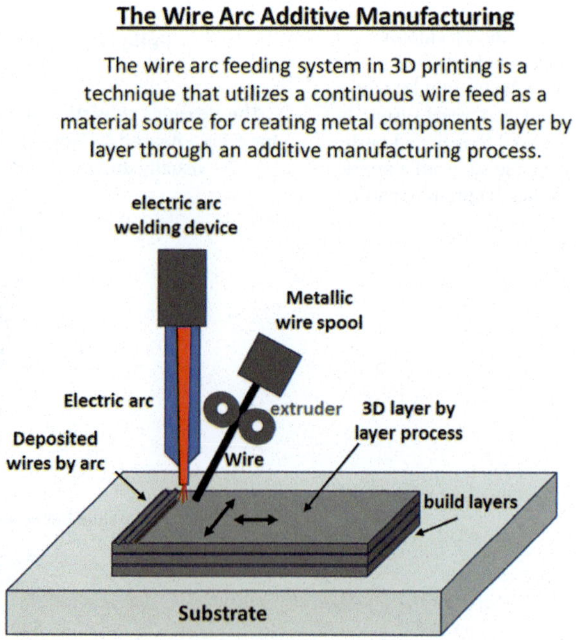

Fig. 15.7 The metallic wire arc feeding system. The heat from the arc can melt the metallic wire material to build the part layer by layer. The wire arc additive manufacturing is also known as WAAM

This electric arc provides the necessary heat to melt the metallic wire, allowing it to be deposited layer by layer to create the desired 3D-printed object. The wire feeding mechanism starts with a spool of metal wire. This wire is typically made of materials like steel, aluminum, titanium, or nickel alloys, depending on the application and desired properties of the final part, but in theory every material that can be welded can also be used for the WAAM process. While WAAM commonly works for metals and their alloys, it can also be used with polymers. However, the term "WAAM" is more commonly used in the context of metal 3D printing.

The wire feeder is responsible for pulling the wire from the spool at a controlled and consistent rate. The rate of wire feed is an essential parameter that affects the deposition rate and quality of the printed part. As the wire is fed out of the feeder, it passes through a contact tip, which maintains electrical contact with the wire. An electrode is also used, usually in the form of a tungsten electrode, which generates the electric arc necessary for melting the wire. Then, an electric arc is established between the wire (anode) and the workpiece (cathode). This high-energy arc generates the heat necessary to melt the wire and causes a molten pool on the substrate. Finally, the metal wire is melted in the arc, and the molten metal is deposited onto the substrate or the previous layer. The motion control system guides the welding torch or robotic arm to precisely position the molten pool to build up the desired 3D shape, layer by layer. To prevent distortion during solidification of molten metal, a cooling system is often used to control the cooling rate and manage thermal stresses. To avoid oxidation, a shielding gas, such as argon or helium, may be needed to avoid the molten metal from oxidation and to create a stable welding environment (Fig. 15.7).

Most WAAM processes require some amount of post processing work because of residual stresses and the surface roughness. The residual stresses are as the result of heating and cooling the metals, just like the heat affected zone (HAZ) after welding, that must be eliminated or reduced. To reduce the residual stresses in the part, a stress relief treatment (i.e., annealing) is applied. The annealing step will be helpful to avoid any premature failure during service usage of manufactured parts when tensile residual stresses will be superimposed with the applied stress.

Surface finishing is another post-processing step in WAAM. Every part is built in layers, which are visible on the surface, and in order to optimize fatigue life, tensile behavior, and corrosion resistance, it is important to adopt a finishing process like milling or grinding.

15.7.4 Powder Feeding Mechanism

Another feeding method in 3D printing is powder feeding mechanisms. It is used in a specific type of 3D printing technology called powder bed fusion. In this feeding process, both laser and electron beam are utilized. The powder feeding mechanism fuses the powdered material layer by layer to create a 3D object (Fig. 15.8). The method of approach and steps related to powder feeding mechanisms are as follows:

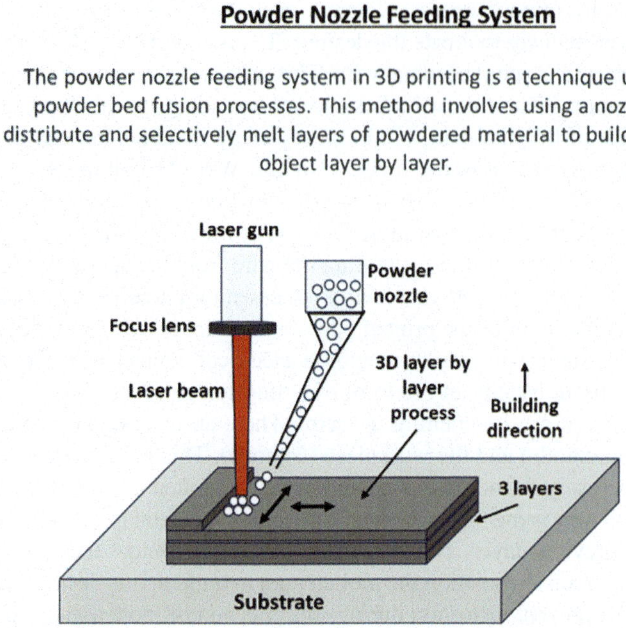

Fig. 15.8 Powder nozzle feeding system. The laser beam (the heat source) can fuse the powder material, melt layers, and build parts layer by layer

(1) the type of powdered material to be used for feed is usually a plastic, metal, or ceramic powder which will be spread as a thin layer across to build layers; (2) the powder bed will be spread after each layer and the thickness of each layer depends on the 3D printing machine, but the powder feeding mechanism guarantees that a consistent and uniform layer of powder is spread for each new layer; (3) a roller or blade is typically involved in spreading a new layer powder evenly across the build platform. By doing that, the powder is uniformly distributed and creating an even surface for the next layer. Aside from roller, a blade is also involved to smooth the surface and prepare for the next layer; (4) to fuse or sinter the powder material an energy source, like high-power laser, is needed in shaping the desired object. Another heat source, such as the electron beam (EBM) is also used for melting the metal powder. The heat source moves according to the 3D CAD model; (5) when the layer is exposed to the heat, it cools, solidifies, and fuses to the layer below it. This process is repeated layer by layer until the entire object is formed; (6) some amount of powder will not be melted or sintered and remains in the bed. The unused powder in the bed must be removed and recycled for future print. Usually, a portion of the unused powder is mixed with fresh powder to maintain the appropriate powder composition, (7) The final object may require post-processing work such as heat treatment or surface finishing to meet the customer's specifications [43–45].

In summary, the powder feeding method in powder-based 3D printing is to uniformly spread a thin layer of powdered material, which will be melted or sintered by an energy source to create the object layer by layer. This process is repeated until the entire 3D object is created. Special importance must be given to powder recycling and reusability of unused powder.

The velocity of the laser beam moving on the powder bed affects the exposure time, heat input, and the thermal efficiency of the fabrication process. Greater scanning speed will affect the porosity amount and oxidation of the parts and decreases the bonding strength. Experimental work on the properties of 3D printing parts indicated that [46] scanning at high rates between 400 and 600 mm/s, which is considered as a high rate, will cause significant pores to form, which significantly reduces tensile strength and stiffness of the final object. Usually, scanning rates depend on material being used, desired resolution and part quality, layer thickness, laser power, and machine capabilities. In addition to the velocity of the laser beam mentioned above, the smaller distance between the two neighboring tracks during the process can affect the microstructure formation of the 3D printed part. Smaller track spacing can improve the surface finish, density, and homogeneity of the parts but also decrease the processing cost and time.

The height of each deposited layer affects the number of layers, dimensional accuracy of the process, and heat conduction. Thinner layer thickness can enhance the resolution, smoothness, and mechanical properties of the parts, but also require more laser passes and energy consumption.

The abovementioned parameters can be interrelated, and the AM process needs to be optimized according to material under consideration and the load environment of AM part. There are other parameters that have also influence on the mechanical properties of 3D printed parts. These are building orientation, preheating temperature, powder size, and post processing techniques.

When fabricating parts through LAM process, material anisotropy can be formed. Anisotropy in LAM material is created by remelting and directional solidification, which are the fundamental forces behind the directional heat transfer phenomenon in LAM processes [43, 44]. Numerous studies have focused on examining anisotropy in the microstructure and mechanical characteristics of sections made utilizing LAM technology [45, 46]. When LAM is used, anisotropy is dependent on the metallic system and the behavior of directional solidification, which may result in possible heterogeneity in the manufactured part and be influenced by processing settings. Additionally, the development of other LAM-related problems/deficiencies such as micropores and absence of fusing [47] may also cause anisotropy and a decline in the mechanical characteristics of the developed sections [48].

The LAM of metallic and polymer composite parts are the primary topic of this chapter, their applications, the DED and subclasses of DED technologies that are now available, the process variables in those technologies, DED technology faults, and their mechanical characteristics. In this review, many classes of materials, steel, titanium and its alloys, polymers, ceramics etc. as well as the repair capabilities of DED process are highlighted.

15.8 Applications and Examples of DED in Laser AM Technology

This section will focus on some of DED's applications in large-scale structural manufacturing, precision repairs, feature addition, multimaterial constructions, alloy design, and coating after creating the part. Also, its advantages and disadvantages will be presented.

Following the fabrication and installation of the components, key factors such as fatigue as a result of heat cycles as well as mechanical load cycles during service usage, local impact, and corrosion that can result in crack formation and growth must be given special attention. For example, in aerospace parts such as turbine blades, corrosion and fatigue can lead to material properties degradation and leading to the initiation of cracks. Most studies have reported that cracks under stress cause high performance materials eventually to fail sooner and being thrown away as a waste part. Recent research has mostly focused on reducing component failure rates and extending component lifespans through repair [49, 50]. However, repairs must be performed when damage has been detected through the NDI inspections. That is, if the size of damage is critical and engineers decided that the repair is necessary, then the repairing through the 3D technique can be helpful.

To increase the number of cycles to failure of parts and to restore their functionality when flaws are present, repairing the damaged areas is required before final failure occurs. The DED repair method is well known not only for producing parts with good metallurgical bonding, but also for the post-repair mechanical qualities. Figure 15.9 shows a damaged turbine blade that was repaired using the DED method after being constructed by the DED procedure. The DED process can be used to manufacture and repair the turbine blades [49, 51], where damage in turbine was successfully repaired and restored by using the DED method. Figure 15.9 shows three different stages of the turbine blade: (1) undamaged turbine blade after the DED manufacturing, (2) damaged turbine blade during operation at their tip, and (3) a fully repaired turbine blade ready to function in the aircraft load environment.

In general, any research related to the DED process must focus on improving the quality and the reliability of the product as well as the decrease in the fabrication time through the process of optimization. Eliminating the residual stresses and porosity, that are the source of crack initiation, propagation, and final failure should be given especial consideration because they all can contribute to the degradation of material properties. After repairing process, due to the rapid cooling rates and significant thermal gradients, a post-DED repair heat treatment is sometimes necessary to reduce the residual stresses. Note that in DED repairing process, the material is deposited in the repaired area by using a computer-controlled deposition head that utilizes the CAD file of the damaged part to complete the repairing task. Finally, as part of the post repair processing, some amount of surface finish may be needed while maintaining the tolerances specified by the customers.

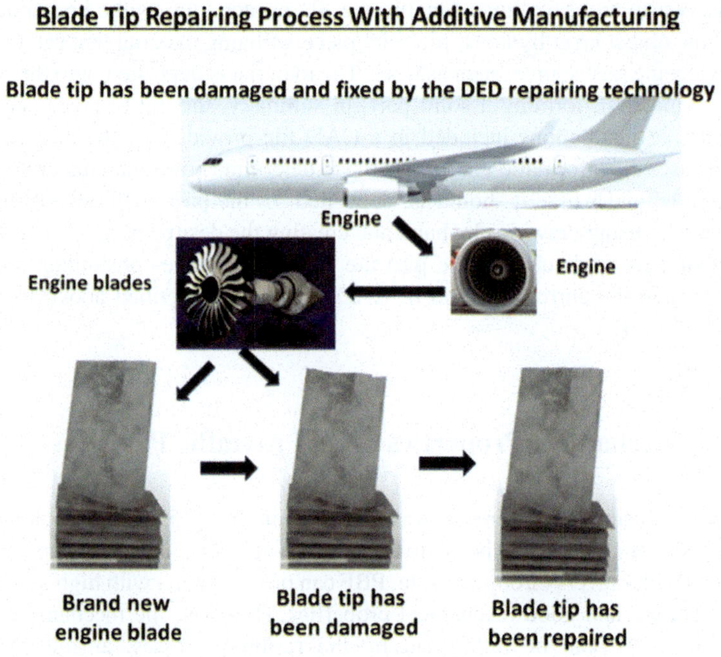

Fig. 15.9 Turbine blades were disassembled from the engine, and blades were checked for cracks or damaged areas. The damaged area was fixed using the DED repairing technology

15.9 Interaction of Digital Model with the 3D Printer

For additive manufacturing a digitized model of the part you wish to create is required. There are two types of software that can interact with the 3D printers to create the desire object for industry, which are (1) computer-aided design (CAD) and (2) computer-aided manufacturing (CAM). The two differ from each other, where the CAD allows users to design new products in 3D, exactly like what engineers have in mind, while the CAM software directs the 3D printer to build the products in accordance with the CAD model. A brief description of how the two files work together is given here. The first step is for the user to create a 3D model of the object under consideration by using the CAD software. The user can choose the shape, size, and material and can modify the appearance of the object as needed. After the task is completed and the object was generated, a file that contains all the information about the geometry of part is now ready for the 3D printing to use.

The next step is for the user to upload the CAD software that is compatible with the 3D printer. On the other hand, the CAM has the capability to transform the CAD file in such a way that instructions can be given to the 3D printer such as how much material should be deposited, where to deposit it, and the speed to move the nozzle. Furthermore, the CAM software has the capability to optimize the printing process to decrease waste and save time. Finally, the 3D printer follows the instructions and creates the object layer by layer, piece by piece, utilizing material that can be melted or cured by the heat source, such as laser. The material is deposited onto the previous layers of material, forming a solid part. In summary, the 3D printer managed by guidelines or instructions included in a CAD file provided by the designers. The machine uses the information in a CAD file to determine how much material needs to be deposited and where it should be deposited. In the past, methods of producing parts were by using drawings by hand and creating the desired parts. As can be seen, the drawing methods used in the past are much more time-consuming and labor-intensive than the current method of CAD and CAM described above for the 3D printing process.

15.10 Mechanical Properties of AM Metallic Parts

Powder bed fusion (PBF) already was discussed in Sect. 15.4. It is a technique that uses a laser or an electron beam to selectively melt or sinter metal, polymer, or ceramic powder layers through nozzle. PBF can produce parts with high density, fine microstructure, and good mechanical properties. However, the mechanical properties of the PBF can be influenced by the residual stresses, anisotropy in the manufactured part, and defects, which can shorten the performance of the 3D printed parts. For example, PBF of Ti-6Al-4 V alloy can reach a tensile strength around 900 MPa (130.4 ksi) and a fracture toughness of about 45 MPa$\sqrt{m}$ (41 ksi$\sqrt{in}$). These values fall in the lower bound of Ti-6Al-4 V alloy when parts are created through forging and at room temperature [52]. In the case of direct energy deposition (DED) method, the process uses metal powder or wire that material fed directly into the focal point of the laser or electron beam. DED manufactured parts can result in poor microstructure, high porosity, and low dimensional accuracy that can lower the mechanical properties of the parts. For example, the DED of Inconel 718 alloy can reach tensile strength close to 800 MPa (116 ksi) and a fracture toughness of 40 MPa$\sqrt{m}$ (36 ksi$\sqrt{in}$). These values are much lower than the 718 Inconel in ST condition [52].

Binder jetting (BJ) technique uses an inkjet printer to deposit a liquid binder onto metal powder layers. BJ can produce parts with low cost, high speed, and full color. However, the technique requires post-processing phases, such as sintering or infiltration to increase the density and strength of the parts. For example, the binder jetting process using the stainless steel 316 L, can reach a tensile strength of approximately 500 MPa (72.5 ksi), and a fracture toughness of around 20 MPa$\sqrt{m}$ (18 ksi$\sqrt{in}$) [52]. The part created through the BJ process with 316 L as the feed material has much lower fracture toughness than the KIc value of 316 L in the cast annealed condition [52]. The sheet lamination (SL) technique uses ultrasonic welding or adhesive bonding to join metal sheets or foils. Note that ultrasonic welding is a technique that joins two pieces of material together using high-frequency ultrasonic vibrations. The SL process can produce parts with low energy consumption, high material utilization, and multi-material capability. However, SL also suffers from low resolution, high surface roughness, and weak interlayer bonding that can impair the mechanical properties of the parts. For example, the sheet lamination process of aluminum alloy 6061, can reach a tensile strength of approximately 200 MPa (29 ksi), and a fracture toughness of around 10 MPa√m (9 ksi√in), which has lower fracture toughness than the 6061-T6 condition with ultimate of 46 ksi and fracture toughness of 36 ksi√in for T-L condition [52].

15.11 Residual Stresses, Distortions, and Porosities (on 3D Parts)

Residual stresses are internal stresses that persist within a manufactured part after its creation. A notable example of residual stresses occurs in metallic parts when two components are welded together through the fusion process. Note that residual stresses can occur in polymers during processes like molding, extrusion, or cooling and for ceramics due to various manufacturing steps like shaping, sintering, or cooling. These stresses can significantly affect the mechanical properties and performance of the printed components. During the DED process, with high heat imposed on the 3D part, the generated object has a very complicated thermal history that involves melting, remelting, and reheating the material follow by cooling [53–55]. The rapid cooling and heating cycles are the primary causes of thermal cycles and residual stresses during AM processing. Both the thermal cycles and residual stresses can contribute to reduction of mechanical properties. Residual strains are produced in the manufactured component, as well as in the substrate, during the part formation from a liquid phase to a solid and during solidification, as well as during several solid-state transformations. The powder feeding rate, laser power and associated temperature, laser transverse speed, and route strategy are design and process variables that may have an impact on a produced part's thermal history, microstructure, and degree of residual stresses [53]. The induced residual stresses have an impact on deposited material and may lead to deformation [56–58], cracking, and delamination, among other problems. Therefore, it is a good idea to find ways to reduce the residual stresses in order to achieve acceptable material properties.

One of the most often employed strategies for decreasing residual stress is preheating the substrate, based on the work provided by [56]. According to a study by [57], it was found that residual stresses dropped by 80.2% and distortions by 90.1%, respectively, when substrate preheating and build chamber heating were coupled. These investigations have demonstrated that while preheating the substrate, printed products, and the chamber at the same time, it can reduce residual stresses, but does not eliminate them totally.

For having an acceptable mechanical property, it is helpful to manage the residual stresses in any LAM process, when manufacturing the 3D printing part. To reduce the residual stresses, there are some factors that can be implemented, which are: (1) thermal gradients, (2) support structures, (3) process parameters, (4) post-processing heat treatment, (5) simulation and modeling, (6) material testing. The following is a brief description of each factor related to managing and reducing the residual stresses in any LAM process.

Thermal Gradients During the LAM process, the rapid heating and cooling of the material can result in significant thermal gradients. This can lead to thermal stresses, which contribute to the development of residual stresses. One way to alleviate this is by optimizing the laser scanning path to control the temperature distribution in the build. Preheating the built platform or the powder bed can help reduce thermal gradients. Post-heating the part after printing can also be used to relieve some of the residual stresses. This is often done in the form of stress relief cycles in a controlled environment. Note that in the case of PBF process, parts are likely to have several issues because of weak bonding between layers and the complicated thermal history. High temperature gradients cause thermal residual stress that accumulates as the layers are built up, which can result in distortion, warpping, delamination, and cracking of the products.

Support Structures The use of support structures is widespread in LAM process to prevent the part from warping or collapsing during the printing process. However, support structures can introduce additional residual stresses because the support structure prevents the part from deformation. Well thought design and placement of supports can minimize the residual stresses.

Process Parameters Experimentation and optimization of laser power and scan speed can influence the temperature distribution and, consequently, the reduction of residual stresses in the printed part.

Post-processing Post-processing techniques such as stress relief annealing, hot isostatic pressing (HIP), or shot peening can be used to reduce residual stresses in the final part. Shot peening will introduce e compressive residual stresses on the surface of the part and prevents the crack initiation at the early stage of operation.

Simulation and Modeling Using advanced simulation and modeling software can predict and know how residual stresses will develop in a printed part. Finite element modeling not only can provide information about the location of high stresses, but also the residual stresses. This information can guide process optimization and design choices.

Material Testing Conducting material testing and characterization, including the measurement of residual stresses, is essential for quality control and to ensure that printed parts meet performance specifications. Conducting proper nondestructive testing can also be helpful to see the presence of any cracking due to the high residual stresses prior to its service usage.

Controlling residual stresses in laser additive manufacturing is a challenge, and it often requires a combination of process optimization, material selection, and post-processing techniques to produce high-quality and reliable components. Engineers continue to develop new methods and strategies to minimize the residual stresses to increase the quality of LAM final products.

15.12 Application of 3D Printing on Fabricating Composites

3D printing technology can generate composite objects by using discontinuous short fibers reinforced polymer matrix composite material using a single extruder 3D printer. Short carbon fibers of diameter around 7.2 μm with length of 150 μm are used as reinforced filaments [59]. Input parameters used in the 3D printing process for obtaining a good-quality short carbon fiber-reinforced polymer specimen are: (1) reinforcement percentage in volume, (2) printing speed, (3) layer thickness, (4) power bed temperature, and (5) print orientation. To examine the quality of 3D printing, the microstructure of fabricated 3D part was examined. Tensile strength, ductility, and toughness were checked to validate the adhesiveness of the matrix and reinforcement and degree of delamination as well as matrix cracking and fiber breakage. It was observed that the hardness properties can be significantly affected when the reinforcement percentage is increased. The results obtained in this study can be effective in fabricating polymer matrix composites for applications in automotive industry, aerospace, and medicine.

Three-dimensional printing or additive manufacturing process is moving forward adding a new dimension in creating complex 3D objects of any desire shape using both metallic and non-metallic materials. In this section, the focus will be on the application of this noble technology on the polymer matrix composites (PMCs) [59]. In addition, this technology has the ability to reinforce the plastic matrix with nanofiber, microfiber, chopped fiber, and continuous fiber where their applications can be employed in many industries. In this section, reinforced fibers used by the commercial 3D printers to develop fiber-reinforced composites and parameters such as uniform distribution of fibers in matrix, as well as fiber volume fraction, that both have direct practical implications in determining and tuning final composites' properties and their applications are investigated. Additionally, the capabilities and limitations of 3D printing software to process materials and control print parameters in relation to print quality, structural integrity, and properties of manufactured composite material are discussed. The discussions aim to present constructive evaluation and criticism of the current technology in order to guide engineers and 3D printing equipment manufacturers on improvements of final 3D printed fiber-reinforced composites products.

Fabricating composite parts using 3D printing technology is a growing area of research, especially when dealing with continuous and long fibers versus the short and chopped one. Recently, this technology has gained enormous thought for building complex geometries with fiber-reinforcement that has brought it into the spotlight and currently has applications in numerous industries [60, 61]. The application of 3D printing is moving forward rapidly ranging from prototyping to high performance products in aerospace, automobile, and many other industries [62]. The idea of applying 3D printing to reinforce plastic with different physical forms of fillers such as nanofibers, microfibers and chopped fibers is of great demand and going forward rapidly. The method of approach is to have fillers mixed with a polymer batch to prepare plastic filament feed for the printer, where the fibers dimensions were small and not continuous [63]. However, reinforcing the polymer with continuous fibers at desired location of a layer to form a part was challenging and did require to have a different feed system unlike simply short fiber contained in the plastic filament [38]. Figure 15.10 illustrates the creation of continuous fiber on the polymer matrix by using the 3D printing technique.

The application of incorporating continuous fibers as the reinforcement was presented by Matsuzaki et al. [64]. Here, continuous carbon fibers were impregnated with the polymer (e.g., polylactic acid) at the extrusion nozzle immediately before printing. This method used a modified nozzle receiving plastic and untouched fiber (fiber as it is) together from two separate feed systems.

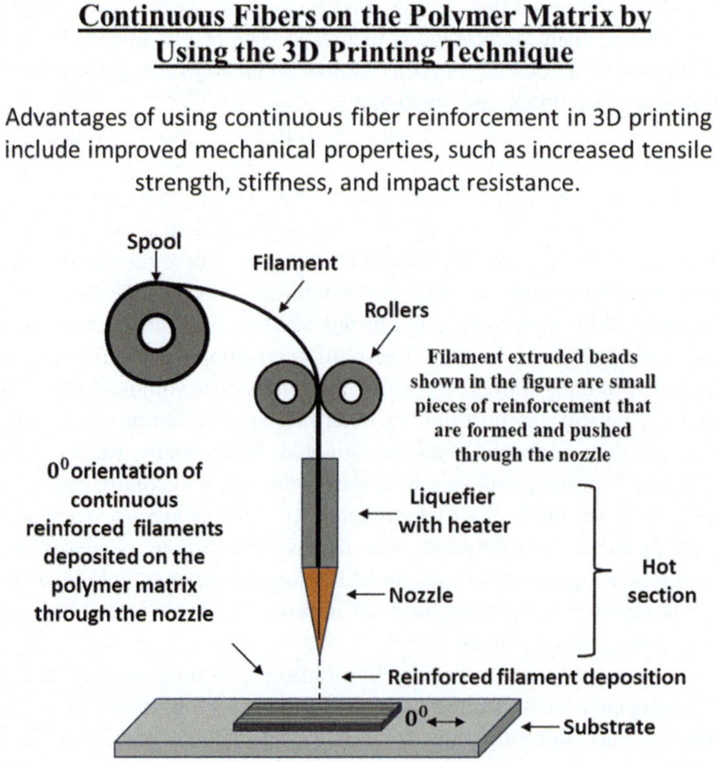

Fig. 15.10 Creation of continuous fiber on the polymer matrix by using the 3D printing technique

The double feed system is referred to as a coaxial extrusion process [65]. Although the double extrusion system (coaxial) might offer the achievement of high fiber content (fiber at each layer), it often suffers from lack of desired fiber placement, poor control of fibers as well as insufficient matrix infusion, resulting in an inferior fiber–polymer interface. The two nozzles deposit feeding system will deposit the plastic and preimpregnated fiber separately, which provides a higher degree of freedom for localized fiber reinforcement and improved fiber–polymer interface [66].

The double nozzle feeder was established by Markforged, a 3D printing company that specializes in producing continuous fiber, and composite parts [66, 67]. For setting print parameters and following the print execution, the printer requires use of supplier-defined software that allows users to import, prepare, and manage their 3D printing projects with ease and efficiently. It also enables users to customize their print settings, optimize their part performance, and access their digital part inventory from anywhere. In brief, the double nozzle feeder printers allow the printer to print both base material and continuous fiber in a single print. The base material is printed by one nozzle and the continuous fiber is laid in layer by a second nozzle.

There are four fiber filaments material including fiberglass, carbon, Kevlar, and high-strength high-temperature fiberglass that are composed of continuous fiber bundles and sized with nylon and Onyx matrices (considered as composite filament) in order to provide strong interfacial bonding between fiber and plastic (matrix). The mechanical properties of these supply materials provided by the manufacturer are available and shown in Fig. 15.11 for short carbon fibers. In some cases, important parameters such as diameters of composite filaments, number of fibers in the

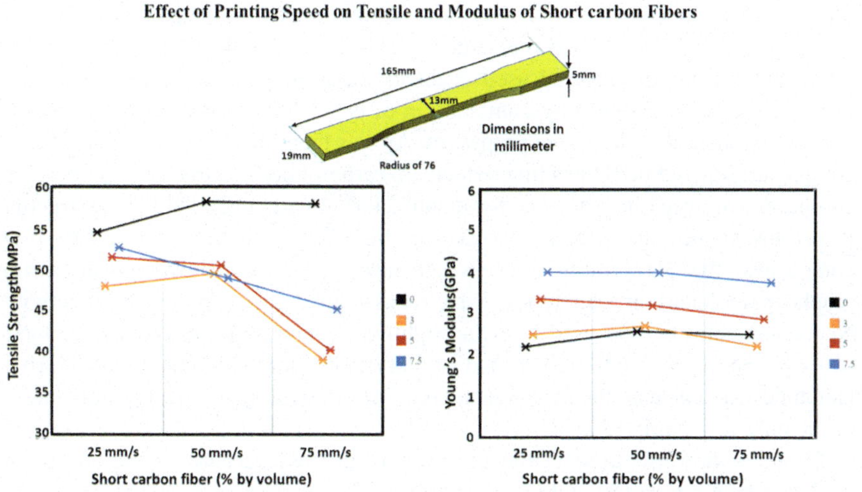

Fig. 15.11 Effect of several printing speed on tensile strength and modulus of short carbon fiber on polycarbonate matrix. The % volume of fibers in the matrix is also shown on the right of the figure (0, 3, 5, 7.5%) [68]

composite filaments, diameter and density of the fiber, and fiber volume fraction, fiber–matrix distribution in the composite filaments as well as printer speed were considered to be key manufacturing parameters that companies do not like to reveal them. But all that information is very important for designing and characterizing a composite part as well as predicting the mechanical properties.

15.12.1 Types of Polymers and Reinforcements Materials Used in AM Composites

Polymers are good for producing pieces that need to be lightweight and flexible. The 3D printing can work with various polymers, including nylon and polypropylene that are not expensive and easy to work with. In addition, it can also employ more sophisticated polymers such as PEEK, which have excellent strength-to-weight ratios and are often used in aerospace and medical applications. One of the key advantages of employing polymers in AM is the ability to make components with complicated interior geometries, making it a preferred choice for constructing medical implants and other specialized devices. Advantages of using polymer in 3D manufacturing are: (1) low cost and being easy to work with, (2) lightweight, (3) ability to use complex geometries, and (4) ability of fabricating the 3D product with different colors and textures. Their disadvantages are: (1) lower strength when compared to metals and composites, and (2) can easily deform under high stress and temperature [69].

In the case of composites, which have great strength to weight ratios (σ_{ul}/ρ) as well as modulus to weight (E/ρ), are used in fabricating 3D parts by using the DED process. There are two types of composites mostly utilized in AM process, which are carbon fiber and fiberglass. Composites, which have excellent strength-to-weight ratios, are a common material used in direct energy deposition (DED). Manufacturers fabricate composites by combination of two or more separate substances to produce a material with unique properties. As was mentioned, two popular types of composites utilized in 3D printing, which are carbon fiber and fiberglass. Fiberglass composites are used in industries for manufacturing parts intended for constructing items that require high-impact resistance. In addition to these frequently used composites, the 3D printing process can also deal with nanocomposites using graphene and carbon nanotubes, which have great strength and conductivity. Nanocomposites can be used to create lightweight and high-performance components for aerospace, automotive, and military applications. Graphene and carbon nanotubes can enhance the mechanical properties (including the fracture toughness) of the polymer matrix [70, 71].

Ceramics are also used in the 3D printing process because of their strength, hardness, heat resistance, and wear resistance. They can be an excellent material to be used in the DED method. Ceramics such as alumina, zirconia, and silicon carbide have high-temperature resistance and are used as matrix or reinforcement in ceramic matrix composites (CMCs). In 3D printing, CMCs can be fabricated through DED using the laser or electron beam as the heat source. One of the main advantages of using ceramics in 3D printing is their ability to withstand extreme high heat environments that other materials cannot.

15.12.2 Parameters Influencing Mechanical Properties of 3D PMCs

The 3D printing process is able to manufacture composite parts with mechanical properties compatible with the traditional approaches described in Chap. 4 for PMCs manufacturing parts. The mechanical properties of fabricated composite parts are a function of several factors, which will be presented here: (1) in additive manufacturing technique, the reinforcing fibers can be arranged in the matrix with proper orientation, which can influence mechanical properties of fabricated laminated composite part, such as strength, stiffness, and toughness significantly; (2) the choice of matrix material and reinforcing agents in 3D printing greatly impacts the mechanical properties of the composite. For instance, combining carbon fibers with a thermoplastic matrix can create a composite with high strength and stiffness, while using glass fibers may emphasize other properties like impact resistance; (3) the volume fraction of the reinforcement in the matrix will influence the mechanical properties of fabricated composite significantly; (4) additive manufacturing allows for complex geometries to be formed and processing the internal structures in such a way that can enhance the mechanical properties; (5) distributing stresses more effectively or reducing material usage while maintaining strength; (6) ensuring proper bonding between layers (layer adhesion) is crucial in additive manufacturing. Strong layer adhesion is essential for achieving desired mechanical properties, as weak interfaces between layers can reduce the overall strength by introducing delamination, (7) adjusting printing parameters such as temperature, speed, and layer thickness can influence the mechanical properties of the final composite part. Fine tuning these parameters is essential to maintain the desired properties.

In item #7 (adjusting printing parameters), the influence of printing speed is demonstrated in [68] on both the tensile strength and modulus of elasticity of short carbon fiber in the polycarbonate matrix. Figure 15.11 shows both the tensile and modulus values for three different printing speeds and different degrees of percent reinforcement. The tests were conducted based on the *ASTM* D638 [72] testing standards with specimen geometry and configuration included in Fig. 15.11.

In Fig. 15.11, the change in the tensile strength and Young's modulus were examined by varying the printing speed and fiber concentrations. There is a substantial increase in the tensile strength and Young's modulus values due to reinforcement can be seen. This figure indicates that the tensile strength increases as reinforcement increases inside the polycarbonate matrix. This happened because short fibers are uniformly distributed inside the matrix material. The uniform distribution of fibers allows the load to be uniformly distributed or transferred from matrix to fibers in the entire 3D printed specimen. However, the tensile strength of short carbon fiber in the matrix is slightly less than pure polycarbonate matrix in all the concentrations shown in Fig. 15.11. Only 3.3% decrease in tensile strength is noticed from the pure polycarbonate. The reason for this decrease is that the short carbon fiber inside the matrix material is not perfectly aligned in one direction or to the loading direction [68]. So, if we stretch the sample, the short carbon fibers inside the matrix materials act like a stress concentrator which eventually reduces the tensile strength of the 3D composite material. But in case of Young's modulus, the composite material reinforced with short fibers has higher stiffness than the pure matrix as shown in the figure, because the presence of short fibers restrains the polymer chain movement in the vicinity of the other chains. 183% increase in Young's modulus is noticed after the short carbon fibers reinforcement were added. Improper melting in the melting zone and slipping of filament in the extruder assembly are some of the reasons for the decrease in tensile and young modulus as we performed the printing on high speed [73].

In general, additive manufacturing of composites offers the potential to create materials with tailored mechanical properties, allowing manufacturers to design components with specific performance characteristics suitable for various applications, including aerospace, automotive, biomedical devices, and more. However, because of parameters involved in this process, it is expected that, reliability, and repeatability in producing these materials with the desired mechanical properties is challenging and will be the focus of future research and development in the 3D printing industry.

15.13 Concluding Remarks

The implementation of 3D printing technology with composite materials marks a shift in the manufacturing methods of composite parts, offering extraordinary opportunities for innovation and customization across many industries. The ability to design complex structures and fabricate them through additive manufacturing, coupled with high customer demand for specific material properties, poses challenges. As research and development continue to push the boundaries of 3D-printing composite manufacturing, achieving enhanced performance, reducing material waste, and eliminating manufacturing-related costs are highly desirable goals for producers of composite parts. While challenges, such as achieving material quality with acceptable properties, are essential for customers, research in 3D printing plays a central role in the evolution of composite materials. The path from concept to manufacturing highlights the transformative potential of this technology, promising a future where the limits of traditional manufacturing are reformed, and various possibilities, such as the adoption of 3D printing technology, will emerge. Finally, engineers should realize that manufacturing the 3D printing parts, just like the traditional methods, need to have material allowables available for analysts to conduct a limited range of damage tolerance analysis to be sure if the 3D-manufactured part has adequate life during the service operation.

References

1. M.K. Thompson, G. Moroni, T. Vaneker, G. Fadel, R.I. Campbell, I. Gibson, et al., Design for additive manufacturing: Trends, opportunities, constructions, and constraints. CIRP Ann. Manuf. Technol. **65**(2), 737–760 (2016)
2. W.E. Frazier, Metal additive manufacturing: A review. J. Mater. Eng. Perform. **23**(2), 1917–1928 (2014)
3. J. Alcisto, A. Enriquez, H. Garcia, S. Hinkson, T. Steelman, E. Silverman, P. Valdovino, H. Gigerenzer, J. Foyos, J. Ogren, J. Dorey, K. Karg, T. McDonald, O.S. Es-Said, Tensile properties and microstructures of laser-formed Ti-6Al-4V. JMEP **20**(2), 203–212 (2011)
4. D.L. Bourell, M.C. Leu, D.W. Rosen (eds.), *Roadmap for additive manufacturing* (University of Texas at Austin, Austin, 2009)
5. W.E. Frazier, *Digital Manufacturing of Metallic Components: Vision and Roadmap*, in Solid Free Form Fabrication Proceedings (University of Texas at Austin, Austin, 2010), pp. 717–732
6. E. Herderick, *Additive Manufacturing of Metals: A Review*, in Proceedings of MS&T11, Additive Manufacturing of Metals, Columbus, 2011
7. D. Ding, Z. Pan, D. Cuiuri, H. Li, Wire-feed additive manufacturing of metal components: Technologies, developments, and future interests. Int. J. Adv. Manuf. **8**(1–4), 465–481 (2015)
8. D.R. Eyers, A.T. Potter, Industrial additive manufacturing: A manufacturing system and perspective. Comput. Ind. **92–93**, 208–218 (2017)
9. D.G. Ahn, Applications of laser assisted metal rapid tooling process to manufacture of molding & forming tools – State of the art. Int. J. Precis. Eng. Manuf. **12**(5), 925–938 (2011)
10. J.J. Lewandowski, M. Seifi, Metal additive manufacturing: A review of mechanical properties. Annu. Rev. Mater. Res. **46**, 151–186 (2016)
11. K. Osakada, M. Shiomi, Flexible manufacturing of metallic products by selective laser melting of powder. Int. J. Mach. Tools Manuf. **46**(11), 1188–1193 (2006)
12. 3D Systems purchased Z Corporation in 2012. Stratasys merged with Object in 2012 and is now incorporated in Israel
13. D.S. Thomas, Economics of the U.S. Additive Manufacturing Industry. NIST Special Publication, 1163 (2013). http://www.nist.gov/manuscript-publication-search.cfm
14. Gartner AM report. Information on http://www.gartner.com/document/2598122. Accessed 16 June 2014
15. Ceramics Additive Manufacturing Markets 2017–2028, SmarTech Analysis, Crozet, VA. http://www.smarTechanalysis.com/reports/ceramics-additive-manufacturing-markets-2017-2028. Accessed 14 Dec 2020
16. Quick 2009, 3D Bio-printer to Create Arteries and Organs. http://www.gizmag.com/3d-bio-printer/13609/
17. K. Vartanian, T. McDonald, Accelerating industrial adoption of metal additive manufacturing technology. J. Occup. Med. **68**(3), 806–810 (2016)
18. L.E. Murr, S.M. Gaytan, D.A. Ramirez, E. Martinez, J. Hernandez, K.N. Amato, et al., Metal fabrication by additive manufacturing using laser and electron beam melting technologies. J. Mater. Sci. Technol. **28**(1), 1–14 (2012)
19. S.M. Thompson, L. Bian, N. Shamsaei, A. Yadollahi, An overview of direct laser deposition for additive manufacturing; part I: Transport phenomena, modeling and diagnostics. Addit. Manuf. **8**, 36–62 (2015)
20. K.M. Hong, Y.C. Shin, Prospects of laser welding technology in the automotive industry: A review. J. Mater. Process. Technol. **245**, 46–69 (2017)

21. N. Venkataraman, S. Rangarajan, M.J. Matthewson, A. Safari, S.C. Danforth, A. Yardimci, *Mechanical and rheological properties of feedstock material for fused deposition of ceramics and metals (FDC and FDMet) and their relationship to process performance,* in Solid Freeform Fabrication Proceedings, University of Texas at Austin, Austin (1999), pp. 351–360
22. A. Singh, S. Kapil, M. Das, A comprehensive review of the methods and mechanisms for powder feedstock handling in directed energy deposition. Addit. Manuf. **35**, 11388 (2020)
23. W.J. Sames, F.A. List, S. Pannala, R.R. Dehoff, S.S. Babu, The metallurgy and processing science of metal additive manufacturing. Int. Mater. Rev. **61**(5), 315–360 (2016)
24. I. Mathoho, E.T. Akinlabi, N. Arthur, M. Tlotleng, B. Masina, *Metallurgical Characteristics of Laser Peened 17–4 PH SS Processed by LENS Technique*, in TMS 2019 148th Annual Meeting & Exhibition Supplemental Proceedings. The Minerals, Metals & Materials Series (Springer, 2019)
25. P. Ahangar, M.E. Cooke, M.H. Weber, D.H. Rosenzweig Current biomedical applications o3D printing and additive manufacturing. Appl. Sci. 2019
26. M. Fateri, A. Kaouk, A. Cowley, S. Siarov, M.V. Palou, F.G. Gonzalez, et al., Feasibility study on additive manufacturing of recyclable objects for space applications. Addit. Manuf. **24**, 400–404 (2018)
27. R. Galante, C.G. Figueiredo-Pina, A.P. Serro, Additive manufacturing of ceramics for dental applications: A review. Dent. Mater. **35**(6), 825–846 (2019)
28. N. Shamsaei, A. Yadollahi, L. Ian, S.M. Thompson, An overview of direct laser deposition for additive manufacturing; part II: Mechanical behavior, process parameter optimization and control. Addit. Manuf. **8**, 12–35 (2015)
29. P. Madeleine, *3D Natives (Your Source of 3D Printing), PBF vs. DED: Which Metal 3D Printing Process Should You Choose?* Published on 14 Mar 2023
30. Elizabeth Engler Modic, *Directed Energy Deposition (DED) vs Powder Bed Fusion (PBF)*, Aerospace Manufacturing and Device, 3 Jul 2018
31. T. Wohlers, *Wohlers Report 2012: Additive Manufacturing and 3D Printing State of the Industry* (Wohlers Associates, 2012)
32. The Additive Manufacturing (TAM), *Binder Jetting Complete Beginner Guide*, By Alireza Jalili, 10 Jan 2020
33. S. Claire, *Everything you need to know about the support Structures*, 3D Natives (Your Source for 3D Printing), Published on 12 Apr 2023
34. M. Langelaar, Topology optimization of 3D self-supporting structures for additive manufacturing. Addit. Manuf. **12**, 60–70 (2016)
35. M. Baumers, *Economic Aspects of Additive Manufacturing: Benefits, Costs, and Energy Consumption*. Doctoral Thesis (Loughborough University, 2012)
36. P.H. Daniel, J.F. Kelly, *The Filament Feeding Mechanism* (Apress, Barkeley, 2011) ISBN 978-1-4302-3443-2
37. C.D. Millholland, *Additive Manufacturing Solutions* (ThermoFisher Scientific, 2021)
38. N. Van de Werken, H. Tekinalp, P. Khanbolouki, S. Ozcan, A. Williams, M. Tehrani, Additively manufactured carbon fiber-reinforced composites: State of the art and perspective. Addit. Manuf. **31**, 100962 (2020)
39. B. O'Neill, *Guide to Additive Manufacturing* (aniwaa.com Company, 2022)
40. *3D Printing with Plastic Pellets – What You Need to Know* (AM Additive Manufacturing, 2023)
41. F. Wang, S. Williams, P. Colegrove, A.A. Antonysamy, Microstructure and mechanical properties of wire and arc additive manufactured Ti-6Al-4V. Metall. Trans. A. **44A**, 968–977 (2013)
42. D. Nelson, *Wire Feed 3D Printing Grows in Popularity* (The Fabricator, 8 May 2019)
43. S. Kelly, S. Kampe, Microstructural evolution in laser deposited multilayer Ti-6Al-4V builds. Part II thermal modeling. Metall. Mater. Trans. **35**(6) (2004)
44. Z. Wang, T.A. Palmer, A.M. Beese, Effect of processing parameters on microstructure and tensile properties of austenitic stainless steel 304L made by directed energy deposition additive manufacturing. Acta Mater. **110**, 226–235 (2016)

45. Y. Kok, X.P. Tan, P. Wang, M. Nai, N.H. Loh, E. Liu, et al., Anisotropy and heterogeneity of microstructure and mechanical properties in metal additive manufacturing: A critical review. Mater. Des. **139**, 565–586 (2018)
46. L. Thijs, M.L.M. Sistiaga, R. Wauthle, Q. Xie, J.P. Kruth, J. Van Humbeeck, Strong morphological and crystallographic texture and resulting yield strength anisotropy in selective laser melted tantalum. Acta Mater. **61**(12) (2013)
47. K. Shah, A.J. Pinkerton, A. Salman, L. Li, Effects of melt pool variables and process parameters in laser direct metal deposition of aerospace alloys. Mater. Manuf. Process. **25**(12), 1372–1380 (2010)
48. Z. Baicheng, L. Xiaohua, B. Jiaming, G. Junfeng, W. Pan, S. Chennan, et al., Study of selective laser melting (SLM) "Inconel 718 part surface improvement by electrochemical polishing". Mater. Des. **116**, 531–537 (2017)
49. A. Saboori, A. Aversa, G. Marchese, S. Biamino, A. Lombardi, P. Fino, Application of directed energy deposition-based additive manufacturing in repair. Appl. Sci. **9**(16), 3316 (2019). https://doi.org/10.3390/app9163316
50. A. Dass, A. Moridi, State of the art in directed energy deposition: From additive manufacturing to materials design, a review. Mech. Aerosp. Eng. (2019)
51. J.M. Wilson, C. Piya, Y.C. Shin, F. Zhao, K. Ramani, Remanufacturing of turbine blades by laser direct deposition with its energy and environmental impact analysis. J. Clean. Prod. **80**, 170–178 (2014)
52. Nasgro Software Overview, Software Research Institute (SwRI), Version 8, 2016
53. D. Svetlizky, M. Das, B. Zheng, A.L. Vyatskikh, S. Bose, A. Bandyopadhyay, J.M. Schoenung, E.J. Lavernia, N. Eliaz, Directed energy deposition (DED) additive manufacturing: Physical characteristics, defects, challenges and applications. Mater. Sci. Eng. **49** (2021)
54. B. Dutta, S. Babu, B. Jarred, *Science, Technology, and Applications of Metals in Additive Manufacturing* (Elsevier, 2019)
55. W.J. Sames, F.A. List, S. Pannala, R.R. Dehoff, S.S. Babu, The metallurgy and processing science of metal additive manufacturing. Int. Mater. Rev. 1–46. D. Schwam, M. Kottman, B. Dutta, NADCA Congress Proceedings (2014), pp. 1–7 (2016)
56. T. DebRoy et al., Prog. Mater. Sci. **92**, 112–224 (2018). https://doi.org/10.1016/j.pmatsci.2017.10.001
57. X. Lu et al., Addit. Manuf. **26**, 166–179 (2018). https://doi.org/10.1016/j.addma.2019.02.001
58. K. Carpenter, A. Tabei, Materials **13**, 255 (2020). https://doi.org/10.3390/ma13020255
59. A. Gupta, S. Hasanov, I. Fidan, *Processing and characterization of 3d-printed polymer matrix composites reinforced with discontinuous fibers* (Department of Manufacturing and Engineering Technology, Tennessee Technological University, Cookeville, TN 38505)
60. G.D. Goh, Y.L. Yap, S. Agarwala, W.Y. Yeong, Recent progress in additive manufacturing of fiber reinforced polymer composite. Adv. Mater. Technol. **4**, 1800271 (2019)
61. S.F. Kabir, K. Mathur, A. Seyam, A critical review on 3D printed continuous fiber reinforced composites: History, mechanism, materials and properties. Compos. Struct. **232**, 111476 (2020)
62. T.D. Ngo, A. Kashani, G. Imbalzano, K.T. Nguyen, D. Hui, Additive manufacturing (3D printing): A review of materials, methods, applications and challenges. Compos. Part B Eng. **143**, 172–196 (2018)
63. P. Parandoush, D. Lin, A review on additive manufacturing of polymer-fiber composites. Compos. Struct. **182**, 36–53 (2017)
64. R. Matsuzaki, M. Ueda, M. Namiki, T.-K. Jeong, H. Asahara, K. Horiguchi, T. Nakamura, A. Todoroki, Y. Hirano, Three-dimensional printing of continuous-fiber composites by in-nozzle impregnation. Sci. Rep. **6**, 23058 (2016)
65. C. Yang, X. Tian, T. Liu, Y. Cao, D. Li, 3D printing for continuous fiber reinforced thermoplastic composites: Mechanism and performance. Rapid Prototyp. J. **23**, 209–215 (2017)
66. S.F. Kabir, K. Mathur, A.F.M. Seyam, Impact resistance and failure mechanism of 3D printed continuous fiber-reinforced cellular composites. J. Text. Inst., 1–15 (2020)

67. K. Mathur, S.M.F. Kabir, A.F.M. Seyam, Tensile properties of 3D printed continuous fiberglass reinforced cellular composites. J. Text. Inst. (2020) under review
68. A. Gupta, S. Hasanov, I. Fidan, *Processing and Characterization of 3d-Printed Polymer Matrix Composites Reinforced with Discontinuous Fibers*, in Solid Freeform Fabrication 2019: Proceedings of the 30th Annual International, Solid Freeform Fabrication Symposium
69. A. Rimkus, M.M. Farh, V. Gribniak, Continuously reinforced polymeric composite for additive manufacturing – Development and efficiency analysis. Polymers **14**(17), 3471 (2022)
70. M. Silva, I.S. Pinho, J.A. Covas, N.M. Alves, M.C. Paiva, 3D printing of graphene-based polymeric nanocomposites for biomedical applications. Funct. Compos. Mater. (2021)
71. X. You, J. Yang, S. Dong, Structural and functional applications of 3D-printed graphene-based architectures. J. Mater. Sci. **56**, 9007–9046 (2021)
72. ASTM (D638-02a), American Society for Testing and Materials. Standard test method for tensile properties of plastics (D 638 – 02a) – SCAN Version. Astm **8**, 46–58 (2003)
73. T.H. Sibisi, M.B. Shongwe, L.C. Tshabalala, I. Mathoho, LAM additive manufacturing: A fundamental review on mechanical properties, common defects, dominant processing variables, and its applications. Int. J. Adv. Manuf. Technol. (2023)

Index

B. Farahmand, *Fundamentals of Composites and Their Methods of Fabrications*,
https://doi.org/10.1007/978-3-031-80201-0

E

F

G

H

I

L

M

The manufacturer's authorised representative in the EU is Springer Nature Customer Service Centre GmbH, Europaplatz 3, 69115 Heidelberg, Germany. If you have any concerns regarding our products, please contact ProductSafety@springernature.com

Printed and bound by CPI Group (UK) Ltd, Croydon, CR0 4YY
30/09/2026
02225593-0011